Springer Texts in Statistics

Advisors:
George Casella   Stephen Fienberg   Ingram Olkin

Springer
New York
Berlin
Heidelberg
Barcelona
Hong Kong
London
Milan
Paris
Singapore
Tokyo

Springer Texts in Statistics

(Continued after index)

George R. Terrell

Mathematical Statistics

A Unified Introduction

With 86 Figures

Springer

George R. Terrell
Department of Statistics
Virginia Polytechnic Institute
Blacksburg, VA 24061
USA

ISBN 978-1-4419-3141-2   e-ISBN 978-0-387-22769-6

Library of Congress Cataloging-in-Publication Data
Terrell, George R.
    Mathematical statistics : a unified introduction / George R.
Terrell.
      p.   cm. — (Springer texts in statistics)
    Includes index.

    1. Mathematical statistics. I. Title. II. Series.
QA276.12.T473   1999
519.5—dc21   98-30565

Printed on acid-free paper.

9 8 7 6 5 4 3 2 1

Teacher's Preface

Why another textbook? The statistical community generally agrees that at the upper undergraduate level, or the beginning master's level, students of statistics should begin to study the mathematical methods of the field. We assume that by then they will have studied the usual two-year college sequence, including calculus through multiple integrals and the basics of matrix algebra. Therefore, they are ready to learn the foundations of their subject, in much more depth than is usual in an applied, "cookbook," introduction to statistical methodology.

There are a number of well-written, widely used textbooks for such a course. These seem to reflect a consensus for what needs to be taught and how it should be taught. So, why do we need yet another book for this spot in the curriculum?

I learned mathematical statistics with the help of the standard texts. Since then, I have taught this course and similar ones many times, at several different universities, using well-thought-of textbooks. But from the beginning, I felt that something was wrong. It took me several years to articulate the problem, and many more to assemble my solution into the book you have in your hand.

You see, I spend the rest of my day in statistical consulting and statistical research. I should have been preparing my mathematical statistics students to join me in this exciting work. But from seeing what the better graduating seniors and beginning graduate students usually knew, I concluded that the standard curriculum was not teaching them to be sophisticated citizens of the statistical community. These able students seemed to be well informed about a set of narrow, technical issues and at the same time embarrassingly lacking in any understanding of more fundamental matters. For example, many of them could discourse learnedly on which sources of variation were testable in complicated linear models. But they became tongue-tied when asked to explain, in English, what the presence of some interaction meant for the real-world experiment under discussion!

What went wrong? I have come to believe that the problem lies in our history. The first modern textbooks were written in the 1950s. This was at the end of the Heroic Age of statistics, roughly, the first half of the twentieth century. Two bodies of magnificent achievements mark that era. The first, identified with Student, Fisher, Neyman, Pearson, and many others, developed the philosophy and formal methodology of what we now call *classical inference*. The analysis of scientific experiments became so straightforward that these techniques swept the world of applications. Many of our clients today seem to believe that these methods *are* statistics.

The second, associated with Liapunov, Kolmogorov, and many others, was the formal mathematicization of probability and statistics. These researchers proved precise central limit theorems, strong laws of large numbers, and laws of the iterated logarithm (let me call these *advanced asymptotics*). They axiomatized probability theory and placed distribution theory on a rigorous foundation, using Lebesgue integration and measure theory.

By the 1950s, statisticians were dazzled by these achievements, and to some extent we still are. The standard textbooks of mathematical statistics show it. Unfortunately, this causes problems for teachers. Measure theory and advanced asymptotics are still well beyond the sophistication of most undergraduates, so we cannot *really* teach them at this level. Furthermore, too much classical inference leads us to neglect the preceding two centuries of powerful but less formal methods, not to mention the broad advances of the last 50 years: Bayesian inference, conditional inference, likelihood-based inference, and so forth.

So the standard textbooks start with long, dry, introductions to abstract probability and distribution theory, almost devoid of statistical motivations and examples (poker problems?!). Then there is a frantic rush, again largely unmotivated, to introduce exactly those distributions that will be needed for classical inference. Finally, two-thirds of the way through, the first real statistical applications appear—means tests, one-way ANOVA, etc.—but rigidly confined within the classical inferential framework. (An early reader of the manuscript called this "the cult of the t-test.") Finally, in perhaps Chapter 14, the books get to linear regression. Now, regression is 200 years old, easy, intuitive, and incredibly useful. Unfortunately, it has been made very difficult: "conditioning of multivariate Gaussian distributions" as one cultist put it. Fortunately, it appears so late in the term that it gets omitted anyway.

We distort the details of teaching, too, by our obsession with graduate-level rigor. Large-sample theory is at the heart of statistical thinking, but we are afraid to touch it. "Asymptotics consists of corollaries to the central limit theorem," as another cultist puts it. We seem to have forgotten that 200 years of what I shall call *elementary asymptotics* preceded Liapunov's work. Furthermore, the fear of saying anything that will have to be modified later (in graduate classes that assume measure theory) forces undergraduate mathematical statistics texts to include very little real mathematics.

As a result, most of these standard texts are hardly different from the cookbooks, with a few integrals tossed in for flavor, like jalapeño bits in cornbread. Others are spiced with definitions and theorems hedged about with very technical conditions, which are never motivated, explained, or applied (remember "regularity conditions"?). Mathematical proofs, surely a basic tool for understanding, are confined to a scattering of places, chosen apparently because the arguments are easy and "elegant." Elsewhere, the demoralizing refrain becomes "the proof is beyond the scope of this course."

How is this book different? In short, this book is intended to teach students to *do* mathematical statistics, not just to appreciate it. Therefore, I have redesigned the course from first principles. If you are familiar with a standard textbook on the subject and you open this one at random, you are very likely to find either a surprising topic or an unexpected treatment or placement of a standard topic. But everything is here for a reason, and its order of appearance has been carefully chosen.

First, as the subtitle implies, the treatment in *unified*. You will find here no artificial separation of probability from statistics, distribution theory from inference, or estimation from hypothesis testing. I treat probability as a mathematical handmaiden of statistics. It is developed, carefully, as it is needed. A statistical motivation for each aspect of probability theory is therefore provided.

Second, I have updated the range of subjects covered. You will encounter introductions to such important modern topics as loglinear models for contingency tables and logistic regression models (very early in the book!), finite population sampling, branching processes, and small-sample asymptotics.

More important are the matters I emphasize systematically. *Asymptotics* is a major theme of this book. Many large-sample results are not difficult and quite appropriate to an undergraduate course. For example, I had always taught that with "large n, small p" one may use the Poisson approximation to binomial probabilities. Then I would be embarrassed when a student asked me exactly when this worked. So we derive here a simple, useful error bound that answers this question. Naturally, a full modern central limit theorem is mathematically above the level of this course. But a great number of useful yet more elementary normal limit results exist, and many are derived here.

I emphasize those methods and concepts that are most useful in statistics in the broad sense. For example, *distribution theory* is motivated by detailed study of the most widely useful families of random variables. Classical estimation and hypothesis testing are still dealt with, but as applications of these general tools. Simultaneously, Bayesian, conditional, and other styles of inference are introduced as well.

The standard textbooks, unfortunately, tend to introduce very obscure and abstract subjects "cold" (where did a horrible expression like $\frac{1}{\sqrt{2\pi}} e^{-x^2/2}$ come from?), then only belatedly get around to motivating them and giving examples. Here we insist on *concreteness*. The book precedes each new topic with a relevant statistical problem. We introduce abstract concepts gradually, working from the special to the general. At the same time, each new technique is applied as widely as possible. Thus, every chapter is quite broad, touching on many connections with its main topics.

The book's attitude toward mathematics may surprise you: We take it seriously. Our students may not know measure theory, but they do know an enormous amount of useful mathematics. This text uses what they do know and teaches them more. We aim for reasonable *completeness*: Every formula is derived, every property is proved (often, students are asked to complete the arguments themselves as exercises). The level of mathematical precision and generality is appropriate to a serious upper-level undergraduate course.

At the same time, students are not expected to memorize exotic technicalities, relevant only in graduate school. For example, the book does not burden them with the infamous "triple" definition of a random variable; a less obscure definition is adequate for our work here. (Those students who go on to graduate mathematical statistics courses will be just the ones who will have no trouble switching to the more abstract point of view later.) Furthermore, we emphasize mathematical *directness*: Those short, elegant proofs so prized by professors are often here replaced by slightly longer but more constructive demonstrations. Our goal is to stimulate understanding, not to dazzle with our brilliance.

What is in the book? These pedagogical principles impose an unconventional order of topics. Let me take you on a brief tour of the book:

The "Getting Started" chapter motivates the study of statistics, then prepares the student for hands-on involvement: completing proofs and derivations as well as working problems.

Chapter 1 adopts an attitude right away: *Statistics precedes probability*. That is, models for important phenomena are more important than models for measurement and sampling error. The first two chapters do not mention probability. We start with the linear data-summary models that make up so much of statistical practice: one-way layouts and factorial models. Fundamental concepts such as additivity and interaction appear naturally. The simplest linear regression models follow by interpolation. Then we construct simple contingency-table models for counting experiments and thereby discover independence and association. Then we take logarithms, to derive loglinear models for contingency tables (which are strikingly parallel to our linear models). Again, logistic regression models arise by interpolation. In this chapter, of course, we restrict ourselves to cases for which reasonable parameter estimates are obvious.

Chapter 2 shows how to estimate ANOVA and regression models by the ancient, intuitive method of least squares. We emphasize geometrical interpolation of the method—shortest Euclidean distance. This motivates sample variance, covariance, and correlation. Decomposition of the sum of squares in ANOVA and insight into degrees of freedom follow naturally.

That is as far as we can go without models for errors, so Chapter 3 begins with a conventional introduction to combinatorial probability. It is, however, very concrete: We draw marbles from urns. Rather than treat conditional probability as a later, artificially difficult topic, we start with the obvious: *All probabilities are conditional*. It is just that a few of them are conditional on a whole sample space. Then the first asymptotic result is obtained, to aid in the understanding of the famous "birthday problem." This leads to insight into the difference between finite population and infinite population sampling.

Chapter 4 uses geometrical examples to introduce continuous probability models. Then we generalize to abstract probability. The axioms we use correspond to how one actually calculates probability. We go on to general discrete probability, and Bayes's theorem. The chapter ends with an elementary introduction to Borel algebra as a basis for continuous probabilities.

Chapter 5 introduces discrete random variables. We start with finite population sampling, in particular, the *negative hypergeometric* family. You may not be familiar with this family, but the reasons to be interested are numerous: (1) Many common random variables (binomial, negative binomial, Poisson, uniform, gamma, beta, and normal) are asymptotic limits of this family; (2) it possesses in transparent ways the symmetries and dualities of those families; and (3) it becomes particularly easy for the student to carry out his own simulations, via urn models. Then the Fisher exact test gives us the first example of an hypothesis test, for independence in the 2×2 tables we studied in Chapter 1. We introduce the expectation of discrete random variables as a generalization of the average of a finite population. Finally, we give the first estimates for unknown parameters and confidence bounds for them.

Chapter 6 introduces the geometric, negative binomial, binomial, and Poisson families. We discover that the first three arise as asymptotic limits in the negative hypergeometric family and also as sequences of Bernoulli experiments. Thus, we have related finite and infinite population sampling. We investigate just when the Poisson family may be used as an asymptotic approximation in the binomial and negative binomial families. General discrete expectations and the population variance are then introduced. Confidence intervals and two-sided hypothesis tests provide natural applications.

Chapter 7 introduces random vectors and random samples. Here is where marginal and conditional distributions appear, and from these, population covariance and correlation. This tells us some things about the distribution of the sample mean and variance, and leads to the first laws of large numbers. The study of conditional distributions permits the first examples of parametric Bayesian inference.

Chapter 8 investigates parameter estimation and evaluation of fit in complicated discrete models. We introduce the discrete likelihood and the log-likelihood ratio statistic. This turns out often to be asymptotically equivalent to Pearson's chi-squared statistic, but it is much more generally useful. Then we introduce maximum likelihood estimation and apply it to loglinear contingency table models; estimates are computed by iterative proportional fitting. We estimate linear logistic models by maximum likelihood, evaluated by Newton's method.

Chapter 9 constructs the Poisson process, from which we obtain the gamma family. Then a Dirichlet process is constructed, from which we get the beta family. Connections between these two families are explored. The continuous version of the likelihood ratio is introduced, and we use it to establish the Neyman–Pearson lemma.

Chapter 10 defines the general quantile function of a random variable, by asking how we might simulate it. Then we may define the expectation of any random variable as the integral of that quantile function, using only elementary calculus. Next, we derive the standard normal distribution as an asymptotic limit of the gamma family. Stirling's formula is a wonderful bit of gravy from this argument. By duality, the normal distribution is also an asymptotic limit in the Poisson family.

Chapter 11 develops multivariate absolutely continuous random variable theory. The first family we study is the joint distribution of several uniform order statistics. We then find the chi-squared distribution and show it to be a large-sample limit of the chi-squared statistic from categorical data analysis. Duality and conditioning arguments lead to bivariate normal distributions and to asymptotic normality of several common families.

Chapter 12 derives the null distributions of the R-squared and F statistics from least-squares theory, on the surprisingly weak assumption that errors are spherically distributed. We notice then that maximum likelihood estimates for normal error models are least-squares. Parameter estimates for the general linear model and their variances are obtained. We show that these are best linear unbiased via the Gauss-Markov theorem. The information inequality is then derived as a first step to understanding why maximum likelihood estimates are so often good.

Chapter 13 begins to view random variables from alternative mathematical representations. First, we study the probability generating function, using the concrete motivation of finding the compound distributions that appear in branching processes. The moment generating function may now be motivated concretely, for positive random variables, by comparison with negative exponential variables. We then suggest (incompletely, of course) how it may be used to derive some limit theorems. We then introduce exponential families, emphasizing how they capture common features and calculations for many of our favorite families. We finish with an introduction to a lively modern topic: probability approximation by small-sample asymptotics. This applies beautifully all the tools developed earlier in the chapter.

Fitting the book to your course. There are, of course, alternative paths through the material if you have different goals for your students. A shorter course in probability and distribution theory may be taught by skipping lightly over those chapters that emphasize data modeling and estimation: Chapters 1, 2, and 8, and 12. Later sections in other chapters, which investigate methods of statistical inference, might also be deemphasized.

At the opposite extreme, a sophisticated sequence in applied statistics may start with this material. Early parts of Chapter 1 could be supplemented by a lecture on statistical graphics and exploratory data analysis. Chapter 8 might be followed by the study of more complicated contingency table models. Then Chapter 12 leads naturally into a fuller treatment of inference in the linear model. The course may be supplemented throughout with tutorials on how to use computer packages to draw better graphs and carry out computations with more elaborate models and larger data sets.

Certain sections, marked with an asterisk (*), may be delayed until later if the instructor wishes at relatively little cost to continuity. The **Time to Review** list at the beginning of each chapter should serve to warn you when to return to these matters.

Acknowledgments

I began this Preface with harsh criticism of earlier texts of mathematical statistics; now I must plead guilty to ingratitude. I learned what I know from books such as these; I am simply exercising here the prerogative of each generation to pass knowledge on to the next in a slightly different and, I believe, improved form.

John Kimmel, my editor, has had the patience to make me do things right, for which I am grateful. Many thanks to the hundreds of students in my statistics classes, for their interest, patience, hard work, occasional enthusiasm, and, above all, for their questions. From them I learned that many matters that were old hat to me could be confusing to a novice. Thanks for the support of the Statistics Department at Virginia Polytechnic Institute and State University, my professional home, where I began this project and where I am finishing it. Thanks, too, to the Statistics Department at Rice University, which welcomed me for a sabbatical in 1994–1995, during which I carried out roughly the middle half of the writing.

Conversations with colleagues, often while standing in the hall, have been a central part of my intellectual development. Those with David Scott, over 20 years, have amounted to a substantial portion of my entire statistics education (casual acquaintances have often assumed, understandably, that I must have been his dissertation student). In addition, he read several chapters of this book and provided detailed and useful comments. No conversations on pedagogical issues have been more useful than those with Don Jensen. In particular, he pointed out to me the central role that spherical symmetry of error distributions plays in classical inference. I.J. Good was a valuable resource, particularly on foundations issues. Marion Reynolds showed me, among other things, how powerful the method of indicators can be. Michael Trosset lent a sympathetic ear and a critical intelligence, often. This list should go on and on.

My own teachers share responsibility, at least when I have gone the right way. In particular, my greatest teacher, Frank Jones, showed me that mathematical clarity and beauty should be the same thing.

My wife, Goldie, has taken in stride the absurd idea that something as dull as a textbook should be allowed to obsess me for many years. Her support has been unwavering, and I am grateful.

George R. Terrell
Virginia Polytechnic Institute

Contents

Getting Started

Why Study Statistics?

We have all been exposed to the popular notion that statistics is about numbers that are deadly-dull, and perhaps intentionally misleading. You will quickly discover in this course that the opposite is the case: Statistics is the science of extracting useful (and therefore interesting) numbers from the world; and the statistician is committed to forcing these numbers to reveal the truth. Therefore, statistics has become an essential tool of modern civilization. For example:

(1) In the early nineteenth century, astronomers observed their first asteroid, Ceres. It then quickly disappeared into the sun's glare, and there was some doubt that it could be found again in the foreseeable future, since it would have moved along in its (unknown) orbit. But the great mathematician Carl Friedrich Gauss managed to compute the orbit of Ceres, using those observations that had been made before it disappeared. He then told observers where to look for it some months hence. The asteroid was found where he had predicted it would be, and Gauss became one of the most respected scientists of his day.

Historians have emphasized Gauss's mathematical achievement in using a few accurately observed positions of Ceres to discover its overall orbit, using the complicated equations of celestial mechanics. But that is not all that Gauss did. He started with a somewhat larger number of not-very-accurate observations of the positions of Ceres. Telescopes, observers, and especially clocks were not as reliable in those days as we would now expect them to be. So the observations he had to work with, if plotted on a chart of the sky, do not show a realistically smooth orbit, but instead bounce around a bit. Fortunately, Gauss was one of the inventors of a marvelous new statistical technique, the method of *least squares*, that takes a number of imperfect observations and reduces them to a few, more precise, numbers characterizing the orbit. So Gauss's technical achievement was twofold; and one aspect of it was a statistical method that has been enormously valuable ever since, throughout science.

(2) In his biography of Richard Feynman, James Gleick observed that we would now be amazed, and perhaps appalled, that after Alexander Fleming's discovery of the first antibiotic, penicillin, it took most of a generation before the drug became a standard treatment for deadly diseases. The process started with Fleming's report about bacteria in petri dishes, which led to an attempt to use penicillin on a sick human being, and evolved into the reports by a number of physicians on how well penicillin seemed to have worked for their patients. Finally, the reputation of the drug in the medical community had become so overwhelmingly favorable that pharmaceutical companies took the risk of gearing up for mass production.

This process was so slow because there was no agreement in the scientific community on what a sensible, orderly way to evaluate new drugs might be. After all, worthless drugs are being invented all the time. Because some people recover spontaneously, while others fail to respond to even the most promising drugs, good and bad drugs are always difficult to tell apart. In the same years medical researchers were studying penicillin, though, statisticians were inventing techniques of *experimental design*, inspired by agricultural research. These were precisely the disciplined, reliable methods that drug-testing needed. Today, new drugs are expected to submit to *controlled*, *randomized* experiments that will, in a reasonably short time, lead to sensible decisions about their clinical value.

(3) Every ten years, the United States carries out a national census. Believe it or not, this process at its heart has very little to do with modern statistics. Since the idea is to collect and organize a basic set of facts about *everybody*, the main skills involved are those of librarians and geographers. However, there are known imperfections in the census: For example, despite its ambitions, it always misses a certain modest percentage of the American population. People would like to have some idea how large this *undercount* is; both so we can estimate the true totals, and also discover how to make future censuses more accurate.

If you think about it, the census itself tells you nothing about its own accuracy (how can it possibly include the information that so-and-so was missed?). But statisticians have developed techniques for parallel, smaller experiments, called *sample surveys*, that can provide such information. These are ways of collecting information about relatively small numbers of people that allow reasonable statements to be made about people in general. Such conclusions are not perfectly accurate, of course, but our statistical methods include ways of estimating just how accurate the conclusions probably are. If you know how good a number is, you can use it with proper care.

One simple way to estimate the undercount would be to do a very thorough recount in a set of small areas chosen somehow to be representative of the country at large. By comparing the results to the original census, you could see what portion of the people were missed the first time. Then you would conjecture that this might be close to the national undercount rate. I am sure you can see problems with this approach; but more sophisticated surveys of this sort have promise, and are in fact used to estimate the undercount.

So statistics today provides a set of valuable tools for dealing with some of the uncertainties of life. You will not be surprised to hear that statistics is a mathematical subject: Mathematics was used to invent these methods and is therefore necessary for any deep understanding of them. Furthermore, new statistical techniques must be developed all the time to deal with new problems. Again, mathematics is required. Statistics courses more elementary than this one often try to avoid such matters, hoping that the student will never encounter a statistical problem that requires novel insights or methods.

But this book is for students who will be the masters of statistical technology, not its slaves. Its subject is "mathematical statistics," or sometimes "theoretical statistics." The methods of mathematics will be in constant use. We assume that you have had a standard calculus sequence, including an introduction to multiple integrals, and the rudiments of matrix algebra. You may find that you do not really know these subjects as well as you thought you did, through lack of interesting applications. Taking this course will solve that problem, since there is no substitute for incisive examples, and for practice. Each chapter begins with some recommendations of topics to review.

How to Read This Book

Now that you have decided to study mathematical statistics, you are probably wondering what you will have to do to master the course. If you have had other applied mathematics courses, you have probably come to realize that the experience is not much like studying history, and even less like studying a foreign language. Let me illustrate:

Example. In 1900, the English mathematical biologist Karl Pearson proposed the formula $\chi^2 = \sum_i \frac{(O_i - E_i)^2}{E_i}$. It is now called Pearson's chi-squared, because, following an old convention, the Greek letter *chi* is to the left of the equal sign. It is a measure of the difference between a set of counts O_i observed in a survey or experiment and a corresponding set of counts E_i expected under some hypothesis about how the survey or experiment should come out. Several years ago, Pearson's formula was on a widely publicized list of the 100 most important scientific discoveries of the twentieth century.

Everything here is useful knowledge, and I would hope that at the end of your statistical education you would know most of the information in the preceding paragraph. But so far this is the sort of thing you get from history classes (the when, where, and who in the first sentence, and the comment about its significance in the last sentence) and from foreign language classes (the formula to memorize, and the definitions of the parts).

But since this is an applied mathematics course, I am sure you realize that there are other things about Pearson's chi-squared that you need to learn. To start with, how do you apply this formula to the real world? For example, I want to know whether a coin that is to be used to choose goals in football games is fair. I toss it 100 times; it lands "heads" 43 times and "tails" 57 times. But my idea of a fair coin would land heads about 50 times and tails 50 times. Were my counts so far from fair that I now have evidence against the coin being balanced? This, you will learn, is a typical application of Pearson's chi-squared; it is the procedure described so abstractly in the sentence about observed and expected counts. The O_i's are 43 and 57, and the E_i's are 50 and 50. You learned in earlier courses that Σ (capital sigma) means "add up the cases," so $\chi^2 = \frac{(43-50)^2}{50} + \frac{(57-50)^2}{50} = 1.96$. In very elementary statistical courses you then learn to consult a table or computer program and report on its authority that this is not a very big value; and so there is little reason to doubt that your coin is fair.

Throughout this book, you will encounter worked numerical examples of what to do with proposed procedures, under the heading **Example**. I have tried to illustrate in this way almost every method discussed, some several times. You should realize that these are not just motivational: They are intended to begin your process of learning how to perform statistical analyses for yourself. Every time you encounter an example you should first read it carefully to try to understand why the given method may be appropriate to the real-world situation. Then you should try to reproduce my mathematics, and my arithmetic, for yourself. (If you find a mistake, please write to me.) In this way, you get the flavor of how the method is applied.

Then you will turn to the **Exercises** at the end of that chapter and try some problems, with numerical data, that use the same method. This may be harder than you expect, because you may not recognize immediately what the new application has to do with the method you are learning. Instead of coin-tossing, it might involve, for example, a consumer survey about lipstick preferences. The fact that this still involves comparing observed to expected counts, and so Pearson's chi-squared applies, is a subtle one. Doing problems on your own is the best way to gain experience at making such judgments.

The exercises, by the way, are in two sections in each chapter. The first set consists of fairly straightforward applications and chances to fill in omitted details. There are some hints and numerical answers to these in an appendix. It is important *not* to look at these answers until you have an answer you are happy with and wish to double-check; or until you are thoroughly stuck. Working backwards from a known answer teaches you much, much less than doing it the right way. The next section, called **Supplementary Exercises**, consists of additional problems of the same kind, for valuable extra practice plus opportunities to develop for yourself interesting and useful extensions of the ideas you have been studying.

If this were a more elementary course, and one that concentrated on applications, this would be all there was to learning the material. But we have ducked some important questions, such as, where in the world did Pearson get that formula? The answer is, he *derived* it, from statistical methods he already knew, using ingenuity and mathematics. You might think that such questions are of mainly historical interest. Remember, though, that it is not obvious why *anyone* would propose the chi-squared method. The question should perhaps be, why would a reasonable person use that formula? In this book you will find not one, but three mathematical derivations of the formula (none of them exactly like Pearson's). That might seem very odd, a waste of time. I suppose it would be, if the purpose of the derivation were just to reassure you that somebody, somewhere (the author, perhaps), knows why we use Pearson's formula. However, the real reason is to learn the ways of thinking that inspire our use of the method. The three derivations show three different aspects of that thinking. My hope is that after studying all three, you will have a pretty good idea of when you might want to use Pearson's formula. (It is a formula that every statistician needs to an important special class.)

So, when you encounter one of the many derivations in the text, read it, slowly and repeatedly, until you believe you understand in detail how it works. Then close your book, and try to carry out that derivation yourself in your own manner. After you have succeeded, turn again to the exercises. There you may be asked to discover for yourself yet another way of obtaining that same method. Or you may be asked to derive a related formula. After you have done all this (it will often take quite a while), you will find that you understand far better than before why statisticians do what they do. In fact, those applied problems involving data and numbers will have become much easier to connect to mathematical methods. Furthermore, you will find that complicated equations, because they are no longer in a foreign language, are much easier to remember than they used to be.

The exercises that require you to derive new formulas give away an important secret: Statisticians do not yet know the answer to every statistical question. Therefore, competent working statisticians spend a good deal of their time inventing new methods, inspired by methods they already know (just as Pearson did). So you should tackle with gusto those exercises that lead you to develop methods new to you, because they give you practice with the creative aspect of statistics.

For example, many Pearson-type problems have the property that the total of all the observed counts in the problem is equal to the total of all the expected counts. In the coin-tossing problem, they both summed to 100. This is usually no accident: When we decided what it meant for a coin to be fair, we split the known total of 100 evenly between heads and tails. The general mathematical statement of this fact says that $\sum_i O_i = \sum_i E_i = n$, where n is just a convenient symbol for the total count. We are going to show that Pearson's chi-squared reduces to a simpler formula in this case. First, we expand the square in the numerator:

$$\chi^2 = \sum_i \frac{(O_i - E_i)^2}{E_i} = \sum_i \frac{(O_i^2 - 2O_i E_i + E_i^2)}{E_i}.$$

Now, remembering that the summation sign just means "add up all the cases," let us sum each of the three terms in the numerator separately (since the order of addition in a finite sum never matters):

$$\chi^2 = \sum_i \frac{(O_i^2 - 2O_i E_i + E_i^2)}{E_i} = \sum_i \frac{O_i^2}{E_i} - 2\sum_i \frac{O_i E_i}{E_i} + \sum_i \frac{E_i^2}{E_i}.$$

In the last two terms, E's in the numerator and denominator cancel, so we get $\chi^2 = \sum_i \frac{O_i^2}{E_i} - 2\sum_i O_i + \sum_i E_i$. But we have decided to concentrate on the case where the total of observed and expected counts are both n, so

$$\chi^2 = \sum_i \frac{O_i^2}{E_i} - 2n + n = \sum_i \frac{O_i^2}{E_i} - n.$$

This last is a new, simplified formula for Pearson's chi-squared, which works in an important special case. (It is a formula that every statistician used to know; but for some reason it is rarely mentioned in modern applied statistics books.)

I hope you have checked my algebra carefully here. The earliest derivations in the book are explained in about this much detail. Later on, as you become more skilled, easy steps are skipped, so that there will be a bit more work for you to do. It will continue to be important that you check all the math for yourself. In fact, omitted steps are often left as exercises.

The last comment I made in working with the coin-tossing experiment was that we would probably decide that 1.96 was not a very large value of chi-squared. Why? This happens to be the hardest question we have yet dealt with. To interpret that number, we will need to investigate deep mathematical properties of the chi-squared statistic. A large percentage of our effort in this course, thoroughly entangled with deriving statistical methods, will be to use mathematics to discover important working characteristics of those methods. When we have found some properties that will be used later in a chapter, we distinguish them as **Propositions**, as is often done in mathematics texts. If the properties are so important that they will be extensively used in later chapters, we call them **Theorems**. We will use here a convention rigorously obeyed by working mathematicians (but not by math books): Theorems are given a name, and are later referred to by that name. (A famous example is Fermat's last theorem.)

Just as we will derive all our methods, we will *prove* all our propositions and theorems. Usually, the proof will be in the discussion leading up to the statement of the result; but sometimes it will be immediately following, labeled **Proof**. Often, students have painful memories of proving things from earlier math courses. You might have come away with the idea that you are supposed to provide a tangle of words like "therefore," "without loss of generality," and "by induction"; then at the end you complete the ritual by invoking the magical formula "QED." Actually, a mathematical proof is nothing more than an explanation of why something is true, which is supposed to be clear enough to convince an intelligent, skeptical listener. We have proofs for the same reason we have derivations of formulas: so you will understand where the theorem comes from and have some idea how to find for yourself similar but novel facts that you may need later.

You should study the mathematical proofs just as you study the derivations. When you encounter a proposition, you should read carefully through my argument until you are convinced that the statement is true. Then close the book, and convince someone else. At that point, turn to the exercises, and work on related problems that say something like "show" or "demonstrate" or "prove" (which all mean the same thing). Your job will again be, first, to persuade yourself that the claim is valid (if it is not, please write to me), and, second, to write down an explanation clear enough to convince other people.

As you begin to tackle the exercises in this book, you will surely begin to wonder how much electronic computing help you should use. The general principle will be this: When you are first learning any subject, you should get your hands very dirty. In the Exercises, I am imagining that you have an ordinary scientific calculator or a fairly low-level mathematics program on your computer handy at all times. A few of the Supplementary Exercises are better tackled by using more sophisticated computing tools—Fortran, Basic, Pascal, C, a spreadsheet program, or Mathematica, for example. At this point you should avoid using any tool that incorporates the statistical procedures you are trying to understand—such as statistical functions in a calculator or spreadsheet, or statistical packages. There will be plenty of time for learning these wonderful timesavers later, after you have mastered mathematical statistics.

You may have noticed that this course has an important characteristic in common with other math and science courses. In many other fields, your job seems to be to believe everything the professor or textbook says; the best student is the most gullible. In this course, the best students are the most skeptical—so long as they are willing to check things for themselves.

So how are you to read this book? As you would read a book on baking bread: If you do not spend much of the time with your hands covered with flour, you are doing it wrong. In the same way, study this book with pencil, pen, paper, calculator, and perhaps computer at your fingertips, and use them to try out every new idea you encounter.

CHAPTER 1

Structural Models for Data

1.1 Introduction

You probably think that statistics has to do with managing lots of numbers. But the basic goal of scientific research (which may well be the reason you collected all those numbers) is to understand them. You will find that statisticians are called in when a scientist, engineer, or planner decides that some survey or experiment has produced too many numbers for a mere human being to comprehend. We statisticians believe that it may still be possible to describe the most important features of those numbers with comparatively simple mathematical *models*. This chapter will give an overview of some of the most useful models that belong in the tool kit of any aspiring statistician.

At least two sorts of models will be required, depending on the experiments we have performed. First, we will study experiments whose results are measured numbers, such as a temperature or pressure. We will try to summarize how those numbers seem to have been affected by experimental conditions. Second, we will consider experiments whose result is a count of how many subjects fell in certain categories, such as male/female or alive/dead. Again, we will want to see how those counts change according to conditions under which the count was taken.

Time to Review

Summation notation
Natural logarithms and exponential functions

1.2   Summarizing Multiple Measurements That Show Variability

1.2.1   Plotting Data

Very often, a scientist finds herself measuring carefully some natural quantity, like a length or weight, in hopes that it will help her understand some phenomenon. But then, showing the care that scientists must show, she takes a second measurement of the same thing. Sometimes the answer will be identical, up to the accuracy of her instruments. In many cases, though, it will be substantially different; and there will be no reason to think a blunder has been made. So she does a series of these comparable measurements, as many as she has time, patience, and resources for. And she may well find that she has obtained an incomprehensible variety of numerical answers to a simple question.

Example. In 1882 Albert Michelson made 23 measurements of the velocity of light in air, in kilometers per second above 299,000:

$$
\begin{array}{ccccc}
883 & 711 & 578 & 696 & 851 \\
816 & 611 & 796 & 573 & 809 \\
778 & 599 & 774 & 748 & 723 \\
796 & 1051 & 820 & 748 & \\
682 & 781 & 772 & 797 & \\
\end{array}
$$

(That is, 711 means he measured a velocity of 299,711 kilometers per second on his sixth try. Do you see what 1051 must mean?)

We need some notation for this situation. Call each of the n observations x_i, where $i = 1, \ldots, n$. Then, for example in the velocity data, $n = 23$ and $x_{17} = (299,)573$. Probably the first thing you would want in this situation is some way of organizing these numbers. Let us try a *geometrical* representation; for example, draw a horizontal number line whose range encompasses our measurements. Then place a thin vertical line at the value representing each of the observations. This is called a **hairline plot** (Figure 1.1).

When two observations are the same, we simply double the thickness of the line. (In other books you may see a similar display called a *dot plot*.)

Strictly speaking, the art of drawing such useful pictures belongs to a field called *statistical graphics*; and that is not the subject of this textbook on mathematical methods. But statisticians find some kinds of pictures so enormously useful that we can hardly imagine doing without them. Besides, there is a mathematical prin-

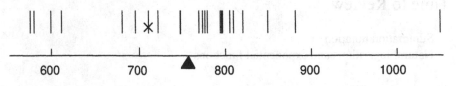

FIGURE 1.1. Measured speed of light in km/s above 299,000

ciple hidden in this diagram: We have represented a numerical measurement by a *coordinate* of a geometrical position on a line. The number did not start out as a point on the line, but we have felt free to put it there. We will see later that this simple step lets all the powerful tools of geometry fall into the statistician's tool kit.

1.2.2 Location Models

In our example, the numbers fell haphazardly in some region of the line. The scientist will tell you that she was trying to measure a constant of nature; but the measurements were so difficult to do well that they vary unpredictably by various amounts above and below the correct value. We have represented the modern accepted value of the speed of light in air, (299,)710.5 km/s, by an $\times$ on the plot.

This is called a (simple) **location model** for how the numbers came about. We hope to simplify the collection down to a single important quantity (that we often denote by the Greek letter μ) that we believe to be the center of our cluster of points. But to be honest, we carefully record the errors that cropped up in each of our observations. These are the n quantities $x_i - \mu$. For example, for observation 17 above, this error is $573 - 710.5 = -137.5$. We have called them errors; but a better word is model **residuals**. After all, with deeper understanding of the science, we may realize *why* some of the measurements were different from μ. The residuals are positive if the measurement is larger than the experimenter thinks it should have been, and negative if it is smaller.

Of course, usually our scientist does not know the value of μ; she did the experiment in order to find out. Perhaps she consulted a statistician, so we could provide her with an intelligent guess that she could report to her fellow scientists. So a statistician needs to be able to determine a number in the middle of the cluster, called an **estimate**, often denoted by $\hat{\mu}$, to report as a plausible value of μ. With luck, this *summary* of many measurements will be better than a single measurement. Of course, you could just stare at the hairline plot and make an educated guess of the center of the data; with practice, this could be a very good method. But it has one fatal flaw as far as a scientist is concerned: It is not *repeatable*—no two statisticians would report the same estimate. This immediately undermines much of the trust her colleagues may have in her proposal. So we ask an important question: What are good ways of making repeatable estimates of unknown quantities, and how good can we expect them to be?

There is one standard method of estimation that is so popular that you should see it right away. Imagine that the hairlines in our plot are equal, physical weights sitting on a (weightless) bar that is our number line. A natural center of those weights would be the point at which we would place a fulcrum so that the bar balances. (Notice the little picture of a fulcrum on the hairline plot of light velocities.) You may remember from high-school physics that the weights times distances must sum to the same value on each side of the fulcrum (so that the torque is zero). This says that the sum of the distances $x_i - \mu$ (their residuals) for observations greater than μ must equal the sum of the distances $\mu - x_i$ (the negatives of their residuals) for observations less than μ, because the weights are the same. If, for example, we number the observations so that the first $i = 1, \ldots, k$ were less than μ and the remaining $i = k + 1, \ldots, n$ were μ or greater, then the balance condition looks like

$$(\mu - x_1) + (\mu - x_2) + \cdots + (\mu - x_k) = (x_{k+1} - \mu) + (x_{k+2} - \mu) + \cdots + (x_n - \mu).$$

If we move the pieces on the left of the equal sign to the right side (changing signs as we do so), then we see that the positive and negative residuals together must sum to zero. We write that condition in *summation* notation (which you should review): $\sum_{i=1}^{n}(x_i - \mu) = 0$.

We will find our estimate $\hat{\mu}$ by solving this equation (called the *normal* equation) for μ. First, we can always split the sum into two pieces around the minus sign: $\sum_{i=1}^{n} x_i - \sum_{i=1}^{n} \mu = 0$. But that second sum just means that you are adding the constant μ to itself n times: $\sum_{i=1}^{n} x_i - n\mu = 0$. Moving it to the other side of the equation and dividing by n, we obtain $\hat{\mu} = \frac{1}{n}\sum_{i=1}^{n} x_i$. This is just the familiar arithmetic *average* of the observations; the summation notation just says that we add them all up, and divide by how many there are: $(x_1 + x_2 + \cdots + x_n)/n$. Statisticians call this the **sample mean**, written $\hat{\mu} = \bar{x}$. (In the speed of light example, $\bar{x} = (299,)756.2$ km/s, as you should check; this is not exactly at the true value, but it is closer than most of the individual measurements.) There are, of course, many other ways to estimate the center of the data μ; one of these is illustrated in your exercises.

I am willing to guess that when you were checking my sample mean calculation, you did not do it precisely the way the formula says to. When I was taking the mean of the speeds of light, I did not calculate $(299,883 + 299,816 + \cdots + 299,723)/23$. Rather, I saved time by calculating $(883 + 816 + \cdots + 723)/23 + 299,000$. To show the mathematical principle, let ν stand for any convenient value on the scale of measurement. Subtract and then add it to each term in the formula for the sample mean: $\bar{x} = \frac{1}{n}\sum_{i=1}^{n} x_i = \frac{1}{n}\sum_{i=1}^{n}(x_i - \nu + \nu)$. Sum those last ν's separately: $\bar{x} = \frac{1}{n}\sum_{i=1}^{n}(x_i - \nu) + \frac{1}{n}\sum_{i=1}^{n} \nu$. When we add a constant to itself n times, that just multiplies it by n, canceling the n in the denominator. We get a new formula, $\bar{x} = \frac{1}{n}\sum_{i=1}^{n}(x_i - \nu) + \nu$. I used $\nu = 299,000$ in our new expression. Some such choice will often be convenient.

1.3   The One-Way Layout Model

1.3.1   Data from Several Treatments

Often a scientist faces a set of measurements obtained in more than one experimental situation.

Example. In 1974 Till reported several samples of the salt content in parts per thousand of three separate water masses in the Bimini Lagoon:

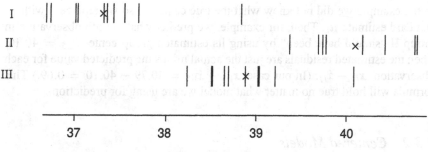

FIGURE 1.2. Salt in parts per thousand in sea water

Mass I: 37.54, 37.01, 36.71, 37.03, 37.32, 37.01, 37.03, 37.70, 37.36, 36.75, 37.45, 38.85
Mass II: 40.17, 40.80, 39.76, 39.70, 40.79, 40.44, 39.79, 39.38
Mass III: 39.04, 39.21, 39.05, 38.24, 38.53, 38.71, 38.89, 38.66, 38.51, 40.08

Figure 1.2 gives hairline plots of these numbers.

If we are lucky, the results in the various situations will be so different that we are obviously measuring completely distinct constants μ. But very often, as in the example, the groups will overlap considerably. Is it just a matter of opinion, or judgment, that one group (the second) seems usually saltier? We would like to say that there are three different typical levels of salt, μ_I, μ_{II} and μ_{III}, and, for example, that $\mu_{II} > \mu_I$. In practice, we have to estimate the salinity in the two masses and check that $\hat{\mu}_{II} > \hat{\mu}_I$. Since these estimates are imperfect, we become more confident of our conclusion as the estimated separation $\hat{\mu}_{II} - \hat{\mu}_I$ becomes larger.

The general setup for this model, called a **one-way layout**, is as follows: We have k **levels of the treatment** numbered $i = 1, \ldots, k$. In our example, the various levels are the different water masses of the lagoon where we found the samples, so $k = 3$. The ith level has n_i separate observations x_{ij}, numbered $j = 1, \ldots, n_i$. In our salinity data, $n_i = 12$; and $x_{II5} = 40.79$, the fifth measurement in the second water mass. We write for the total number of observations $n = \sum_{j=1}^{k} n_i$ ($n = 30$ measurements in our data set). Our model then says that the true value for the ith level is μ_i. We call these unknown but important constants the *parameters* of the model. If our estimates are $\hat{\mu}_i$, then the *estimated* residuals, representing the failure of our estimated model to describe the observations completely, are $x_{ij} - \hat{\mu}_i$.

We have standard estimates for our parameters: just take the sample mean of the observations in each level of the treatment: $\hat{\mu}_i = \bar{x}_i = \frac{1}{n_i} \sum_{j=1}^{n_i} x_{ij}$.

Example (*cont.*). Though the measurements at the sites overlap considerably, there seem to be characteristic salinities at each. The group means are $\hat{\mu}_I = 37.31$, $\hat{\mu}_{II} = 40.10$, and $\hat{\mu}_{III} = 38.89$; these are marked $\times$ on the plot.

We often think of a statistical model as making *predictions* of some future observation taken under conditions similar to some of the old ones; in the one-way layout, the prediction would just be the center for that level, $\hat{x}_{ij} = \mu_i$. Of course, in the example we did not know what the true center is, so we replace it with its standard estimate $\hat{\mu}_i$. Then, for example, we predict what the 5th observation in group II "should have been" by using its estimated group center $\hat{x}_{II5} = 40.10$. Then the estimated residuals are just the actual minus the predicted value for each observation: $x_{ij} - \hat{x}_{ij}$. (In our case, $x_{II5} - \hat{x}_{II5} = 40.79 - 40.10 = 0.69$.) This formula will hold true no matter what model we are using for prediction.

1.3.2   Centered Models

Since comparisons between the treatment levels are usually our primary interest, we have a different way to parametrize our model, called the **centered** model. With two levels, we start with a common center μ for all our observations and then compute how much the higher group is above center: $b_1 = \mu_2 - \mu$. Similarly, we compute the (negative) amount by which the second group is below the center by $b_2 = \mu_2 - \mu$. Now we can write the predictions for each of the two groups as $\mu_1 = \mu + b_1$ and $\mu_2 = \mu + b_2$. This is the first of many examples of **linear** models: We start our prediction with a common value, then *add* an adjustment corresponding to the particular treatment level (see Figure 1.3).

Generally, the centered model for the one-way layout looks like $\hat{x}_{ij} = \mu_i = \mu + b_i$. You might have noticed a problem with this: It is ambiguous. You could use any value of μ at all and then calculate the b's by subtraction. For example, if our level means are 30 and 40, we might use a common μ of 20, then add b's of 10 and 20. On the other hand, we could let μ be 35 and the b's be -5 and 5. To limit ourselves to one possibility, we need a restriction on the parameters.

We will borrow the restriction from a nice property of sample means, which are the most common estimates. Let μ have the obvious estimate, the overall sample mean of all the measurements $\hat{\mu} = \bar{x} = \frac{1}{n} \sum_{i=1}^{k} \sum_{j=1}^{n_i} x_{ij}$ (a double summation tells us to add the values for all possible combinations of the indices i and j). Then we would just estimate the b's by $\hat{b}_i = \hat{\mu}_i - \hat{\mu} = \bar{x}_i - \bar{x}$.

Example (*cont.*).   For the three sections of Bimini Lagoon, we find $\hat{\mu} = \bar{x} = 38.58$ for the typical salinity in our sample. Then $\hat{b}_1 = \bar{x}_1 - \bar{x} = 37.31 - 38.58 =$

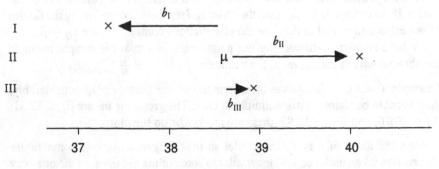

FIGURE 1.3. A centered model for salinity

-1.27 parts per thousand measures how atypical the sample from section I is. Similarly, $\hat{b}_{\mathrm{II}} = 1.52$ and $\hat{b}_{\mathrm{III}} = 0.31$.

Now I want to ask, what is the average value of these predicted adjustments b? It will, of course, just be the difference of the average of all the $\bar{x}_i$ and the average of the $\bar{x}$. Obviously, the average of all the $\bar{x}$, because they are all the same, is still $\bar{x}$. To average the level means, we calculate $\frac{1}{n} \sum_{i=1}^{k} \sum_{j=1}^{n_i} \bar{x}_i$. But this way of writing the double summation means that we should do the second, inner, sum first. This inner sum $\sum_{j=1}^{n_i} \bar{x}_i$ just tells us to add the same number n_i times, to get $n_i \bar{x}_i$. But $n_i \bar{x}_i = n_i \frac{1}{n_i} \sum_{j=1}^{n_i} x_{ij} = \sum_{j=1}^{n_i} x_{ij}$. Then going to the outer sum, the average of the level means is $\frac{1}{n} \sum_{i=1}^{k} \sum_{j=1}^{n_i} x_{ij} = \bar{x}$, the same as the overall average. By subtraction, $\bar{x} - \bar{x}$: The average of the b's is zero. Our adjustments from the common mean are on average the same in the positive and negative directions. (Remember the related fact, that the sum of residuals about a sample mean is zero.) This is such a plausible property that we will require it of any centered model:

Definition. A location model for the one-way layout $\hat{x}_{ij} = \mu_i = \mu + b_i$ is **centered** if the average of the b's over all observations is zero.

Then our algebra gives us the following mathematical result:

Proposition. *The sample mean estimates for the one-way layout parameters create a centered model.*

You should check that this is actually true for the salinity estimates.

1.3.3 Degrees of Freedom

Now we should stop and do a little bookkeeping. We prefer simple models, when we can get away with them; so we need an index of how complicated our model is. An obvious criterion is, the more parameters, the more complicated the model. In the one-way layout, we measure n observations, then try to predict them as well as we can with only k treatment means. We say that the model has k *degrees of freedom*. For example, in the saltwater problem we try to represent 30 measurements by just 3 water-mass averages.

At first glance, it may seem that in the centered model we must estimate a single μ and k different b_i's, for a total of $k + 1$ parameters. But remember that the b's average is 0, which means that the grand total of the b's for all observations is zero: $\frac{1}{n} \sum_{i=1}^{k} n_i b_i = 0$. This means that after computing the first $k - 1$ parameters b_i, we can compute the last one without doing any more estimating by just solving this equation: $b_k = -\frac{1}{n_k} \sum_{i=1}^{k-1} n_i b_i$. So we really have only one μ and $k - 1$ *algebraically independent* b's to estimate. For the salinity data, this comes to 1 overall average μ, plus the fact that 2 (out of 3) adjustments b are algebraically independent. In a similar manner, as an exercise you should discover that the n estimated residuals $x_{ij} - \hat{x}_{ij}$ actually involve only $n - k$ algebraically independent quantities (27 independent residuals in the salinity data).

The way statisticians say this is that the original experiment has n **degrees of freedom**, and we have broken them down into 1 degree of freedom for the center μ, $k - 1$ degrees of freedom for the adjustments b_i, so that the model has a total of k degrees of freedom. Then we are left with $n - k$ degrees of freedom for the estimated residuals. That is, $n = 1 + (k - 1) + (n - k)$. We blame the loss of those k degrees of freedom on the fact that we had to estimate k parameters using our n pieces of data. This check-sum bookkeeping will turn out to be increasingly important as our models and their analyses become more complicated.

1.4   Two-Way Layouts

1.4.1   Cross-Classified Observations

Very often our scientist will want to allow for the possibility that some further distinction among the measurements affects the comparisons he is primarily interested in.

Example. Educational psychologists are excited about a new way of teaching arithmetic to third graders. Obviously, we would test whether it is really an improvement by trying it out on a collection of children, while at the same time having a similar sample of children use the old lessons (this second group is called a *control* group). At the end, we give both groups a test to see how they do; this is just the sort of one-way layout we talked about earlier.

But some teachers claim that the new curriculum seems to work better with girls than with boys. From our own experience, we do not believe this claim, but if we are to convince our fellow teachers, we must allow for this possibility somehow. We clearly want to give each of the curricula to both boys and girls. The results may be displayed in a table of test scores:

Arithmetic Test Scores

	Boys	Girls
New	15 18 26	13 17 21
	28 30	25 29
Old	11 14 16	9 10 18
	22 23	19 24

This is an example of a **two-way layout**. It will require an impressive triple-index notation, but which fortunately will be easy to decode. Generally, we have a collection of observations denoted by x_{ijk} where $i = 1, \ldots, l$ keeps track of the levels of the first (row) **factor**, and $j = 1, \ldots, m$ keeps track of the levels of the second (column) factor. Then the pair of indices ij determine a particular **cell**, a box in a table like the one in the example, in which all subjects receive the same levels of the treatments. That third index just keeps track of the observations in the ijth cell, so that $k = 1, \ldots, n_{ij}$, where we had n_{ij} observations in that cell. Then the total number of subjects receiving the ith level of the first factor must be $n_{i\bullet} = \sum_{j=1}^{m} n_{ij}$ (summing over columns); and the number receiving the jth level from the second factor is $n_{\bullet j} = \sum_{i=1}^{l} n_{ij}$ (summing over rows). The dot keeps track of the missing index, so we can tell whether the letter is a row or column index. Then the total number of subjects for the experiment must be $\sum_{i=1}^{l} n_{i\bullet} = \sum_{j=1}^{m} n_{\bullet j} = n_{\bullet\bullet} = n$. In the example above, $x_{213} = 16$, $n_{21} = 5$, $n_{\bullet 2} = 10$, and $n = 20$.

As usual, we want to summarize these results so we can tell people simple and useful things about the treatments we have carried out. The easiest model to construct just ignores the table organization and lets every pair of factor levels, every cell, be a single level of treatment. Then the location model prediction just says $\hat{x}_{ijk} = \mu_{ij}$; presumably, the estimate of the typical value for, say, girls learning arithmetic the old way will be based only on the result for the five girls in that part of the experiment. This is called the **full** model, because we are making the finest distinctions possible among our subjects. The model has, of course, $l \times m$ degrees of freedom, one for each cell.

The standard estimate will be simply the sample mean of the observations in that cell: $\hat{\mu}_{ij} = \bar{x}_{ij} = \frac{1}{n_{ij}} \sum_{k=1}^{n_{ij}} x_{ijk}$.

Example (*cont.*).   In the arithmetic-teaching example, we estimate $\bar{x}_{11} = 23.4$, $\bar{x}_{12} = 21.0$, $\bar{x}_{12} = 17.2$, $\bar{x}_{22} = 16.0$. That is complicated enough that a picture should help (see Figure 1.4).

Hairlines are individual test scores, and they show that, as usually happens in experiments with people as subjects, the peculiarities of children and tests seem to matter much more than the groups we are distinguishing. We can still see possible patterns: The solid lines show that for each gender, the new teaching method

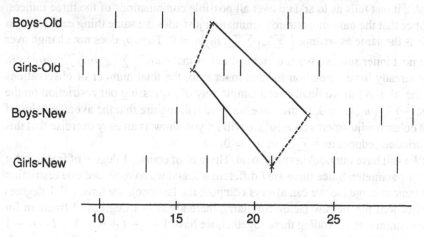

FIGURE 1.4. Arithmetic test scores: full model

averaged higher scores than the old. The dotted lines show that in each curriculum group, the boys' scores were on average slightly higher than the girls'.

1.4.2   Additive Models

What about the complaint that led to this analysis, that the curriculum is more of an improvement for girls than for boys? Actually, in our little experiment, the boys' average improvement (6.2) was slightly more than the girls' improvement (5.0); so our results provide no evidence for the claim. The similarity of these two improvements supports the idea that the two improvements were in fact the same. We can write a simple model for this situation: We imagine that there is an overall test-performance center, then add or subtract some amount for each curriculum; next we add or subtract some other amount for each gender. The sample mean estimates are easy to get: For the center, the overall mean is just 19.4. Since the mean for the new curriculum is 22.2, then its improvement is $22.2 - 19.4 = 2.8$ on average. The boys' mean is 20.3; so their edge is $20.3 - 19.4 = 0.9$. The disadvantages of the old curriculum and of being a girl are expressed by adding the negatives of these differences. Such numbers answer the most obvious questions about test performance.

What does this model say, for example, about girls who take the new curriculum? We predict a score of $19.4 + 2.8 - 0.9 = 21.3$. This is clearly not the same as the prediction of the full model using cell estimates, 21.0 (though in this particular experiment they chanced to be very close).

Our new model is called the **additive** model for the two-way layout, and the notation is as follows: $\hat{x}_{ijk} = \mu + b_i + c_j$, where b_i is the adjustment for the ith level of the row factor, and c_j is the adjustment for the jth level of the column factor. These were estimated by adding or subtracting from the overall mean; so once again we want a centered model. We impose the restriction that on average, the b's must be zero: $\frac{1}{n} \sum_{i=1}^{l} \sum_{j=1}^{m} \sum_{k=1}^{n_{ij}} b_i = 0$. As threatening as a triple summation looks, it just tells us to add up over all possible combinations of the three indices. Notice that the innermost (third) summation just adds the same thing each time, so this is the same as writing $\frac{1}{n} \sum_{i=1}^{l} \sum_{j=1}^{m} n_{ij} b_i = 0$. Then b_i does not change over the next inner sum, so we can factor it out of that sum: $\frac{1}{n} \sum_{i=1}^{l} b_i \sum_{j=1}^{m} n_{ij} = 0$. We already have a notation for that inner sum, the total number of observations in the ith row; so we finally get a simple way of expressing our restriction on the b's: $\frac{1}{n} \sum_{i=1}^{l} n_{i\bullet} b_i = 0$. In the same way, we will require that the average value of the column adjustments be zero; we will let you show as an easy exercise that this restriction reduces to $\frac{1}{n} \sum_{j=1}^{m} n_{\bullet j} c_j = 0$.

We still have our bookkeeping to do. There is, of course, 1 degree of freedom for the μ parameter. Since there are l different b's, and we have placed one restriction on their average (so we can always compute the last one), we have $l - 1$ degrees of freedom for the row factor. Similarly, there are $m - 1$ degrees of freedom for the column factor. Adding these together, we have $1 + l - 1 + m - 1 = l + m - 1$ degrees of freedom for the additive model for the two-way layout. The residuals in our predictions of how each child will do, $x_{ijk} - \hat{x}_{ijk}$, of which there are n, must then have $n - l - m + 1$ degrees of freedom, because we had to estimate our $l + m - 1$ parameters from the n observations. That is, we have a checksum $n = 1 + (l - 1) + (m - 1) + (n - l - m + 1)$.

Standard estimates of the parameters are obtained just as in our example. The overall center may be estimated using the mean of everybody, $\hat{\mu} = \bar{x} = \frac{1}{n} \sum_{i=1}^{l} \sum_{j=1}^{m} \sum_{k=1}^{n_{ij}} x_{ijk}$. Then we estimate the column adjustment b_i by finding the column sample mean $\bar{x}_{i\bullet} = \frac{1}{n_{i\bullet}} \sum_{j=1}^{m} \sum_{k=1}^{n_{ij}} x_{ijk}$ and then subtracting the overall mean: $\hat{b}_i = \bar{x}_{i\bullet} - \bar{x}$. In the same way, we estimate the column adjustments by $\hat{c}_j = \bar{x}_{\bullet j} - \bar{x}$. The estimated prediction of the model then looks like $\hat{x}_{ijk} = \hat{\mu} + \hat{b}_i + \hat{c}_j = \bar{x} + (\bar{x}_{i\bullet} - \bar{x}) + (\bar{x}_{\bullet j} - \bar{x}) = \bar{x}_{i\bullet} + \bar{x}_{\bullet j} - \bar{x}$.

1.4.3   Balanced Designs

Our standard estimate of the additive model seems quite reasonable; but that is a little bit of an accident, because in our example we had the same number of observations in each cell. The additive model would still be interesting in other cases. But if the numbers of observations in the cells of different rows vary, our estimates of the column adjustments $\hat{c}_j$ using the sample average of each column are no longer entirely convincing. For example, if the counts of observations are $\begin{array}{|cc|} 3 & 6 \\ 5 & 2 \end{array}$, the sample average of the first column is based mostly on the second row (5 observations versus 3); but in the second column the average is based mostly on the first row (6 observations versus 2). Intuitively, this is not fair; so we will single out a class of designs that do not have this problem:

Definition. A two-way layout has a **balanced design** whenever the numbers of observations in the cells of each row are proportional; that is, $(n_{ij}/n_{i\bullet}) = (n_{\bullet j}/n)$ for each $i = 1, \ldots, m$ and each $j = 1, \ldots, l$.

Any design (like our example) in which all the n_{ij}'s are the same, is of course, balanced. Another example of a balanced design is one where the counts of observations are $\begin{array}{|cc|} 1 & 2 \\ 2 & 4 \end{array}$, since $\frac{1}{3} = \frac{2}{6} = \frac{3}{9}$. You should prove as an exercise that we could equally as well have said that the cells of each *column* are proportional.

The only significance of balanced designs is that the standard estimates of parameters make sense. Lazy statisticians have made themselves very unpopular with scientists by telling them that their experiments were bad if they were not balanced. This is false; we can, with slightly more sophisticated estimates, extract just as much information from an unbalanced experiment. We will see how in Chapter 2.

We will let you show off your skill with summation signs by proving the following as an exercise:

Proposition. *The standard estimates for the additive model are centered.*

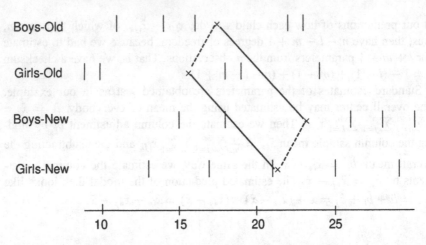

FIGURE 1.5. Arithmetic test scores: additive model

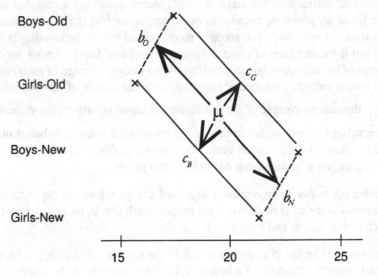

FIGURE 1.6. Parallelogram of additive model

We draw a picture of the additive model for the math test in Figure 1.5. Because in this instance the additive model is similar to the full model, you may have to stare at Figures 1.4 and 1.5 a moment to see the difference. In the additive model, opposite edges of the quadrilateral go over and down by the same amount (when you add your row or column corrections); therefore, opposite edges are parallel and of the same length. The figure is now a *parallelogram*, and not just any quadrilateral (see Figure 1.6).

Generally, the graph for any two-by-two experiment with an additive model will be a parallelogram. If there are more than two levels of a factor, the picture is more complicated; but the solid lines connecting equivalent levels of the first factor are still parallel. In the same way, dotted lines connecting the equivalent levels of the second factor are parallel.

1.4.4 Interaction

Just how different are the full and the additive models for the two-way layout? Our geometrical analysis suggests that additive models are more restricted in what they can predict—they must form parallelograms, while full models may (or may not) form parallelograms. This suggests that the full models have the freedom to follow the sample observations better, leading to generally smaller residuals. Let us quantify the difference by subtracting the degrees of freedom for the additive model from those for the full model: $l \times m - (l + m - 1)$. Factor that expression to conclude that the latter requires us to estimate $(l - 1)(m - 1)$ more parameters than the former $((2 - 1) \times (2 - 1) = 1$ more parameter in the case of our 2 rows by 2 columns experiment).

Now let us quantify the difference in the predictions made by the full model and the additive model: Of course, the standard estimated prediction for the full model was just $\hat{x}_{ijk} = \bar{x}_{ij}$. The difference between the two is then $\bar{x}_{ij} - \bar{x}_{i\bullet} - \bar{x}_{\bullet j} + \bar{x}$. In our example, for boys in the old curriculum, it is -0.3. You should notice that for every cell in our example, it is either plus or minus that same quantity. This is what we meant when we said that the full model had exactly one more degree of freedom; only that one amount is available to improve the predictions. In general, these quantities measure a very important feature of the full model, the **interaction**. It is the amount by which you *cannot* say that the result of a two-way experiment is just a common value plus a column adjustment plus a row adjustment. In our example, it is the amount by which the girls in the class were helped more than the boys by the new curriculum.

There is no reason for interactions to be small; In Figure 1.7 are plots of the cell averages (full models) for three different two-by-two experiments

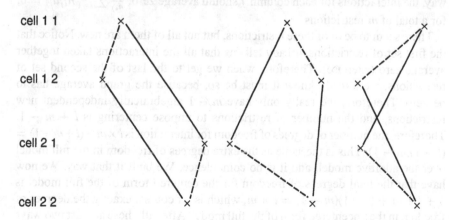

FIGURE 1.7. Three degrees of interaction

The horizontal axis (whatever you measured in each experiment) and the raw data have been left out so that you can see the qualitative features of the models. In the leftmost example, the figure is just about a parallelogram; this means that an additive model seems to explain the cell centers satisfactorily.

In the middle example, there are consistent and perhaps noteworthy row and column adjustments; row 1 is higher than row 2, and column 2 is higher than column 1. But these adjustments are enough different in the different cells that we have nothing like a parallelogram. In this case, interaction will be substantial.

In the rightmost example, we see no common row or column adjustments; the factors seem to lack any consistent effects. This time, there is a great deal of interaction, and little else going on. We might see such a picture, for example, when experimenting with one of those drugs that is a tranquilizer when given to children and a stimulant when given to adults; therefore, its effect on level of activity is opposite for the two groups.

1.4.5   Centering Full Models

We can now provide a centered parametrization of the full model. We just append an interaction term to the additive model: $\hat{x}_{ij} = \mu = b_i + c_j + d_{ij}$. The d's are just those corrections whose standard estimates were $\hat{d}_{ij} = \bar{x}_{ij} - \bar{x}_{i\bullet} - \bar{x}_{\bullet j} + \bar{x}$, calculated above. The restrictions that make this a centered model are as before: $\frac{1}{n} \sum_{i=1}^{l} n_{i\bullet} b_i = 0$ and $\frac{1}{n} \sum_{j=1}^{m} n_{\bullet j} c_j = 0$.

What restrictions do the interaction terms require? Of course, as corrections we want them to be zero on average; but even more, we want the set of corrections to *each level* of the row factor to average zero. This is because if the average interaction in that row is not zero, we should have added that average adjustment to the corresponding row adjustment b_i in the first place. Then the additive part of the model would be that much more accurate in its predictions. So our restriction for row i looks like $\frac{1}{n_{i\bullet}} \sum_{j=1}^{m} n_{ij} d_{ij} = 0$. There are l of these restrictions. In the same way, the interactions for each column j should average zero: $\frac{1}{n_{\bullet j}} \sum_{i=1}^{l} n_{ij} d_{ij} = 0$, for a total of m restrictions.

There seem to be m of these restrictions, but not all of them are new. Notice that the first set of restrictions already tells us that all the interactions taken together average zero (exercise). Therefore, when we get to the last of the second set of restrictions, we already know it must be so, because the grand average has to be zero. Therefore, we really only have $m - 1$ algebraically independent new restrictions, and the number of restrictions to impose centering is $l + m - 1$. Therefore, the number of degrees of freedom for interaction is $l \times m - (l + m - 1) = (l - 1)(m - 1)$. This is the same as the extra degrees of freedom in the full model over the additive model, and it is no coincidence: We built it that way. We now have that the total degrees of freedom for the centered form of the full model is $1 + l + m + (l - 1)(m - 1) = l \times m$, which is, of course, exactly the degrees of freedom in the uncentered form of the full model. After all, these are just two ways of writing the same thing. You should now prove the following as an exercise:

Proposition. *The standard estimates of* μ*, the b's, and the c's plus the standard estimates of the interactions* $\hat{d}_{ij} = \bar{x}_{ij} - \bar{x}_{i\bullet} - \bar{x}_{\bullet j} + \bar{x}$ *form a centered full model.*

Of course, our standard estimates of the full centered model are satisfactory only if the experiment is balanced. The cell averages still give the right predictions for any full two-way layout, though, because they come from the one-way layout, where there was never a problem with balance.

The style of statistical analysis we have been studying in this chapter was first explored in depth in the 1920s by R. A. Fisher; and it has revolutionized scientific research throughout biology, medicine, and the social sciences. You may explore its many variations in advanced courses called something like "experimental design." For example, a number of new possibilities arise when there are three factors.

Of course, we have not yet addressed a fundamental issue: How do we tell how well a model matches (statisticians say *fits*) the data? It is perfectly possible to estimate the parameters of a truly stupid model, such as an additive model in cases where a great deal of interaction seems to be present. In other cases, it may seem to the eye that an additive model is adequate in a particular application, or even that we can ignore one of the factors. But is there some more objective way to decide whether we are doing the right thing? We will tackle such matters later in this book.

1.5   Regression

1.5.1   Interpolating Between Levels

Sometimes, if the levels of our treatment have a numerical meaning, we can extract still more information from the observations in even a one-way layout.

Example. Twelve subjects whose blood pressure is disturbingly high are given an eight-week regimen of a new pressure-lowering drug. At the end of that time, the *change* in their diastolic pressures is measured (a negative number is good). The patients were arbitrarily divided into two groups: One got 100 milligrams a day, the other, 200 milligrams. The results were

$$100 \text{ mg} : -40, -30, -25, -10, 0, 15;$$
$$200 \text{ mg} : -50, -35, -30, -20, -15, 10.$$

You might draw parallel hairline plots to see what is going on here. The sample means of the two dosage groups are -15 and -23.33, with an overall mean -19.17. Then the standard estimates of the centered model are $\hat{\mu} = -19.17$, $\hat{b}_{100} = 4.17$, and $\hat{b}_{200} = -4.17$. On average, the group who received the larger dose did better.

There is nothing new here, but what if the investigators notice something else: The higher-dose group are just beginning to show signs of an unpleasant (but not deadly) side effect? The lower-dose group has no problems. From experience with similar drugs, it is suggested that a relatively modest drop in the dosage may alleviate the side effects. So a new series of experiments is proposed, with doses like 175 mg per day included. Since these new experiments take time and money, it would be nice to make intelligent guesses in advance of their effect on blood pressure, using what we have already learned. Unfortunately, we did not give anybody 175 mg per day. You will probably have thought of a reasonable thing to do: *interpolate*. The halfway point between the doses, 150 mg, should correspond in this case to the overall mean, -19.17 mm. A dose of 175 mg is $(175-150)/(200-150)$ of the way from the middle to the upper dose, which corresponded to an increase blood pressure of -4.17 mm. So our predicted response to a dose of 175 mg is $-19.17 - 4.17(175-150)/(200-150) = -21.25$ mm. That was certainly easier than doing the whole experiment again.

Notice that this interpolation procedure works for *any* new dose:

$$\hat{p} = -19.17 - 4.17\frac{d-150}{200-150} = -19.17 - 0.0833(d-150)$$

(where p is change in blood pressure and d is drug dose). You should check that this is just a novel way of writing the usual one-way model—it makes the same predictions at 100 and at 200 mg. It is called the *linear regression* model for this experiment.

Let us draw a picture of our situation (Figure 1.8).

We have turned the picture on its side; this is the conventional way to draw a regression model. The $\times$'s represent the sample means of the changes for our two dosage groups. Notice that a linear regression model was the equation of a straight

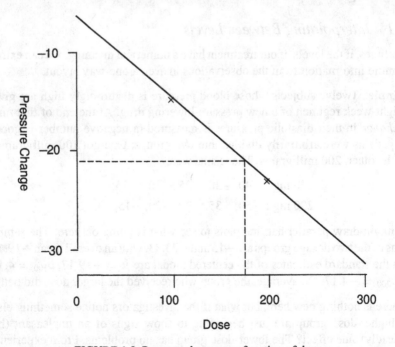

FIGURE 1.8. Pressure change as a function of dose

line, which we have drawn on the graph. This sloped line represents our various possible interpolations. The dotted line shows how to make such a prediction: start at 175 mg, go up until you hit the solid line, then go across to read off the prediction on the vertical scale.

How seriously should we take such predictions at interpolation points like 175? There are two limitations to this method:

(1) The predictions are unlikely to be much better than the means at the original doses. Remember that the 6 people in the 100 mg dose group varied from -40 to 15, and the 6 people in the 200 mg dose group varied from -50 to 10; so the predictions at 100 and 200 mg are not likely to be wonderfully accurate anyway. In between, at, say, 150 mm, there may be a slight improvement because 12 people rather than 6 contributed to the calculation. But notice that outside the actual experimental range, at, say, 0 or 300 mg, the prediction would likely be quite a bit worse: Errors in one sample mean or the other will swing the line wildly by a sort of lever effect (see the graph in Figure 1.9). That is why we should rarely trust such *extrapolated* rather than merely interpolated estimates.

(2) Are we at all sure that the actual pattern of response to the various doses is a straight line? Laws of nature can take a great many mathematical forms. Since pharmacology provides no helpful general theory about what sort of equation to use, we guessed the simplest continuous function we knew of, a straight line. If the line in our picture should really be curved, our predictions will be systematically wrong (*biased* is the statistician's word). Furthermore, they are likely to be, again, even worse for extrapolated than for interpolated doses.

Example. If the true connection between dose and blood pressure follows the dotted line in Figure 1.9, so that our estimates were only slightly off at the experimental doses, notice how far off our extrapolations are near 0 and 300 mg. On the other hand, if the true connection is the dashed, curved line, our experimental estimates were just about right; but our extrapolated straight line still goes quickly wildly wrong for extreme doses. In the exercises you will see an example of how to make predictions with curved models (if you know you need one).

1.5.2   Simple Linear Regression

If we remember to be cautious, regression can be a widely useful tool. Generally, a **simple linear regression** model works as follows: We measure the numerical responses of our subjects, y_i, for $i = 1, \dots, n$. The responses to the experiment are values of the *dependent* variable (the blood pressure changes in our example). For each subject we have a numerical value describing the conditions of the experiment, x_i, which are values of the *independent* variable (in our example, drug dosages of 100 or 200 mg). Then we make predictions $\hat{y}_i = \mu + b(x_i - \bar{x})$ where $\bar{x}$ is the average independent variable value at all the observations (here, 150 mg). (This is a centered model, as you will check in an exercise.) You should remember from analytic geometry that b is the slope of the line we have drawn. The model possesses two degrees of freedom, one each for μ and for b.

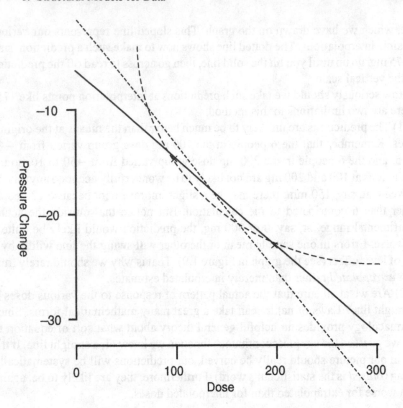

FIGURE 1.9. Erroneous and nonlinear regression

Example. Our example had only two values of the dependent variable, the drug dosage; but a simple linear regression model allows for any number. Figure 1.10 shows the weights of purebred beagles at four different ages, 6, 8, 10, and 12, with four puppies of each age.

The diamonds mark the cell-mean estimates of a one-way layout; the crosses, the weights of individual dogs. To interpolate for other ages, the obvious device is to connect the crosses with straight segments, as in our dotted path. This is an example of a *nonparametric regression* estimator, which you may see again in advanced courses.

In our example, it is interesting how the crosses fall near a single straight line (though not exactly); a possible line is the solid segment. Such a simple linear regression prediction has the advantage of being much simpler than the broken line. (2 degrees of freedom instead of the 4 for the one-way-layout estimates). The predictions are obviously nearly the same. Of course, we do not expect the curve to continue to follow closely a straight line, or we would have 50-pound beagles at the end of a year. On the other hand, our prediction for a puppy age 7 weeks (about 6.5 pounds) is quite plausible.

You have no doubt noticed a problem. Since I did not find the line by interpolation of level means, how do I draw that straight line, that is, estimate μ and b? We are

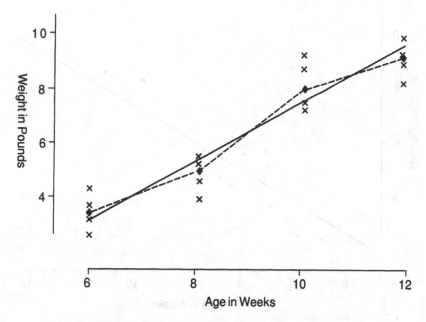

FIGURE 1.10. Weight as a function of four ages

stuck: There is no longer an obvious choice for the standard estimator. A powerful general method for obtaining such estimates will be introduced in the next chapter.

Simple linear regression models may be useful for summarizing the results of many other experiments. For example, instead of selecting puppies of a few specific ages, we might have simply taken a variety of puppies, recorded each of their ages, then weighed them. There might then be as many independent variable values (ages) as there are dogs. The results are captured in the Figure 1.11.

We use ×'s to mark the points whose coordinates are the age and weight of a particular dog. This kind of diagram, one of the most useful in all of statistical graphics, is called a *scatter plot*. We use it to compare any two distinct measurements we take on each of a number of different subjects. In this example, though the ×'s for the puppies are widely scattered, we see a pattern that might be stated as follows: The average weights of the puppies of approximately the same age follow a linear upward trend. The solid line is a proposed simple linear regression model, $\hat{w}_i = \mu + b(a_i - \bar{a})$ (w is a weight and a is an age). Once again, we shall have to wait until Chapter 2 to find good estimates of μ and b.

1.6   Multiple Regression*

1.6.1   Double Interpolation

In factorial experiments, we split up our subjects among several levels of two or more treatments. We successfully interpolated numerical levels in the one-

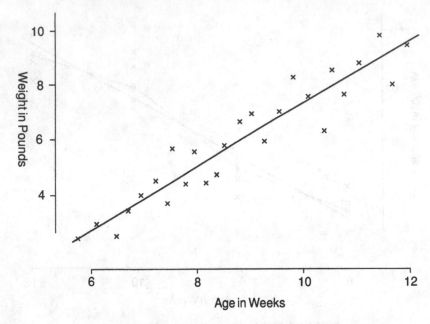

FIGURE 1.11. Weight as a function of many ages

way layout; perhaps something similar might work when each of the factors has numerical levels.

Example. We study the effect of cooking time and temperature on a standard cake recipe. Three cakes are baked at each of 350 and 375 degrees, and for 20 and 25 minutes. At the end we measure the percentage of the original moisture that remained in the cake:

		Time	
		20	**25**
Temperature	**350**	40 36 41	28 27 32
	375	32 37 30	19 24 25

When we compute the standard estimates of an additive model, we get $\hat{\mu} =$ 30.917 and that the adjustment for going to the higher temperature is -3.083 and the increment for going to the longer time is -5.083. (You should check my calculations as an exercise.) A graph looks like that shown in Figure 1.12.

The two baking times correspond to the lines that go from lower left to upper right, and the two temperatures to the lines at right angles to them. You can see from the observations that the additive model works fairly well.

Now we can carry out a *double* interpolation to predict, for example, how much moisture will remain in a cake left in a 360 degree oven for 23 minutes. The center of the experiment is at 22.5 minutes and 362.5 degrees. We would, of course, predict that the percentage of moisture in cakes cooked in that way would be the overall average of all our cakes, 30.917%. Now adjust for the distance from that

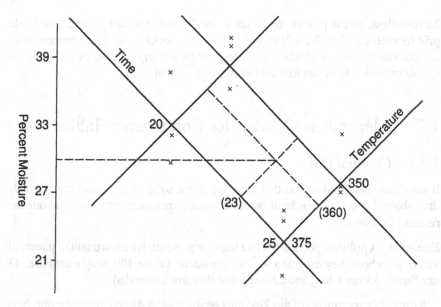

FIGURE 1.12. Cake moisture as a function of time and temperature

center by computing

$$\hat{m} = 30.917 - 5.083\frac{23 - 22.5}{25 - 22.5} - 3.083\frac{360 - 362.5}{375 - 362.5} = 30.517.$$

You can read this in a rough way off the plot: Interpolate between 20 and 25 to get one dotted line, and between 350 and 375 to get the other; then find their intersection. That position on the vertical scale gives an estimate of their moisture level. (We felt free to use the standard estimates of the parameters in this model because it was based on a balanced two-way layout.)

1.6.2 Multiple Linear Regression

Generally, a linear regression model for a dependent variable y using two independent variables x_1 and x_2 looks like

$$\hat{y}_j = \mu + (x_{1j} - \bar{x}_1)b_1 + (x_{2j} - \bar{x}_2)b_2$$

in centered form, where j keeps track of the settings for a single observation. The model has 3 degrees of freedom, one each for μ, b_1, and b_2. We noticed from our example that it corresponds to a two-by-two additive factorial model when there are two levels of each independent variable. Therefore, the standard estimates could be obtained in the obvious way from row and column means.

If there are more than two levels of either variable, the regression model is no longer equivalent to a factorial design, as you may see by counting degrees of freedom. The regression model is a simplification of the factorial model, and we do not yet know a standard estimator for it, whether the design is balanced or not.

Nevertheless, we can plot the model just as we did above, with a parallel coordinate grid for each variable. We will let you graph one as an exercise. Furthermore, there are obviously *multiple linear regression* models for any number of independent variables, which look just like the two-variable model.

1.7   Independence Models for Contingency Tables

1.7.1   Counted Data

It may have occurred to you that there are other sorts of statistical experiments than those that provide us with repeated, varied measurements. What about the results of surveys?

Example. A political polls asks a (we hope) representative assortment of potential voters for whom they expect to vote for President. Of the 100 people they ask, 43 say Smith, 35 say Chan, and 22 insist that they are undecided.

Results of experiments of this kind may be summarized as counts of the numbers of subjects who fall into various categories. The most common model for these counts is the *proportions* model, which is what we are doing when we summarize our survey as 43% Smith, and so forth.

Formally, we have a set of counts of the numbers of subjects falling in distinct categories x_i for $i = 1, \ldots, k$, where $\sum_{i=1}^{k} x_i = x_\bullet = n$. In the example above, $k = 3$, $n = 100$, and, for example, $x_2 = 35$. We imagine that these subjects are representative of a much larger class of potential subjects, called a *population*. The **multinomial proportions model** asserts that a true proportion p_i of potential subjects from that population falls into the ith category, so that $\sum_{i=1}^{k} p_i = p_\bullet = 1$ (as we expect proportions to behave). The predicted counts in the category for our experiment are then, of course, $\hat{x}_i = np_i$.

Example. Genetic theory predicts that in a third-generation crossbreeding experiment there should be population proportion of 25% individuals of type AA, 50% of type AB, and 25% of type BB. In the notation for the multinomial proportions model, $p_{AA} = 0.25$, $p_{AB} = 0.50$, and $p_{BB} = 0.25$. If we do the experiment with 40 individuals arising in the third generation, then our predicted counts (we sometimes say *expected* counts) are $\hat{x}_{AA} = 0.25 \times 40 = 10$, $\hat{x}_{AB} = 20$, and $\hat{x}_{BB} = 10$. But of course, when the experiment is carried out, the recombinations are not precisely predictable, and we get actual counts like $x_{AA} = 11$, $x_{AB} = 22$, and $x_{BB} = 7$ (called the *observed* counts). Later in the book we will learn something about just how large a difference between observed and expected counts might reasonably be accepted as ordinary variation.

Of course, in the political polling example we do not know the true proportions to expect. You will surely have guessed the standard estimates of the population proportions: $\hat{p}_i = x_i/n$, the *sample proportions*. In our example, we estimate that candidate Chan has 0.35 of the vote, since $n = 100$. If we do use the sample proportion estimate for this model, notice that the actual and estimated counts always coincide: $\hat{x}_i = n\hat{p}_i = x_i$. This time, we have nothing like residuals with which to evaluate the quality of the model.

As with measurement experiments, counting experiments become much more interesting when the subjects are classified by the levels of two or more factors:

Example. A Hollywood studio is test-marketing a new film; and viewers are simply asked whether or not they liked the movie enough to recommend it to friends. An executive voices concerns that its market may be limited if substantially smaller proportions of either men or women like it; so responders are classified by gender:

| | Observed Counts | | |
	Male	Female	
Like	51	83	134
Dislike	42	24	66
	93	107	200

The survey counts appear in the middle of the table. The other numbers are row and column totals, and the grand total of 200 subjects. This is called a *contingency table*.

Generally, we will denote a two-way classification by an array of counts x_{ij}, for $i = 1, \ldots, k$ and $j = 1, \ldots, l$; then write $\sum_{i=1}^{k} x_{ij} = x_{\bullet j}$, $\sum_{j=1}^{l} x_{ij} = x_{i\bullet}$ and

$$\sum_{i=1}^{k}\sum_{j=1}^{l} x_{ij} = \sum_{j=1}^{l} x_{i\bullet} = \sum_{i=1}^{k} x_{\bullet j} = x_{\bullet\bullet} = n.$$

In our movie example, $x_{12} = x_{LF} = 83$, $x_{2\bullet} = x_D = 66$, and $n = 200$.

The multinomial, or *saturated*, model consists of population proportions for the individual cells p_{ij}, with column proportions $\sum_{i=1}^{k} p_{ij} = p_{\bullet j}$, row proportions $\sum_{j=1}^{l} p_{ij} = p_{i\bullet}$, and, of course,

$$\sum_{i=1}^{k}\sum_{j=1}^{l} p_{ij} = \sum_{j=1}^{l} p_{i\bullet} = \sum_{i=1}^{k} p_{\bullet j} = p_{\bullet\bullet} = 1.$$

It corresponds to the full model for a two-way layout.

The standard estimates of these parameters are again the sample proportions $\hat{p}_{ij} = x_{ij}/n$, and, of course, $\hat{p}_{i\bullet} = x_{i\bullet}/n$ and $\hat{p}_{\bullet j} = x_{\bullet j}/n$. In our example, the proportion of moviegoers we wanted to survey who are female fans of the movie we estimate to be $\hat{p}_{LF} = 83/200 = 0.415$. The proportion of females in the survey population is about $\hat{p}_F = 107/200 = 0.535$.

1.7.2   Independence Models

In our example, $51/93 = 0.548$ of the men liked the movie, whereas $83/107 = 0.776$ of the women did. This suggests that it is more of a women's movie; but of course, we have no idea whether this is an accident of our sample and perhaps not a characteristic of people in general. To get a better idea, let us see how consistent our survey is with another model, in which gender makes no difference at all.

If that were the case, then the important parameters would be a population proportion of males p_M and a proportion p_L of people who would like the movie. If gender and taste are unrelated, then of the np_M males you would expect to find in the survey, a proportion p_L would like it, for a predicted count of favorable male viewers $np_L p_M$. We may estimate this by $n\hat{p}_L \hat{p}_M = 200 \frac{134}{200} \frac{93}{200} = 62.3$ men in the survey who might be expected to like the movie, if gender is irrelevant to taste. Then we may ask ourselves whether this is different to an important degree from the 51 men who actually liked it in our survey, and whether such a difference might have been an accident of who we happened to pick for our sample. (Of course, we do not know enough yet to come up with a sensible answer.) This sort of model, in which row and column classification are assumed irrelevant to each other (and so we calculate proportions of proportions by multiplication), is called an **independence** model. The concept is one of the most useful in all of statistics. The row and column proportions become the key parameters of the model, and we predict counts by $\hat{x}_{ij} = np_{i\bullet} p_{\bullet j}$.

In Figure 1.13, we have represented the moviegoing population by a square of area one. The vertical subdivisions represent the proportions of males and females in that population; the horizontal subdivisions represent the proportions of the population who like and dislike the movie. Therefore, our model predicts that the shaded area, $p_L p_M$, will be the proportion of moviegoers who are male enthusiasts for our movie.

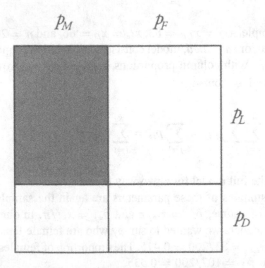

FIGURE 1.13. The independence model

You might notice (exercise) that if the independence model is exactly true, we get a table of counts that, if it represented the numbers of observations in each cell in a two-way layout, would be balanced. Therefore, when we design a two-factor experiment to be balanced, we are arranging that the factors be independent of one another.

To evaluate the model, we estimate the row and column proportions, then use them to create a table of the counts we would have *expected* to see. For example,

	Expected Counts		
	Male	Female	
Like	62.3	71.7	134
Dislike	30.7	35.3	66
	93	107	200

We called the original table, with the raw data, the *observed* counts; comparing the two tables should tell us how good the independence model is. Notice, by the way, that the difference between observed and expected counts, a sort of residual, is plus or minus 11.3 in each of our four cells. Notice also that the row and column totals are exactly the same in the two tables. As an exercise, you should check that this is always true for independence models.

1.7.3 Loglinear Models

You probably noticed that our two-way contingency tables and two-way layouts may both be displayed in rectangular tables. The similarity goes deeper. The additive model for the layout involved *adding* adjustments for the row and column factors, whereas the independence model for a contingency table required us to *multiply* row and column proportions. But we can make the parallel clearer by turning multiplication into addition. You know how to do that: take *logarithms*, and use the standard fact that $\log ab = \log a + \log b$. Starting with the multinomial proportions model $\hat{x}_i = np_i$, we get $\log \hat{x}_i = \log n + \log p_i$. (Time to start getting used to a convention: In statistics, logarithm always means *natural* logarithm [base e] unless you clearly state otherwise.) Read this as a linear predictive model for the logarithms of cell counts.

So far nothing interesting has happened; but we found earlier that it helped to create a *centered* version of the model, with a middle value plus a correction for the particular category. This would look like $\log \hat{x}_i = \mu + b_i$, much like a one-way layout. Then we required that the level effects, averaged over the observations, be zero. In this model the individual numerical observations are cell counts; so we will require that the averages of the b's over *cells* be zero: $\frac{1}{k} \sum_{i=1}^{k} b_i = 0$. Now let us connect the two ways we have written our models. Sum both versions over all categories to get

$$\sum_{j=1}^{k} \log \hat{x}_j = k \log n + \sum_{j=1}^{k} \log p_j = k\mu + \sum_{j=1}^{k} b_j = k\mu.$$

The sum of the b's disappeared because of the centering condition. Therefore, $\mu = \log n + \frac{1}{k}\sum_{j=1}^{k}\log p_j$. Now substitute this back into the centered version and solve for $b_i = \log \hat{x}_i - \mu = \log n + \log p_i - \log n - \frac{1}{k}\sum_{j=1}^{k}\log p_j = \log p_i - \frac{1}{k}\sum_{j=1}^{k}\log p_j$.

Example. In the genetics example above with $n = 40$ individuals and $k = 3$ genotypes, we obtain $\mu = 2.534$, $b_{AA} = b_{BB} = -0.231$, and $b_{AB} = 0.462$ (so the adjustments do sum to zero).

Sample estimates of the μ's and b's can be gotten by using sample proportions in the same way. We count degrees of freedom by starting with k categories and letting μ have 1 and the b's have only $k - 1$, because we force them to average 0.

But what do the parameters in these new models mean? The parameter μ is just an average log count, but we can say more about the b's. In the case where there are only two categories, as in Like/Dislike (or Yes/No, or Male/Female) the formula reduces to $b_L = 1/2(\log p_L/p_D) = 1/2(\log p_L/(1 - p_L))$, by familiar facts about logarithms and the fact that $p_L + p_D = 1$. The quantity $p_L/(1 - p_L)$ is called the *odds ratio* for someone liking the movie; and $\log p_L/(1 - p_L)$ is called the *log-odds*, or the *logit*. This is an alternative way of measuring the proportion of a population. For example, 10% of Americans are left-handed; we might as easily say that the odds ratio for being left-handed is $0.1/0.9 = \frac{1}{9}$. In horse-racing parlance, this is 9:1 against a typical person being left-handed. The statistician turns it into the logit for left-handedness $\log(\frac{1}{9}) = -2.197$. Since a proportion of $\frac{1}{2}$ is an odds-ratio of 1 and so a logit of $\log(1) = 0$, we conclude that a positive logit refers to better than even odds, and a negative one to worse than even.

Definition. Corresponding to a population proportion p where $(0 < p < 1)$, we have its **odds** $o = \frac{p}{1-p}$ and its **logit** $l = \log o = \log \frac{p}{1-p}$.

In a case like this in which we have divided the population into two categories such as Like/Dislike, notice that the odds ratio for disliking the movie $p_D/(1 - p_D) = (1 - p_L)/p_L$ is one over the odds for liking it. But $\log(1/a) = -\log(a)$. So the logit for disliking the movie is the negative of the logit for liking it, and similarly, for Male versus Female and any other division of a population into two parts. This is just another way of remembering our centering condition $b_L + b_D = 0$.

For more than two categories, the b's are called *multiple logits*; you may see them again in advanced courses.

1.7.4   Loglinear Independence Models

Our problem becomes more interesting when we construct linear versions of our independence models for two-way contingency tables. In the movie example

$$\log \hat{x}_{LM} = \log n p_L p_M = \log n + \log p_L + \log p_M.$$

The centered version is $\log \hat{x}_{LM} = \mu + b_L + c_M$. We will require the row and column effects each to average 0 over cells; so that in this case $b_L + b_D = 0$ and $c_M + c_F = 0$.

We again need to connect the two models with the different parameters, for each of the four cells:

$$\log n + \log p_L + \log p_M = \mu + b_L + c_M,$$
$$\log n + \log p_L + \log p_F = \mu + b_L + c_F,$$
$$\log n + \log p_D + \log p_M = \mu + b_D + c_M,$$
$$\log n + \log p_D + \log p_F = \mu + b_D + c_F.$$

Add together the four cell predictions under each of the two forms of the model to get

$$4 \log n + 2 \log p_L + 2 \log p_D + 2 \log p_M + 2 \log p_F$$
$$= 4\mu + 2b_L + 2b_D + 2c_M + 2c_F = 4\mu,$$

since by the centering conditions, the b's and c's cancel out. This gives us

$$\mu = \log n + \frac{1}{2}(\log p_L + \log p_D + \log p_M + \log p_F).$$

Now sum just the first row of predictions:

$$2 \log n + 2 \log p_L + \log p_M + \log p_F = 2\mu + 2b_L.$$

Substitute what we got for μ in the previous expression and solve to get $b_L = \frac{1}{2}(\log p_L - \log p_D)$. By a similar argument (exercise), $c_M = \frac{1}{2}(\log p_M - \log p_F)$; and of course, $b_D = -b_L$ and $c_F = -c_M$.

Example (*cont.*).   We will use the sample proportions to estimate the parameters in our movie example:

$$\hat{\mu} = 5.298 + (-0.400 - 1.109 - 0.766 - 0.625)\frac{1}{2} = 4.473,$$

$$\hat{b}_L = [-.400 - (-1.109)]\frac{1}{2} = 0.355, \quad \hat{c}_M = [-0.766 - (-0.625)]\frac{1}{2} = -0.071$$

Wonderfully enough (though perhaps not surprisingly, given our motivation for it), the row and column adjustments in this independence model are half the separate logits for the row treatments and the column treatments. The μ parameter, though, has a slightly different meaning.

Generally, the *loglinear* independence model for a two-way contingency looks like $\log \hat{x}_{ij} = \mu + b_i + c_j$ with centering constraints $\sum_{i=1}^{k} b_i = 0$, and $\sum_{j=1}^{l} c_j = 0$. As an exercise, you should derive general formulas for the μ, b's, and c's in terms of the row and column p's. We can do a degrees-of-freedom calculation identical to the one for the additive two-way model: The saturated model has kl degrees of freedom, and the independence model has $k + l - 1$. Therefore, the residuals in the cell counts have $kl - (k + l - 1) = (k - 1)(l - 1)$ degrees of freedom. The simple differences between raw counts and expected counts in our 2-by-2 table had only one value, 11.3, because the saturated model had only one extra degree of freedom.

1.7.5 Loglinear Saturated Models*

Inspired by our success, we propose a loglinear form of the saturated model: $\log \hat{x}_{ij} = \mu + b_i + c_j + d_{ij}$, with the additional constraints $\sum_{i=1}^{k} d_{ij} = 0$ for each j and $\sum_{j=1}^{l} d_{ij} = 0$ for each i. The d's are called *measures of association*, or sometimes just interactions, as in the measurement models. We count the free parameters just as we did for the corresponding argument for the full measurement model, and the total kl is the same as for the saturated contingency table. Therefore, we expect to be able to solve for the parameters using n and the cell proportions p_{ij}.

For example, in our movie experiment, the two versions look like $\log \hat{x}_{LM} = \log n p_{LM} = \log n + \log p_{LM} = \mu + b_L + c_M + d_{LM}$. Now add these up over all four cells to get $4\mu = 4\log n + \log p_{ML} + \log p_{FL} + \log p_{MD} + \log p_{FD}$ (the centering conditions have canceled all the b's, c's, and d's).

Then sum the first row and substitute for μ to get $b_L = \frac{1}{4}(\log p_{LM} + \log p_{LF} - \log p_{DM} - \log p_{DF})$. Similarly, for the first column, $c_M = \frac{1}{4}(\log p_{LM} - \log p_{LF} + \log p_{DM} - \log p_{DF})$.

Something should strike you here: Unlike our measurement models for balanced two-way layouts, these estimates are not the same as the ones for the independence model. In fact, you might notice (exercise) that they are equal only if the independence model is exactly true. The interpretation of b and c, as adjustments in the predicted log-count as we change row or column, is still the same; but the amount of that adjustment depends on the model.

Now back-substitute to get

$$d_{LM} = \frac{1}{4}(\log p_{LM} - \log p_{LF} - \log p_{DM} + \log p_{DF})$$
$$= \frac{1}{4} \log \left(\frac{p_{LM} p_{DF}}{p_{LF} p_{DM}} \right).$$

The quantity $p_{LM} p_{DF}/(p_{LF} p_{DM})$ is called the *relative odds ratio*, and it is perhaps the most widely quoted measure of association in two-by-two tables. We may rewrite it $(p_{LM}/p_{DM})/(p_{LF}/p_{DF})$. The numerator p_{LM}/p_{DM} is just an odds ratio for liking the movie, when we restrict the population to men only; we call it a *conditional* odds ratio. Similarly, the denominator p_{LF}/p_{DF} is the conditional odds for liking the movie when we consider only women. The ratio compares the two; the farther it is from 1, the more different are the tastes of men and women, and the less appropriate the independence model must be. In our survey we estimate the relative odds ratio to be $(0.255/0.21)/(0.415/0.12) = (1.214)/(3.458) = 0.351$. Then $\hat{d}_{LM} = \log(0.351)/4 = -0.262$. The fact that our relative odds ratio was less than one (and so d was negative) says that in our sample, more women than men liked the movie.

You should notice that as a reflection of the one degree of freedom available to the d's, their logarithms are all same size with varying sign. Whenever the d's are all close to zero, we should probably conclude that we did not need them and that the simpler independence model is appropriate.

There are, of course, 3-way and higher contingency tables, with loglinear models including various sorts of association with which to summarize them. We will study some of these in exercises, and later in the book.

1.8 Logistic Regression*

1.8.1 Interpolating in Contingency Tables

You will recall that linear regression allowed us, whenever independent variables corresponded to numerical settings, to predict what a measurement might be at other settings. When our responses are counts, we can still, with ingenuity, do something of the same thing.

Example. A studio wonders whether the popularity of its latest movie has more to do with the age of the audience than anything else. They do a special screening for a number of subjects, some of approximately age 20 and some of approximately age 40; at the end they are each asked whether they like the movie.

		Opinion	
		Like	Dislike
Age	20	42	19
	40	13	51

All the methods of the last section apply. As an exercise, you should estimate the independence model. When I did so, I was led to the conclusion that it was not very appropriate here; there is indeed probably some association. This means that age does have something to do with opinion: Younger people liked the movie better.

We can put this as a prediction: If you know the ages of a collection of people, what proportion of them will like the movie? Express this in terms of the *saturated* loglinear model (since the *independence* model assumes that age makes no difference to opinion). Now, we have already noted that the natural quantity to predict in a loglinear model is the logit for liking the movie, in particular, the *conditional* logits $l_{20} = \log(p_{L20}/p_{D20})$ and $l_{40} = \log(p_{L40}/p_{D40})$, each of which refers only to the patrons of one age.

$$l_{20} = \log \frac{p_{L20}}{p_{D20}} = \log \frac{np_{L20}}{np_{D20}} = \log np_{L20} - \log np_{D20}$$

$$= \log \hat{x}_{L20} - \log \hat{x}_{D20}$$

$$= \mu + b_L + c_{20} + d_{L20} - (\mu + b_D + c_{20} + d_{D20})$$

$$= (b_L - b_D) + (d_{L20} - d_{D20}).$$

In the same way, $l_{40} = \log p_{L40}/p_{D40} = (b_L - b_D) + (d_{L40} - d_{D40})$. But going back to the last section,

$$d_{L20} - d_{D20} = \frac{1}{2} \log \frac{p_{L20} p_{D40}}{p_{D20} p_{L40}}$$

and

$$d_{L40} - d_{D40} = -\frac{1}{2} \log \frac{p_{L20} p_{D40}}{p_{D20} p_{L40}}.$$

We have managed to write our predictions of a conditional logit as a centered model with a middle liking level

$$b_L - b_D = \frac{1}{2} \log \frac{p_{L20} p_{L40}}{p_{D20} p_{D40}},$$

to which we add or subtract a correction proportional to the log of the relative odds ratio.

There are no new conclusions here; but what if you wanted to predict how popular the movie would be in other age groups, besides those in the survey? We already tried linear interpolation in the regression problem; that should work here, too. Let the new age be x, and write its predicted logit as $\hat{l} = \log(p_{Lx}/p_{Dx}) = \mu + (x - \bar{x})b$, where $\bar{x} = (20 + 40)/2 = 30$ is the average level of the independent variable. Match this to one of the prediction equations in the last paragraph, to get $\mu = b_L - b_D$ and $b = (d_{L40} - d_{D40})/(40 - 30)$.

Using the standard estimates, the cell proportions, we have $\hat{p}_{L20} = \frac{42}{125} = 0.336$, $\hat{p}_{D20} = 0.152$, $\hat{p}_{L40} = 0.104$, and $\hat{p}_{D40} = 0.408$. Then $\hat{\mu} = \frac{1}{2} \log(0.336 \times 0.104)/(0.152 \times 0.408) = -0.287$ and $\hat{b} = -\frac{1}{20} \log(0.336 \times 0.408)/(0.152 \times 0.104) = -0.108$. Then we have a regression equation for predicting the logit, $\hat{l} = \log \frac{p_{Lx}}{p_{Dx}} = -0.287 - 0.108(x - 30)$. If this model is reasonable, what proportion of 25-year-olds would we expect to like our movie? The predicted logit, conditional on age $x = 25$, is $\hat{l}_{25} = \log(p_{L25}/p_{D25}) = 0.253$.

The slashes in Figure 1.14 show the estimated logits at the two survey ages, 20 and 40. The dotted line shows how the regression equation estimates the logit at age 25 by interpolation.

This does not answer our question about the proportion of favorable reactions; but fortunately, that information can always be extracted from the logit. Notice that $(p_{Lx})/(p_{Lx} + p_{Dx}) = (p_{Lx}/p_{Dx})/(p_{Lx}/p_{Dx} + 1)$. The logit l is the logarithm of these fractions; but we know that $e^{\log a} = a$; so $p_{Lx}/p_{Dx} = e^l$. Then the proportion favorable is $p = e^l/(e^l + 1)$.

Proposition. *Given a proportion p and its odds o and logit l, $p = o/(1 + o)$ and $p = e^l/(e^l + 1) = 1/(1 + e^{-l})$.*

Our estimate of the proportion of favorable patrons of age 25 would be $e^{0.253}/(e^{0.253} + 1) = 0.563$. This is between the 69% of 20-year-olds and the 20% of 40-year-olds, as we intended.

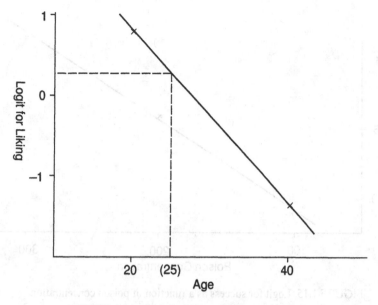

FIGURE 1.14. Logit for liking as a function of age

1.8.2 Linear Logistic Regression

The method illustrated above is an example of *logistic regression*, which may be used to predict the proportion of "successes" in some experiment when there are numerical settings to the independent variables that we can interpolate. It possesses all the powers of linear regression and requires the same care—interpolate with caution, extrapolate doubly so. We certainly need not restrict ourselves to the case of only two settings for the independent variable.

Example. Three different concentrations of a new ant poison are applied to a number of fire ant nests, and we record whether or not the nests are destroyed:

		Concentration		
		100 mg/l	200 mg/l	300 mg/l
Destroyed	Yes	15	20	25
	No	17	11	8

We can estimate the conditional logits just from the ratios of the counts in each column and plot them against the concentration (see Figure 1.15).

The ×'s show the estimated logit at each concentration. They are, of course, not exactly on a straight line, but they are plausibly close to the one we have drawn. So a logistic regression equation of the form $\log p_{Yx}/p_{Nx} = \hat{l} = \mu + (x - \bar{x})b$ is a plausible summary of our experiment, where x is the concentration of poison, and so $\bar{x} = 200$ mg/l is the center of our three concentrations.

Let us estimate this equation by the line drawn (by eye) on the plot, which happens to be $\hat{l} = 0.538 + 0.00632(x - 200)$. We had a good bit of success with

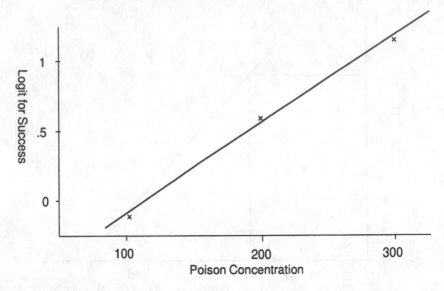

FIGURE 1.15. Logit for success as a function of poison concentration

300 mg/l; so we are tempted to try 400. Before we buy the poison, we may as well use logistic regression to predict the result. Of course, this is *extrapolation* (see Section 5.1), so we would be foolish to take the conclusion too seriously. Anyway, $\hat{l} = 1.802$, and we translate that to a proportion of successful kills $\hat{p} = e^{1.802}/(e^{1.802} + 1) = 0.858$. You will have to decide whether that is a good enough success rate to justify the experiment.

Of course, we have not told you how to find the line on the plot. Reliable methods for estimating logistic regression equations will have to await a later chapter. There are, of course, logistic regression models for far more complicated experiments. Just as in ordinary regression of measured data, our experimental results may consist of any number of values of one or several independent variables, so long as the dependent variable records simply whether that experiment was a "success" or a "failure" (like/dislike, male/female, or any other *dichotomous* outcome).

1.9   Summary

In this and subsequent chapter summary sections we will briefly review the key technical terms and the most important mathematical expressions that should now have meaning for you after studying the chapter. If any of these are at all fuzzy, it is time for you to study those sections more carefully. When you see a notation like (3.4) it will mean section 3, subsection 4 of the current chapter.

First we studied *linear* models for experiments where we try to measure some important numbers (such as people's blood pressure), but for some reason our measurements are not all the same. We can estimate the "true" value μ using the *sample mean* $\hat{\mu} = \frac{1}{n} \sum_{i=1}^{n} x_i = \bar{x}$ (2.2). Often, different subjects of your experiment will undergo different *levels* of a *treatment* (such as types of drug). In that case, the model that describes the experiment is called a *one-way layout* (3.1). We try to discover whether the different levels lead to consistent differences in our measurements, and we express the result as $\hat{x}_{ij} = \mu_i = \mu + b_i$ so that the b's tell us how different the ith level is from the average level μ (3.2). If the observations were subjected to more than one sort of treatment at the same time (for example, bed rest or not, as well as drugs), we have a *two-way* (or more) layout (4.1). Sometimes, these data may be described well enough by an *additive* model $\hat{x}_{ijk} = \mu + b_i + c_j$, where the c's tell us the effect of levels of the second treatment (4.2). Often, though, that will not be sufficient, and we will need to add *interaction* terms d_{ij} that tell how differently the j levels affect the individual i levels (4.4). When the experimental levels correspond to numerical settings (such as dosages of a single drug), we may be able to predict the results of future measurements using *regression* models (5.1). For a single predictor x of a measurement y, we may start with a *simple linear* regression model that looks like $\hat{y}_i = \mu + b(x_i - \bar{x})$ (5.2). The extension to several predictor variables gives us a *multiple regression* model, such as $\hat{y}_j = \mu + (x_{1j} - \bar{x}_1)b_1 + (x_{2j} - \bar{x}_2)b_2$ (6.2).

On the other hand, our data may consist of categorized counts (as from a political poll); we summarize the results with *population proportions* p_i, which predict the count in the ith category by $\hat{x}_i = np_i$. We usually estimate these by the *sample proportion* $\hat{p}_1 = x_i/n$. When we have two ways of categorizing counts (such as gender and party preference), we construct *contingency tables* (7.1). When it may be that certain classifications have nothing to do with each other, *independence* models provide an important simplification. These look like $\hat{x}_{ij} = np_{i\bullet}p_{\bullet j}$ (7.2). A powerful way to express many models for counted data will be as *loglinear* models (7.3), for example in a two-way contingency table $\log \hat{x}_{ij} = \mu + b_i + c_j + d_{ij}$. The d_{ij} measure the failure of the independence model, which we call the *association* between the two kinds of categories (7.5). When we want to predict proportions from numerical experimental settings x, we often use (*simple linear*) *logistic* regression, which looks like $\log(p_{Yx}/p_{Nx}) = \hat{l} = \mu + (x - \bar{x})b$ for the case of Yes or No categorization (8.2).

1.10 Exercises

1. *Science* magazine in 1978 announced that various American lunar probes had obtained the following values for the ratio of the mass of the Earth to that of the Moon: 81.3001, 81.3015, 81.3006, 81.3011, 81.2997, 81.3005, and 81.3021.

   a. Draw a hairline plot or similar graphical display of these measurements.
   b. Compute the sample mean $\hat{\mu} = \bar{x}$ for these numbers, and mark it clearly on your plot.

c. Compute the residuals from this location model. Now compute the sum of these residuals. Did you get the answer you were supposed to?

2. In 1982, Sternberg et al. reported in *Science* on the level of an enzyme called DBH in the bloodstream of a number of schizophrenia patients. The patients were separated into groups that were judged by clinicians to be either psychotic or nonpsychotic:

psychotic: 0.0150, 0.0204, 0.0208, 0.0222, 0.0226, 0.0245, 0.0270, 0.0275, 0.0306, 0.0320

nonpsychotic: 0.0104, 0.0105, 0.0112, 0.0116, 0.0130, 0.0145, 0.0154, 0.0156, 0.0170, 0.0180, 0.0200, 0.0210, 0.0230, 0.0252

a. Draw parallel hairline plots of the DBH levels for the two clinical groups. What does this suggest to you about the effect of clinical status on enzyme level?
b. Find the standard (sample mean) estimates of a one-way layout model. Mark the group centers on your plot.
c. Find the standard estimates of a centered model for this experiment.

3. Four different shrimp nets are under consideration for use on your shrimp boat. On 16 days with acceptable weather conditions, you note the yield in hundreds of pounds, using each net on 4 randomly chosen days:

InSein	75	82	91	93
Crusty	51	58	62	76
Hample	90	53	56	84
NetProfit	112	78	104	97

a. Draw parallel hairline plots of the performance of each net. Mark the sample means on each.
b. Construct standard estimates of a centered one-way layout model for this experiment.

4. We claimed that the centered model $\hat{x}_{ij} = \mu_i = \mu + b_i$ is determined unambiguously if we know the group centers μ_i, so long as we impose the centering condition $\frac{1}{2} \sum_{i=1}^{k} n_i b_i = 0$. Show that we can always determine what μ and b_i are if we know the μ_i, and vice versa.

5. Show that the collection of all residuals in the standard estimate of the one-way layout model, $x_{ij} - \hat{\mu}_i$, has $n - k$ degrees of freedom. That is, even though there are n residuals, you can specify $n - k$ of them that would allow you to compute the remaining k residuals.

6. Nine 20-year-olds who are classified as moderately overweight are recruited into a three-month weight-loss program. Some will go on a 2000-calorie diet, some will enter a 30-minute-a-day vigorous aerobics program, and some will be "controls." At the end of the program, each weight loss in pounds is recorded:

	none	diet
none	2	2
		7
exercise	4	6 10
	8	13 14

a. Is this experiment balanced? Why or why not?

b. Use the standard estimates to find values for the parameters of an additive model. Plot the resulting model, and interpret it.

c. Find standard estimates for the parameters of the full centered model. Plot the resulting model. Explain why you do or do not believe this model substantially superior to the additive model.

7. Show that the standard estimates for an additive model turn out to be centered in a two-way layout.

8. Assume that a two-way layout has equal numbers of observations (call it r) in each cell. Show that the standard estimates of the parameters in a full model for this two-way layout meet the centering conditions.

9. You would like to know how much money a higher thermostat setting saves you during a Houston summer. So for six years in a row you flip a coin to decide whether to set the thermostat to 72°F or 78°F for all of August, with the following bills:

72°: $178, $195, $201

78°: $180, $153, $164

a. Write down and estimate a simple linear regression model for predicting monthly bills, given your thermostat setting.

b. If you set your thermostat to 76°F next August, use your model to predict what your electric bill will be. Do you find this prediction plausible? Why or why not?

c. You decide that air conditioning is bad for you, so next August you set your thermostat to 86°F. Use your model to predict your electric bill. Do you find your prediction plausible? What practical aspects of the problem might lead you to doubt your prediction?

10. A sociologist suspects that crowding and heat contribute to violent crime rates, so she locates medium-size cities near 32 and 40 degrees latitude and with population densities approximately 2000 and 6000 people per square mile. Her 8 representative cities had the following crime rates in 1990 (in crimes per 1000 population):

	32 degrees N	40 degrees N
2000/sq mile	80 48	60 35
6000/sq mile	97 79	63 83

a. Construct and estimate a multiple regression model for predicting crime rate from density and latitude, using the standard estimates for an additive two-way layout. Plot your model.

b. I live in a town that is 37 degrees, 20 minutes north latitude, with a population density of 2400 people per square mile. Use your model to predict its crime rate.

11. Without telling them what you are doing, you issue some (arbitrarily selected) soldiers a 25-pound backpack for a strenuous field exercise: 13 out of 49 complain afterwards of muscle or joint pain. The other soldiers on the same exercise have a 30-pound pack: 23 out of 52 complain of muscle or joint pain. If in fact there is no connection between pack size and complaints, how many soldiers in each group would you expect to complain?

12. A political polling organization would like to know whether upper, middle, or lower socioeconomic status (SES) has anything to do with whether a voter considers himself or herself libertarian, conservative, or liberal in political philosophy. Two hundred voters picked at random were classified on standard scales into the possible combinations; the counts were as follows:

SES\Phil.	Libertarian	Conservative	Liberal
Upper	17	20	17
Middle	12	45	17
Lower	5	18	49

Under the hypothesis that status and philosophy are independent of one another, construct a table of the predicted counts for each table entry.

13. For the expected table in an independence model, you of course compute $\hat{x}_{ij} = n\hat{p}_{i\bullet}\hat{p}_{\bullet j}$, where you use the standard estimate for the p's. Show that the row and column sums in this table are always the same as the row and column sums $x_{i\bullet}$ and $x_{\bullet j}$ in the observed table.

14. For the political poll data of Section 7.1, estimate the parameters of a centered loglinear model.

15. **a.** For a general two-way contingency table, derive formulas for μ, the b's, and the c's of a centered parametrization of the **independence** model, in terms of n and the p's.

    b. Derive formulas for μ and the b's, c's, and d's of the **saturated** model, in terms of n and the p's.

16. For the experiment of Exercise 12 (political philosophy),

    a. Compute standard estimates for μ, the b's, and the c's of a centered parametrization of the **independence** model.

    b. Compute standard estimates for μ and the b's, c's, and d's of the **saturated** model. Interpret the values you get in words.

17. For the experiment of Exercise 11 (soldier's backpacks),

    a. Compute standard estimates for μ, the b's, and the c's of a centered parametrization of the **independence** model.

    b. Compute standard estimates for μ and the b's, c's, and d's of the **saturated** model. Interpret the values you get in words.

18. In Exercise 11, use linear logistic regression to predict the proportion of soldiers who would complain with a 28-pound pack.

1.11 Supplementary Exercises

19. A common alternative to the sample mean to estimate μ in a location model is the **sample median**: Sort the observations in ascending order $x_{(1)} \leq x_{(2)} \leq \cdots \leq x_{(n)}$. The median is then in the middle of that list: (i) if n is an odd number, then the median is the middle number $\hat{\mu} = x_{\left(\frac{n+1}{2}\right)}$; and (ii) if n is an even number, the median is conventionally the average of the two numbers flanking the middle $\hat{\mu} = (x_{(n/2)} + x_{(n/2+1)})/2$.
   Find the sample median of the mass ratios from Exercise 1. How does it compare to the sample mean?

20. Three long-distance telephone companies, BSS, CMI, and DWP, are competing for your business. To evaluate the impacts of their rates, you test them on 15 quite similar branch offices of your company, randomly assigning 5 offices to each carrier. Here are their phone bills for the same month, in thousands of dollars:

BSS	20	23	25	32	21
CMI	39	21	22	36	23
DWP	50	33	46	42	38

   a. Draw parallel hairline plots of the observations for the three carriers. Mark on them the sample means for each level.
   b. Estimate the parameters of a centered one-way layout model.

21. a. Use the sample median of each group to estimate the one-way layout model in the schizophrenia data from Exercise 2.
    b. Use the results from (a) to estimate a centered model for this experiment. Compare your estimates to what you got in Exercise 2 (b) and (c).

22. Demonstrate that we could just as well have defined a balanced design to be one in which the numbers of observations in each cell in each *column* were proportional to those in the other columns.

23. You want to compare, over the year 1995, how the three locations of your identically sized pizza restaurant are doing. Somebody points out that because of weather, school, and so forth, the time of year affects sales. So you record the total dollar sales (in units of $10,000) at each location in each season to get the following data: for Price's Fork, Sp(ring) 34, Su(mmer) 30, Au(tumn) 34, Wi(nter) 34; for North Main, Sp 34, Su 14, Au 26, Wi 21; and South Main, Sp 44, Su 27, Au 37, and Wi 30.

   a. Estimate the parameters of an additive model in this two-way design.
   b. Estimate the parameters of the full model in this design. Comment on the differences between the two.

24. Show that in any **balanced** two-way layout, the standard estimates for the parameters of the full model are centered.

25. An example of a **balanced incomplete block design** for a two-way layout is

	1	2	3
1	x_{11}	x_{12}	
2	x_{21}		x_{23}
3		x_{32}	x_{33}

where we have taken only six observations, yet we can still estimate a centered additive model $\hat{x}_{ij} = \mu + b_i + c_j$. We might wish to do this if observations are very expensive.

The standard estimates are $\hat{\mu} = \bar{x}$, $\hat{b}_1 = \frac{1}{3}(x_{11} + x_{12}) - \frac{1}{6}(x_{21} + x_{23} + x_{32} + x_{33})$, and $\hat{b}_2 = \frac{1}{3}(x_{21} + x_{23}) - \frac{1}{6}(x_{11} + x_{12} + x_{32} + x_{33})$. Find the corresponding estimate for b_3. Assuming column corrections are estimated just as row corrections are, find standard estimates for the c's.

26. For a balanced incomplete block experiment (see Exercise 25) to estimate the breaking strength of three beam cross-sections (A, B, C) made of three steel alloys (I, II, III), we got, in thousands of pounds,

	I	II	III
A	35.2	28.1	
B	18.7		40.3
C		31.6	60.5

What does an additive model predict for the typical breaking strength of a beam with cross-section B made from alloy I? Compare it to the actual result. How many degrees of freedom for residuals does this model have? What does your model predict for the untried case of cross-section A and alloy III?

27. There is a more complicated linear regression problem for which a standard estimate is easy to guess. We will assume that there are three distinct values of the independent variable, equally spaced (for example, 10, 20, 30). Furthermore, the number of observations at the highest and lowest levels of the independent variable must be the same. Then the average of all observations should give you a predicted value for the middle level of the independent variable. Furthermore, the slope of the regression line should be the slope of the line connecting the averages of the observations at the highest and lowest levels (because the slope does not affect your middle-level prediction, which is at the fulcrum around which the line is free to rock).

   a. Write down a precise notation for such an experiment and for a simple linear regression model for predicting it.

   b. Write down the standard estimate of your regression model.

28. The highest-volume item at your beach supply store is a certain brand of sunscreen lotion. You would like to know how your price affects your weekly sales volume. You try three different prices for various weeks during the summer, with the following unit sales:

$2.50: 82, 74, 83

$3.00: 55, 54, 61, 58

$3.50: 40, 46, 37

   a. Construct a plot of these numbers, marking also the sample mean of each group.

   b. Calculate the standard estimate of a simple linear regression model for predicting unit sales from price (see Exercise 27). Draw the prediction line on your plot. Does the model seem plausible? Why or why not?

   c. Predict unit sales for a week in which your price is $2.79. Now predict the number of units you would get rid of if you gave sunscreen away for free. Comment on the plausibility of your predictions.

29. You have surely noticed that in our two-by-two examples of regression we insisted on using an additive model. What would have happened if we had used the full model instead?

   a. Write down a model that looks like

$$\hat{y}_i = \mu + (x_{i1} - \bar{x}_1)b_i + (x_{i2} - \bar{x}_2)b_2 + (x_{i1} - \bar{x}_1)(x_{i2} - \bar{x}_2)b_{12}$$

in Exercise 10, and estimate the new parameter b_{12} by setting the last term equal to one of the interactions in the full model. Recalculate the prediction in (b). (The new model, which makes sense for any number of levels of each of the independent variables, is called a *bilinear* model because it is linear in each independent variable if the other is held fixed. Here it has four degrees of freedom.)

   b. Write down what a multilinear model in some larger number of independent variables would look like.

30. Young people on the lookout for prospective husbands or wives often claim that certain cities have more women or more men. To study this issue, you sample the voter rolls in three cities looking for people who are between 20 and 30 years of age and single. Here are the numbers of those you find, by gender:

	New York	Chicago	Houston
Males	230	211	297
Females	312	225	255

Your question might be addressed in the following way: An independence model would mean that the proportions of men and women did not depend on which city you looked in. So you should define and find standard estimates for an independence model. Then build a table of expected values. Comment on what the comparison between the two tables says about the question you began with.

31. For the survey of Exercise 30,

a. Compute standard estimates for μ, the b's, and the c's of a centered parametrization of the **independence** model.

b. Compute standard estimates for μ and the b's, c's, and d's of the **saturated** model. Interpret the values you get in words.

32. Sometimes in a two-by-two contingency table experiment, the count in one of the cells is *unobservable*. We believe that there is a count, but we do not know what it is:

	1	2
1	n_{11}	n_{12}
2	n_{21}	?

a. It is still possible in this experiment to estimate the parameters of an independence model $\hat{n}_{ij} = np_{i\bullet}p_{\bullet j}$. Then we could, with a little ingenuity, predict the unknown count $\hat{n}_{22}$. Find standard estimates, using all the available information, of the parameters of the independence model in this experiment. (Do not forget that n is also an unknown parameter in this case.)

b. This method may be used to correct census undercounts. The people in a census tract are counted by two methods we believe to be independent (say, mail and visit). Then n_{11} = people counted by both methods, n_{12} = people counted by mail but not by visit, n_{21} = people counted by visit but not by mail, and n_{22} = people counted by neither method (obviously unobservable). Use the model from (a) to estimate the **total** population of a certain census tract if $n_{11} = 12{,}384$, $n_{12} = 589$, $n_{21} = 1466$.

33. Ultrapasteurization of cream requires it to be heated to a very high temperature for a short time. We count how many pints have spoiled under refrigeration for two weeks after ultrapasteurization at two temperatures:

	170°F	180°F
Spoiled	9	3
Good	21	27

a. Write down and estimate a linear logistic regression model for the rate of spoilage at various temperatures. Plot your equation.

b. Use your model to predict the proportion of pints of cream that would spoil within two weeks if they were originally heated to 176°F. Do the same for a temperature of 160°F. How confident are you about these two predictions?

34. A three-way contingency table consists of counts resulting from an experiment x_{ijk}, where there are $i = 1, \ldots, l$ levels of the first treatment, $j = 1, \ldots, m$ levels of the second treatment, and $k = 1, \ldots, q$ levels of the third treatment. The **complete independence** model of this experiment looks like $\hat{x}_{ijk} = np_{i\bullet\bullet}p_{\bullet j\bullet}p_{\bullet\bullet k}$.

a. What does this model say about your experiment? Write down standard estimates of the parameters in the complete independence model.

b. Invent a notation for the centered, loglinear parametrization of this model. Be sure to specify your centering conditions. **Hint:** You need four kinds of parameters.

35. You want to find out how many people in various walks of life still smoke cigarettes. You note during your poll whether the responder is male or female, and whether he or she lives in a rural or urban area. Your results are as follows:

	Rural	Urban
Male	23	43
Female	27	52

Smokers

	Rural	Urban
Male	43	135
Female	32	118

Nonsmokers

a. Define and estimate a complete independence model for this experiment.
b. Write down a table of expected counts under this model. How well does the model match the facts?

36. With three-way contingency tables we can propose a great variety of models for the results of an experiment. For example, a **conditional independence** model would be one that says something like this: the second and third treatments are independent of each other, for each level of the first treatment. That would require us to say, about our proportions, $p_{ijk}/p_{i\bullet\bullet} = (p_{ij\bullet}/p_{i\bullet\bullet})(p_{i\bullet k}/p_{i\bullet\bullet})$. After cancellation, we see that our predictions must be $\hat{x}_{ijk} = n(p_{ij\bullet}p_{i\bullet k})/p_{i\bullet\bullet}$.

a. Write down standard estimates for the parameters in this model.
b. Write down a centered loglinear version of this model, including centering conditions. **Hint:** There should be six kinds of parameters.

37. Estimate the p's of a conditional independence model for the survey in Exercise 36, where you assume that gender and location are conditionally independent of one another for each of smokers and nonsmokers. Construct a table of expected counts under this model. In words, what does this model say about your experiment? How well does it match the facts?

38. Linear regression can be generalized to *polynomial* regression by making terms that involve the square, the cube, etc. of the independent variable into additional independent variables. To illustrate this, estimate a model for the case of Exercise 27 (three equally spaced design points) with

$$\hat{y} = \mu + b(x - \bar{x}) + c(x - \bar{x})^2,$$

by interpolating the sample means at each design point. Apply it to the data of Exercise 28 and redo part (c) with your new model. Do you find the results more or less convincing than before?

CHAPTER 2

Least Squares Methods

2.1 Introduction

In the last chapter we considered models that summarized the measurements that we obtained in several kinds of experiments. We ran into two sorts of difficulties. First, we had nothing but our practical intuition to tell us how good a job we had done when we summarized our data. Sometimes our averages and our regression lines nearly equaled each data point; the difference could be attributed to measurement "noise." At other times our numbers were all over the plot, and only our faith in the simplicity of nature led us to take our elementary mathematical models seriously. We need some sort of index to score how well we do when we reduce the data to these expressions.

Second, we found for most of our regression models no good way to estimate the parameters. We need reasonable, repeatable estimators for regression models.

Fortunately, in 1805 the French mathematician Adrien Marie Legendre proposed a beautiful solution for both of our problems: the method of *least squares*. This simple idea based on coordinate geometry will give us a powerful, unified way to deal with all the measurement problems discussed in the last chapter (and many more).

Time to Review

Vector algebra
Matrix algebra

2.2   Euclidean Distance

2.2.1   Multiple Observations as Vectors

We pointed out at the beginning that our measured responses x_i could be thought of as points on a number line. In a similar way, our regression *scatter plots* were graphs of pairs of coordinates (x_i, y_i) for points in the plane; we again translated numbers into geometrical objects. We can take this idea one radical step further and pretend that an entire sample of observations x_i for $i = 1, \ldots, n$ are the coordinates of a single vector in n-dimensional space, this despite the fact that we cannot readily visualize figures or plot points in a space of more than three dimensions. Nevertheless, it will turn out that we can use methods from analytic geometry to work with these *sample vectors*.

We need to translate our measurements into vector and matrix notation. First of all, we will follow the convention that a vector is written as a boldface, lowercase letter, such as **x**. When we expand the vector into its component coordinates, we will use *matrix notation*. A vector is conventionally an $n \times 1$ matrix, a *column*, of coordinates:

$$\mathbf{x} = \begin{pmatrix} x_i \\ \vdots \\ x_n \end{pmatrix}.$$

This is a bit inconvenient when we are writing text in a line, so we will often use the *transpose* operator (which interchanges rows and columns of a matrix) to change a row vector to a column vector: $\mathbf{x} = (x_i, \ldots, x_n)^T$.

Example. On Monday through the following Sunday, I note how long I have to wait for my hamburger at my favorite local lunch counter. The answers, in minutes, are $\mathbf{x} = (12, 15, 9, 10, 14, 16, 14)^T$.

The usual situation when we are analyzing multiple measurements of the same sort is that we have some theory that says that the ith number ought to be μ_i; but when we actually did our error-prone experiment, we got x_i. So we ask how *far apart* the sample vector **x** and the theoretical vector $\mu = (\mu_1, \ldots, \mu_n)^T$ are. Analytic geometry suggests that we find the *length* of the vector $\mathbf{x} - \mu$ from the hypothesis to the experiment, called the *Euclidean distance* from **x** to μ. Notice that the ith coordinate of this vector is $x_i - \mu_i$, the residual defined in 1.2.2 (when we say this, we mean that you can look for the earlier discussion in Chapter 1, Section 2.2).

Example (*cont.*). The manager of the lunch counter announces that typically one should have to wait about 10 minutes on weekdays and 15 minutes on weekends. His theory (we usually call it a model, or hypothesis) says that $\mu = (10, 10, 10, 10, 10, 15, 15)^T$. Then the residual between our data and his model is $\mathbf{x} - \mu = (2, 5, -1, 0, 4, 1, -1)^T$.

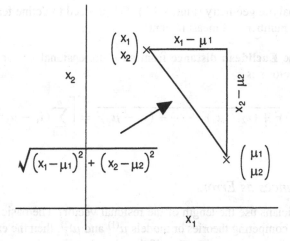

FIGURE 2.1. Pythagorean theorem

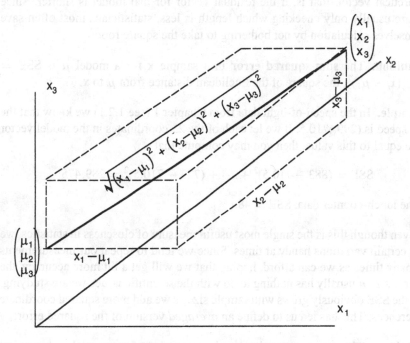

FIGURE 2.2. 3-D Pythagorean theorem

To remind you how to calculate this length. Let us look at the graphable case of two measurements (Figure 2.1): The Pythagorean theorem tells us that the length of the residual vector, the hypotenuse of the triangle, is $\sqrt{(x_1 - \mu_1)^2 + (x_2 - \mu_2)^2}$. You will probably have seen the corresponding expression, with three squared coordinate differences under the square root, for the length of a vector in three-dimensional analytic geometry (Figure 2.2): We proceed to define fearlessly, for the case of any number n of measurements:

Definition. The **Euclidean distance** from an n-dimensional vector μ to an n-dimensional vector x is

$$\sqrt{(x_1 - \mu_1)^2 + (x_2 - \mu_2)^2 + \cdots + (x_n - \mu_n)^2} = \left[\sum_{i=1}^{n} (x_i - \mu_i)^2 \right]^{1/2}.$$

2.2.2 Distances as Errors

How do statisticians use the length of the residual vector? The basic idea is that if we have two competing theories or models $\mu^{(1)}$ and $\mu^{(2)}$, then the experimental results tend to favor one or the other if the observed vector x is closer to the theoretical vector, that is, if the residual vector for that model is shorter. Since we are usually only checking which length is less, statisticians most often save themselves calculation by not bothering to take the square root:

Definition. The **sum-squared error** in a sample x for a model μ is SSE $= \sum_{i=1}^{n} (x_i - \mu_i)^2$, the square of the Euclidean distance from μ to x.

Example. In the speed-of-light data from Chapter 1 (see 1.2.1) we know that the true speed is (299,)710.5. If we let each of the 23 coordinates in the model vector μ be equal to this value, then you may compute that

$$\text{SSE} = (883 - 710.5)^2 + \cdots + (723 - 710.5)^2 = 289{,}478.$$

In the lunch-counter data, SSE = 48.

Even though this is the single most useful measure of closeness in statistics, we find certain variations handy at times. Since we tend to repeat our measurements as many times as we can afford, hoping that we will get a bit more accuracy, the sample size n usually has nothing to do with the scientific issues we are studying. But the SSE obviously grows with sample size as we add more squared coordinate differences. This has led us to define an *averaged* version of the squared error:

Definition. The **mean-squared error** in a sample x for a model μ (proposed before the experiment is carried out) is MSE $= \frac{1}{n} \sum_{i=1}^{n} (x_i - \mu_i)^2$.

Example (*cont.*). In the speed-of-light data, MSE $= 12{,}586$. In the lunch-counter data, MSE = 6.86.

This gives us a rough idea of the quality of a typical observation from the point of view of the model. It has, however, one obvious failing that is clearly our fault: If the measurements are in some units such as, say, grams, then the MSE is in units of grams-squared. These are likely to have no meaning for us. So we sometimes repair an earlier adjustment and take the square root of the mean-squared error:

Definition. The **root-mean-squared error** in a sample **x** for a model μ is

$$\text{RMSE} = \sqrt{\text{MSE}} = \left[\frac{1}{n} \sum_{i=1}^{n} (x_i - \mu_i)^2 \right]^{1/2}.$$

Example (*cont.*).   In the speed-of-light data, RMSE $= 112.2$ km/sec. In the lunch-counter data, RMSE $= 2.62$ minutes.

The RMSE, and its many special cases depending on the sort of model we are studying, is perhaps the single most intuitively useful summary of how well our experimental setup seems to be matching the model. It is a sort of typical absolute difference between an observed and a predicted value.

2.3   The Principle of Least Squares

2.3.1   Simple Proportion Models

Often we have only a partial idea about what sort of simple model does the best job of matching our data approximately. We noted earlier that Euclidean distance could be used to pick from among several alternative models, according to how close they are to the observations.

Example.   In the lunch-counter problem, my personal opinion was that it takes about 15 minutes to be fed every day. Therefore, I proposed another model, $\mu^{(2)} = (15, 15, 15, 15, 15, 15, 15)^T$. Its SSE is 73. The manager's claim looks slightly better, because its SSE is smaller.

But can we apply this approach when there is an infinity of choices?

Example.   In the early decades of the twentieth century, astronomers had found that they could tell how fast objects in the sky were moving toward us or away from us by using the Doppler shift in the color of their light (just like a traffic cop catching speeders using radar). With much more difficulty, they had also found ways to tell how far away some objects were. In 1927, Edwin Hubble juxtaposed those two facts about 24 galaxies:

velocity (km/sec)	distance (1,000,000 parsecs)	velocity (km/sec)	distance (1,000,000 parsecs)
170	0.032	650	0.9
290	0.034	150	0.9
−130	0.214	500	0.9
−70	0.263	920	1.0
−185	0.275	450	1.1
−220	0.275	500	1.1
200	0.45	500	1.4
290	0.5	960	1.7
270	0.5	500	2.0
200	0.63	850	2.0
300	0.8	800	2.0
−30	0.9	1090	2.0

He of course drew a scatter plot (Figure 2.3):

After staring at this a while, you will probably come to the same conclusion Hubble did: The faster a galaxy is moving away from us, the farther off it is (with quite a bit of variation in the peculiar motions of each galaxy). If this is a general law, then we see a way to exploit it: Since it is easy to observe the outward velocity of a distant galaxy, we can use some simple law like $d = kv$ to estimate roughly the distance d, where k is our hypothesized proportionality constant. (One possible such relation is given by the sloped line on the plot.) To this day, this is the most common way to estimate the distance of newly discovered galaxies.

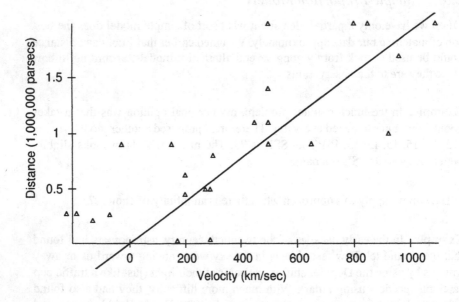

FIGURE 2.3. Distance as a function of velocity

Astronomers soon suggested an implication of our model: Perhaps the universe is expanding. The expansion rate is measured by the Hubble constant $1/k$. This eventually led to the famous big bang hypothesis for the evolution of our universe.

2.3.2 Estimating the Constant

But what is k? If we knew some physical mechanism for expansion of the universe, maybe that would tell us; but at this time we do not. Instead, we shall try to *estimate* our k by assuming a *regression* model $\hat{d} = kv$ similar to those of the last chapter (see 1.5.2), but with only the one parameter k. Unfortunately, Chapter 1 gave us no clue as to how to estimate k, except by eye. Now to Legendre's great step: We may phrase the problem as one of Euclidean distance. We want to choose k such that the vector of distances **d** is as close as possible to the vector predicted by the Hubble model $\hat{d} = kv$. Equivalently, we want somehow to pick out a k that makes SSE $= \sum_{i=1}^{n}(d_i - \hat{d}_i)^2$ as small as possible (since making the *squared* distance small is just the same as making the distance small, if all we want is the right k). We have a name for this:

Definition. If we choose the parameters of a model for predicting observed measurements by making the Euclidean distance from the observed vector to the predicted vector as small as possible, we are applying the **method of least squares** (because we are minimizing the SSE).

How is it possible to find k, since there is an infinite number of possible values to compare? We shall use some ingenuity: Let l stand for *any* other possible value of the proportionality constant in the Hubble model, besides k. Then if k is the least-squares estimate, we know that always $\sum_{i=1}^{n}(d_i - lv_i)^2 \geq \sum_{i=1}^{n}(d_i - kv_i)^2$. Here comes the first trick: Subtract and add k to l on the left-hand side to get

$$\sum_{i=1}^{n}(d_i - lv_i)^2 = \sum_{i=1}^{n}(d_i - kv_i + kv_i - lv_i)^2.$$

Now expand the square in the second expression, using the first two and last two terms:

$$\sum_{i=1}^{n}(d_i - lv_i)^2 = \sum_{i=1}^{n}(d_i - kv_i)^2 + 2\sum_{i=1}^{n}(kv_i - lv_i)(d_i - kv_i) + \sum_{i=1}^{n}(kv_i - lv_i)^2$$

$$\geq \sum_{i=1}^{n}(d_i - kv_i)^2.$$

We can cancel out the identical sums on the two sides of the inequality and factor out some constants from sums to get

$$2(k - l)\sum_{i=1}^{n}v_i(d_i - kv_i) + (k - l)^2\sum_{i=1}^{n}v_i^2 \geq 0.$$

To review, this inequality must always be true, no matter what l is, if k is the least-squares estimate. But the second term must always be at least zero, because it is a sum of squares. The first term is more of a problem: Since l is free to be anything, the term can obviously be either positive or negative. One more bit of ingenuity: We can make the first term zero, and therefore never negative, without paying attention to l, by setting $\sum_{i=1}^{n} v_i(d_i - kv_i) = 0$. This is called the *normal equation* for this least-squares problem. To solve it, split the sum and move the minus sign to the other side of the equation to get $\sum_{i=1}^{n} v_i d_i = k \sum_{i=1}^{n} v_i^2$. We can solve this for k whenever we do not have to divide by zero, that is, when all the v's are not zero. In that case we have an estimate $\hat{k} = \sum_{i=1}^{n} v_i d_i / \sum_{i=1}^{n} v_i^2$.

This estimate gets rid of the middle term in the big equation above, leaving $\sum_{i=1}^{n}(d_i - lv_i)^2 = \sum_{i=1}^{n}(d_i - kv_i)^2 + (k - l)^2 \sum_{i=1}^{n} v_i^2$. Now we know we have succeeded; since our k meets the normal equation, it always has the smallest SSE: In any other case of l, we have to add that positive last term, which makes the SSE larger.

This equation has a practical application; if we are curious about what happens if we use another value of k than the least squares value, we may use it to calculate how much further away the prediction vector is from the observation vector. Another use of it comes about when l (which, remember, can be anything) is set equal to zero. Then $\sum_{i=1}^{n} d_i^2 = \sum_{i=1}^{n}(d_i - lv_i)^2 + k^2 \sum_{i=1}^{n} v_i^2$. The Pythagorean theorem has appeared once again: You can read this as a relationship between the squared length of the observed vector $\mathbf{d}$, the squared length of the vector of residuals, and the squared length of the vector of predictions $\mathbf{v}k$. We do this so often in statistics that we have names for the terms: $\sum_{i=1}^{n} d_i^2$ is called the *(total) sum of squares*, TSS; the next term we already know as the sum of squares for error, SSE; and $\sum_{i=1}^{n} v_i^2$ is called the *sum of squares for regression*, SSR.

Example (*cont.*).   For Hubble's model, you should check that $\hat{k} = 0.001922$ where SSE $= 5.469$. This is the slope of the line we drew on the scatter plot. Therefore, if we observe that a galaxy is moving away from us at 600 km/sec, we would expect it to be about $600 \times 0.001922 = 1.15$ million parsecs distant.

Let us summarize all our mathematics as follows:

Proposition. *To predict a vector of dependent variables* $\mathbf{y}$ *from a vector of independent variables* $\mathbf{x}$ *using the regression model* $\mathbf{y} = \mathbf{x}b$,

(i) *the least squares estimate* $\hat{b}$ *is a solution of the normal equation* $\sum_{i=1}^{n} x_i y_i = b \sum_{i=1}^{n} x_i^2$, *because then*

(ii) $\sum_{i=1}^{n}(y_i - cx_i)^2 = \sum_{i=1}^{n}(y_i - bx_i)^2 + (b - c)^2 \sum_{i=1}^{n} x_i^2$ *for any parameter value* c;

(iii) *in particular, if we choose* $c = 0$, $\sum_{i=1}^{n} y_i^2 = \sum_{i=1}^{n}(y_i - bx_i)^2 + b^2 \sum_{i=1}^{n} x_i^2$, *which we conventionally write* TSS $=$ SSE $+$ SSR.

All that I have done here is to use generic letters for the special symbols from the Hubble problem: y for d, x for v, and b for k.

2.3.3   Solving the Problem Using Matrix Notation

The result above is so important that anything we can do to understand it better will be useful. First we will translate it into matrix notation. Remember that $\mathbf{x}a$ where $\mathbf{x}$ is a vector and a is a constant is the vector we get by multiplying each coordinate of $\mathbf{x}$ in turn by a. Second, an *inner product* of any two vectors $\mathbf{x}$ and $\mathbf{y}$, expressed in terms of their coordinates, is $\mathbf{x} \cdot \mathbf{y} = \sum_{i=1}^{n} x_i y_i$. This can also be written in terms of matrix *products*, which you should review:

$$(x_1 \cdots x_n) \begin{pmatrix} y_1 \\ \vdots \\ y_n \end{pmatrix} = \mathbf{x}^T \mathbf{y} = \sum_{i=1}^{n} x_i y_i.$$

In particular, this means that the squared length of a vector may be written $\mathbf{x}^T\mathbf{x} = \sum_{i=1}^{n} x_i^2$.

Now we retackle our problem, to find the b that makes $(\mathbf{y} - \mathbf{x}b)^T(\mathbf{y} - \mathbf{x}b)$, the sum of squares of residuals, as small as possible. Again, let c be any possible value of the slope, and subtract and add $\mathbf{x}b$ to get

$$(\mathbf{y} - \mathbf{x}c)^T(\mathbf{y} - \mathbf{x}c) = (\mathbf{y} - \mathbf{x}b + \mathbf{x}[b - c])^T(\mathbf{y} - \mathbf{x}b + \mathbf{x}[b - c]).$$

Now we can expand this "square" just as before, because matrix multiplication and addition distribute and associate just like the ordinary operations:

$$\begin{aligned} (\mathbf{y} - \mathbf{x}c)^T(\mathbf{y} - \mathbf{x}c) = (\mathbf{y} - \mathbf{x}b)^T(\mathbf{y} - \mathbf{x}b) + [b - c]\mathbf{x}^T(\mathbf{y} - \mathbf{x}b) \\ + (\mathbf{y} - \mathbf{x}b)^T\mathbf{x}[b - c] + [b - c]\mathbf{x}^T\mathbf{x}[b - c]. \end{aligned}$$

This is not quite the same as before, because there are two middle terms. However, these happen to be the same (they are just the inner product of two vectors, listing the vectors in different orders). Our middle term is then just $2[b - c]\mathbf{x}^T(\mathbf{y} - \mathbf{x}b)$. The new normal equation to get rid of this term is $\mathbf{x}^T(\mathbf{y} - \mathbf{x}b) = 0$, which can be solved whenever $\mathbf{x} \neq 0$ to get $\hat{b} = \mathbf{x}^T\mathbf{y}/\mathbf{x}^T\mathbf{x}$. Our decomposition has become

$$(\mathbf{y} - \mathbf{x}c)^T(\mathbf{y} - \mathbf{x}c) = (\mathbf{y} - \mathbf{x}b)^T(\mathbf{y} - \mathbf{x}b) + [b - c]^2\mathbf{x}^T\mathbf{x}.$$

(You should decode these last three expressions to check that we got the same thing before, when we were using summation signs.)

Why have we done the same derivation twice? Because much later the matrix notation will be essential for similar but harder derivations; and we have given you some practice with it while you kept in mind what it really meant in terms of summation. But there is something deeper here: Remember from vector geometry that if two vectors $\mathbf{x}$ and $\mathbf{y}$ are both not zero, then their inner product $\mathbf{x}^T\mathbf{y} = 0$ exactly when they are at *right angles* to each other. In fact, in n-dimensional analytic geometry, this is the *definition* of a right angle. Therefore, our normal equation (leaving the b in but with c chosen to be zero) $(\mathbf{x}b)^T(\mathbf{y} - \mathbf{x}b) = 0$ may be restated as follows: Choose the parameter b such that the vector of predictions $(\mathbf{x}b)$ is at right angles to the vector of residuals $(\mathbf{y} - \mathbf{x}b)$. (In fact, this is the meaning of *normal* in geometry.) You can see from Figure 2.4 where the theorem of Pythagoras comes in. Further, you can see that our whole argument is just a familiar theorem

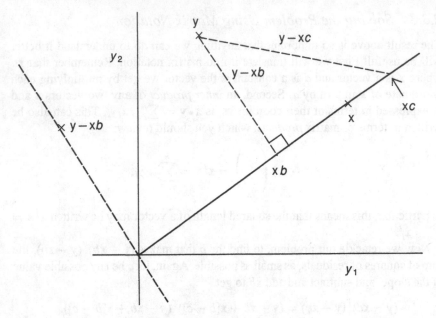

FIGURE 2.4. Geometry of least squares

from Euclidean geometry: To find the shortest distance from a point (y) to a line (xc for any number c), drop a *perpendicular*. It hits the line at some point (xb), and we call that value of the constant b our least-squares estimate $\hat{b}$.

2.3.4 Geometric Degrees of Freedom

Now we will use our geometric pictures to reinterpret the idea of degrees of freedom (see 1.3.3). Imagine that we have not carried out our regression experiment yet, but we know which n values of the independent variable x we will use as settings when we later observe our dependent y's. Here is what we already know: The vector $y - xb$, whatever it turns out to be, will be perpendicular to the predictions xb. With two observations, it may turn out to be any point on a certain line through the origin (imagine sliding the dotted line $y - xb$ down to where the coordinate axes intersect, as we have done in Figure 2.4).

With three observations, $y - xb$ may be any point in a whole *plane* perpendicular to the vector xb, that is, a two-dimensional *subspace* of our coordinate space (Figure 2.5).

Generally, our residual vector will be a point in the $(n - 1)$-dimensional *hyperplane* through the origin and perpendicular to the vector of possible predictions. (We need n coordinates to determine a point in the space of sample vectors. Let one coordinate axis be at an angle, in the direction x. The remaining $n - 1$ coordinates are needed to determine any vector at right angles to this one.) This turns out to be the geometrical way of looking at an issue we discussed in the previous chapter: When we say that the predictions have 1 degree of freedom, we mean that they

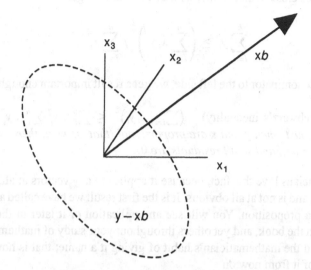

FIGURE 2.5. 3-D geometry of least squares

lie in a 1-dimensional subspace (varying according to possible values of b). When we say that this leaves $n - 1$ degrees of freedom for the errors, we mean that the residual vectors lie in an $(n - 1)$-dimensional subspace of the data space. We will greatly exploit this interpretation later.

We would now say that SSE $= \sum_{i=1}^{n}(y_i - bx_i)^2$ is the squared length of the vector of residuals $y - xb$, which lies in a known $(n - 1)$-dimensional subspace. Somewhat conventionally, when we average the squared errors, we average over the number of dimensions (degrees of freedom) rather than the number of observations to get the mean squared error MSE$= \frac{1}{n-1} \sum_{i=1}^{n}(y_i - bx_i)^2$. At the beginning of this chapter (see 2.2) we divided by n because we assumed that the predictions μ were given in advance of observation, and so the residual vector $y - \mu$ could lie anywhere in n-dimensional space.

Example. In Hubble's problem, MSE $= 5.469/23 = 0.2378$. Then the RMSE $= \sqrt{0.2378} = 0.4876$; in this data set, we typically misestimated the distance by not quite half a million parsecs.

2.3.5 Schwarz's Inequality

One more interesting fact comes out of the least-squares method: Remember that when we let $c = 0$ in our proposition, we got $\sum_{i=1}^{n} y_i^2 = \sum_{i=1}^{n}(y_i - bx_i)^2 + b^2 \sum_{i=1}^{n} x_i^2$. We can conclude from this that since the first term on the right is at least zero, then $\sum_{i=1}^{n} y_i^2 \geq b^2 \sum_{i=1}^{n} x_i^2$. Now substitute our least-squares estimate $\hat{b} = \sum_{i=1}^{n} x_i y_i / \sum_{i=1}^{n} x_i^2$, which makes the sum of squares of residuals $(\sum_{i=1}^{n}(y_i - bx_i)^2$, the term we threw away) as small as possible. Then the inequality is as close to an equality as it can be, and we get

$$\sum_{i=1}^{n} y_i^2 \geq \left(\sum_{i=1}^{n} x_i y_i\right)^2 \bigg/ \sum_{i=1}^{n} x_i^2.$$

Moving the denominator to the left side, we get a result important enough to name:

Theorem (Schwarz's inequality). $\left(\sum_{i=1}^{n} x_i y_i\right)^2 \leq \sum_{i=1}^{n} x_i^2 \sum_{i=1}^{n} y_i^2$; and we have equality just when **y** and **x** are proportional (that is, when there is a b such that each $y_i = bx_i$, and so all residuals are 0).

Mathematicians love this fact, because it applies to any vectors at all, is amazingly simple, and is not at all obvious. It is the first result we have called a theorem, and not just a proposition. You will see an application of it later in the chapter, others later in the book, and yet others throughout your study of mathematics. We have followed the mathematician's habit of giving it a name; that is how we will remind you of it from now on.

2.4   Sample Mean and Variance

2.4.1   Least-Squares Location Estimation

Our first summary model for measurements in the last chapter was the location model: We imagined that our n repeated measurements were unimportant errors in measuring a common constant μ. We can estimate μ by least squares: Let **x** be the vector of measurements; then our vector of predictions is $(\mu \cdots \mu)^T$, since every prediction is the same. To write this as a regression problem, we use the notation $(1 \cdots 1)^T = \mathbf{1}$ for a vector of all ones. Then $(\mu \cdots \mu)^T = \mathbf{1}\mu$ just multiplies each 1 by the constant μ. Now we have a regression equation like Hubble's: $\hat{\mathbf{x}} = \mathbf{1}\mu$, where **y** has been replaced by **x**, b has been replaced by μ, and **x** has been replaced by **1**. Our least-squares estimate is then $\hat{\mu} = \sum_{i=1}^{n} 1 x_i / \sum_{i=1}^{n} 1^2 = \frac{1}{n} \sum_{i=1}^{n} x_i = \bar{x}$. Interestingly enough, the least-squares estimate for the location model is just the sample mean, our standard estimate from the last chapter. So we see another reason that the sample mean is important. Let us, as promised, list some of its properties:

Proposition (properties of the sample mean).

(i) $\bar{x}$ is the least-squares location estimate for the sample vector **x**.
(ii) Add a constant to every observation: $x_i + a$. Then $\overline{x + a} = \bar{x} + a$.
(iii) Multiply every observation by a constant: bx_i. Then $\overline{bx} = b\bar{x}$.
(iv) The sum of the residuals $\sum_{i=1}^{n}(x_i - \bar{x}) = 0$.

We will let you show why (ii) and (iii) are true, as an easy exercise. We discovered (iv) in Chapter 1 (see 1.2.2).

2.4.2 Sample Variance

To measure how well the mean describes our observations we have SSE $= \sum_{i=1}^{n}(x_i - \bar{x})^2$; and adjusted for the number of degrees of freedom, MSE $= \frac{1}{n-1}\sum_{i=1}^{n}(x_i - \bar{x})^2$. This last quantity tells us how spread out the results typically are from their center. Statisticians have found it to be so enormously useful that they have given it a special name and notation:

Definition. The **sample variance** of a sample vector **x** is the mean-squared error about the sample mean $s_x^2 = \frac{1}{n-1}\sum_{i=1}^{n}(x_i - \bar{x})^2$. The **standard deviation** is its square root $s_x = \sqrt{s_x^2}$, the root-mean-squared error about $\bar{x}$.

Our Pythagorean law for location becomes $\sum_{i=1}^{n}(x_i - v)^2 = \sum_{i=1}^{n}(x_i - \bar{x})^2 + n(\bar{x} - v)^2$ for any number v. Dividing by $n - 1$ and solving for the sample variance, we have $s_x^2 = \frac{1}{n-1}\left[\sum_{i=1}^{n}(x_i - v)^2 - n(\bar{x} - v)^2\right]$. Letting $v = 0$, we get a famous formula for simplified computation of the variance: $s_x^2 = \frac{1}{n-1}\left(\sum_{i=1}^{n}x_i^2 - n\bar{x}^2\right)$. Judicious use of other values of v will often do much better; letting it be a round number that is fairly close to μ will lead to a calculation of the variance that is easier for pencil-and-paper computing, and less subject to round-off error in electronic computing.

Example. You want to know how far it is from your apartment to your college. You count your paces on five successive days, getting 1007, 998, 1023, 1025, and 1002 paces. You will use the sample mean as a summary measurement. To make the calculation easy (see 1.2.2), subtract $v = 1000$ from each number, and average: $(7 - 2 + 23 + 25 + 2)/5 = 11$. Then it is about $\bar{x} = 1000 + 11 = 1011$ paces to school. To get the sample variance, use this same value of v in the equation above:

$$s_x^2 = \left[7^2 + (-2) + 23^2 + 25^2 + 2^2 - 5 \times 11^2\right]\frac{1}{4} = 166.5.$$

Then the sample standard deviation is $s_x = \sqrt{166.5} = 12.9$. It appears that you varied about ± 13 paces from day to day as you walked to school.

We can use the mean and standard deviation to provide another kind of simple summary of a set of measurements. Add and subtract *twice* the standard deviation from the sample mean to get an interval in which a large majority of the numbers should fall. In the walking example, the interval is $985 \leq x_i \leq 1037$. We call this a 2-s *interval*. Our definition looks somewhat arbitrary, but we will see some sort of justification later.

Let us summarize our results:

Proposition (properties of the sample variance).

(i) *For any number* v, $s_x^2 = \frac{1}{n-1}\left[\sum_{i=1}^{n}(x_i - v)^2 - n(\bar{x} - v)^2\right]$.

(ii) $s_{x+a}^2 = s_x^2$ *and* $s_{x+a} = s_x$ *for any constant a (location invariance).*

(iii) $s_{bx}^2 = b^2 s_x^2$ *and* $s_{bx} = |b|s_x$ *for any constant b (scale equivariance).*

You should discover the last two as an exercise. Together they say that the standard deviation has nothing to do with where your measurements were centered but is directly proportional to how spread out they are.

2.4.3   Standard Scores

These measures of location and scale of our variable samples give us a way to compare "atypicality" of observations that were originally evaluated in quite different ways:

Example. On the first midterm exam in a statistics class, you make an 82; but on the second you make only 65. However, the professor grades on the "curve," by which she seems to mean that your score will be compared to how your classmates scored on the same test. You learn that on the first test, the class average was 75 with a standard deviation of 15. On the second test, the average was 51 with a standard deviation of 12. On which one is your professor likely to conclude that you did better?

We will, as in the 2-s interval, describe each observation as some number of standard deviations above or below the mean. Letting t_i denote that number, we write $x_i = \bar{x} + t_i s_x$; solving for t_i, we get the following:

Definition. For a sample of n observations x_i, the **standardized** measurements (or **standard scores**) are $t_i = \frac{x_i - \bar{x}}{s_x}$.

For example, 1007 paces becomes $(1007 - 1011)/12.9 = -0.31$. In words, 1007 is 0.31 standard deviations below the mean. Notice that the 2-s limits are always at $t = \pm 2$. A standardized measurement has lost the scale on which it was originally measured:

Proposition (properties of standard scores).

  (i) *Under the changes of variable $x + a$ and bx (for $b > 0$), t does not change.*
 (ii) *$\bar{t} = 0$ and $s_t = 1$.*

You should show these as exercises.

Example (*cont.*).   On that first exam, your standard score was $(82 - 75)/15 = 0.47$. On the second test, your standard score was $(65 - 51)/12 = 1.17$. It turns out that you did relatively better on the second test, in the sense of being farther above the class average if the test scores were similarly variable. Your professor should be quite a bit more impressed with you the second time.

2.5   One-Way Layouts

2.5.1   Analysis of Variance

Remember that a one-way layout experiment splits up a number of observations x_{ij} among the levels i of a treatment (see 1.3.1). It may have occurred to you that we have now established that the standard estimates for the one-way layout $\hat{x}_{ij} = \hat{\mu} = \bar{x}_i$ are actually the least-squares estimates for the parameters μ_i of that model. This is because the SSE is just the sum of squared deviations of each measurement about the center of its cell; and we have discovered that these are made smallest for each cell in turn by using the cell means as centers.

What about the centered model $\hat{x}_{ij} = \mu + b_i$? The least-squares estimates only make the *residuals* small, and these are determined by the cell estimates $\hat{x}_{ij}$. The centered model has exactly these same standard cell predictions (we just wrote them in terms of different parameters), so the residuals $x_{ij} - \hat{x}_{ij}$ are still the same, and as small as possible. Therefore, the standard estimates $\hat{\mu} = \bar{x} = \frac{1}{n} \sum_{i=1}^{k} \sum_{j=1}^{n_i} x_{ij}$ and $\hat{b}_i = \bar{x}_i - \bar{x}$ are also least-squares estimates.

The fact that the standard estimates are least-squares will teach us something important. The sum-squared error is SSE $= \sum_{i=1}^{k} \sum_{j=1}^{n_i} (x_{ij} - \hat{x}_{ij})^2 = \sum_{i=1}^{k} \sum_{j=1}^{n_i} (x_{ij} - \bar{x}_i)^2$. But the inner sum $\sum_{j=1}^{n_i} (x_{ij} - \bar{x}_{ij})^2$ is just the SSE of the location model for the n_i observations in the ith level by themselves. Then the Pythagorean law in (4.2) letting $v = \bar{x}$, the overall mean, gives us $\sum_{j=1}^{n_i} (x_{ij} - \bar{x}_i)^2 = \sum_{j=1}^{n_i} (x_{ij} - \bar{x})^2 - n_i (\bar{x}_i - \bar{x})^2$. Putting this back in the double sum for the SSE, we get

$$\sum_{i=1}^{k} \sum_{j=1}^{n_i} (x_{ij} - \bar{x}_i)^2 = \sum_{i=1}^{k} \sum_{j=1}^{n_i} (x_{ij} - \bar{x})^2 - \sum_{i=1}^{k} n_i (\bar{x}_i - \bar{x})^2.$$

Moving the negative part over to the other side yields $\sum_{i=1}^{k} \sum_{j=1}^{n_i} (x_{ij} - \bar{x})^2 = \sum_{i=1}^{k} \sum_{j=1}^{n_i} (x_{ij} - \bar{x}_i)^2 + \sum_{i=1}^{k} n_i (\bar{x}_i - \bar{x})^2$. Now remembering what these had to do with the parameters of the centered model, $\hat{\mu} = \bar{x}$ and $\hat{\mu} + \hat{b}_i = \bar{x}_i$, we can rewrite this last expression:

$$\sum_{i=1}^{k} \sum_{j=1}^{n_i} (x_{ij} - \hat{\mu})^2 = \sum_{i=1}^{k} \sum_{j=1}^{n_i} (x_{ij} - \hat{\mu} - \hat{b}_i)^2 + \sum_{i=1}^{k} n_i \hat{b}_i^2.$$

Proposition. *In the centered model for a one-way layout, with least-squares estimates $\hat{\mu} = \bar{x}$ and $\hat{b}_i = \bar{x}_i - \bar{x}$, we have*

$$\sum_{i=1}^{k} \sum_{j=1}^{n_i} (x_{ij} - \bar{x})^2 = \sum_{i=1}^{k} n_i (\bar{x}_i - \bar{x})^2 + \sum_{i=1}^{k} \sum_{j=1}^{n_i} (x_{ij} - \bar{x}_i)^2,$$

or

$$\sum_{i=1}^{k} \sum_{j=1}^{n_i} (x_{ij} - \hat{\mu})^2 = \sum_{i=1}^{k} n_i \hat{b}_i^2 + \sum_{i=1}^{k} \sum_{j=1}^{n_i} (x_{ij} - \hat{\mu} - \hat{b}_i)^2.$$

This is so important that we have a shorthand notation to help us remember it. The rightmost term was SSE. The term on the left is called the *corrected* sum of squares and is denoted by SS. We call $\sum_{i=1}^{k} n_i \hat{b}_i^2$ the *sum of squares for treatment* (SST) (or sometimes the *between-groups* sum of squares). It is the total of the squares of all the adjustments we have made for the level of treatment in the individual observations. Therefore, our result may be written SS = SST + SSE.

The expansion can go one step further: Since SS $= \sum_{i=1}^{k} \sum_{j=1}^{n_i} (x_{ij} - \bar{x})^2$ is just the error sum of squares for a simple location model with only one location μ for all the observations, apply the result from (4.2) with $\nu = 0$ to get

$$\sum_{i=1}^{k}\sum_{j=1}^{n_i}(x_{ij} - \bar{x})^2 = \sum_{i=1}^{k}\sum_{j=1}^{n_i} x_{ij}^2 - n\bar{x}^2.$$

Plugging this into the proposition yields an impressive result:

Theorem (analysis of variance for the one-way layout).   *In the centered model for a one-way layout, with least-squares estimates $\hat{\mu} = \bar{x}$ and $\hat{b}_i = \bar{x}_i - \bar{x}$, we have*

$$\sum_{i=1}^{k}\sum_{j=1}^{n_i} x_{ij}^2 = n\bar{x}^2 + \sum_{i=1}^{k} n_i(\bar{x}_i - \bar{x})^2 + \sum_{i=1}^{k}\sum_{j=1}^{n_i}(x_{ij} - \bar{x}_i)^2,$$

or

$$\sum_{i=1}^{k}\sum_{j=1}^{n_i} x_{ij}^2 = n\hat{\mu}^2 + \sum_{i=1}^{k} n_i\hat{b}_i^2 + \sum_{i=1}^{k}\sum_{j=1}^{n_i}(x_{ij} - \hat{\mu} - \hat{b}_i)^2.$$

We have now decomposed the total sum of squares of the measurements TSS $= \sum_{i=1}^{k}\sum_{j=1}^{n_i} x_{ij}^2$ into three pieces: The new one, $n\bar{x}^2$, is called the *sum of squares for the mean* SSM. We then remember the analysis of variance theorem symbolically as TSS $=$ SSM $+$ SST $+$ SSE.

2.5.2   Geometric Interpretation

Looking at this model geometrically, let $\hat{\mu} = \mathbf{1}\hat{\mu}$,

$$\hat{\mathbf{b}} = \left(\underbrace{\hat{b}_i \cdots \hat{b}_i}_{n_i \text{entries}} \quad \cdots \quad \underbrace{\hat{b}_k \cdots \hat{b}_k}_{n_k \text{entries}} \right)^{\mathrm{T}},$$

and the residual vector $\hat{\mathbf{e}} = \mathbf{x} - \hat{\mu} - \hat{\mathbf{b}}$, where the observation vector is

$$\mathbf{x} = \left(x_{11} \cdots x_{1n_1} x_{21} \cdots x_{2n_2} \cdots x_{k1} \cdots x_{kn_k} \right)^{\mathrm{T}}.$$

Each vector is n-dimensional. You should check as an exercise that our theorem may be written $\mathbf{x}^{\mathrm{T}}\mathbf{x} = \hat{\mu}^{\mathrm{T}}\hat{\mu} + \hat{\mathbf{b}}^{\mathrm{T}}\hat{\mathbf{b}} + \hat{\mathbf{e}}^{\mathrm{T}}\hat{\mathbf{e}}$. But then we note some important facts:

Proposition.

(i) $\hat{\mu}^{\mathrm{T}}\hat{\mathbf{b}} = 0.$
(ii) $\hat{\mu}^{\mathrm{T}}\hat{\mathbf{e}} = 0.$
(iii) $\hat{\mathbf{b}}^{\mathrm{T}}\hat{\mathbf{e}} = 0.$

These also should be verified, as an exercise. We say the vectors are orthogonal to one another.

Perhaps now you can imagine the geometry of the theorem, which is a three-dimensional version of the ubiquitous theorem of Pythagoras. Imagine a

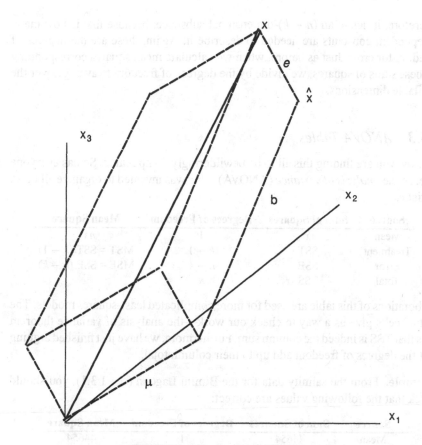

FIGURE 2.6. Geometry of ANOVA

rectangular box whose length, width, and height are our three estimated vectors, which you have checked are at right angles to each other (Figure 2.6). Then the observation vector **x** is the diagonal of that box. The various sums of squares are the squared lengths of the edges, which sum to the squared length of the diagonal.

Once again we can use our picture (Figure 2.6) to interpret the degrees of freedom in the one-way layout model. The vector $\hat{\mu}$ lies in a one-dimensional subspace, those vectors proportional to **1**, which corresponds to the single degree of freedom for the mean. The vector $\hat{\mathbf{b}}$ is determined by the k different level adjustments, which may each have any value at all (at least until you make your observations), except of course for the centering constraint, which requires them to average zero. This last statement, by the way, is just what our result $\hat{\mu}^T \hat{\mathbf{b}} = 0$ tells us: Our adjustments must be at right angles to the constant vector. Therefore, $\hat{\mathbf{b}}$ is determined by $k - 1$ independent constants and necessarily lies in a $(k-1)$-dimensional subspace of our data space. This matches the degrees of freedom for the b's. The residuals vector may take on any n values at all, except that our results $\hat{\mu}^T \hat{\mathbf{e}} = 0$ and $\hat{\mathbf{b}}^T \hat{\mathbf{e}} = 0$ say that it must be perpendicular to any mean vector and any adjustments vector.

Therefore, it lies in an $(n - k)$-dimensional subspace, because that is how many independent constants are needed to describe it. Again, these are the degrees of freedom for error. Just as before, when we calculate mean squares corresponding to these sums of squares, we divide by the degrees of freedom to average over the available dimensions.

2.5.3 ANOVA Tables

By now you are finding this all to be bewilderingly complicated. So has everyone else, so the *analysis of variance* (ANOVA) *table* was invented to organize all these statistics:

Source	Sum of Squares	Degrees of Freedom	Mean Square
Mean	$n\bar{x}^2$	1	$n\bar{x}^2$
Treatment	SST	$k - 1$	MST = SST/$(k - 1)$
Error	SSE	$n - k$	MSE = SSE/$(n - k)$
Total	TSS	n	

Elaborations of this table are used for more complicated least-squares models. The "total" cells give us a way to check our work—the analysis of variance theorem says that TSS is indeed the column sum. Furthermore, we have just finished arguing that the degrees of freedom add up to their column total.

Example. From the salinity data for the Bimini Lagoon (see 1.3.1), you should check that the following values are correct:

Source	Sum of Squares	Degrees of Freedom	Mean Square
Mean	44654	1	44654
Water Mass	38.80	2	19.40
Error	7.934	27	0.2938
Total	44700.7	30	

How shall we interpret the quantities in this table? In this problem (and often in other ANOVA problems) we find ourselves uninterested in the overall mean and its table entry. It is so large because the ocean is salty, and that is where the water comes from. We are interested rather in the differences among samples. We retreat to the proposition SS = SST + SSE, and the table simplifies to this more commonly seen form:

Source	Sum of Squares	Degrees of Freedom	Mean Square
Treatment	SST	$k - 1$	MST = SST/$(k - 1)$
Error	SSE	$n - k$	SSE/$(n - k)$
Total	SS	$n - 1$	

To quantify the relative importance of the treatment level, we may compute the following statistic:

Definition. The **coefficient of determination** is given by $R^2 = \frac{SST}{SST+SSE} = \frac{SST}{SS}$.

In our salinity example, the corrected sum of squares is 46.734, so $R^2 = 0.83$. We might interpret R^2 as the proportion of the sample variance that is "explained" by systematic differences among the levels. As its name is a mouthful, most statisticians just call it "R-squared." You might remember from trigonometry that R^2 is the square of the cosine of the angle between the vectors $\hat{b}$ and $x - \hat{\mu}$.

2.5.4   The F-Statistic

We might ask instead a somewhat harder question: Are the apparent differences among treatment means just an accident? That is, did we just by bad luck pick saltier samples in area II and fresher samples in area I? Given the variability of our measurements, that certainly seems possible; but we can never tell with reasonable certainty without doing a much more extensive set of measurements.

Since we are using the principle of least squares, we must think that the most important fact about our random errors is the *length* of the error vector. Therefore, if we *rotate* that error vector in any direction whatsoever, keeping it the same length, we should get the same least-squares estimates of our model parameters. This suggests that if least squares is indeed the right way to look at errors in our experiment, the following assumption about what those errors look like is plausible:

Assumption of Spherical Distribution.   If we repeat the whole experiment many times, the scatter of sample vectors in n-dimensional space is much the same in any direction from the vector of "true" values.

This says that the error, or residual, vectors tend to be of similar lengths in any direction. In one dimension, this means that the scatter of numbers above the true value looks much the same as the scatter of numbers below the true value, reversed as if in a mirror. In two dimensions, this pattern is called *circular symmetry*; an example is shown in Figure 2.7, where each triangle marks the error vector for one repetition of the experiment. If you rotate this scatter plot through any number of degrees, it still looks much the same. In three-dimensional space, the scatter plot would look like what astronomers call a globular star cluster. The mathematical word for such a pattern is *spherical symmetry*, hence the name of our assumption.

One implication of this assumption is that the *order* of the observations, the indices $j = 1, 2, \ldots$ that we gave them, is not scientifically important. This is because changing the order just involves switching coordinate axes around; that obviously has no effect on the general appearance of our spherical cloud of sample vectors. This is often a desirable property of fair sampling practices. Much later in the book you will discover that certain very common statistical models will imply that our assumption is true.

Now assume in our centered model that if we actually knew the deep scientific truth about what is going on, $b = 0$, so that the treatments should not matter, then the vector whose squared length is SST, $\hat{b}$, would consist of irrelevant peculiarities about our data. Much later in the book we will discover the mathematical reasons

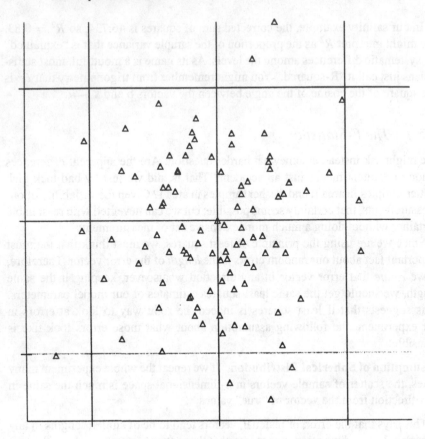

FIGURE 2.7. Observations with circular symmetry

for an amazing and wonderful fact: If in the one-way layout model $\mathbf{b} = \mathbf{0}$, and the assumption of spherical distribution is true for this sort of experiment, then MST and MSE will often be similar in size.

You have no obligation to believe me about this yet, or even understand what it means. But it tells us why we like to calculate the following:

Definition. The **F-statistic** is given by $F_{k-1,n-k} = \frac{\text{MST}}{\text{MSE}}$.

Our justification suggests that when the true adjustments due to the experimental levels should be zero, the F-statistic is somewhere near 1. On the other hand, if the adjustments for level are substantially different from zero, MST increases, as you may see by looking at its formula, and so does the F-statistic. In our salinity example $F_{2,27} = 66.03$; this is so much greater than 1 that we are fairly confident that the salinity does vary from site to site. If our statistic had been, say 0.7, we would have to say that the evidence for the treatment mattering was weak, since some number like this might have arisen by routine accident in an experiment with no real treatment effect.

We will see other F-statistics with which to evaluate the evidence for experimental treatment effects in other least-squares models. Of course, we have nothing but experience to guide us in how much bigger than 1.0 an F must be before we jump to any conclusions about nontrivial effects; this will come later.

2.5.5 The Kruskal–Wallis Statistic

Another simple way to see whether several levels of a treatment show different measurement values is to **rank** all the measurements from smallest to largest. For example, in the salinity data, the value 36.71 gets a rank of 1, 36.75 gets a rank of 2, and the two 37.01s are tied for third, so we conventionally give each a rank of 3.5. We continue until 40.80 gets a rank of 30. The complete rankings are

Mass I: 10 3.5 1 5.5 7 3.5 5.5 11 8 2 9 17
Mass II: 27 30 24 23 29 28 25 22
Mass III: 19 21 20 12 14 16 18 15 13 26

as you should check.

It seems reasonable to perform an analysis of variance on these ranks. The notation we shall use is R_{ij} for the rank in the whole sample of the jth observation in the ith level; for example, $R_{II3} = 24$. In our example, the level means are $\bar{R}_I = 6.917$, $\bar{R}_{II} = 26.0$, and $\bar{R}_{III} = 17.40$. This tells us much the same thing as the level means of the original salinities: Mass II is a bit saltier than III and much saltier than I. (Traditionally, if we are interested in only the question of whether the ith level is peculiar, we compute its **Wilcoxon rank-sum statistic** $W_i = \sum_{j=1}^{n_i} R_{ij} = n_i \bar{R}_i$. For example, $W_{III} = 174$. Of course, this is harder to interpret than the level means.)

The new way of comparing the levels has two important disadvantages: first, it no longer says anything at all about just how salty the water actually is. Second, it loses some distinctions that were present in the original observations; for example, the distinction between 19 and 20 was only 0.01%, but the difference between 11 and 12 is fully 0.54%.

On the other hand, the new statistic has an important advantage: If we attach very little importance to the actual values on the scale of measurement, but only trust it usually to tell us which sample has a larger value, then these comparisons based on ranks seem plausibly to capture what we want to know. For example, our salinity gauge might be poorly calibrated, so that the only thing we are sure of is that it reads higher with saltier water. Or our scale may have been an arbitrary one, designed just for this one experiment. The arithmetic test from (1.4.1) was a collection of problems the teacher invented on the spur of the moment. A grade of 26 means nothing in itself; but the student who scored 26 is likely doing better in the class than the one who scored 17. Therefore, analyzing this problem by ranks might well tell us almost as much as using the grades.

The obvious statistic to summarize differences between water masses is the sum of squares for treatment, which, remember (Section 5.2), compares these level means to the overall mean $\bar{R} = 15.5$. Then SST $= \sum_{i=1}^{k}(\bar{R}_i - \bar{R})^2$; in our water example, SST $= 1802.18$. Notice that some simplification will turn out to be possible, because $\bar{R}$ is just the average of the ranks $1, \ldots, n$; this is exactly the same, no matter how the experiment came out. In fact, the average of the first and last ranks is $(1 + n)/2$; the average of the second and next to last is $(2 + (n - 1))/2 = (1 + n)/2$; in fact, all such low and matching high pairs average the same, $(1 + n)/2$. Therefore, it is always the case that $\bar{R} = (1 + n)/2$.

It gets better; our corrected sum of squares SS depends only on all the ranks, so it will be the same however the experiment comes out (if we ignore ties). You will figure out a formula for SS in the next chapter as an exercise. But this fact has an important implication: Earlier in this section we had to invent R-squared and F to compare SST and SSE, because they were independent pieces of information. Now SS = SST + SSE, with SS known in advance, says that they are no longer independent; we need calculate only SST, and interpret it.

Definition. The **Kruskal–Wallis statistic** is $K = 12/(n(n + 1))$SST, where SST is the sum of squares for treatment when the ranks of the observations are used as the data.

In the water example, $K = 23.25$. The larger this is, the more different are the water masses. In an exercise in a later chapter, you will discover that if there are in fact no systematic differences among the levels, so there is no pattern to which ranks are where, a typical value of K is somewhere in the neighborhood of $k - 1$, the degrees of freedom for treatment. (This is why it is usual to multiply by $12/(n(n + 1))$; the interpretation will no longer depend on our sample size.) In our example, $3 - 1 = 2$ is so much smaller than 23.25 that we suspect we have spotted a real salinity difference.

The Kruskal–Wallis statistic is an important example of a *rank statistic*, which are of considerable historical interest in applied statistics. You will see another example in a later chapter.

2.6   Least-Squares Estimation for Regression Models

2.6.1   Estimates for Simple Linear Regression

Finally, we come to an important estimation problem from the last chapter that the method of least squares can solve for us. Remember the simple linear regression model $\hat{y}_i = \mu + (x_i - \bar{x})b$? (See 1.5.2.) We were able to suggest standard estimates of the parameters μ and b in only the simplest case, where exactly two distinct values of the independent variable **x** appeared in the data, so we could interpolate between them. The method of least squares would suggest that we choose our parameters to make $\sum_{i=1}^{n}[y_i - \mu - (x_i - \bar{x})b]^2$ as small as possible. This looks harder than the problem we solved in Section 3; but fortunately, we have already done most of the work.

First, pretend we already knew the correct value of b. Then the least-squares problem just asks what constant value μ makes $\sum_{i=1}^{n}\{[y_i - (x_i - \bar{x})b] - \mu\}^2$ smallest. That is, what single μ is closest to the known numbers $[y_i - (x_i - \bar{x})b]$? We already solved this problem in Section 4: The least-squares estimate is just their average

$$\hat{\mu} = \frac{1}{n}\sum_{i=1}^{n}[y_i - (x_i - \bar{x})b] = \frac{1}{n}\sum_{i=1}^{n}y_i - \frac{b}{n}\sum_{i=1}^{n}(x_i - \bar{x}).$$

The last term is zero, from a property of the sample mean, so $\hat{\mu} = \bar{y}$. This works out so nicely because we used a centered model.

We get the same result for any b; to get the best b, we are left with the problem of minimizing $\sum_{i=1}^{n}[y_i - \bar{y} - (x_i - \bar{x})b]^2$. That is, we want a least-squares prediction of the values $y_i - \bar{y}$ from the model $(x_i - \bar{x})b$. This is the simple proportion model from Section 3; so $\hat{b} = \sum_{i=1}^{n}(y_i - \bar{y})(x_i - \bar{x})/\sum_{i=1}^{n}(x_i - \bar{x})^2$. This is important enough to make into a theorem.

Theorem (linear regression by least squares). *Given a vector of independent variable settings* $\mathbf{x}$ *and a vector of dependent measurements* $\mathbf{y}$, *then the least-squares estimates of the prediction model* $\hat{y}_i = \mu + (x_i - \bar{x})b$ *are given by* $\hat{\mu} = \bar{y}$ *and*

$$\hat{b} = \sum_{i=1}^{n}(y_i - \bar{y})(x_i - \bar{x})/\sum_{i=1}^{n}(x_i - \bar{x})^2$$

whenever not all values of $\mathbf{x}$ *are the same.*

(Why did I have to put in that last quibble?) You should check as an exercise that the estimates in the theorem are the same as our standard estimates from the last chapter, in case there are only two different values of the independent variable.

Example. Mapes and Dajda in 1976 collected data on the percentage of the time that ill British children of various ages were taken to the doctor:

age	0	1	2	3	4	5	6	7
percentage	70	76	51	62	67	48	50	51
age	8	9	10	11	12	13	14	
percentage	65	70	60	40	55	45	38	

It is plausible that a very crude prediction of a child's likelihood of being taken to the doctor might be made by a linear regression model: If p stands for the percentage of time an age group has gone to the doctor, and a for their age, then we predict $\hat{p} = \mu + (a - \bar{a})b$. Actually, since the raw data were individual cases of a child either going or not going, I should be using *logistic* regression here (see 1.8.2); but I have no access to the raw data. We shall do the best we can with a least-squares estimate of a linear regression model. We calculate $\bar{a} = 7$ years, $\hat{\mu} = \bar{p} = 56.53\%$, $\sum_{i=1}^{n}(a_i - \bar{a})^2 = 280$, and $\sum_{i=1}^{n}(a_i - \bar{a})(p_i - \bar{p}) = -440$. Then $\hat{b} = -440/280 = -1.5714$. (You should check my arithmetic.) We arrive

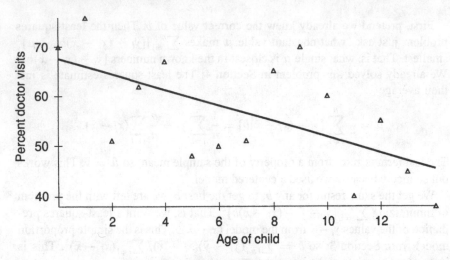

FIGURE 2.8. Doctor visits as a function of age

at a prediction equation

$$\hat{p} = 56.53 - 1.5714(a - 7).$$

This line is displayed on the scatter plot in Figure 2.8. For example, we predict that a child of 9.5 years of age will be taken to the doctor about 52.6% of the time. From looking at the graph, this is a very crude estimate; on the other hand, I think I would trust it better than just the data values for 9 and 10 years.

2.6.2   ANOVA for Regression

We partition the sum of squares as in Section 3 to get

$$\sum_{i-1}^{n}(y_i - \bar{y})^2 = \sum_{i-1}^{n}\left[y_i - \bar{y} - \hat{b}(x_i - \bar{x})\right]^2 + \hat{b}^2 \sum_{i-1}^{n}(x_i - \bar{x})^2,$$

and then decompose the left-hand side as in Section 4:

Theorem (analysis of variance for simple linear regression). *For the least-squares estimates for simple linear regression,*

$$\sum_{i=1}^{n} y_i^2 = n\hat{\mu}^2 + \hat{b}^2 \sum_{i-1}^{n}(x_i - \bar{x})^2 + \sum_{i-1}^{n}\left[y_i - \bar{y} - \hat{b}(x_i - \bar{x})\right]^2.$$

As an exercise, you should interpret this as a statement about vectors at right angles to each other. The new term we call the sum of squares for regression: SSR= $\hat{b}^2 \sum_{i-1}^{n}(x_i - \bar{x})^2$; it has one degree of freedom. So now we can write down an analysis of variance table:

Source	Sum of Squares	Degrees of Freedom	Mean Square
Mean	$n\bar{y}^2$	1	$n\bar{y}^2$
Regression	SSR	1	MSR = SSR
Error	SSE	$n-2$	MSE = SSE/$(n-2)$
Total	TSS	n	

Example (*cont.*).   In the problem of rates of going to the doctor, we have the following:

Source	Sum of Squares	Degrees of Freedom	Mean Square
Mean	47940.0	1	47940.0
Age	691.429	1	691.429
Error	1202.305	13	92.485
Total	49834	15	

That gives us $R^2 = 691.43/(691.43 + 1202.3) = 0.3651$. Only about 37% of the variability in our rates of going to the doctor is explained by the linear trend we have proposed. On the other hand, $F_{1,13} = \frac{691.43}{92.485} = 7.4761$ is much bigger than one, so that even though our predictions do not accomplish a great deal, the downward trend may be real.

2.7   Correlation

2.7.1   Standardizing the Regression Line

To see some qualitative features of the least-squares regression equation, divide both the numerator and denominator of the slope estimate by $n - 1$,

$$\hat{b} = \frac{\frac{1}{n-1}\sum_{i=1}^{n}(y_i - \bar{y})(x_i - \bar{x})}{\frac{1}{n-1}\sum_{i=1}^{n}(x_i - \bar{x})^2},$$

so that the denominator is just the sample variance of **x**. Let us give the numerator a name:

Definition. The **sample covariance** of sample measurement vectors **x** and **y** is

$$s_{xy} = \frac{1}{n-1}\sum_{i=1}^{n}(y_i - \bar{y})(x_i - \bar{x}).$$

Then we can write compactly $\hat{b} = s_{xy}/s_x^2$. Now our regression equation, with $\hat{\mu} = \bar{y}$ moved back to the other side of the equation, looks like $\hat{y} - \bar{y} = (x - \bar{x})s_{xy}/s_x^2$. These subtractions may remind you of *standard scores*; we can force them to appear by dividing both sides by s_y and rearranging to get

$$(\hat{y} - \bar{y})/s_y = ((x - \bar{x})/s_x)(s_{xy}/(s_x s_y)).$$

Let us play a standard mathematician's game by giving the messy part a name:

Definition. The **sample correlation** between x and y is

$$r_{xy} = \frac{s_{xy}}{s_x s_y} = \frac{\sum_{i=1}^{n}(y_i - \bar{y})(x_i - \bar{x})}{\sqrt{\sum_{i=1}^{n}(y_i - \bar{y})^2 \sum_{i=1}^{n}(x_i - \bar{x})^2}}.$$

We have canceled out the $(n-1)$'s. For example, in the age/doctor–visit problem, $r = -0.604$.

Giving obvious names to the parts that are standard scores, we have a remarkably compact formulation of simple least-squares regression:

Proposition. $t_{\hat{y}} = r_{xy} t_x$.

2.7.2   Properties of the Sample Correlation

This last equation is not terribly useful for doing predictions, and it will help our understanding only if we develop some insight into what the correlation means. It will turn out to be a dimensionless measure of the degree to which the two variables change together. First, let us apply the Schwarz inequality (see Section 3.5) to $x_i - \bar{x}$ and $y_i - \bar{y}$ to get that

$$\left[\sum_{i=1}^{n}(y_i - \bar{y})(x_i - \bar{x})\right]^2 \le \sum_{i=1}^{n}(y_i - \bar{y})^2 \sum_{i=1}^{n}(x_i - \bar{x})^2$$

always holds, where all the quantities are familiar from earlier in this section. Dividing by the right-hand side, we find

$$\frac{\left[\sum_{i=1}^{n}(y_i - \bar{y})(x_i - \bar{x})\right]^2}{\sum_{i=1}^{n}(y_i - \bar{y})^2 \sum_{i=1}^{n}(x_i - \bar{x})^2} \le 1.$$

This is just the square of the correlation, so always $r_{xy}^2 \le 1$, which gives us the first part of the following:

Proposition (properties of the correlation).

 (i) $-1 \le r_{xy} \le 1$.
 (ii) $r_{xy} = r_{yx}$.
 (iii) $r_{x+a,y} = r_{xy}$ *for any constant a.*
 (iv) $r_{cx,y} = r_{xy}$ *for any constant c > 0.*
 (v) $r_{cx,y} = -r_{xy}$ *for c < 0.*

Notice that (ii) is true because x and y may be switched in the defining formula. You should prove (iii)–(v) as exercises.

Parts (iv) and (v) are what we mean by calling a quantity *dimensionless*: Think of c as the conversion factor that you need to change one of the variables from feet into meters, for example. In the process r does not change.

Now go back to the statement of the Schwarz inequality: It becomes an equality just when the vector of quantities $y_i - \bar{y}$ is exactly proportional to the vector of quantities $x_i - \bar{x}$. That is, there is some constant b such that $y_i - \bar{y} = b(x_i - \bar{x})$.

But then $y_i = \bar{y} + b(x_i - \bar{x})$, and the regression prediction is exactly true. The points in the scatter plot are lined up perfectly along this straight line, and SSE = 0. In this case, because the inequality has become an equality, necessarily $r_{xy}^2 = 1$; so $r_{xy} = 1$ (if $b > 0$) or $r_{xy} = -1$ (if $b < 0$).

Now summarize what we can conclude from knowing the correlation:

1. If $r_{xy} = 1$, then all pairs (x, y) fall on an upward-sloping line.
2. If $r_{xy} > 1$, there is an upward-sloping regression line; the larger it is, the more tightly the pairs cluster about the line (we call this a *positive association* between x and y).
3. If $r_{xy} = 0$, a regression line is flat, and it does not help you predict one variable from the other (we say x and y are *uncorrelated*).
4. If $r_{xy} < 0$, there is a downward-sloping regression line; the more negative it is, the more tightly the pairs cluster about the line (x and y have a *negative association*).
5. If $r_{xy} = -1$, then all pairs (x, y) fall on a downward-sloping line.

You might notice that because of our properties of the correlation, it simply does not matter in Figure 2.9 where the origin is, or what units our axes are in, or which axis is x and which is y.

For the example where $r = -0.604$, there is a moderate degree of negative association. You might notice that in this example $r^2 = R^2$. You should show as an exercise that this is always true for simple linear regression. Of course, r may be either positive or negative, and so tell us also the direction of the association. On the other hand R^2 makes sense for any model estimated by least squares.

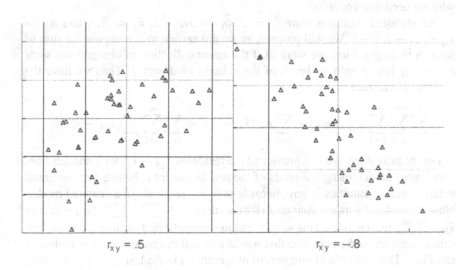

$r_{xy} = .5$ $r_{xy} = -.8$

FIGURE 2.9. Examples of correlation

2.7.3 Regression to the Mean

The regression equation $t_{\hat{y}} = r_{xy}t_x$ tells us something interesting right away. Since r is always no bigger in size than one, it follows that $|t_{\hat{y}}| \le |t_x|$: The standard score of the prediction is no bigger in size than the independent-variable standard score. We always predict that our experimental result will be closer to average than our experimental setting. This is called *regression to the mean*; it was so named by the pioneering mathematical biologist Francis Galton in the late nineteenth century, and is the origin of the statistical use of the word regression. His example was that the sons of tall fathers tend to be taller than average, but less so than their fathers; the reverse is true for sons of short fathers. This correlation is about 0.5; so on average, children regress halfway to the mean height of their generation, by our equation.

2.8   More Complicated Models*

2.8.1 ANOVA for Two-Way Layouts

The method of least squares should tell us how to estimate the parameters of models for more elaborate experiments. For example, what about two-way layouts? In the full model $\hat{x}_{ijk} = \mu_{ij}$, we know what to do; as before, we get a least-squares estimate for each cell separately: $\hat{\mu}_{ij} = \bar{x}_{ij}$. This is the standard estimate. But now consider the centered parametrization $\hat{x}_{ijk} = \mu + b_i + c_j + d_{ij}$. What are the least-squares estimates for the parameters, and do we have an analysis of variance to rate their importance? In Chapter 1, we claimed that the standard estimates were appropriate only for *balanced* designs, when the numbers of observations of the cells of each row were proportional to each other (see 1.4.3). Now we shall see why we need that condition.

The standard estimates were $\hat{\mu} = \bar{x}$, $\hat{b}_i = \bar{x}_{i\bullet} - \bar{x}$, $\hat{c}_j = \bar{x}_{\bullet j}$, and $\hat{d}_{ij} = \bar{x}_{ij} - \bar{x}_{i\bullet} - \bar{x}_{\bullet j} + \bar{x}$. We will proceed, as we did earlier, to decompose the sum of squares in stages. First, we work as if the entire collection of observations were a one-way layout split by levels of the column treatment j. Then we have the analysis of variance

$$\sum_{i=1}^{l}\sum_{j=1}^{m}\sum_{k=1}^{n_{ij}} x_{ijk}^2 = n\bar{x}^2 + \sum_{j=1}^{m} n_{\bullet j}(\bar{x}_{\bullet j} - \bar{x})^2 + \sum_{i=1}^{l}\sum_{j=1}^{m}\sum_{k=1}^{n_{ij}} (x_{ijk} - \bar{x}_{\bullet j})^2.$$

For the next stage, we will predict all the residuals $x_{ijk} - \bar{x}_{\bullet j}$ with another one-way layout model, using the levels of the *row* treatment i. Notice that the grand mean of these quantities is zero, because they are residuals in a centered model. Now we need to figure out their mean for the ith row: $\frac{1}{n_{i\bullet}}\sum_{j=1}^{m}\sum_{k=1}^{n_{ij}}(x_{ijk} - \bar{x}_{\bullet j}) = \bar{x}_{i\bullet} - \sum_{j=1}^{m}(n_{ij}/n_{i\bullet})\bar{x}_{\bullet j}$. This would lead to a complicated decomposition of the sum of squares, and worse, one that would turn out different if we had looked at rows first. But that ratio of numbers of observations in the last term, $n_{ij}/n_{i\bullet}$ does not depend on i, because we are talking about balanced designs. Substituting its constant value $n_{\bullet j}/n$, we get

$$\frac{1}{n_{i\bullet}}\sum_{j=1}^{m}\sum_{k=1}^{n_{ij}}(x_{ijk} - \bar{x}_{\bullet j}) = \bar{x}_{i\bullet} - \sum_{j=1}^{m}\frac{n_{\bullet j}}{n}\bar{x}_{\bullet j} = \bar{x}_{i\bullet} - \bar{x}$$

(as an easy exercise, check my claim that $\sum_{j=1}^{m}(n_{\bullet j}/n)\bar{x}_{\bullet j} = \bar{x}$). This, then, is the predicted value of these residuals $x_{ijk} - \bar{x}_{\bullet j}$ by row. The sum of squares of residuals can then be expanded, again by the analysis of variance theorem:

$$\sum_{i=1}^{l}\sum_{j=1}^{m}\sum_{k=1}^{n_{ij}}(x_{ijk} - \bar{x}_{\bullet j})^2 = \sum_{i=1}^{l}n_{i\bullet}(\bar{x}_{i\bullet} - \bar{x})^2 + \sum_{i=1}^{l}\sum_{j=1}^{m}\sum_{k=1}^{n_{ij}}(x_{ijk} - \bar{x}_{\bullet j} - \bar{x}_{i\bullet} + \bar{x})^2.$$

The last stage in the decomposition will see us predicting the current residuals $x_{ijk} - \bar{x}_{\bullet j} - \bar{x}_{i\bullet} + \bar{x}$ with a *full* model. The average residual over all the observations in the ijth cell is obviously $\bar{x}_{ij} - \bar{x}_{\bullet j} - \bar{x}_{i\bullet} + \bar{x}$, because only the first term changes inside that cell. This is, of course, the standard estimate of interaction. We get a third decomposition of sum of squares

$$\sum_{i=1}^{l}\sum_{j=1}^{m}\sum_{k=1}^{n_{ij}}(x_{ijk} - \bar{x}_{\bullet j} - \bar{x}_{i\bullet} + \bar{x})^2 = \sum_{i=1}^{l}\sum_{j=1}^{m}n_{ij}(\bar{x}_{ij} - \bar{x}_{\bullet j} - \bar{x}_{i\bullet} + \bar{x})^2$$

$$+ \sum_{i=1}^{l}\sum_{j=1}^{m}\sum_{k=1}^{n_{ij}}(x_{ijk} - \bar{x}_{ij})^2.$$

Combining the three stages, we get a result that is impressive-looking, but easy to interpret:

Theorem (analysis of variance for a balanced two-way layout). *If the design is balanced, then*

$$\sum_{i=1}^{l}\sum_{j=1}^{m}\sum_{k=1}^{n_{ij}}x_{ijk}^2 = n\bar{x}^2 + \sum_{j=1}^{m}n_{\bullet j}(\bar{x}_{\bullet j} - \bar{x})^2 + \sum_{i=1}^{l}n_{i\bullet}(\bar{x}_{i\bullet} - \bar{x})^2$$

$$+ \sum_{i=1}^{l}\sum_{j=1}^{m}n_{ij}(\bar{x}_{ij} - \bar{x}_{\bullet j} - \bar{x}_{i\bullet} + \bar{x})^2 + \sum_{i=1}^{l}\sum_{j=1}^{m}\sum_{k=1}^{n_{ij}}(x_{ijk} - \bar{x}_{ij})^2.$$

We see the familiar TSS term, the SSM term, and the final SSE term. Since we now have two treatment sums of squares, we will name them sum of squares for columns, SSC = $\sum_{j=1}^{m}n_{\bullet j}(\bar{x}_{\bullet j} - \bar{x})^2$; and sum of squares for rows, SSR = $\sum_{i=1}^{l}n_{i\bullet}(\bar{x}_{i\bullet} - \bar{x})^2$. (We will not be confused by the latter, because it is not a regression problem.) Finally, we need the sum of squares for interaction,

$$\sum_{i=1}^{l}\sum_{j=1}^{m}n_{ij}(\bar{x}_{ij} - \bar{x}_{\bullet j} - \bar{x}_{i\bullet} + \bar{x})^2.$$

Our complicated theorem just says TSS = SSM + SSC + SSR + SSI + SSE. Notice that nothing in our result depends on the fact that we decomposed by columns, and then rows. We are ready to put the terms into an ANOVA table:

Source	Sum of Squares	Degrees of Freedom	Mean Square
Mean	SSM	1	MSM
Rows	SSR	$l - 1$	MSR
Columns	SSC	$m - 1$	MSC
Interaction	SSI	$(l - 1)(m - 1)$	MSI
Error	SSE	$n - lm$	MSE
Total	TSS	n	

Once again, because most applications are not concerned with the overall mean, we commonly reduce it to a decomposition of the corrected sum of squares SS = SSC + SSR + SSI + SSE:

Source	Sum of Squares	Degrees of Freedom	Mean Square
Rows	SSR	$l - 1$	MSR
Columns	SSC	$m - 1$	MSC
Interaction	SSI	$(l - 1)(m - 1)$	MSI
Error	SSE	$n - lm$	MSE
Total	SS	$n - 1$	

Example. Returning to the third-grade arithmetic test (see 1.4.1), we compute the ANOVA table for the full model:

Source	Sum of Squares	Degrees of Freedom	Mean Square
Curriculum	156.8	1	156.8
Gender	16.2	1	16.2
Interaction	1.8	1	1.8
Error	774.8	16	48.425
Total	949.6	19	

We find ourselves interested in several different F-statistics here. Comparing the mean square for interaction to that for error, we get a ratio of 0.037. This is much *less* than 1 (in fact, surprisingly so; you will rarely encounter such a small value in practice). This suggests that there is no evidence that the change of curriculum treats boys and girls differently.

Now we know that it is at least plausible to imagine that we had two separate experiments: one that looked at differences in the scores for different curricula and the other that looked at the scores of girls versus boys. Comparing the gender mean square to error, we get an F-statistic of 0.335; still less than 1. We have no evidence that boys really tended to do better. Comparing the curriculum to error, we get a ratio of 3.24. Experience will teach you that this is not amazingly larger than 1; still, it is some evidence that the students using the new curriculum are really doing better.

2.8.2 Additive Models

What about additive models like $\hat{x}_{ijk} = \mu + b_i + c_j$ (that is, which neglect interactions) for balanced two-way layouts? Going back to the ANOVA for full models, simply combine the first two stages, skipping the decomposition involving the interaction:

$$\sum_{i=1}^{l}\sum_{j=1}^{m}\sum_{k=1}^{n_{ij}} x_{ijk}^2 = n\bar{x}^2 + \sum_{j=1}^{m} n_{\bullet j}(\bar{x}_{\bullet j} - \bar{x})^2 + \sum_{i=1}^{l} n_{i\bullet}(\bar{x}_{i\bullet} - \bar{x})^2$$
$$+ \sum_{i=1}^{l}\sum_{j=1}^{m}\sum_{k=1}^{n_{ij}} (x_{ijk} - \bar{x}_{\bullet j} - \bar{x}_{i\bullet} + \bar{x})^2.$$

This tells us that the decomposition of the observations

$$x_{ijk} = \bar{x} + (\bar{x}_{\bullet j} - \bar{x}) + (\bar{x}_{i\bullet} - \bar{x}) + (x_{ijk} - \bar{x}_{\bullet j} - \bar{x}_{i\bullet} + \bar{x})$$

is, by the Pythagorean theorem, orthogonal. That is, the four n-dimensional vectors consisting of each of the four terms on the right-hand side are at right angles to one another. Remember that the additive model has standard estimates $\hat{\mu} = \bar{x}$, $\hat{b}_i = \bar{x}_{i\bullet} - \bar{x}$, $\hat{c}_j = \bar{x}_{\bullet j} - \bar{x}$. Therefore, our prediction is the sum of the first three vectors, and it is at right angles to the fourth, residual, vector. Apparently, the standard estimate consists of a *perpendicular projection* into the subspace of additive predictions; therefore, the residual vector is as short as it could be. This means that our estimate is least squares.

Proposition. *The standard estimates of the centered, additive model for a balanced two-way layout are least squares.*

The ANOVA table looks just like the one for the full model, except that the interaction and error rows have been summed into a single, error, row.

Example. We concluded earlier that the additive model worked quite adequately in the arithmetic curriculum problem. Its ANOVA table for its corrected sum of squares is as follows:

Source	Sum of Squares	Degrees of Freedom	Mean Square
Curriculum	156.8	1	156.8
Gender	16.2	1	16.2
Error	776.6	17	45.68
Total	949.6	19	

The method of least squares will still find the parameters for a centered, additive model from an unbalanced experiment, but the answer is more complicated and raises some questions better left for advanced courses. Furthermore, least-squares estimation may be applied to estimating multiple-regression models. You will do some important cases as exercises.

Unfortunately, the method of least-squares is not really appropriate for estimating loglinear contingency table models and logistic regression models, which must wait for a later chapter.

2.9 Summary

We first suggested that the ordinary idea of geometrical distance, applied to sample vectors and their model predictions, gives us a way to tell a good model from less good ones (2.1). Therefore, the failure of a model μ to fit the data may be measured by SSE $= \sum_{i=1}^{n}(x_i - \mu_i)^2$ (2.2). When we choose our model by making this quantity as small as possible, we are applying the *principle of least squares*. We then used this principle to find the best estimate in a *simple proportionality* regression model $\mathbf{y} = \mathbf{x}b$ and concluded that we must solve a *normal equation* $\sum_{i=1}^{n} x_i y_i = b \sum_{i=1}^{n} x_i^2$ for $\hat{b}$ (3.2). This had an intriguing consequence: The standard estimates, based on sample means, for the measurement models from Chapter 1 are really least-squares estimates (4.1). The natural measure of how well these means described a sample was the *sample variance* $s_x^2 = \frac{1}{n-1} \sum_{i=1}^{n}(x_i - \bar{x})^2$ (4.2). This led to a method for evaluating how well more general models are doing, called the *Analysis of Variance* (*ANOVA*), based on generalizations of the theorem of Pythagoras. For example, in a one-way layout we get

$$\sum_{i=1}^{k}\sum_{j=1}^{n_i} x_{ij}^2 = n\hat{\mu}^2 + \sum_{i=1}^{k} n_i \hat{b}_i^2 + \sum_{i=1}^{k}\sum_{j=1}^{n_i}(x_{ij} - \hat{\mu} - \hat{b}_i)^2,$$

so that the second term on the right measures how important the levels of the treatment were, and the last term is the SSE again (5.2). This allowed us to interpret degrees of freedom geometrically, as the dimension of a subspace. We then applied least squares to simple linear regression models $\hat{y}_i = \mu + b(x_i - \bar{x})$; the estimates are $\hat{\mu} = \bar{y}$ and

$$\hat{b} = \frac{\sum_{i=1}^{n}(y_i - \bar{y})(x_i - \bar{x})}{\sum_{i=1}^{n}(x_i - \bar{x})^2}. \tag{6.1}$$

To interpret these, we introduced the idea of the *correlation* between two measurements,

$$r_{xy} = \frac{\sum_{i=1}^{n}(y_i - \bar{y})(x_i - \bar{x})}{\sqrt{\sum_{i=1}^{n}(y_i - \bar{y})^2 \sum_{i=1}^{n}(x_i - \bar{x})^2}}. \tag{7.1}$$

Finally, we showed that several more sophisticated measurement models, involving cross-classification, may also be estimated by least squares (8.2).

2.10 Exercises

1. The Fahrenheit boiling point of water is 212 degrees at sea level. You measure the boiling point of water from six cheap thermometers, all from the same manufacturer, getting 214.4, 211.8, 210.6, 212.4, 212.0, and 210.8. What are the SSE and Euclidean distance of this sample from the correct value? What are the MSE and RMSE?

2. Draper and Smith in 1981 reported a study of the relationship between concentration of aflatoxin (parts per billion) and percentage of contaminated nuts in batches of peanuts:

toxin	% bad	toxin	% bad	toxin	% bad
3.0	0.029	18.8	0.058	46.8	0.189
4.7	0.021	18.9	0.068	58.1	0.123
8.3	0.018	21.7	0.092	62.3	0.202
9.3	0.029	21.9	0.030	70.6	0.145
9.9	0.043	22.8	0.015	71.1	0.212
11.0	0.039	24.2	0.067	71.3	0.179
12.3	0.044	25.8	0.142	83.2	0.170
12.5	0.028	30.6	0.013	83.6	0.282
12.6	0.111	36.2	0.042	99.5	0.358
15.9	0.039	39.8	0.091	111.2	0.342
16.7	0.018	44.3	0.141		
18.8	0.025	46.8	0.137		

a. Draw a scatter plot relating percentage of contaminated peanuts to concentration of aflatoxin.

b. Since measuring the concentration of aflatoxin is much easier than counting contaminated peanuts, we would like to predict the percentage contaminated, using the aflatoxin concentration, perhaps by simply multiplying the concentration by some constant. Specify and estimate the parameter of such a model, by the method of least squares, and graph the line on your scatter plot.

c. You measure a 50.0 parts per billion aflatoxin in a new batch of peanuts. What prediction does your model provide for the percentage of contaminated peanuts in that batch? To get some idea of the accuracy of your prediction, estimate the root-mean-squared error for predictions in general.

3. Compute both sides of the Schwarz inequality for the toxin and percentage of bad peanut vectors in Exercise 2 and note how close it is to an equality.

4. Prove properties (ii) and (iii) of the sample mean $\bar{x}$.

5. For the 7 measured ratios of the mass of the earth and moon from Exercise 1 of Chapter 1:

a. Calculate the sample variance and sample standard deviation using the defining formula $s_x^2 = \frac{1}{n-1}\sum_{i=1}^{n}(x_i - \bar{x})^2$.

b. Now redo your calculation of the sample variance using the computational formula $s_x^2 = \frac{1}{n-1}\sum_{i=1}^{n}(x_i - v)^2 - \frac{n}{n-1}(\bar{x} - v)^2$, first using the traditional value $v = 0$, then using an intelligent choice $v = 81.3$. **Be sure to use exactly six significant figures for every step in your calculations.** Compare your answers to each other and to (a).

6. In recent years, many alternative methods of estimating the center of a sample of measurements have been proposed. For a newly discovered subatomic particle, 15 measurements of its mass have been carried out. Being old fashioned,

you find the sample mean, 124 Mev, and its sum-squared error, SSE = 1570. Three new methods have been proposed: From the same data, Larry comes up with a center estimate for which he claims SSE = 1625; Moe suggests one for which he claims SSE = 1528; and Curly proposes one for which he claims SSE = 1591.

a. At least one of the three has made an arithmetic error. Which one, and why?

b. Assuming that the other two made no mistakes, what are the possible values of the estimates they might have made of the particle's mass?

7. Prove properties (ii) and (iii) of the sample variance s_x^2 and the sample standard deviation s_x.

8. Prove the properties of standardized measurements.

9. Show that for the one-way layout model, the vector form of the analysis of variance for the one-way layout indeed says exactly the same thing as the theorem. Then prove the proposition about the mutual orthogonality of the three vectors $\hat{\mu}$, $\hat{\mathbf{b}}$, and $\hat{\mathbf{e}}$.

10. Construct the analysis of variance table for the one-way-layout model for the DBH level data from Exercise 2 from Chapter 1. Calculate the F-statistic for treatment. Does it suggest that clinical state made a real difference in patient DBH level?

11. Construct the analysis of variance table for the one-way-layout model for the shrimp-net data from Exercise 3 of Chapter 1. Calculate the F-statistic for brand of net. What do you conclude about the importance of which net you use?

12. Calculate the Kruskal–Wallis statistic K for the shrimp-net data from Exercise 3 of Chapter 1. What do you conclude about the importance of which brand of net to use?

13. Prove that our least-squares estimates for a simple linear regression model are exactly the same as the standard estimates, in case (as in 1.5.1) there are exactly two different values of the independent variable.

14. In the data of Exercise 2 estimate a two-parameter simple linear regression model $\hat{p} = \mu + (t - \bar{t})b$, where p is the percentage of bad peanuts and t is the parts per billion of aflatoxin. Predict once again the percentage of bad peanuts you would expect to find in a batch with 50.0 parts per billion aflatoxin.

a. Construct the ANOVA table for this regression problem. Compute the RMSE for predictions under this model. Compare it to the RMSE for the simpler model of Exercise 2. What do you conclude?

b. Calculate the correlation r between percentage of contaminated nuts and concentration of aflatoxin.

15. Prove parts (iii)–(v) of the properties of sample correlations.

16. Show that for least-squares estimates of simple linear regression we always have $r^2 = R^2$

17. a. For the experimental data of Exercise 6 in Chapter 1, construct the ANOVA table for the additive model. Now do the same for the full model.

b. Compute F-statistics for the presence of interaction, a diet effect, and an exercise effect. What do you conclude?

2.11   Supplementary Exercises

18. You extract a sample of 25 resistors from a batch that are supposed to be 100 ohms. Here are their actual resistances:

83	85	109	100	89
82	97	83	107	87
105	107	94	96	85
96	97	100	83	96
92	91	89	97	84

a. Find the sample mean and sample standard deviation for these numbers.

b. Construct a 2-s interval for this sample. Find the standard score for a resistance of 83 ohms.

19. One alternative to using the principle of least squares to estimate linear models is the **principle of least total error**, which just says to choose parameter values that make the sum of the absolute values of the residuals as small as possible. We will do this for the simple location model, which finds a center μ for a collection of n measurements x_i by minimizing $\text{TE} = \sum_{i=1}^{n} |x_i - \mu|$. We will proceed in stages, for the special case that n is odd. First, sort your observations in ascending order, and write the results $x_{(1)} \leq x_{(2)} \leq \cdots \leq x_{(n)}$. Now write the total error as the sum of the first and last term, then the second and next-to-last, and so forth, until only the middle term is unpaired:

$$\text{TE} = \sum_{i=1}^{(n-1)/2} \left(|x_{(i)} - \mu| + |x_{(n+1-i)} - \mu| \right) + |x_{[(n+1)/2]} - \mu|.$$

a. Prove the **triangle inequality** $|a - b| + |c - b| \geq |c - a|$ for any three numbers a, b, c, noting that it is an equality exactly when b is between a and c.

b. Use (a) to conclude that $\text{TE} \geq \sum_{i=1}^{(n-1)/2}(x_{(n+1-i)} - x_{(i)}) + |x_{[(n+1)/2]} - \mu|$. For what value of μ is this an equality, which also makes TE as small as possible? This is our least total error location estimator $\hat{\mu}$; have you seen it before?

c. Compute $\hat{\mu}$ and TE for the mass ratios of Exercise 5. (Notice that you found a formula for TE in (b) that does not directly mention the value $\hat{\mu}$).

20. As yet another way of measuring the error in a collection of n measurements x_i, perhaps we should just average the squared differences between them, $(x_i - x_j)^2$. Using the algebraic fact that there are $n(n - 1)/2$ pairs of different observations, this would be $d^2 = \frac{2}{n(n-1)} \sum_{i<j}(x_i - x_j)^2$.

a. Compute d^2 using this formula for the water temperatures of Exercise 1.

b. Show that always $d^2 = \frac{2}{n-1}\sum_{i=1}^{n}(x_i - \bar{x})^2 = 2s^2$; so we have nothing very new here. (However, this provides our first insight into why we usually divide by $n - 1$ in computing variances. It comes from the formula for counting pairs, to which we will return.)

21. In Exercise 20 from Chapter 1, the telephone bill problem, construct an ANOVA table. Now compute the F-statistic for the effect of choice of carrier. What do you conclude?

22. In Exercise 23 from Chapter 1, the pizza problem:

   a. Construct an ANOVA table for the additive model. Calculate an F-statistic for the importance of location. What do you conclude?
   b. Is it possible to carry out (a) for the full model? Why or why not?

23. Show that for the situation of Exercise 27, Chapter 1 (three equally spaced values of the independent variable, equal numbers of observations at the smallest and largest value), the standard estimate you proposed for the simple linear regression model was in fact the least-squares estimate.

24. The pressure and volume of a fixed mass of an ideal gas follow the law $PV^\gamma = C$ under adiabatic (insulated) compression, where C and γ are constants. We get the following results for a quantity of real gas:

P (lb/sq. in)	V (cu. in.)
212	10
111	15
64	20
46	25
36	30
25	35

   a. Estimate the constants C and γ by simple linear regression by predicting pressure from the volume to which you have compressed your gas. **Hint:** our law is not linear, so you will have to take logarithms of both sides first to make it so.
   b. Though we do not like to extrapolate, our apparatus will not let us compress the gas to 5 cubic inches. Use the results in (a) to estimate the pressure in that case.

25. Using the theorem of the analysis of variance for simple linear regression, define the three mutually orthogonal vectors that sum to y, and prove that they are indeed orthogonal.

26. Find the parameter estimate for the simple proportions regression model $\hat{y}_i = bx_i$ using the principle of least total error (Exercise 19).

27. Use the method of Exercise 26 to estimate the Hubble parameter k.

28. Given an observation vector x and a model vector μ:

   a. Find an inequality connecting the SSE and the total error TE defined in Exercise 19. **Hint:** Apply the Schwarz inequality to the vector **1** (all ones) and the vector whose coordinates are $|x_i - \mu_i|$.

b. Translate it into a more useful relationship between the RMSE and the mean absolute error $\text{MAE} = \text{TE}/n$.

29. To estimate a multiple regression model $\hat{y} = \mu + (x_1 - \bar{x}_1)b_1 + (x_2 - \bar{x}_2)b_2$, we might naively hope that the estimates would be $\hat{\mu} = \bar{y}$, $\hat{b}_1 = s_{x_1 y}/s_{x_1}^2$, $\hat{b}_2 = s_{x_2 y}/s_{x_2}^2$. This is usually false, but for one important sort of experiment it works. We say that the design is **orthogonal** if $s_{x_1 x_2} = 0$. Show, by reasoning in stages, that in this case the naive estimates are the least-squares estimates.

30. We measure the efficiency of a polymerization reaction for various vessel temperatures and pressures:

efficiency (%)	temperature (F)	pressure (lb/sq in.)
74	250	100
81	300	100
85	350	100
76	250	120
85	300	120
88	350	120
76	250	140
82	300	140
91	350	140

a. Using the method of Exercise 29, show that this design is orthogonal, and find a linear prediction equation for efficiency in terms of temperature and pressure.

b. Plot your model, using the method of Chapter 1, Section 6. How well does the linear equation seem to describe your data?

c. At 320 degrees and 115 pounds per square inch, what would you expect the percent efficiency of this reaction to be? Find the RMSE , to get some idea how good your prediction is likely to be.

31. a. Since we already know that the least-squares estimate for centered simple linear regression is $\hat{\mu} = \bar{y}$, estimate b instead by calculus: That is, minimize $\sum_{i=1}^{n}[y_i - \bar{y} - b(x_i - \bar{x})]^2$ as a function of b by *differentiating* to find an extremum and differentiating again to see whether you have found a minimum.

b. Do the same thing to estimate the slopes (b's) in the multiple regression model $\hat{y} = \mu + (x_1 - \bar{x}_1)b_1 + (x_2 - \bar{x}_2)b_2$, where still $\hat{\mu} = \bar{y}$. Do *not* assume that the design is orthogonal. Take *partial* derivatives of the sums of squares for each b in turn to get a *system* of normal equations, two linear equations in two unknowns. (You need not take second derivatives here.)

32. Use calculus as in Exercise 31 to find the normal equations for estimating the model $\hat{y} = \mu + b(x - \bar{x}) + c(x - \bar{x})^2$ in a regression problem with one independent variable. (This is called *polynomial* regression. You can imagine how to generalize it to polynomials of higher degree.) Solve them for the aflatoxin data of Exercise 2. Repeat the prediction and error estimate of part (c), and compare.

CHAPTER 3

Combinatorial Probability

3.1 Introduction

We have seen useful ways of summarizing complicated data sets in the last two chapters. We have taken that process about as far as we can without developing ways of deciding whether our models are reasonable and how accurate our parameter estimates are, a process called *statistical inference*. The great breakthrough on this problem came about when people realized that we needed mathematical models for the origin of our *variability*, as well as for the important natural processes they were studying. The statistician's favorite mathematical tool for doing this is *probability*. An example will introduce one application of probability to statistical inference:

Example. The great statistician R. A. Fisher described a party he attended in which the hostess was serving tea with milk (this was England). She claimed that she could tell whether her maid had poured tea or milk into the cup first, just by tasting. Fisher was skeptical. He proposed an experiment to test her claim: He would put the tea first in some cups, and the milk first in the others, stir up the contents, scramble the cups, then let her taste them all and announce which ones had tea poured first. The more she got right, the more impressed he would be with her claim. This is a statistical experiment because we use *replication*; we pour a number of cups. After all, few of us would be impressed if she guessed correctly what had happened with a single cup.

How do we interpret the results? Fisher's approach, called *classical* or *frequentist* inference, starts before the experiment. We specify all possible outcomes. For example, with six cups we might write the numbers 1, 2, 3 on the bottom of those cups that are to get tea first and 4, 5, 6 on those that will get milk first. Then we pour the beverages: Let the lady taste and tell us which three she believes got tea first. Here are her possible choices of the cups that perhaps got tea first:

123	three correct	145	
124		146	
125		156	
126		245	one correct
134	two correct	246	
135		256	
136		345	
234		346	
235		356	
236		456	none correct

Fisher suspected that she was just guessing; so just by accident any of these possibilities might have arisen. If she gets all three cups right, that would happen only one time in twenty; because we have listed twenty different things she could have said. The statistician would conclude that either she had been fairly lucky, or there is some substance to her claim. On the other hand, if she gets two cups out of three, she might say that this supported her claim. But Fisher would point out that fully ten of our twenty cases, or half the time, she would get at least some two of the three cups right by luck. No doubt he would remain a skeptic.

In the next several chapters, this kind of reasoning will help us evaluate some of our models for counted data. Eventually, it will do the same for measurement models.

Time to Review

Set notation
Integration

3.2   Probability with Equally Likely Outcomes

3.2.1   What Is Probability?

In the example above, we invented a measure of how rare or surprising various possible results of our experiment are, in light of an opinion about what is really going on. Intuitively, the *probability* will be the proportion of times we expect the results to come out in some particular way, when the experiment has yet to be done. The calculation in this case was particularly simple but widely useful. When we believe that a number of possible outcomes are *equally likely*, then the probability of some event is

$$\text{probability of event} = \frac{\text{number of outcomes leading to event}}{\text{number of outcomes possible}}$$

Therefore, the lady's probability of two of three cups or better was 10/20, or 0.5. Let us turn this into some formal notation:

Definition. An **event** is a set whose elements are distinct **outcomes**.

Intuitively, an event is a collection of interest to us of the individual things that we believe might happen in some experiment not yet performed.

At this point, you should review the basic concepts and the notation of mathematical set theory. Events are often represented by capital letters (A, B, ...). The number of outcomes in a finite event will be denoted by $|A|$. We will talk about the probability that an event A will happen when the set of outcomes we believe possible is B, calling it *the probability of* A *relative to* B (or *given* B, or *conditioned on* B); we denote it by $P(A \mid B)$. Remembering that $A \cap B$, the intersection of A and B, is the set of outcomes in B that are also in the event A, our ratio above suggests the following:

Definition. A **probability space** with **equally likely** outcomes, has

$$P(A \mid B) = \frac{|A \cap B|}{|B|},$$

where A and B are events, and B is not empty and has a finite number of outcomes.

If it is obvious what the set of possibilities B should be in a particular problem, we will often use the shorthand $P(A)$ for $P(A \mid B)$, called an *unconditional* probability. Notice that in a way, *equally likely* is being defined here; it is any circumstance in which the probability of an event may be determined by the simple proportion of outcomes from that event.

3.2.2 Probabilities by Counting

Probabilists (mathematicians who study probability) traditionally use *urns*, which are just opaque jars containing a number of marbles of the same size, weight, and surface texture, to construct probability models. Our favorite urn, which will appear through much of the rest of the course, will contain some number W of white marbles and some number B of black marbles (Figure 3.1).

Our experiment is performed by stirring up the marbles so well that we have no idea which marbles are where. Then someone reaches in without looking and removes a marble. Is it black or white? This procedure matches our intuitive notion that all the marbles are equally likely to be chosen. The probability that the marble will be white is then

$$P(\text{white marble} \mid \text{from jar}) = \frac{|\text{white marble and from jar}|}{|\text{from jar}|} = \frac{W}{W + B}.$$

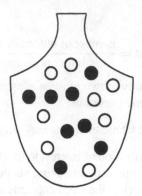

FIGURE 3.1. An urn

Even though we will use urns mainly as simple models for probabilistic experiments, they have practical applications. For example, what if we had decided to test a new medical procedure on a certain number of patients? It is considered good policy also to use the standard medical regimen on a similar set of patients, called *controls*. A simple way to help ensure that the controls are a group of patients similar to the ones who get treated, *randomization*, might work as follows: If we decide to test the procedure on W patients and have B controls, simply put those numbers of white and black marbles in the urn and stir it up. Now as each qualified patient appears at the hospital, we draw out a marble. If it is white, the patient gets the new treatment, if black, the old treatment. By the time the urn is empty, we have our full complement of subjects. The very unpredictability of patient assignments is the great virtue of this method: It makes it very difficult for experimenters, consciously or unconsciously, to bias the choice of patients either for or against the new procedure.

One nice feature of the basic urn experiment is that it can be arranged so that the probability of a white marble is any fraction (*rational* number) between 0 and 1. However, as we shall see, there is a famous geometrical experiment (the Buffon needle problem) in which the probability of an event is $2/\pi$. (This number is known to be irrational; so it is not a fraction, and the decimal representation begins $0.63661977...$.) We cannot construct an urn to give this exact probability. However, we can construct a sequence of urn models that gives probabilities as close as we please to $2/\pi$: 6 white marbles and 4 black marbles gives probability 0.6 for drawing a white marble; 64 and 36 gives probability 0.64; 637 and 363 gives probability 0.637; 6366 and 3634 gives probability 0.6366; and so forth. For reasons we shall discover, it would take several million sets of draws from the urn before we were likely to notice that even the third of our sequence of models had the wrong probability. This process, constructing a sequence of models whose probabilities approach that of another experiment, will be one of our most important mathematical tools. (It will be called *convergence in distribution*.)

So calculating probabilities is trivial so far, because all we have to do is count. But that is not as easy as it sounds. In Fisher's actual tea-tasting experiments, there were four cups with tea first and four with milk first. To proceed with our analysis we would have to list all sets of four out of eight cups his hostess might guess: 1234, 1235, 1236, ... This would take much longer than before—you should do it as an exercise. Fortunately, there is a branch of mathematics, called *combinatorics*, that studies counting. Some of its results will make life much easier for us.

3.3   Combinatorics

3.3.1   Basic Rules for Counting

The counting methods we need will be based on only two simple principles. The first notes that if you want a complete count of the outcomes in two events that do not overlap, you may count them separately and add the two counts. In our formal notation, $A \cap B = \phi$, where ϕ is the event with no outcomes, means that the two events have no outcome in common. The *union* of A and B, $A \cup B$, is, of course, the event that the outcome is either in A or in B.

Addition Rule. In the case $A \cap B = \phi$, $|A \cup B| = |A| + |B|$.

This rule is obvious enough, though we will use it very often. For example, in a poll of candidates for a political office, candidate DiBiasi might drop out of the race between the time of the poll and the time of the statistical analysis. Then it would make sense to combine the formerly distinct categories DiBiasi and Undecided into a single category and sum the numbers of subjects in the two old categories.

The second of our two principles is less obvious. We will illustrate with an example:

Example. The menu for a Chinese restaurant has on it three appetizers: hot and sour soup, egg rolls, and steamed dumplings. There are four main courses: pepper beef, lemon chicken, sweet and sour pork, and shrimp stir-fry. A meal consists of one appetizer and one main course; how many meals are possible? It would be easy to list them, but there is a shortcut: Construct a table.

		Main Course			
		beef	chicken	pork	shrimp
	soup				
Appetizer	egg rolls			×	
	dumplings				

Each *cell* (rectangle) corresponds to a distinct meal; for example, the marked cell corresponds to a lunch of egg rolls followed by sweet and sour pork. The number of cells is just rows×columns, $3 \times 4 = 12$ meals.

This should remind you of the number of distinct treatment levels in a two-way layout with l rows and m columns, which was of course lm (see 1.4.1). To formalize this idea, recall from mathematics that $A \times B$, the *Cartesian product* of the sets

A and B, is the set of all ordered pairs (a, b) in which $a \in A$ and $b \in B$. In the restaurant example, we would write our meal (egg rolls, sweet and sour pork).

Multiplication Rule. $|A \times B| = |A| \cdot |B|$

Example. Your daughter's best friend was assigned to the most popular teacher in their elementary school grade level, supposedly by random assignment, in each of the first four grades. This makes you suspicious that the assignments were not done honestly. There are five teachers in each grade. You reason, by using the multiplication rule three times, that there are $5 \times 5 \times 5 \times 5 = 625$ different teacher assignments possible, one factor per grade. Therefore, the probability that the girl would be this lucky is $1/625 = 0.0016$, which sounds very lucky indeed.

3.3.2   Counting Lists

We will now use these principles to derive three special formulas that will, with ingenuity, solve most of the counting problems faced by statisticians. Imagine that we have an urn with n marbles in it; but now all the marbles have labels, so we can tell them apart once they are out of the jar.

Example. Let the 26 marbles correspond to the letters of the Roman alphabet. We could create all six-letter "words" by removing letters from the jar, such as GXNGEK. Notice that we allowed G to appear twice, as often happens with real words, by replacing its marble after it was used the first time.

We could potentially make all such words by the following procedure:

Urn Problem 1. Remove a marble, write down its label, and *put it back*. Now remove a second marble, write down its label below the first one, and put it back. Continue until the list has k entries in it. How many lists are possible? We call this counting *ordered lists with replacement*.

The teacher assignment problem was an instance of this; the same solution technique works. We have n choices for each of the k stages, so the multiplication rule tells us that we have

$$\underbrace{n \cdot n \cdot n \cdots n}_{k \text{ copies}} = n^k.$$

We have established the following result:

Proposition. *The number of ordered lists of k objects taken with replacement from a set of n objects is n^k.*

Example (*cont.*).   In the six-letter word problem, $n = 26$ and $k = 6$; therefore, we could get $26^6 = 308,915,776$ different words.

Example. Eight swimmers are about to race in the Olympic games. The first to finish will get a gold medal, the second a silver medal, and the third a bronze. How many distributions of medals are possible? The gold medal can go to one of eight competitors. But then the silver medal can go to only one of seven swimmers, because no one may receive two. Finally, the bronze can go to one of the six remaining swimmers. By the multiplication rule, there are $8 \cdot 7 \cdot 6 = 336$ placing orders.

Our most recent formula does not apply; this is an instance of

Urn Problem 2. Choose a marble, write it down, *leave it out*, and repeat until you have a list of k marbles. How many lists are possible? This is counting *ordered lists without replacement*; we call them **permutations** *of n taken k at a time*. The mathematical symbol for the number of lists is $(n)_k$.

The Olympic example, which is counted by $(8)_3$, shows us how to do this:

Proposition. $(n)_k = n \cdot (n-1) \cdot (n-2) \cdots (n-k+1)$.

The last factor appears because before the selection of the last marble, we have removed $k-1$ of the n marbles, leaving $n - (k-1)$ to choose among.

Example. Of the 50 United States, 15 have an Atlantic coastline. A researcher picks 6 states at random for a detailed study of their emergency preparedness for severe wind storms. Obviously, it would be a poor sample group that did not include any Atlantic coastal states, which are subject to hurricanes and nor'easters. What is the probability that her sample, by accident, will include no Atlantic coastal states?

First notice that if she picks her states in some sequence, then she essentially has Urn Problem 2, and there are $(50)_6 = 50 \cdot 49 \cdot 48 \cdot 47 \cdot 46 \cdot 45 = 11,441,304,000$ possible sequences of choices. That will be the denominator, if we assume that they are all equally likely. If we consider the event that they are all chosen from among the $50 - 15 = 35$ non-Atlantic states, these peculiar sample sequences may be chosen in

$$(35)_6 = 35 \cdot 34 \cdot 33 \cdot 32 \cdot 31 \cdot 30 = 1,168,675,200$$

ways. Therefore, the probability of getting a bad sample is

$$P(6 \text{ non-Atlantic}|6 \text{ states}) = \frac{1,168,675,200}{11,441,304,000} = 0.102.$$

Unfortunately, this is rather likely; about one time in 10.

Example. A product testing lab wants to evaluate 5 new automobiles. Each driver will try all the cars. There may be an order effect; for example, there may be an unconscious bias in favor of the first car driven. Therefore, different drivers are to test the 5 cars in different orders. How many such orders are possible? This is like drawing the names of the cars from a jar without replacement; so we have $(5)_5 = 5 \cdot 4 \cdot 3 \cdot 2 \cdot 1 = 120$ sequences.

This last should be familiar: $(n)_n = n!$ (*n factorial*), which we call simply the permutations of n things. This leads to a useful alternative formula for permutations: To find the total number of complete lists ($n!$), we arrange the first k

marbles in $(n)_k$ ways, then the remaining $n - k$ in $(n - k)!$ ways. Therefore, by the multiplication rule, $n! = (n)_k(n - k)!$. We may then solve for the unknown term:

Proposition. $(n)_k = n!/(n - k)!$

For example, in the medal problem, we could have calculated $(8)_3 = 8!/5! = 40320/120 = 336$. Notice that our new formula is rarely convenient for computation: The numbers stay much smaller if we use the original formula. It will be useful, however, for algebraic manipulation.

3.3.3 Combinations

You may have complained that the Atlantic states problem was not explained realistically. We talked about selecting our sample in order; but you may know that for purposes of the study of emergency planning, the order of choice simply did not matter. It was just a set of 6 states. Therefore, we have counted far too many samples, because we have counted (Maine, Oregon, Nebraska, Rhode Island, Texas, West Virginia) separately from (Oregon, Texas, Nebraska, West Virginia, Rhode Island, Maine). We need another counting formula:

Urn Problem 3. Remove a handful (set) of k marbles from a jar containing n. How many sets are possible? This is counting *unordered sets, without replacement*; we call them **combinations** *of n things taken k at a time*. The mathematical symbol for the number of sets is $\binom{n}{k}$, sometimes read "n choose k."

Some ingenuity will be required to find the number of combinations. I propose that we do it by counting the number of permutations in Urn Problem 2 by a slightly different procedure: (1) Remove a handful of k marbles from the jar of n; then (2) place the unordered handful in an ordered row on the table. We can construct every permutation in this way. The multiplication rule says that we multiply the number of ways each of the two steps was performed to get the total number of possible lists. Therefore, $(n)_k = \binom{n}{k} \cdot (k)_k$. The first and third counts are known, so once again we may solve for the unknown term:

Theorem (combinations).

$$\binom{n}{k} = \frac{n!}{k!(n - k)!}.$$

Staring at this formula, we see some equivalent ways of writing it:

Proposition.

(i)
$$\binom{n}{k} = \binom{n}{n - k}.$$

(ii)
$$\binom{n}{k} = \frac{(n)_k}{k!} = \frac{(n)_{n-k}}{(n - k)!}.$$

The first fact just notices that removing k marbles from a jar is the same as leaving $n - k$ marbles behind in the jar.

Example. There are $\binom{50}{6} = 50!/(6!44!)$ samples of 6 states. No one would calculate 50! by choice; but we consider also the two equivalent formulas in (ii) of the proposition, and computing $(50)_6/6! = (50 \cdot 49 \cdot 48 \cdot 47 \cdot 46 \cdot 45)/(6 \cdot 5 \cdot 4 \cdot 3 \cdot 2 \cdot 1) = 15,890,700$ requires by far the least arithmetic.

We now solve a counting problem that will be of repeated interest from here on. From an urn with W white marbles and B black marbles, remove all the marbles, one at a time without replacement, and make an ordered list of their colors. For example, if there are 3 white and 4 black marbles, one such list is $\underset{1}{\bullet}\,\underset{2}{\circ}\,\underset{3}{\bullet}\,\underset{4}{\circ}\,\underset{5}{\bullet}\,\underset{6}{\circ}\,\underset{7}{\bullet}$. How many lists are possible?

The trick here will be to translate the problem into a second urn problem, as follows: Obtain $W + B$ additional marbles, number them, and place them in a second urn. Now number the positions in your ordered list, also from 1 to $W + B$. Reach into the second urn and select an unordered handful of W of the numbered marbles. Put white marbles into the numbered list positions you have chosen, and black marbles in all the others. In our example, we must have picked marbles numbered 2, 4, and 6. This process uniquely determines all possible lists, so the number of lists is $\binom{W+B}{W}$.

When all these lists of black and white marbles are picked from a well-stirred urn, we might assume each to be equally likely. Then choosing a list is called a *hypergeometric process*. It is the first important example of a *stochastic process*. We will see several other important examples in this course; but we will construct them all as ways to approximate hypergeometric processes.

3.3.4 Multinomial Counting

Example. A professor has a peculiar grading curve, so that she expects to assign grades to the 12 students in her new graduate seminar as follows: 5 A's, 4 B's, 2C's, and 1 D. She has graded no work, so she knows nothing as of yet about her student's performance. In how many ways will she be able to assign grades at the end of the term?

We can assign grades one at a time; she may choose the students to receive A's in $\binom{12}{5}$ ways. Then she has 7 students left; she may give 4 of them B's in $\binom{7}{4}$ ways. The remaining 3 students may get the 2 C's in $\binom{3}{2}$ ways, and the last student automatically gets the D. The multiplication rule says that the grades have $\binom{12}{5} \cdot \binom{7}{4} \cdot \binom{3}{2} = 83,160$ distributions.

Notice that when we write out the calculation from the theorem, we get several very convenient cancellations: $(12!/(5!7!)) \cdot (7!/(4!3!)) \cdot (3!/(2!1!)) = 12!/(5!4!2!1!)$.

Definition. The number of ways of assigning n_1 objects to category 1, n_2 other objects to category 2, ..., and finally the last n_k objects to category k, where $\sum_{i=1}^{k} n_1 = n$, is the **multinomial symbol**. It is denoted by $\binom{n}{n_1 n_2 \cdots n_k}$.

Proposition. $\binom{n}{n_1 n_2 \cdots n_k} = n!/(n_1! n_2! \cdots n_k!)$.

You should prove this as an exercise (perhaps by cancellation as in the example above; or you might imitate the proof of the theorem about combinations). Since choosing a set of l from n is the same as grouping objects into a selected set of l and another set of $n - l$ that was not selected, then $\binom{n}{l} = \binom{n}{l\ n-l}$. Notice generally that any rearrangement (permutation) of our k categories leads to the same multinomial symbol, because we just multiply our denominator factorials in a different order.

3.4   Some Probability Calculations

3.4.1   Complicated Counts

Our small tool kit of counting methods will already allow us to calculate a great many interesting probabilities.

Example. We can test a new drug for lowering blood pressure in the following plausible way: Of the next 40 patients that might benefit from the drug, match them so that each pair of patients is as similar as possible in blood pressure, sex, age, health, and other relevant matters. We now have 20 pairs; for each, flip a coin, and give the standard treatment to one patient and the new drug to the other. We will evaluate them after six weeks and, for each pair, decide which patient has lower blood pressure. Thus, we will end up with a count of how many times out of 20 the new drug was the winner. We might decide to advocate use of the new drug if it wins, for example, 14 or more comparisons. If, in fact, the new drug is no better than the old, what is the probability we will (unfortunately) advocate it anyway? (This is another example of the *frequentist* style of inference.)

There would be 2^{20} sequences of wins by either the new or old treatment; we are presuming these equally likely. In $\binom{20}{14}$ of these sequences, the new drug was superior exactly 14 times similarly for 15, 16, ..., 20. Therefore,

$$P(14 \text{ or more}|20 \text{ pairs}) = \frac{\binom{20}{14} + \binom{20}{15} + \cdots + \binom{20}{20}}{2^{20}} = \frac{60,460}{1,048,576} = 0.05766.$$

If this chance of making a foolish claim is too large for us, we might require 15 or more wins; we easily check that $P(15 \text{ or more}|20 \text{ pairs}) = 0.0207$, which is safer.

Example. Remember that Fisher's tea-tasting experiment was actually bigger than our example suggested; to make it more informative, he had his hostess taste 8 cups of tea, in which 4 had tea poured first. She then tried to determine which 4. Our first step previously was to list all her possible sets of guesses; we noticed that the list is too long to be fun to write down. But now we are more sophisticated: There are $\binom{8}{4} = 70$ lists. The probability that she will get all four guesses right is $1/70 = 0.0143$. Most of us would be very impressed, and perhaps modify our opinion that she was just guessing. Should we be surprised if she gets 3 out of 4 correct? Enumerate these lists by noting that she must choose 3 of the 4 cups that had tea first; and then 1 of the 4 that had milk first. We get

$$P(3 \text{ of } 4 | 4 \text{ of } 8 \text{ tea first}) \frac{\binom{4}{3}\binom{4}{1}}{\binom{8}{4}} = \frac{16}{70} = 0.229.$$

Our total probability for a result this good, 3 or 4 out of 4 correct, is $(16+1)/70 = 0.243$, or about 1 in 4. We are not likely to be impressed with her skill.

Example. Scientists suspect that initial handling of patients with a certain form of acute mental illness may have something to do with chances of recovery. Therefore, when a long-term drug therapy is proposed, they are careful to create a patient pool for a study that has exactly 5 patients who were first seen by each of the 16 participating clinics (for a total of 80 patients).

For a small substudy, 7 patients from this pool are selected at random. What is the probability that it will be found that 2 patients in this substudy came from the same clinic (while the other 5 came from 5 additional clinics)?

Of course, there are a total of $\binom{80}{7}$ equally likely samples for the substudy. We need to count the samples that duplicate a clinic, in stages. First, which clinic appears twice and which clinics appear once? We may decide this in $\binom{16}{1 \ 5 \ 10}$ different ways; the 1 refers to the duplicated clinic, the 5 to the clinics represented by one patient each, and the 10 to clinics not represented. We can pick the patients from that duplicated clinic in $\binom{5}{2}$ ways, and from each of the other 5 clinics we may pick the patient in $\binom{5}{1}$ ways. Therefore,

$$P(\text{one clinic duplicated} | 7 \text{ patients}) = \frac{\binom{16}{1 \ 5 \ 10}\binom{5}{2}\binom{5}{1}^5}{\binom{80}{7}} = 0.473.$$

This coincidence will happen almost half the time.

3.4.2 The Birthday Problem

Example. In a class with 35 students, what is the probability that no two of them will have a birthday on the same day of the year? We will assume, not quite correctly, that all birthdays are equally likely, and that there are 365 of them.

Most people will find the answer surprising; to understand why, let us first ask what one might expect the answer to be like:

Naive Intuition. if the number of people is small compared to the number of birthdays, then the probability of having any two the same is small, since then the average time between birthdays is certainly large.

Since 35 is fairly small compared to 365, people's birthdays have plenty of room to be scattered over the year; we expect that the probability of a coincidence is fairly small.

This is an example of an *occupancy problem*: Let there be n slots in a board (possible birth dates). Throw k marbles at the board (people) so that they fall in slots at random; each slot can potentially hold all the marbles. What is the probability that no two marbles fall in the same slot (no two people have the same birthday)?

The denominator is easy: By the first urn problem, throwing a marble at a slot is like picking a slot number out of a jar. Since more than one marble can fall in a slot, we are choosing slots with replacement. There are n^k ways this can be done; presumably they are equally likely. On the other hand, for the numerator we want to count the number of ways slots can be chosen no more than once; that is, without replacement. By the second urn problem, we can do that in $(n)_k$ ways. We state our conclusion as a proposition:

Proposition. *The probability that no two objects occupy the same category, from among k assigned at random to n categories, is* $(n)_k/n^k$.

Example (*cont.*).   In the birthday problem, $n = 365$ and $k = 35$, so that the probability of no coincidences is just about 0.186. It would actually be a bit surprising if no two people in the class had the same birthday. A laborious calculation shows that in any class with at least 23 students, there is a less than even chance (0.5 probability) that no two will share a birthday.

3.4.3   General Principles About Probability

Now that we find ourselves capable of a calculating a number of complicated probabilities, it might be worth our time to stop and notice some general facts about equally likely probability.

Since our definition says that $P(A|B) = |A \cap B|/|B|$, we notice that the numerator was defined so it would be a *subset* of the denominator: $A \cap B \subset B$. But then the numerator set is always no bigger than the denominator, so when we count them, $0 \leq |A \cap B| \leq |B|$. (Counts are, of course, never negative.) We insisted that B was not an empty set, so we can divide by its count $|B|$ in this inequality to get $0 \leq |A \cap B|/|B| \leq 1$, But this tells us that for any equally likely probability, $0 \leq P(A|B) \leq 1$. Certainly, all our example calculations fell between 0 and 1; and our intuitive idea that probabilities are the proportion of the time something will happen says this ought to be true.

A couple of special cases are worth noting. If A cannot happen at the same time as B, then $A \cap B = \phi$, and so $|A \cap B| = 0$. Then we have $P(A|B) = 0$. In English, the probability of an event impossible under the circumstances is 0. On the other hand, $P(B|B) = |B|/|B| = 1$, and we say that the probability that anything possible will happen is 1.

Let us see what our addition and multiplication rules for counting tell us generally about probability. Assume we know that C will happen, and we have any two events, A and B (for example, two ways that an experiment might be considered successful). Then the probability that one or the other will happen is

$$P(A \cup B|C) = \frac{|(A \cup B) \cap C|}{|C|}.$$

Informally, we might count the outcomes in A and then count the remaining outcomes in B. In set notation, this idea of remaining cases is written as the set *difference*:

$$B - A = \{\text{outcomes in B and not in A}\}.$$

Then clearly,

$$P(A \cup B|C) = \frac{|A \cap C| + |(B - A) \cap C|}{|C|} = \frac{|A \cap C|}{|C|} + \frac{|(B - A) \cap C|}{|C|}$$

because the counts in A and B − A do not overlap. We conclude that there is an addition rule for our equally likely probabilities: $P(A \cup B|C) = P(A|C) + P(B - A|C)$.

This general principle has two important special cases. First, what if, as above, A and B cannot happen at the same time, so that $A \cap B = \phi$? Then $B - A = B$, and our formula simplifies to $P(A \cup B|C) = P(A|C) + P(B|C)$. In the example of testing a blood-pressure drug, we could have combined the cases of 14 through 20 wins by summing probabilities of each, instead of adding counts in the numerator:

$$P(14 \text{ or more}|20 \text{ pairs}) = \frac{\binom{20}{14}}{2^{20}} + \frac{\binom{20}{15}}{2^{20}} + \cdots + \frac{\binom{20}{20}}{2^{20}} = 0.03696 + 0.01479 + \cdots$$
$$= 0.05766.$$

For the second case, let $B = C$. Then $P(A \cup C|C) = 1$, because the same cases are in the numerator and the denominator. But then $1 = P(A \cup C|C) = P(A|C) + P(C - A|C)$. Rearrange to get $P(C - A|C) = 1 - P(A|C)$. This says that the probability something will *not* happen under experimental conditions C is just 1 minus the probability that it *will* happen.

For such a simple result, this equation is amazingly useful. For example, in the birthday problem, people most usually ask, What is the probability that there are *any* birthday coincidences in a group? To tackle that question directly, I would need to figure out the probability that exactly two have the same birthday, then the probability that three do, then that two have one birthday and two another, and so forth. Each calculation is hard, and there are very many of them. But now I know what to do (with 35 students):

$$P(\text{any coincidences}|35 \text{ students}) = 1 - P(\text{no coincidences}|35 \text{ students})$$
$$= 1 - 0.186 = 0.814.$$

Very often, this *complementary* question is much easier to answer.

Does the multiplication rule for counting tell us something similarly useful about probability computations? Indirectly, it does. If you will, recall the study of disaster-preparedness in the states; let me ask what is the probability that of two states chosen, they are both Atlantic states? This is just like the original problem: $(15)_2/(50)_2 = (15 \cdot 14)/(50 \cdot 49) = 3/35$. Notice, though, that the calculation can be factored as a product and the factors each interpreted as probabilities:

$$\frac{15}{50} \cdot \frac{14}{49} = P(\text{Atlantic}|15 \text{ of } 50) \cdot P(\text{Atlantic}|14 \text{ of } 49).$$

This says that when we chose the first state, we had 15 chances in 50 of succeeding, but to choose the second state we had to get one of the remaining 14 Atlantic states from among the remaining 49 states.

We have hit upon a general feature of probabilities that is obvious so long as we think of them as *proportions* of the possibilities: The probability that two things will both happen is the proportion of the time the first will happen multiplied by the proportion of *those* times in which the second also happens.

We can look at this generally for intersections of two events, because we are concerned with whether both will happen. Multiply both numerator and denominator by $|A \cap C|$ (the number of cases when the first thing has happened, which must not be zero):

$$P(A \cap B|C) = \frac{|A \cap B \cap C|}{|C|} \cdot \frac{|A \cap C|}{|A \cap C|} = \frac{|A \cap C|}{|C|} \cdot \frac{|A \cap B \cap C|}{|A \cap C|}.$$

We can interpret each of the factors as a probability to get

$$P(A \cap B|C) = P(A|C) \cdot P(B|A \cap C).$$

This just says, as before, that proportions of proportions are gotten by multiplication.

Now assemble the results of this section:

Proposition (properties of equally-likely probability).

(i) $0 \le P(A|B) \le 1$.
(ii) If $A \cap B$, then $P(A|B) = 0$.
(iii) $P(B|B) = 1$.
(iv) $P(A \cup B|C) = P(A|C) + P(B - A|C)$; *and if* $A \cap B = \phi$, *then* $P(A \cup B|C) = P(A|C) + P(B|C)$.
(v) $P(C - A|C) = 1 - P(A|C)$.
(vi) If $A \cap C \neq \phi$, *then* $P(A \cap B|C) = P(A|C) \cdot P(B|A \cap C)$.

Not only are these useful now; when later we study other forms of probability, they will continue to be true.

3.5   Approximations to Coincidence Probabilities

3.5.1   An Upper Bound

Let us return to some issues raised by the surprising results of the birthday problem (see Section 3.4.2). It is a bit disturbing that our naive intuition about birthday coincidences was so wrong. The formula is sufficiently obscure that it contributes little to our intuitive understanding, and if you compute it multiplication by multiplication, it is time-consuming. We will look at some *approximations* to the answer that may teach us more.

It would be nice to have an easy-to-calculate *maximum* value for our birthday probability. If we do a good job, perhaps it will be close to the exact value of our probability for a wide range of cases. First, expand our formula for the probability of no birthday coincidences:

$$\frac{(n)_k}{n^k} = \frac{n(n-1)(n-2)\cdots(n-k+1)}{n \cdot n \cdots \cdot n} = \left(\frac{n-1}{n}\right)\left(\frac{n-2}{n}\right)\cdots\left(\frac{n-k+1}{n}\right)$$

$$= \left(1 - \frac{1}{n}\right)\left(1 - \frac{2}{n}\right)\cdots\left(1 - \frac{k-1}{n}\right).$$

This product may be interpreted as the probability that the second birthday was different from the first, multiplied by the probability that the third was different from the first two, and so forth. Long products are difficult to work with, so we use the fact about the logarithm function that $\log(ab) = \log(a) + \log(b)$ to turn it into a sum:

$$\log \frac{(n)_k}{n^k} = \log\left(1 - \frac{1}{n}\right)\left(1 - \frac{2}{n}\right)\cdots\left(1 - \frac{k-1}{n}\right) = \sum_{i=1}^{k-1} \log\left(1 - \frac{i}{n}\right).$$

Reviewing our calculus, there is a maximum that comes from a simple property of the (natural) logarithm function (in fact, it is sometimes *defined* this way): $\log(1+x) = \int_1^{1+x} \frac{dt}{t}$. Whenever $x \geq 0$, since under the integral sign $1 \leq t \leq 1+x$, we have $\frac{1}{t} \leq 1$. But then

$$\log(1+x) = \int_1^{1+x} \frac{dt}{t} \leq \int_1^{1+x} dt = x.$$

On the other hand, if $x < 0$, then $1 + x \leq t \leq 1$, and we have $\frac{1}{t} \geq 1$. Then

$$\log(1+x) = \int_1^{1+x} \frac{dt}{t} = -\int_{1+x}^1 \frac{dt}{t} \leq -\int_{1+x}^1 dt = -(-x) = x.$$

The inequality is the same for both positive and negative x (see Figure 3.2).
We summarize:

Proposition. $\log(1 + x) \leq x$ *for all* $x > -1$.

Apply this result to our expansion of the log-probability:

$$\log \frac{(n)_k}{n^k} = \sum_{i=1}^{k-1} \log(1 - \frac{i}{n}) \leq -\sum_{i=1}^{k-1} \frac{i}{n}.$$

Now you should show as an exercise that the sum of the first m integers is $\sum_{i=1}^{m} i = (m(m + 1))/2 = \binom{m+1}{2}$. Replacing our rightmost term, we get $\log(n)_k/n^k \leq -\binom{k}{2}/n$. Now, to get our probabilities back we need to undo the logarithm. The exponential function is the inverse function to the natural logarithm; that is, $e^{\log(x)} = x$ and $\log(e^x) = x$. Furthermore, the exponential function, like the logarithm, is a *nondecreasing* function (a function f is nondecreasing if whenever $a \leq b$, we also have $f(a) \leq f(b)$). Therefore, they both preserve inequalities. Apply the exponential function to both sides of our inequality:

Proposition. P(*no coincidence*) $= (n)_k/n^k \leq e^{-\binom{k}{2}/n}$.

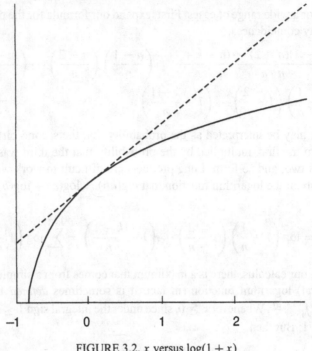

FIGURE 3.2. x versus $\log(1 + x)$

(We have used the shorthand probability notation here (see 2.1). What condition $(\,|B)$ are we assuming that you know?)

We have the upper limit we wanted. In the case of the class in which $k = 35$, this says that $0.186 \leq 0.196$, which is fairly close, with much less arithmetic. Remember that $\binom{k}{2}$ is the number of *pairs* of people (choose 2 from the class of k). Furthermore, as an exponent becomes a large negative number, the exponential function approaches zero. Then the inequality says that the probability of no coincidences is even smaller. This gives us an improved intuition.

Improved Intuition 1. Coincidences become highly probable when the number of pairs of people is large compared to the number of birthdays.

This will not be hard to remember, since of course individuals do not have birthday coincidences, two people at a time do.

3.5.2   A Lower Bound

We have a useful answer to the question, When are coincidental joint occupations likely? But an inequality tells less than half the story. We would also like to know when coincidences are unlikely. Therefore, we need a convenient *minimum* value for the probability; when the minimum is close to one, then so must be the exact probability.

A strategy remarkably parallel to the last one will work here, too. First note that since $\log(1) = 0$, then for any positive number a, $0 = \log(1) = \log(a \cdot 1/a) =$

$\log(a) + \log(1/a)$. Then $\log(1/a) = -\log(a)$. Apply this to each term in our sum for the log-probability $\log(n - i/n) = -\log(n/(n - i)) = -\log(1 + i/(n - i))$. Then our simple inequality for the logarithm yields $\log((n - i)/n) \geq -i/(n - i)$, where our inequality has reversed, as we wanted it to, because of the minus sign. So our log-probability is

$$\log \frac{(n)_k}{n^k} = \sum_{i=1}^{k-1} \log\left(\frac{n - i}{n}\right) \geq - \sum_{i=1}^{k-1} \frac{i}{n - i}.$$

We do not have a convenient sum formula, because the denominators $(n - i)$ are not constant; therefore, we will replace them by their smallest value, $n - k + 1$. This makes the right side even smaller, so our inequality is still true:

$$\log \frac{(n)_k}{n^k} \geq - \sum_{i=1}^{k-1} \frac{i}{n - i} \geq - \sum_{i=1}^{k-1} \frac{i}{n - k + 1} = -\frac{\binom{k}{2}}{n - k + 1}.$$

Again, taking the exponential of both sides, we get a reversed inequality:

Proposition. P(*no coincidence*) $= (n)_k/n^k \geq e^{-\binom{k}{2}/(n-k+1)}$.

In our example with $k = 35$, we compute $0.166 < 0.186$.

We conclude that when the exponent is small, that is, $\binom{k}{2}$ is small compared to $n - k + 1$, then the probability of no coincidence is close to one. In that case, of course, k itself is small compared to $n - k + 1$, which is therefore little different from n. Thus $\binom{k}{2}$ is small compared to n. Then we have another improvement for our intuition:

Improved Intuition 2. When the number of pairs of people is small compared to the number of birthdays, coincidences are rare.

For example, among 6 students sharing a house, there are 15 pairs of birthdays, out of 365 possible birthdays. We conjecture that coincidences are unlikely. Our inequality says that the probability of no coincidences is at least 0.9592. In fact, it is 0.9595.

When we say that a number a is small compared to a number b, we mean more precisely that the fraction a/b is close to zero, and in particular is much less than one. In our example, $15/365 = 0.04$.

3.5.3 A Useful Approximation

It will be convenient to combine our inequalities into a single fact:

Theorem (the birthday inequality).

$$e^{-\binom{k}{2}/(n-k+1)} \leq \frac{(n)_k}{n^k} \leq e^{-\binom{k}{2}/n}.$$

In the case $k = 6$, we now bracket our answer rather tightly: $0.9597 \geq 0.9595 \geq 0.9592$. Either bound could be used as a nice quick approximate probability. When can we get away with this? When the upper and lower bound are close together, then we may be sure that either approximation is good. To see how close together the two exponents are, we first compare denominators: $\frac{1}{n-k+1} - \frac{1}{n} = \frac{k-1}{n(n-k+1)}$. After rearrangement, the exponents are related by $\binom{k}{2}/(n-k+1) = \binom{k}{2}/n + ((k-1)\binom{k}{2})/(n(n-k+1))$. Now using the fundamental fact about exponents that $e^{a+b} = e^a e^b$, we are able to rewrite the birthday inequality as

$$e^{-\binom{k}{2}/n} e^{-((k-1)\binom{k}{2})/(n(n-k+1))} \leq \frac{(n)_k}{n^k} \leq e^{-\binom{k}{2}/n}.$$

If the second exponent on the left is close to zero, then its exponential is close to one, because $e^0 = 1$ and the exponential function is continuous. Therefore, the upper and lower bounds are within a factor hardly different from one of each other. We have established a practically useful approximation that works when $((k-1)\binom{k}{2})/(n(n-k+1))$ is close to 0. But it is easy to translate this into a condition easier to remember by looking at the highest powers when this is multiplied out:

Proposition. $(n)_k/n^k \approx e^{-\binom{k}{2}/n}$ when k^3 is small compared to $2n^2$.

In the $k = 6$ example, we are saying that $0.9595 \approx 0.9597$ because our relative error estimate 0.00048 is small. We will see a number of other important uses for this approximation later in the book.

Trying to find simple bounds and approximations when probability calculations become complicated will be fundamental to our progress through mathematical statistics. We call these *asymptotic* methods.

3.6   Sampling

One way of looking at statistical experimentation is that we are trying to find out something about a great many potential subjects of a survey or repetitions of a measurement. We call the collection of these potential subjects or measurements the *population* of interest. Of course, because of our limited resources, we can usually only study relatively few subjects, or carry out only a few replications, from among the population. We call the subjects actually studied, or the measurements actually carried out, a *sample*.

A survey, such as a political poll, can be thought of as removing a random collection of n subjects (sample) from among the pool of m potential subjects (population) without replacement (it would be stupid to survey anybody twice) so that they may be asked certain questions. Statisticians call this a *simple random sample from a finite population*. There are, of course, $(m)_n$ possible ordered samples: Our probability calculations will use this as the denominator.

If we had drawn the subjects with replacement, risking repeated interviews, there would be m^n ordered samples; notice that this is now the denominator in probability calculations and is a much simpler number to work with. The solution to the birthday problem says that the probability that nobody gets interviewed twice is the ratio $(m)_n/m^n$. The inequality $(m)_n/m^n \geq e^{-\binom{n}{2}/(m-n+1)}$ then tells us that this probability is close to 1 so long as the number of pairs of subjects in a sample $\binom{n}{2}$ is small compared to the remaining population size $m - n + 1$. When this is so, though we sample without replacement, we sometimes do the easier arithmetic for the case of sampling with replacement because it is unlikely we would have interviewed anybody twice. We will see later that in such cases the errors we have introduced are usually small.

For example, in a small city with 100,000 voters, a sample with replacement of 100 would have better than probability $e^{-\binom{100}{2}/(100,000-100+1)} \approx 0.95$ of having no duplications. As the population size goes up for a given size sample, the probability of no duplication approaches 1. Therefore, if we are willing to pretend that there is no chance of duplication, we say that we are sampling from an *infinite* population.

3.7 Summary

Whenever we reason about uncertain things, such as experiments not yet performed, by trying to measure the proportion of times various things would happen, we are applying *probability* theory. In simple situations we may count *equally likely* outcomes, so that a probability is $P(A|B) = |A \cap B|/|B|$ (2.1). This counting is easy until the number of outcomes becomes numerous; then we invoke the science of counting, called *combinatorics*, to help us. Most counting problems of interest to statisticians may be solved with the aid of *permutations*, the number of ordered lists of k things from n, which is $(n)_k = n!/(n - k!)$ (3.2), or with *combinations*, the number of sets of k from n, given by $\binom{n}{k} = n!/(k!(n - k)!)$ (3.3). An amazing number of complicated probabilities may be calculated using these. For example, the *occupancy* problem, which asks how probable it is that there will be no duplicate assignments to n categories by k observations, has solution $P(\text{no duplicates}|k \text{ assigned to } n) = \frac{(n)_k}{n^k}$ (4.2). Then we discover an approximate or *asymptotic* method for calculating this probability when the number of pairings of k objects is small compared to n, $\frac{(n)_k}{n^k} \approx e^{-\binom{k}{2}/n}$ (5.3). Finally, we use this approximation to investigate when the distinction between *finite* and *infinite* population sampling becomes important (6).

3.8 Exercises

1. You awaken in the middle of the night because a truck has backfired. You glance at your lighted bedside clock, and as always, to the nearest minute the minute hand points to some number between 00 and 59. What is the probability that the minute hand points nearest to a number divisible by 7?

2. A student has 5 clean shirts (white, brown, blue, green, and maroon) and 5 clean pants of the same colors in his closet. He has to dress before dawn without waking his roommate, so he grabs a pair of pants and a shirt without being able to see them and puts them on. What is the probability that the two are **not** the same color?

3. List all the ways Fisher's hostess could choose the 4 out of 8 cups that she believed had tea poured first. How long is your list?

4. How many nonnegative integers with at most 3 decimal digits are there? Solve the problem first by ordinary arithmetic, then using the solution to Urn Problem 1.

5. You intend to go to two of the Grand Canyon, the Smithsonian, Disney World, and Niagara Falls, one this summer and the other next summer. List all possible vacation plans. Now check that your count is right by applying the formula for permutations.

6. You are going to spend a month each studying the penal systems of 12 of the country's 50 states. Count how many different ways (in sequences of states) you can spend your year.

7. A deck of playing cards consists of 52 cards = {4 suits} × {13 ranks}. A poker hand consists of five different cards, chosen so that any five are equally likely. A spade is one of the suits, so there are 13 of them in the deck. What is the probability that a poker hand will consist of five spades?

8. To keep control of my time, I decide this semester to be active in only 3 of bowling, volleyball, softball, basketball, and rugby. How many choices are possible? List all the possibilities and then count again using the combinations formula.

9. Show that $\binom{n}{k} = \binom{n-1}{k-1} + \binom{n-1}{k}$ by algebra. Now show it again, in a completely different way, by interpreting the symbols as counts in Urn Problem 3.

10. Use Exercise 9 and the fact that $\binom{n}{0} = \binom{n}{n} = 1$ (since there is only one set with no marbles and one set with all the marbles) to construct the table of combination symbols $\binom{n}{k}$,

$n \backslash k$	0	1	2	3	4	5	6	7
1	1	1						
2	1	2	1					
3	1	3	3	1				
4	1	4	6	4	1			
5	1	5	10	10	5	1		
6	1	6	15	20	15	6	1	
7	1	7	21	35	35	21	7	1

etc. (*Pascal's triangle*) by repeated addition.

11. I walk to work through a section of town where all streets are either north–south or east–west, and I must go 6 blocks west and 4 blocks south. Of course, I never take a path that would take me farther away from work. How many possible complete routes from home to work do I have to choose from?

12. Prove that $\binom{n}{n_1 n_2 \cdots n_k} = n!/(n_1! n_2! \cdots n_k!)$.

13. A police department has 10 detectives in the homicide division. In how many ways can the supervisor assign 4 detectives to the Coors case and 3 other detectives to the Hard case?

14. In the 5000 meter women's Olympic finals there are 4 Americans, 2 Canadians, and 2 Jamaicans, plus one runner each from Great Britain, Korea, Ukraine, and Japan.

    a. How many finishing orders, by nationality and **not** the name of the individual, are possible in this race?
    b. If as far as you know any finishing order is as likely as any other, what is the probability that the first two finishers will come from the same country?

15. Of the last 10 students who came from a certain small town, 7 finished above the middle of their classes at the University of Minnesota. If you believe that students from that small town are really typical of all UM students, how probable is this result? Assume that by "typical" we mean that all possible sequences like ABAAABBAAA of the arriving students finishing above (A) and below (B) the middle are equally likely.

16. Of 40 engineering majors in an engineering statistics class, 12 are mechanical engineers and 15 are industrial engineers. The instructor chooses 10 students to represent the class in a statistics contest.

    If major should have no effect on who is chosen, what is the probability that 3 mechanical engineers and 5 industrial engineers will be chosen for the contest?

17. You are playing a version of poker in which all cards are dealt from a 52-card deck. The four cards in your hand include one ace. Some of your opponents' cards are face up: You see among them one ace and 3 other cards. You are about to be dealt two more cards. What is the probability that at least one of them will be an ace?

18. Male and female chicks are very difficult to distinguish without expert examination. Eight of 12 chicks in a batch are female. You casually select 5 chicks from the batch.

    a. What is the probability that they are all female?
    b. What is the probability that there are 3 males and 2 females?

19. The 9 sororities on a certain campus form a sorority senate consisting of 7 representatives from each sorority. The president is then supposed to choose an executive committee of 8 senators. Unfortunately, 4 of the executive committee turn out to be from one sorority and 4 from another, and the president is accused of favoring these sororities. She claims it was an accident, that they were chosen without regard to the sorority they came from. Find the probability that this would have happened by chance.

20. There are sixteen well-hidden cameras, each of which is triggered by a moose wandering into its range; as far as we know, all are equally well placed for observing moose. If we wait until 9 pictures have been taken, what is the probability that 9 different cameras will have been involved? Assume that separate triggering events are independent.

21. In Exercise 20, what is the probability that exactly 7 cameras will have been involved?

22. The 150 voters in a small town are to be chosen for a panel of 12 jurors by lot, that is, chance. Of course, their names should be removed from the voter list as they are chosen, so there will be no duplications; unfortunately, the county clerk is not that smart. What is the probability that some people will be chosen twice for the panel? Also, calculate simpler upper and lower bounds for your answer, using the results of this chapter.

23. Prove that $\sum_{i=1}^{m} i = (m(m + 1)/2) = \binom{m+1}{2}$.

24. Prove that $e^x \geq 1 + x$ for all numbers x.

25. A bag of candy is supposed to contain 20 chocolates and 20 caramels. After you have eaten your way through 5 pieces, you realize suddenly that they were all caramels.

    a. If the bag was well mixed, what is the probability that this would have happened?

    b. An easier, approximate, version of this calculation follows from the approximation for the probability of birthday coincidences. Find it, and compare.

26. Show that if k^3 is small compared to $2n^2$, then $(k - 1)\binom{k}{2}/(n(n - k + 1))$ is close to zero.

3.9   Supplementary Exercises

27. The Virginia Lottery Pick 4 game draws 4 digits (from 0 though 9) each from an urn containing all ten digits.

    a. A player wins by having selected the same 4 digits in the same order, in advance of the drawing. What is the probability of winning?

    b. A lesser prize is offered for getting any three of the digits correct including order, but not a fourth. What is the probability of winning this prize?

28. a. More generally than in Exercise 23, show that $\sum_{i=m}^{n} \binom{n}{i} = \binom{n+1}{m+1}$ for any integers $n \geq m \geq 1$.

    b. Use (a) to show that $\sum_{i=1}^{n} i^2 = (n(n + 1)(2n + 1))/6$.

29. In the game of poker, the hand called a *pair* consists of 2 cards of the same rank, plus 3 cards of ranks different from the first and different from each other. If the deal is from a well-shuffled deck, what is the probability that a hand will be a pair?

30. The Virginia Association of Triplets has 9 sets of triplets as members (for a total of 27 individuals). Four individual members are picked at random to go to a national convention. What is the probability that some two of the delegates will be from the same set of triplets (but the other two delegates are from two other sets)?

31. You are a federal narcotics agent, and you have gotten a reliable tip that 6 one-kilogram packets of cocaine have been placed, one to a locker, among the 100 rental lockers at the local airport. You have gotten a search warrant

to search the lockers, but time is very tight. Your partner has searched nine lockers and found two packets. You have searched eight lockers and found one packet. What is the probability that among the next three lockers you open, there will be at least one package of cocaine?

32. You are thinking of installing a robot inspector to spot defective products at the end of an assembly line. To test it you run 6 good and 6 bad items through the inspector, in random order, and ask it to select the 6 that it judges are bad. If it finds 5 or 6 of the 6 bad ones in its list of 6, you will pass it. If the robot labels defective products purely by chance, what is the probability that you will pass it anyway?

33. A publisher sends one copy each of 25 new books to every large newspaper. The editors of the 6 large newspapers in the state each pick completely randomly one book from that list to have reviewed in next Sunday's papers. What is the probability that there will be more than one review of at least one book next Sunday?

34. It is 1944, and soldiers are building two runways, at the north end and at the south end of a Pacific atoll. There are 25 foxholes near the south runway and 20 foxholes near the north runway. One evening, 8 soldiers are working on the south runway and 6 soldiers are working on the north runway, so late at night that they can no longer see each other. The air-raid siren sounds, and each soldier independently chooses a foxhole and leaps into it.

   a. What is the probability that in some foxholes, a soldier lands on top of another soldier at the south runway? at the north runway?
   b. What is the probability that somewhere on the atoll, a soldier lands on top of another soldier?

35. Four different digits from among the digits 1, 2, ..., 9 are picked at random, one at a time.

   a. What is the probability that they are selected in *increasing* numerical order? (That is 2, 3, 7, 9 is a success, but 4, 8, 1, 3 is a failure.)
   b. If 3 is the first digit selected, what is now the probability that the four digits selected will be in increasing numerical order?

36. An absent-minded grandfather hands out 7 pieces of candy among his 12 grandchildren. He gives each piece to a randomly chosen child, without regard to whether that child has already received candy.

   a. What is the probability that 7 different children will get candy?
   b. What is the probability that exactly 6 different children will get candy?

37. There is an obvious Urn Problem Four: How many *unordered* sets of k marbles can be chosen *with replacement* from among n distinct marbles?
   Hint: Each such set is determined by knowing how many 1's, how many 2's, and so forth, up to how many n's you got in your set of k marbles. You might keep track of these as follows: Put a movable marker on your table to separate the 1's from the 2's, one to separate the 2's from the 3's, ..., and one

to separate the $(n - 1)$'s from the n's. There will always be $n - 1$ markers on the table. Now write down the marbles in the appropriate place as they come in. For example, $11||3|4444|5$ might keep track of the set of 2 ones, no twos, 1 three, 4 fours, and 1 five, in the case $n = 5$ and $k = 8$. The vertical bars are your markers. Now count the possible strings of numerals *and separating markers*.

38. A millionaire intends to give seven identical, perfect ten-carat blue diamonds to his four children. They only care how many, not which ones, they get. In how many ways can he distribute the diamonds?

    Hint: Use the results of Exercise 37.

39. In Urn Problem 4 (Exercise 37) you established that the number of ways of drawing k unordered objects, with replacement, from among n objects is $\binom{n+k-1}{n-1}$. Prove (that is, convince me you know why it is true) that this count $\binom{n+k-1}{n-1}$ is always less than or equal to $(n^k e^{\binom{k}{2}/n})/k!$.

40. In fact, the second expression in Exercise 39 may be shown to be an *asymptotic approximation* to the first when k/n is close to zero: That is, the ratio between the count and the approximate count is close to one. We will illustrate this by example:

    A computer arithmetic program for children picks 4 integers between 1 and 20, arranges them in ascending order, and presents them as an addition problem; for example, $7 + 9 + 9 + 13$. How many different problems can it generate? Now calculate the approximate answer from Exercise 39 and compare.

41. Dice are cubes (6 sides) in which the sides are numbered 1, 2, 3, 4, 5, 6. When one of these cubes is rolled across a table, it is believed to be equally likely that each of the sides will end face up; the number facing up is the result of that roll. In the game of Yahtzee, a player rolls 5 dice at once; the 5 numbers that result are a hand.

    A full house is a hand in which one number comes up three times and a second, different, number comes up twice. What is the probability that a Yahtzee hand will be a full house?

42. A consumer group claims that heavy-metal music causes cancer. As a fan of the music, I doubt this, but I will do an experiment with rats anyway, to check. I expose 8 rats to no music, 8 rats to a low dose of music, and 8 rats to a high dose of music. Eventually, 3 of the rats with no music exposure get cancer, 2 of the rats with low doses get cancer, and 5 of the rats with high doses get cancer.

    In my opinion, those rats who got cancer were destined to do so, and all possible assignments of cancerous and cancer-free rats to the three treatment groups could just as easily have happened. In that case, what was the probability of the results we actually observed?

43. A **runs test** is a way to tell whether or not there may be "serial dependence" in a sequence of experiments, that is, whether each experiment is affecting later results. Imagine that in our study of headache remedies, pill A did better in a cases and pill B did better in the remaining b cases. We count the runs, that

is, the number of sets of adjacent cases with the same results. (For example ABBAAABA has 5 runs: A, BB, AAA, B, and A.) If there are too few (or too many) runs, each result may be influencing later results.

a. Find the probability that there are exactly k runs, where k is an **even** number, if all sequences are equally likely.
(**Hint:** If there are k runs, then you already know where $k/2$ A's and $k/2$ B's are. You just have to count the ways of placing the rest.)
b. Find the probability of 4 runs if aspirin was better 5 times and Tylenol was better 6 times.

44. Now find the answer to Exercise 43 in the case where k is an **odd** number. Apply your formula to find the probability of 5 runs in the aspirin/Tylenol problem.

45. When we were defining the Kruskal–Wallis statistic K (see 2.5.5), we applied analysis of variance to the ranks $1, \ldots, n$ of a collection of measurements. Assuming that there were no ties, use Exercise 28 to show that the corrected sum of squares SS (see 2.5.3) is always $(n(n + 1)(n - 1))/12$, and therefore $R^2 = \text{SSE}/\text{SS} = \frac{K}{n-1}$.

CHAPTER 4

Other Probability Models

4.1 Introduction

We think of probability as measuring our degree of uncertainty in the results of experiments not performed yet. But in general, there is no reason to believe that each of our possible outcomes would be equally likely, as we assumed in the last chapter. Can we still come up with a science of probabilities in other cases? Some examples will suggest directions in which the concept might be extended.

Example. The weather forecast asserts that the probability of rain for tomorrow is 20%. What can be meant by that? We could imagine consulting extensive weather records, until we find 100 days in the past that were as much like today as possible. Then we assume that tomorrow is equally likely to be most similar to each of the 100 days that followed. Now, simply count how many of those days reported rain; if the answer is 20, we have our forecast. The procedure is laborious and fraught with difficult decisions; but presumably a computer could be programmed to do it. However, meteorologists of my acquaintance assure me that it is not done this way.

Example (Buffon needle problem). Consider a striped flag with all stripes of equal width, such as the stripe field of the U.S. flag. Throw a needle of the same length as the width of a stripe at random onto the field (see Figure 4.1). What is the probability that it will cross the boundary of a stripe?

It sounds as if all positions and orientations are "equally likely"; but since there are an *uncountable infinity* of these, we cannot answer the question directly from combinatorics. It was claimed in the last chapter that the probability is $2/\pi$. Since this number is *irrational*, we cannot hope to transform it to any combinatorial problem; another approach will be necessary.

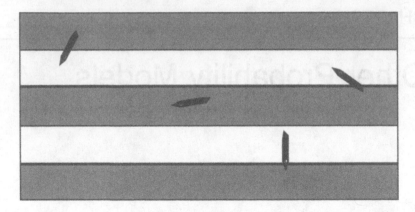

FIGURE 4.1. Buffon needle problem

The strategy of this chapter will be to describe a general probability theory, of which combinatorial probability is only one special case. We will try as we go to preserve as much as possible of the essential character of our work so far, without mentioning equal likelihood. Then we will develop some general tools for working with probabilities, however these arise.

Time to Review

Algebra of sets
Calculus of trigonometric functions
Geometric series

4.2   Geometric Probability

4.2.1   Uniform Geometric Probability

We gave an example in the introduction, the Buffon needle problem, of the probability of a sort of geometric outcome; unfortunately, none of the techniques for deriving probabilities discussed so far will help with it: It is in no sense a combinatorial probability. This particular problem is a bit hard to start with, so let us first tackle an easier one.

Example. I throw darts at a simple dart board, which consists of a 10-inch circular disk with a 3-inch circular disk called the bull's eye at its center (Figure 4.2). If a dart does chance to hit the board, what is the probability that it will hit the bull's eye?

To study this problem realistically, you would have to know a great deal about my skill at darts. Fortunately, there is very little to know. I would be lucky to hit the board at all; therefore, I am presumably just as likely to hit anywhere on the board,

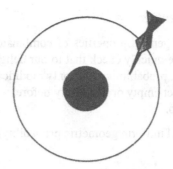

FIGURE 4.2. Dart board

if I do hit it. Intuitively, therefore, the chances of hitting a spot are proportional to the size of the spot; the relative *area* of the bull's eye to the area of the whole board is the issue. So, using the familiar formula for the area of a circular disk, we get P(bull's eye|board)$= 2.25\pi/25\pi = 0.09$.

In general, we see that events of interest on two-dimensional surfaces are usually regions that we think of as possessing *area*. Similarly, events in three-dimensional space are usually regions that possess *volume*. (What is the probability that a surface-to-air missile will explode in a certain volume of space?) And even if you do not usually think of one-dimensional problems, on lines, as being geometrical, it seems reasonable to measure the size of a segment by its *length*:

Example. My pocket calculator has a command on it called Ran#, or something like it, that produces an unpredictable nine-digit number somewhere between zero and one (most computer languages, spreadsheets, and mathematical and statistical packages have something similar). If we think of this as the coordinate of a random point on the number line between zero and one, then its probabilities are intended to be *uniform* on the event (0,1). The probability the random number will fall in the interval from 0.15 to 0.40 is then just the length of that interval, 0.25 (since the denominator is 1, the length of the whole interval).

These are related ideas: lengths in one dimension, areas in two dimensions, volumes in three dimensions, and in fact, hypervolumes in more than three dimensions. We call all of these concepts **volume** with respect to the appropriate dimensional space, and write the volume of A as V(A) for an event A. Our dart board example suggests one simple kind of probability assignment that is sometimes useful.

Definition. A geometric probability space is **uniform** if given events A and B such that $0 < V(B) < \infty$, probabilities are given by $P(A|B) = V(A \cap B)/V(B)$ whenever the numerator exists.

As in the darts example, this model applies to cases in which any point in B seems as likely as any other.

4.2.2   General Properties

Going back to our list of general properties of combinatorial probability in the last chapter (see 3.4.3), we quickly check that to our delight, they all are equally true for uniform geometric probabilities. The only modification we might make is that where we had some set empty or not empty before, we now ask only that its volume be zero or not zero.

Proposition (properties of uniform geometric probability).

(i) $0 \leq P(A|B) \leq 1$.

(ii) *If* $V(A \cap B) = 0$, *then* $P(A|B) = 0$.

(iii) $P(B|B) = 1$.

(iv) $P(A \cup B|C) = P(A|C) + P(B - A|C)$, *and if* $V(A \cap B) = 0$, *then* $P(A \cup B|C) = P(A|C) + P(B|C)$.

(v) $P(C - A|C) = 1 - P(A|C)$.

(vi) *If* $V(A \cap C) \neq 0$, $P(A \cap B|C) = P(A|C) \cdot P(B|A \cap C)$.

You should use familiar properties of length, area, and volume that you learned in geometry and in calculus to prove these facts. You can use the analogous proofs from Chapter 3 as models.

As similar as these are to properties of combinatorial probability, the one small difference has interesting implications. An event on an interval does not now have to be empty to have probability equal to zero: For example, a single point has length zero, so its probability conditioned on the whole interval is zero. Thus $P(\{1/\pi = 0.318309886\ldots\}|(0,1)) = 0$; the chances that I will get one exact number when I hit Ran# is vanishingly small. If I think I have hit it, there is a very good bet that if I measure my answer to another few decimal places of accuracy, I will find I just barely missed. Nevertheless, I could conceivably hit that number. So in this version of probability, "impossible" and "zero probability" have subtly different meanings.

In fact, sets do not have to be small to have zero volume and therefore zero probability. Consider a square dart board C and an interval B that cuts across it Figure 4.3. Since this is a problem in two dimensions, probability is in terms of area; and the area of that segment B is zero. Therefore, even though B is much

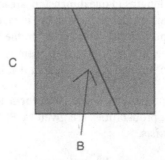

FIGURE 4.3. A line interval inside a square

more than a single point, it must still be that $P(B|C) = 0$. If you think that your dart has hit B, it is almost certain that if you looked a little closer, you would see that you have hit just to one side or another of the line segment.

4.3 Algebra of Events

4.3.1 What Is an event?

Now we know that probability may be usefully applied both to counting problems and to geometrical problems, and have remarkably similar properties in these very different situations. We are inspired to talk about a general concept of probability, in which our two types so far would be only two special cases among many.

As before, we will be interested in probabilities of *events*, which will still be sets of individual *outcomes*. In combinatorial probability, any finite set at all was a plausible candidate to be an event, even if it is hard to imagine why we would be interested in a particular set for a practical application. In uniform geometric probability problems, it is obvious that only events that have volume (whether that means length, area, ordinary volume, or whatever) are candidates to be events. In advanced real analysis courses, you will discover that certain sets (though not any you would be likely to guess) can never be assigned a volume, no matter how good you are at computing volumes. These can never be events in geometric probability problems. So each application of probability may require a different definition of what constitutes an event.

We need to know when we have done a satisfactory job of defining the events in a probability problem. Our strategy will be to write down some simple rules for which other sets of outcomes ought to be events, if we know which ones we certainly want.

For example, there might be two sorts of results of an experiment that we would call successes; we could write them down as two collections A and B of successful outcomes. If these are each to be events, we would also be interested in the event of simply succeeding. This event would be given in set theory by $A \cup B$, the outcomes in either A or B or both. We will generalize this and insist that if you wish to study any two events, their union must also be an event.

If B is the possible outcomes of a certain experiment and A is the event of *succeeding* at that experiment, then surely *failing* at the same experiment is also an event of interest. In set notation, $B - A = \{x \in B \text{ and } x \in A\}$, the set of failing outcomes. We shall insist generally that if A and B are any events, then $B - A$ is an event as well.

4.3.2 Rules for Combining Events

To summarize our requirements:

Definition. An **algebra of events** is a nonempty collection of events such that

(i) if A and B are events, then A ∪ B is also an event (*unions*); and

(ii) if A and B are events, B − A is also an event (*complements*).

From now on, we will expect the collections of events to which we assign probabilities to be algebras. You might be surprised that we have not required the presence of certain other events, such as intersections, that we talked about when computing equally likely probabilities. It turns out that the two requirements given are enough.

Proposition. (i) ϕ (*the empty set*) *is an event; and*

  (ii) *if* A *and* B *are events, then* A ∩ B *is also an event.*

PROOF.   (i) B − B = ϕ is an event; (ii) exercise.   □

Notice that already we have one easy example of an algebra. When we did combinatorial probability, we had a finite list of all possible outcomes. The events included *any* subset of that list. But the rules for an algebra just insist on a minimum collection of events, and since we are using all possible subsets of that list as events, it must be an algebra.

When we do uniform geometric probability, we start with the biggest event in which we may be interested U, which must have finite volume in whatever dimension we are working, $0 < V(U) < \infty$. (Think of a dart board.) Now, I will propose an algebra whose events are all the subsets of U that have a volume (possibly 0). Then it is plausible that for two events A and B that each have a volume, A ∪ B and B − A will also have a volume (for one thing, we know immediately that they can be no bigger than V(U)). We will come back to this issue later in the chapter, when we will describe more carefully the algebra needed for geometrical probability.

4.4   Probability

4.4.1   In General

Now we will try to say what all sorts of probability should be like, guided by our experience with combinatorial and uniform geometric probability. These share a common intuition that the probability of a future event is something like the proportion of times we might reasonably expect it to happen if we did the same experiment many times. Certainly, then, we should have an addition rule of some sort—for example, the proportions of the time one event or another would happen, if they cannot both happen, must surely just add. Surely, too, there must always be a multiplication rule:

Example. What is the probability that an entire weekend will be rained out in September, precluding a picnic? The weather service is unlikely to have this question already answered, but they might be able to tell us that the probability of a rainy day is 20% this time of year. With further research, they might tell us that on a typical rainy day, the probability that rain will recur the next day is 50% (because many storms last longer than a day). Our answer is the probability that it will rain Saturday, and then also the next day; which will come about 50% of 20% of the time, or 10% of the time.

This just uses the familiar principle that proportions of proportions simply multiply. So general probability theory will be founded on those two requirements.

4.4.2   Axioms of Probability

The two requirements from the last section will be the most important statements in an *axiom system* for probability; their purpose is to summarize the general features we will look for in any possible application of probability theory. This approach was first popularized by the Russian mathematician Kolmogorov in the 1930s (though our choice of axioms is somewhat different from his). The axioms are contained in the following:

Definition. A **(finitely additive) probability space** is an algebra of events, together with a real-number-valued function $P(A|B)$ defined on pairs of events with $B \neq \phi$ such that

(i) $P(A|B) \geq 0$ (*nonnegativity*);
(ii) $P(B|B) \neq 0$ (*nontriviality*); under a condition C,
(iii) $P(A \cup B|C) = P(A|C) + P(B - A|C)$ (*additivity*); and
(iv) $P(A \cap B|C) = P(A|C) \cdot P(B|A \cap C)$ whenever $A \cap C \neq \phi$ (*multiplicativity*).

Comments: Our motivating examples of probabilities are proportions, which are certainly never negative; therefore, I cannot imagine what a negative probability would mean, and I put in rule (i). Rule (ii), certainly true in our examples, is a simple device to make sure that there are *some* positive probabilities; a probability system that is always zero, and so completely useless, meets all the other rules. The last two are just our addition and multiplication computing rules.

You may have seen in other books what are called *unconditional* probabilities, written something like $P(A)$. As mentioned in the last chapter (see 3.2.1), this is simply a shorthand notation for our usual $P(A|B)$, whenever you feel free to assume that your audience knows which condition B is meant. When discussing dart throwing, we felt free to assume that a common general condition would be that you have hit somewhere on the dart board. Now let us see what the shorthand does to the appearance of our axiom (iv) when we assume that everybody is aware of the general condition $C : P(A \cap B) = P(A) \cdot P(B|A)$. You have to remember that a subtle convention is hidden here. Not only have we written $P(A)$ for $P(A|C)$ and $P(A \cap B)$ for $P(A \cap B|C)$; we have also written $P(B|A)$ for $P(B|A \cap C)$. The only way you can tell about that last substitution is to see that it appears in the same formula as the unconditional probabilities. Nevertheless, many people find this simplified form easier to remember.

The shorthand form of the axiom of additivity is $P(A \cup B) = P(A) + P(B - A)$. You may find that it helps you remember the two axioms to notice the remarkably parallel form they take. Interchange $\cup$ and $\cap$, addition and multiplication, and $-$ and $|$, and you find that one axiom has been transformed into the other.

While we are at it, let us solve for the second factor in the axiom of multiplicativity to get a famous formula.

Proposition (conditioning). *If* $P(A) \neq 0$, *then* $P(B|A) = \frac{P(A \cap B)}{P(A)}$, *where all probabilities are with respect to a common condition.*

In older texts, this is sometimes used as the definition of a conditional probability. We will use it whenever we want to introduce a new condition, because we have learned something relevant to the question.

Example. Your ornithology group is capturing and attaching location finders to predatory birds in a large wildlife preserve. Only 25% of the birds you catch are eagles, and only 6% of the birds are golden eagles, which you are studying. Your colleague Susan, who is surveying eagles in general, comes running in and announces "We caught an eagle today!" What is the probability that it is a golden eagle?

We calculate $P(\text{golden}|\text{eagle}) = \frac{P(\text{golden eagle})}{P(\text{eagle})} = \frac{0.06}{0.25} = 0.24$.

4.4.3 Consequences of the Axioms

You may be wondering where all those common properties of combinatorial and uniform geometrical probabilities went to. Axioms are supposed to be short lists of the most critical properties; so now let us check that our list is long enough. With a little ingenuity, we can extract from our axioms all the other usual properties of probability.

Let $A \supset B$ so that every outcome from B is also in A. Then we know that $A \cap B = B$. Calculate

$$P(B|B) = P(A \cap B|B) = P(A|B) \cdot P(B|A \cap B) = P(A|B) \cdot P(B|B),$$

where the second equality just uses axiom (iv). Axiom (ii) says that $P(B|B) \neq 0$, so we can divide the first and last terms of the equality by it:

Proposition. (i) $P(A|B) = 1$ *whenever* $A \supset B$.
 (ii) $P(B|B) = 1$ (*because* $B \supset B$)

The second fact is often given as an alternative to our axiom (ii).

If we know the probability that something will happen, what is the probability that it will not happen, that is, $P(B - A|B)$? We know what the answer should be from combinatorial probability; in fact, when we solved this problem in (3.4.3), we used only additivity and the proposition above. Therefore, it is true for all kinds of probability. We summarize our results as follows:

Proposition. (i) $P(B - A|B) = 1 - P(A|B)$.
 (ii) *Always* $P(A|B) \leq 1$.

The first result says, for example, that the probability of success with an experiment is one minus the probability of failure. You should check (ii) as an exercise.

4.5 Discrete Probability

4.5.1 Definition

So far, we have nothing new, and our purpose in writing down the axioms was to allow for new applications of probability theory. The weather forecasting example in the introduction suggests another sort of model: Tomorrow's weather consists of two outcomes, rain and dry (T = {r, d}). We assign somehow (in this case, by expert opinion), P[{r}|T] = 0.2. The previous proposition shows that P[{d}|T] = 0.8; this is all we need to say about the probabilities in this situation. To summarize, we want a type of probability space that consists of a complete list of possible outcomes and such that we have some way of assigning a positive probability to each. We will want all of these probabilities to sum to one, by our addition rule for probabilities and the fact that P(All|All) = 1.

Sometimes we will need to say even more. Imagine an outcome to be the number of Atlantic hurricanes during the next season. The possible outcomes are {0, 1, 2, 3, . . .}, the nonnegative integers. I know of no natural law that places an upper limit on this number (certainly not 26, the available first letters for the annual names list), so even though I do not take seriously the possibility of a million hurricanes, I include all these integers among my outcomes. Now, the case of exactly three hurricanes is an event of interest, written {3}. Might I also be curious (do not ask why) about the probability of an *odd* number of hurricanes? If so, that event could be written {1, 3, 5, . . . , 2k − 1, . . .}. (We are now certainly not in the world of equally likely probability. We do not know how to do arithmetic with infinite counts.) We need some restriction on the sizes of such collections of outcomes:

Definition. A **countable** collection is one whose elements can be numbered, that is, can have a different positive integer assigned to each.

Example. Any finite collection is countable, since you can just write down the assigned numbers: {A_1, A_2, A_3, A_4}.

Example. For an infinite collection like the odd positive integers, we will need a rule for numbering the elements, since we would fail to finish numbering them by hand before our species becomes extinct. Notice that 1 is the first odd number, 3 is the second, 5 is the third, and by a leap of ingenuity, k is the (k + 1)/2 odd number. For example, 1793 is the 897th odd number. We can number them all, so our collection is countable.

Let us formalize this sort of probability space:

Definition. A **discrete** probability space consists of a countable event $U = \{x_i\}$, the algebra consisting of all subsets of U, numbers $p_i > 0$ associated with each outcome x_i such that $\sum_i p_i = 1$, and probabilities $P(A|B) = \sum_{i \in A \cap B} p_i / \sum_{i \in B} p_i$.

The idea is that $P(\{x_i\}|U) = p_i$; the general probability formula was inspired by the proposition on conditioning. To see that this special, but important, concept is consistent with what has gone before, we need to see that it is consistent with our axioms.

4.5.2   Examples

Proposition. *Any discrete probability space is also a probability space.*

PROOF.   Check the axioms:

(i) $P(A|B) \geq 0$ because neither numerator nor denominator is ever negative;

(ii) $P(B|B) = \sum_{i \in B \cap B} p_i / \sum_{i \in B} p_i = 1 \neq 0$ because B is not allowed to be empty;

(iii) The secret of verifying this axiom is to be unafraid of our complicated notation:

$$P(A \cup B|C) = \frac{\sum_{x_j \in (A \cup B) \cap C} p_j}{\sum_{x_j \in C} p_j}$$

$$= \frac{\sum_{x_j \in (A \cap C)} p_j + \sum_{x_j \in [(B-A) \cap C]} p_j}{\sum_{x_j \in C} p_j} = P(A|C) + P(B - A|C),$$

where the first equality just uses the definition, and the second works because A and B − A do not have any outcomes in common. Finally, split the fraction in two, and we are done.

(iv) Exercise. When you have done it, our proof will be complete.   □

You should check, as an easy exercise, that equally likely (combinatorial) probability (where the events are any subsets of some finite set of outcomes, and probabilities are gotten by counting outcomes) is an example of a discrete probability space.

The shorthand notation is particularly useful with discrete probabilities, if your audience agrees in advance on the complete list of outcomes U (for Universe). Then, almost always, $P(A) = P(A|U)$. But notice that

$$P(A|U) = \frac{\sum_{i \in A \cap U} p_i}{\sum_{i \in U} p_i} = \frac{\sum_{i \in A} p_i}{1} = \sum_{i \in A} p_i;$$

we have learned the following fact:

Proposition. $P(A) = \sum_{i \in A} p_i$.

Of course, we intended this to be true when we first defined discrete probability.

Example. In the example of the number of hurricanes in a season, we had U = $\{0, 1, 2, \ldots\}$. I do not know enough meteorology to assign realistic probabilities to the various numbers of hurricanes; but let me propose the following simple rule: $P(\{0\}) = p_1 = \frac{1}{2}$, $P(\{1\}) = p_2 = \frac{1}{4}$, $P(\{2\}) = p_3 = \frac{1}{8}$, and generally $P(\{i\}) = p_{i+1} = 2^{-i-1}$. Since we have assigned all outcomes a positive probability, then we will have a discrete probability space if only the grand total is right: $\sum_i p_i = \frac{1}{2} + \frac{1}{4} + \frac{1}{8} + \cdots$. This infinite series is one of a very important class, called *geometric* series; it will be useful, now and later, to recall from calculus how to sum it:

Proposition. $\sum_{i=0}^{\infty} a \cdot r^i = a + a \cdot r + a \cdot r^2 + \cdots = a/(1-r)$ *whenever* $|r| < 1$.

You can see why this ought to be the right sum by multiplying both sides by $1 - r$. Our series is of this form if $a = \frac{1}{2}$ and $r = \frac{1}{2}$, so that the sum of all our probabilities is $\frac{1}{2}/(1 - \frac{1}{2}) = 1$, as it should be.

Now we may do various calculations with hurricane probabilities. For example,

$$P(\text{odd number}) = P(1) + P(3) + \cdots = \frac{1}{4} + \frac{1}{16} + \frac{1}{64} + \cdots$$

This is another geometric series, with $a = \frac{1}{4}$ and $r = \frac{1}{4}$; so the probability of an odd number of hurricanes is, peculiarly enough, $\frac{1}{3}$.

Now you can see why we restricted our attention to countable collections of outcomes (yes, there are bigger sets, which you may study in classes in real analysis). We learned in calculus how to sum certain infinite series, which just involve adding up a countable sequence of terms. This is just what we needed to do in this example.

4.6   Partitions and Bayes's Theorem

4.6.1   Partitions

Now that we have a richer variety of examples of probability spaces, we can show off some more powerful computing tools. One important idea is that when we want the probability of an event under complex conditions, it may be useful to split the conditions into simpler special cases.

Example. What proportion of undergraduates at a certain college might be expected to drop out in a given year? Well, the situation is presumably different for freshmen, sophomores, juniors, and seniors; the youngest students presumably are less committed, and more likely to quit. Furthermore, they have different advisors, who have completely separate data bases of information about the different years. You find that 30% of freshmen, 15% of sophomores, 10% of juniors, and 8% of seniors drop out each year; presumably the answer is some sort of average of these. But it cannot be a simple average, because presumably there are more freshmen than there are students in any of the other classes, so the 30% who dropout represent proportionally more students. You go to the registrar and find that of all undergraduates, 35% are freshmen, 25% are sophomores, 20% are juniors, and 20% are seniors. Now you can reason as follows: 30% of 35% of students, or 10.5%, are freshman dropouts (using the intuition behind our multiplication law). Now sum the proportion of dropouts over all classes:

$$0.3 \times 0.35 + 0.15 \times 0.25 + 0.1 \times 0.2 + 0.08 \times 0.2 = 0.1785$$

We state this as a result in probability: If you pick an arbitrary student in September, the probability that he or she will drop out by the end of the year is 0.1785.

We need to formalize this idea of dividing the condition into special cases.

Definition. A (finite) **partition** of an event B is a finite collection of events $\{C_i\}$ such that

(i) $C_i \cap C_j = \phi$ for $i \neq j$ (*mutually exclusive*).
(ii) $\underset{i}{\cup} C_i = B$ (*exhaustive*).

The notation in (ii) just says to take the union over all values of j; it is a relative of summation notation. A Venn diagram should make this definition easy to remember (Figure 4.4):

Example. (1) Freshman, sophomore, junior, senior is a partition of undergraduates.

(2) Male, female is a partition of people.

(3) Given $A \subset B$, then $\{A, B - A\}$ is a partition of B (exercise).

4.6.2 Division into Cases

Partitions are useful because we can sum probabilities over them.

Proposition (finite additivity). *Given a finite collection of events $\{A_i\}$ that are mutually exclusive, $A_i \cap A_j = \phi$ for $i \neq j$, $P(\underset{j}{\cup} A_j) = \sum_j P(A_j)$, where the probabilities are taken with respect to a common condition.*

PROOF. We showed in (3.4.3) that for any two mutually exclusive events (in shorthand), $P(A \cup B) = P(A) + P(B)$, as a direct consequence of the additivity

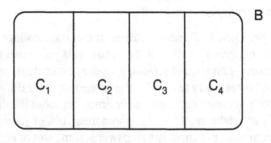

FIGURE 4.4. A partition

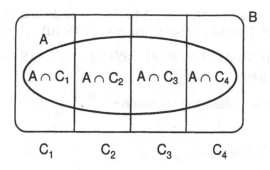

FIGURE 4.5. Division into cases

axiom. Repeat this, taking the union with one additional event at a time, until you have the union of the entire collection (as mathematicians say, by induction).   □

Now let us see what a partition can tell us about a probability:

$$P(A|B) = P(A \cap B|B)$$

(which you should verify, as an exercise)

$$= P[A \cap (\cup_i C_i)|B] = P[\cup_i (A \cap C_i)|B]$$

using a famous identity from set theory, which you should check for yourself.

Therefore, $P(A|B) = \sum_i P(A \cap C_i|B)$ by the proposition of finite additivity (see Figure 4.5). So a partition does indeed allow us to break up a probability as a sum. But

$$P(A \cap C_i|B) = P(C_i|B) \cdot P(A|C_i \cap B) = P(C_i|B) \cdot P(A|C_i)$$

from the multiplicative axiom. Let us summarize:

Theorem (division into cases). *Let $\{C_i\}$ be a finite partition of B. Then*

$$P(A|B) = \sum_i P(C_i|B) \cdot P(A|C_i).$$

Note that our calculation of the dropout probability took this form.

Example. A city is thought to have about 1% of its population carrying the HIV virus, which is believed to cause the deadly AIDS syndrome. There exists a good inexpensive blood test for the HIV virus whose performance may be summarized as follows:

 (i) If a patient does have HIV, 90% of the time the test will say so; and
 (ii) If a patient does not have HIV, 96% of the time the test will say so.

The number in (i) is called the *sensitivity* of a test; the number in (ii) is called the *specificity* of the test. In practice, they will not both be 100%. Usually, there is a trade-off; the more sensitive a test is, the less specific, and vice versa.

What is the probability that a randomly chosen person from this city will test positive for HIV? Our partition formula will work here: Let C be residents of the city, H be those who have HIV, and D be those who do not. Then {H, D} is a partition of C. Let T be the event of testing positive for HIV. Then we want

$$P(T|C) = P(H|C) \cdot P(T|H) + P(D|C) \cdot P(T|D) = 0.01 \cdot 0.9 + 0.99 \cdot 0.04$$
$$= 0.0486.$$

Not quite 5% of our patients will test positive.

4.6.3   Bayes's Theorem

You may find the result above rather disturbing, if you imagine that a program to test everybody in the city for HIV would be a good idea. You would get far more positives than you had HIV patients and run the risk of scaring many healthy patients to death. To further quantify this difficulty, we might ask; What is the probability that someone who tests positive actually has the virus? In symbols, we want $P(H|T)$. Notice that this is the reverse conditional probability of the $P(T|H)$ that we were given; is there a way to exchange the roles of event of interest and condition?

Let E, F, G be events, and compute $P(F|E \cap G) = P(E \cap F|G)/P(E|G)$ using our formula for introducing a condition. The event that E and F happen is just the event that F and E happen, so long as we treat events as sets, since $E \cap F = F \cap E$. Then

$$\frac{P(E \cap F|G)}{P(E|G)} = \frac{P(F \cap E|G)}{P(E|G)} = \frac{P(F|G)P(E|F \cap G)}{P(E|G)}$$

by another use of the multiplication axiom. But notice that G was a common condition in every probability in this formula; so it is natural to use shorthand and leave it out. We have proved a famous fact:

Theorem (Bayes's theorem). $P(F|E) = P(F)P(E|F)/P(E)$ *whenever* $P(E) \neq 0$, *where all probabilities are with respect to a common condition.*

This is attributed to Thomas Bayes, an eighteenth-century Presbyterian minister. (His example was a problem in the game of billiards.)

In our AIDS example, we notice that we have already computed the quantities we need. $P(H|T) = (0.01 \cdot 0.9)/0.0486 = 0.185$. Fewer than 20% of the people positive on our test really have HIV.

This seems to suggest that the blood test described, which we thought was a good one, is really terrible. But that is not entirely fair; notice that at one time we thought that the chances a patient would have HIV was 1%. After the same patient is positive on the test, the chances leap to 18.5%, or almost 20 times greater. As an exercise, calculate the probability that the patient has HIV after testing negative on the blood test. You will find that it is many times smaller than before. If our goal was to screen out a high-risk group from among, for example, blood donors, it seems that the test could be very useful indeed.

This illustrates an important style of statistical reasoning, called *Bayesian inference*. We start with some state of knowledge about some important question (new patient has probability 0.01 of having HIV). We perform an experiment (give the blood test) that is relevant to the question, in the sense that the probabilities of various events are different for different answers to the question. We then use Bayes's theorem to compute new probabilities for the possible answers to the question (a patient positive on the blood test has probability 0.185 of having HIV). Good experiments can make us ever more confident, though never quite certain, of the truth. The probabilities we knew before the experiment are called *prior* probabilities; those we compute after the experiment are called *posterior* probabilities.

4.6.4 Bayes's Theorem Applied to Partitions

When we calculated the probability in our example using Bayes's theorem, we found that both numerator and denominator were quantities that had appeared in our division into cases theorem for probabilities. This suggests that we might use Bayes's theorem to find the probability of one of the partition events C_i once the event of interest has happened:

$$P(C_i | A \cap B) = \frac{P(C_i|B)P(A|C_i \cap B)}{P(A|B)} = \frac{P(C_i|B)P(A|C_i)}{\sum_j P(C_j|B) \cdot P(A|C_j)},$$

where the first equality is Bayes's theorem, and the second just uses the partition theorem. As before, people often prefer the shorthand notation. The common condition B does not appear in every term, but it is, in effect, there because the $\{C_j\}$ are subsets of B. This is a nice enough formula that we mark it:

Theorem (Bayes's theorem for partitions).   *Let* $\{C_j\}$ *be a (finite) partition of an event* B, *and* A *an event. Then if* $P(A) \neq 0$, *we have*

$$P(C_i|A) = \frac{P(C_i)P(A|C_i)}{\sum_j P(C_j) \cdot P(A|C_j)},$$

where all probabilities share the common condition $B = \bigcup_j C_j$.

I think of this case of Bayes's theorem as a sort of detective's equation. Imagine that the $\{C_j\}$ are the cases of various suspects being guilty of a crime, and A the crime actually taking place. Then $P(C_i|B)$ is the probability that suspect i would commit such a crime (*motive*), and $P(A|C_i)$ the probability that were he to commit such a crime, it would be the particular one being investigated (*opportunity*). So now we see that when detectives evaluate their suspects for motive and opportunity, they should really multiply the two and compare the product to the corresponding products for all the other suspects.

4.7   Independence

4.7.1   Irrelevant Conditions

We exemplified the multiplicativity axiom in Chapter 3 (see 3.4.3) by choosing two states without replacement for a survey and asking whether they were both Atlantic states. How much did it matter that we drew without replacement? We might instead draw one name from a jar, write down which state we got, put it back in the jar, stir up the names, and draw a second state (draws *with* replacement). How has the probability of two Atlantic states changed? This is back to what we called Urn Problem 1; the answer is $15^2/50^2 = 15/50 \cdot 15/50 = 0.09$, which is slightly larger than before. We once again interpret the product to mean something like P(Atlantic|15 of 50)P(2nd Atlantic|1st Atlantic and 15 of 50). But by putting the first state back in the jar, we have made the jar equivalent to what it was before, and the probability that the second state is Atlantic is the same as the probability that the first was. When, as in Chapter 1, we do an experiment repeatedly in hopes of making our overall conclusion more accurate, we often work very hard to make sure that each repetition of the experiment is unaffected by what happened in previous runs. Here, we have done this by putting the removed state back in the jar. This is an example of an important phenomenon in probability: Some conditions that you may consider (previous experiments) may have no effect on the probability of a certain event.

Definition. An event B is **independent** of an event A relative to a condition C if $P(B|A \cap C) = P(B|C)$.

Example. Let B be the event that it rains tomorrow in Blacksburg, Virginia, and A be the event that it rains later today in Athens, Greece. I cannot imagine much of a connection over so short a period between two places so far apart; so I assume that B is independent of A. Under current conditions, using shorthand, I say $P(B|A) = P(B)$. If the weather report gives a 20% chance of rain tomorrow in Blacksburg, I will not expect that to change if a few minutes later I hear on television that a shower is falling on the Parthenon.

Our motivating problem was reduced to a rather simple multiplication by re-placing a state, and thereby making our two choices independent. The general idea is

$$P(A \cap B|C) = P(A|C) \cdot P(B|A \cap C) = P(A|C) \cdot P(B|C)$$

by the multiplication axiom, if B is independent of A. We summarize, using shorthand:

Proposition. *If* B *is independent of* A *relative to* C, *then* $P(A \cap B) = P(A) \cdot P(B)$, *where all probabilities are relative to* C.

Example. In the darts problem in Section 2, what is the probability that I will hit the bull's eye 3 times in a row? I presume that a little practice will do me little good, so that each throw is independent of previous throws. Therefore, $P(3 \text{ bull's eyes}|3 \text{ hits}) = 0.09^3 = 0.000729$. It will not happen very often.

Example. I hope to have a successful picnic on Labor Day or Memorial Day next year. What is the probability that at least one of these days will be rainless? The weather service says the probability of rain on Memorial Day is 20%, and on Labor Day is 15%. They are so far apart in time that I presume that Labor Day rain is independent of Memorial Day rain; so the probability of being rained out on both days is $0.2 \times 0.15 = 0.03$. My probability of success is therefore $1 - 0.03 = 0.97$.

4.7.2 Symmetry of Independence

Notice that our product formula for independent events does not care whether B is independent of A, or vice versa. In fact, when B is independent of A, we may apply Bayes's theorem to check that

$$P(A|B) = \frac{P(A)P(B|A)}{P(B)} = \frac{P(A)P(B)}{P(B)} = P(A),$$

where C is the common condition.

Proposition. *If* B *is independent of* A, *then* A *is independent of* B, *relative to the same condition.*

Because of this symmetry, we usually just say that A and B are independent relative to C. If your audience knows the condition C, it is a common shorthand not to mention it; we just say that A and B are independent of one another.

Example. A certain scholarship is given to a Tech junior each year, without regard to gender. Yet for the past five years, it has gone to women. We learn that 42% of Tech juniors are women. If we imagine that the scholarship was given by picking a student completely at random, what is the probability that the next five recipients will also be women? Presumably, the annual choices are independent, so we simply use our multiplication result repeatedly: $P(5 \text{ women}|5 \text{ students}) = 0.42^5 = 0.013069$. I did not need to know how many juniors there were, even though the number of people involved is known and finite.

4.7.3 Near-Independence

Example. Another scholarship is given to five Tech juniors each year, without regard to gender. What is the probability all five will go to women this year? This is a draw without replacement (nobody gets two scholarships), so independence does not apply; we need to find out from the registrar that there are 4850 juniors, of whom 2037 are women (exactly 42%). This is another finite population sampling calculation, so

$$P(5 \text{ women}|5 \text{ students}) = \frac{(2037)_5}{(4850)_5} = \frac{2037 \cdot 2036 \cdot 2035 \cdot 2034 \cdot 2033}{4850 \cdot 4849 \cdot 4848 \cdot 4847 \cdot 4846}$$
$$= 0.013032.$$

It is noteworthy that this answer and the answer to the last problem differ only in the fourth decimal place. The reason is easy to see; even after four people have been removed from the pool, the proportion of women that remain is $\frac{2033}{4846} = 0.4195$, which hardly differs from 42%. Thus, the calculations of the two answers are practically the same. This is an example of the phenomenon we noticed in Chapter 3.6, where sampling from a finite population was almost the same as sampling from an infinite one. Apparently, sometimes we can get away with assuming that we are doing draws with replacement (which lets us do the easy, independence, calculations) when we are in fact not replacing our draws. This presumably works when the number of draws is small compared to the number of marbles in our urn, so we are not changing the proportion of available choices much.

We can say something about when the number of draws is small enough. If we draw k marbles from an urn with W whites and B blacks, then the probability of getting all white marbles with and without replacement is approximately the same when $(W)_k/(W + B)_k \approx W^k/(W + B)^k$. This is true when $(W)_k/W^k \approx 1$ and $(W + B)_k/(W + B)^k \approx 1$. But we already know from the last chapter (see (3.5.3), the birthday problem) when we can count on this to be true. Using the inequalities established there, $e^{-\binom{k}{2}/W-k+1} \le (W)_k/W^k \le e^{-\binom{k}{2}/W}$. This says that the ratio is practically 1 when $\binom{k}{2}$ is very small compared to $W - k + 1$, and therefore also to $W + B - k + 1$ (which is obviously bigger). In our problem, we had $W = 2037$ women and $k = 5$ scholarships, so $\binom{k}{2} = 10$; so we are not surprised that the approximation to the draw without replacement by the easier calculation of the draw with replacement (assuming independence) was rather good.

4.8 More General Geometric Probabilities

4.8.1 Probability Density

Uniform geometric probabilities can sometimes help us solve more complicated geometric probability problems.

Example. On our circular dart board (2.1), what is the probability for a dart falling in a certain vertical strip? (See Figure 4.6.)

To make the math easier, center the board on the origin of a coordinate system, and let the board be of radius 1. Then our strip of interest is those points with x-coordinates between a and b. The total area of the board is now π. The parts of the strip above and below the x-axis have the same area, and the upper half of the entire dart board is the area under the curve $y = \sqrt{1 - x^2}$, the equation for the unit circle. Areas under a curve may be obtained by integration. Dividing by the total area of the board π, we get $P\{x$ between a and $b\} = \int_a^b \frac{2}{\pi}\sqrt{1 - x^2}dx$. This often happens: A geometric probability can be expressed as the integral of a relatively simple function, in this case $\frac{2}{\pi}\sqrt{1 - x^2}$, which we will call the probability *density* of the x-coordinate. Here the density has a simple geometrical interpretation as being proportional to the height of the strip above a given x. Now we can reason

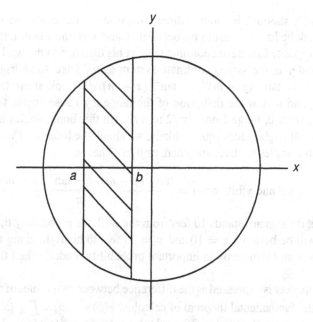

FIGURE 4.6. A strip inside a circle

backwards, solving the integral (exercise) to get

$$P\{x \text{ between } a \text{ and } b\} = \frac{1}{\pi}\left[\sin^{-1}(b) + b\sqrt{1-b^2} - \sin^{-1}(a) - a\sqrt{1-a^2}\right].$$

For example, the probability of hitting between 60% to the left of center and 20% to the left of center is P$\{x$ between -0.6 and $-0.2\} = 0.231$.

Example (Great Wall of China problem).   The Great Wall of China is a stone wall 1500 miles long, but not very high. Imagine a guard standing before a long straight and level stretch of the wall. He is very inebriated, so he shoots his rifle completely at random. Occasionally, by chance, a bullet hits the wall. What are the probabilities that it lands in various places along the wall?

Since the wall is very long but low, I will pay no attention to how high on the wall the bullet lands; just to where horizontally. The first thing we notice is that there are so many points along the wall the bullet could hit that the probability of hitting any one point is negligible. The best we can do is figure the probability of hitting in a stretch of wall, for instance, between x and y (see Figure 4.7).

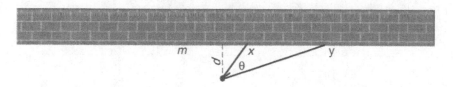

FIGURE 4.7. The Great Wall of China

If the guard is shooting in random directions, it seems reasonable that the angle θ within which he has to shoot to hit between x and y is important. Let the point on the wall opposite him have coordinate m; let his distance to the wall be d, and measure x and y in the same coordinate system as m. Then some trigonometry tells us that $\theta = \tan^{-1}(y - m)/d - \tan^{-1}(x - m)/d$. (Look at the triangles in the diagram, and review the definition of the tangent.) Those angles that hit the wall, starting from 0, range from $-\pi/2$ to $\pi/2$. (In this book, angles are always in radians.) If all angles seem equally likely, we should be looking at what portion of the available angles we have included, or θ/π. That is,

$$P(\text{between } x \text{ and } y|\text{hits wall}) = \frac{\tan^{-1}(y - m)/d - \tan^{-1}(x - m)/d}{\pi}.$$

For example, if the guard stands 10 feet from the wall, the probability that his next bullet hole will be between $x = 10$ and $y = 20$ feet to his right along the wall is 0.1024. This is an example of an important probability model, called the *Cauchy law*.

Since our answer is expressed as the difference between two values of a function, we can use the fundamental theorem of calculus, $g(b) - g(a) = \int_a^b g'(x)dx$, to rewrite the Cauchy law. Remember from calculus that $(d \tan^{-1}(z))/dz = 1/(1 + z^2)$. Therefore,

$$P(\text{between } x \text{ and } y|\text{hits wall}) = \int_x^y \frac{dz}{\pi[(1 + \{(z - m)/d\}^2)]}.$$

This may seem a peculiar thing to do, but notice that the expression under the integral sign, the *density* again, does not involve the transcendental arc tangent function. It is in a sense simpler when written this way. In the case $m = 0$ and $d = 1$, the Cauchy density function looks like $f(z) = 1/(\pi(1+z^2))$, and its graph looks like the graph in Figure 4.8.

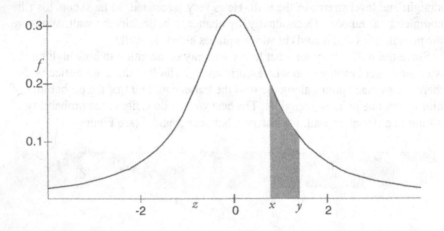

FIGURE 4.8. The Cauchy density

The shaded area is the probability that a bullet will hit between the points x and y along the wall if it hits the wall at all. We will discover later many other uses for densities.

4.8.2   Sigma Algebras and Borel Algebras*

It is time to tackle the problem of what sort of algebra of events we need for geometric-based probability problems. This has become a more important question, because now we know how to tackle geometric problems whose probabilities are not necessarily uniform. We will do it by analogy with how areas are found.

Remember that when studying probabilities of an outcome falling along a line, we are usually interested in the probabilities of it falling in *intervals*. These are, after all, the sets whose lengths are easy to measure (an interval (a, b) has length $b - a$). So we need an algebra that incorporates our idea that we need events on the line based on intervals. By custom, statisticians start building their events in one dimension by insisting that all half-open intervals $(a, b]$ (which include the point b but exclude a) are events.

But that may not be enough intervals to satisfy us. Is the entire line $(-\infty, \infty)$ an event? It would seem relevant in Cauchy probability spaces, for example. We could build the line out of our half-open intervals in the following way: $(-\infty, \infty) = (-1, 1] \cup (-2, 2] \cup \cdots \cup (-k, k] \cup \cdots$. That is, we combine bigger and bigger intervals until, somewhere, every real number is included. Unfortunately, in our definition of algebras of sets, we did not say that you necessarily included such an infinite, but *countable*, union of events.

Furthermore, are single numbers, like $\{b\}$, events in geometric probability problems? It seems silly not to include them; they have a known area (zero). Imagine that the following (countably infinite) intersection is an event:

$$(b - 1, b] \cap (b - \tfrac{1}{2}, b] \cap (b - \tfrac{1}{3}, b] \cap \cdots (b - \tfrac{1}{2}, b] \cap \cdots .$$

Obviously, b is in this event. Also obviously, any number $c > b$ is not in this event. Now think about any number $c < b$. Then $b - c$ is a positive number, and I can always find an integer n big enough that $1/n \le b - c$. So $c \le b - 1/n$, and c is not in the interval $(b - 1/n, b]$. So c is not in the infinite intersection event. We conclude that b must be the only point in that event. So we could argue that a point is indeed an event, if only *countable* intersections of events were necessarily events.

The same approach may be used to assign probabilities on the plane. We start with events that are certain *rectangles*, because the definition of area starts with that of a rectangle. Again, we conventionally start by declaring that all rectangles $(a, b] \times (c, d]$ for any numbers $a < b$ and $c < d$ are events (see Figure 4.9).

In p-dimensional space we include all hyper-rectangles $\times_{i=1}^{p} (a_i, b_i]$. (Can you figure out this fancy notation?)

Then if we want to find the probability of an irregular area, we might partition the conditioning event with a grid of rectangles. The dark line bounds an event of

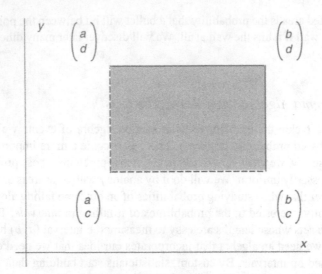

FIGURE 4.9. A rectangle in the plane

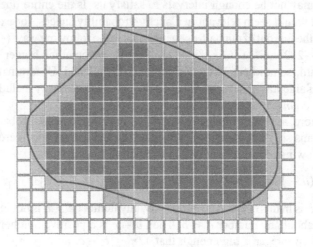

FIGURE 4.10. Approximating an irregular region

interest (Figure 4.10). The probability of the event of interest could then be calculated rectangle by rectangle from the division-into-cases formula. Much more easily, we can get a *lower* limit on the probability by simply summing the probabilities of those (darkly shaded) rectangles that are entirely within the event. Then we can get an *upper* limit on the probability by summing the probabilities of those rectangles (shaded at all) that intersect the event in any way. With ever smaller rectangles, we could then pin down the probability as accurately as we wish. But the lower limit corresponds to a countable union of an ever-growing combination of rectangles, and the upper limit to an ever-shrinking countable intersection.

We will decide that we always want to be able to do things like this, so we strengthen our definition of an algebra from Section 3.2:

Definition. A σ-**algebra (sigma algebra)** is an algebra of events such that if $\{A_i\}$ is a countable collection of events, then $\bigcup_i A_i$ is also an event (*countable unions*).

Proposition. *If* $\{A_i\}$ *is a countable collection of events in a* σ*-algebra, then* $\bigcap_i A_i$ *is also an event.*

You should check this as an exercise. This makes no difference to equally likely probability spaces and to discrete probability spaces, of course. In those examples, all subsets of a large set were events, so we certainly had a σ-algebra.

Now we are ready to apply this to real numbers.

Definition. The **Borel algebra** on the real line is the smallest σ-algebra that contains all the intervals of the form $(a, b]$.

By "smallest" we mean that there are no extra events; we have, of course, the events that can be gotten by applying the σ-algebra rules (complements and unions) to the half-open intervals. Furthermore, if we remove any events, either some of them will be those we can build out of half-open intervals (which is bad) or we will discover that we no longer have a sigma algebra.

Proposition.

   (i) *Any single point* $\{b\}$ *is an event.*
  (ii) (a, b), $[a, b]$, *and* $[a, b)$ *are events.*
 (iii) *The entire line as well as all possible half-lines (*$[a, \infty)$*, etc.) are events.*

The point and the line we already took care of. The rest are exercises.

The last several paragraphs claim that to assign probabilities on the real line, all we need to be able to do is assign probabilities to intervals. Thus, the formula we derived for the hit probability for any stretch of the Great Wall of China potentially tells us anything we want to know about hit probabilities.

Definition. The Borel algebra on the plane is the smallest σ-algebra that includes the rectangles $(a, b] \times (c, d]$ for any numbers $a < b$ and $c < d$, and the Borel algebra in p-dimensional space is the smallest σ-algebra that includes the hyper-rectangles $\times_{i=1}^{p} (a_i, b_i]$.

So now our probability spaces whose outcomes are in several dimensions can potentially tell us how probable all sorts of irregular areas are.

4.8.3 Kolmogorov's Axiom*

When we restrict the idea of probability space to σ-algebras, does that have any consequences for computing probabilities? Presumably, we must be able to calculate the probabilities for those new events imposed on us by the requirement of countable unions and intersections. In each of our examples of a union in the last

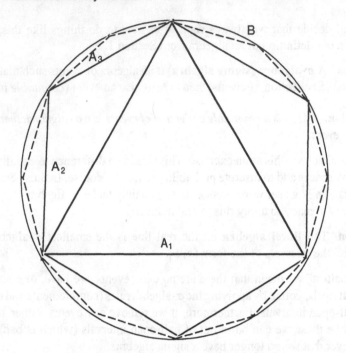

FIGURE 4.11. Polygons approximating a circle

section, we defined a growing sequence of events $A_1 \subset A_2 \subset \cdots \subset A_k \subset \cdots$ whose union was the event of interest, $\underset{k}{\cup} A_k = B$ (see Figure 4.11). A common notation for this union of a list of growing sets is $\lim_{k \to \infty} A_k = B$.

Obviously, the probability of B should be the limit of the probabilities of the events A_k.

Definition. Kolmogorov's axiom states that for a countable sequence of events $A_1 \subset A_2 \subset \cdots A_k \subset A_k \subset \cdots$, $P\left(\underset{k}{\cup} A_k | C\right) = P\left(\lim_{k \to \infty} A_k | C\right) = \lim_{k \to \infty} P(A_k | C)$.

Example. Let me check this for the probability of an odd number of hurricanes. Let $A_1 = \{1\}, A_2 = \{1, 3\}, A_3 = \{1, 3, 5\}$, and so forth; this is clearly an increasing sequence of sets. The limit of the A_k's is the event of an odd number of hurricanes. From calculus, you might remind yourself about the sum of a *finite* geometric series; this says that $P(A_k) = \frac{1}{4}(1 - (\frac{1}{4})^k)/(1 - (\frac{1}{4}))$. Then $\lim_{k \to \infty} P(A_k) = \frac{1}{3}$, which matches our earlier result.

The new axiom is then obviously true for equally likely probability spaces, because any union of events is only a union of a finite number of events. It is also clearly true for discrete probability spaces: We find ourselves adding an always-convergent countable sum of those probabilities p_j in order to take any such limit. It is certainly true for uniform geometric probability problems: The axiom imitates a valid way of computing areas of events by filling the region up from inside.

We are now ready to amend the definition of a probability space to include countable unions and a sensible rule for computing their probabilities:

Definition. A **probability space** meets conditions i–iv for a finitely additive probability space, and further the set of events form a σ-algebra and (v) Kolmogorov's axiom holds.

All our theorems about probability in general are still true, because we have only placed new restrictions on possible probability spaces.

The calculation in the example suggests that we may now be able to generalize the proposition about finite additivity (see 6.2). Consider a *countable* collection of events $\{A_i\}$ that are mutually exclusive, $A_i \cap A_j = \phi$. Let $B_1 = A_1, B_2 = A_1 \cup A_2$, and generally $B_k = \bigcup_{i=1}^{k} A_i$. Then $B_1 \subset B_2 \subset B_3 \subset \cdots$, and $\lim_{k \to \infty} B_k = \bigcup_i A_i$. Finite additivity says that $P(B_k) = \sum_{i=1}^{k} P(A_i)$, and so using Kolmogorov's axiom,

$$P(\bigcup_i A_i) = P(\lim_{k \to \infty} B_k) = \lim_{k \to \infty} P(B_k) = \lim_{k \to \infty} \sum_{i=1}^{k} P(A_i) = \sum_{i=1}^{\infty} P(A_i).$$

Proposition (countable additivity). *Consider a countable collection of events* $\{A_i\}$ *that are mutually exclusive,* $A_i \cap A_j = \phi$. *Then* $P(\bigcup_i A_i | C) = \sum_{i=1}^{\infty} P(A_i | C)$.

Now we can state more general versions of other things in Section 6. A countable partition is just one with a countable list of events in it, and the theorem on division into cases and Bayes's theorem for partitions are true as well for these countable partitions.

You may well be wondering why we bothered to go back and require probability spaces to be σ-algebras and to obey Kolmogorov's axiom. After all, each of the types of probability we discussed—equally likely, discrete, geometric—already meet these restrictions. The problem is that we can invent some finitely additive probability spaces that do not. Imagine a probability space whose outcomes are all the nonnegative integers, but where the events include only finite sets of integers. Define probability as in the equally likely case, by counting: $P(A|B) = |A \cap B|/|B|$ when B is not empty. This space meets all axioms (i)–(iv), so we might imagine that it is a perfectly reasonable probability space. However, the set of events is obviously not a σ-algebra: We can piece together by countable union events with an *infinite* number of members and so cannot calculate probabilities involving them from our definition.

Should this strange space be allowed to be a probability space? Probabilists are not in general agreement. Some would say yes, because mathematical uses have been found for it. Others point out that it is quite impossible to imagine any experiment that would lead to these probabilities, even approximately—there are just too many integers to have them all be equally likely. You may see both points of view in advanced courses. We will choose to keep Kolmogorov's axiom for the rest of this book, since we emphasize here experiments that one can actually carry out.

4.9   Summary

In this chapter we analyzed certain geometrical experiments, using *uniform geometric* probability, which says that if the outcome is any point in a region, all equally likely, then $P(A|B) = V(A \cap B)/V(B)$ (2.1). Then we gave general rules for what sorts of sets *events* in any probability problem must be: if A and B are events, then so are $A \cup B$ and $A - B$. They will then belong to an *algebra* of events (3.2). Next we stated a short list of *axioms* that all probability models must follow; the ones that tell us how to calculate are $P(A \cup B) = P(A) + P(B - A)$ and $P(A \cap B) = P(A) \cdot P(B|A)$ (4.2). Then we demonstrated new sorts of probability that meet our axioms, such as *discrete* probability spaces. In this case, $P(A) = \sum_{i \in A} p_i$, where the p's are probabilities of individual outcomes (5.2).

From these rules we extracted several useful formulas, such as the *division into cases* formula $P(A|B) = \sum_i P(C_i|B) \cdot P(A|C_i)$, where $\{C_i\}$ *partition* B (6.2). Then we derived the famous *Bayes's theorem* $P(C_i|A) = P(C_i)P(A|C_i)/\sum_j P(C_j) \cdot P(A|C_j)$ (6.4). When certain conditions turned out to be unimportant to the probability of an event, we concluded that the events must be *independent* of each other, which simplified such calculations as $P(A \cap B) = P(A) \cdot P(B)$ (7.1). Then we explored more general geometric probability problems, which suggested the important idea of a probability *density*, a function f such that

$$P(\text{outcome between } a \text{ and } b) = \int_a^b f(x)dx \quad (8.1).$$

It turned out that geometrical probability problems required us to invent the *Borel algebra* of events, which essentially says that geometric events have length, area, or volume. These algebras are *sigma* algebras, which include countable unions of events (8.2), and we need an additional axiom, *Kolmogorov's axiom*, $P(\bigcup_k A_k|C) = \lim_{k \to \infty} P(A_k|C)$ whenever $A_1 \subset A_2 \subset \cdots \subset A_k \subset \cdots$, to compute necessary probabilities (8.3).

4.10   Exercises

1. Prove the six properties of uniform geometrical probability.
2. List all the events that could conceivably be built out of the collection of outcomes $\{1, 2, 3, 4, 5\}$.
3. Prove that if A and B are events, then $A \cap B$ is also an event.
4. If $\{2, 3\}$, $\{3, 4\}$, and $\{4, 5\}$ are events in an algebra, prove (that is, convince me, using only the definition) that $\{3, 4, 5\}$ must also be an event in that same algebra.
5. You are playing a game in which you toss two coins, and if they *both* land heads, you win. A friend who is watching has a side bet with someone else that she will win if *at least one* of your coins lands heads. You toss the coins, but they roll behind a chair. Your friend races ahead of you, looks behind the

chair, sees both coins, and announces "I won!" What is now the probability that you will win?

6. Prove that $P(A \cap B \cap C|D) = P(A|D) \cdot P(B|A \cap D) \cdot P(C|A \cap B \cap D)$.

7. Prove that the multiplication axiom $P(A \cap B|C) = P(A|C) \cdot P(B|A \cap C)$ whenever $A \cap C \neq \phi$ is always true for a discrete probability space.

8. Prove that if you have an equally likely rule for probabilities on some set of possible results C (that is, all probabilities are gotten by counting), then that probability rule is also an example of a discrete probability space.

9. Prove that if $A \subset B$, then $\{A, B - A\}$ is a partition of B.

10. Prove that always $P(A|B) = P(A \cap B|B)$.

11. In the AIDS example (see Section 6.3), find the probability that a patient has HIV, given that the patient has tested negative on the blood test.

12. As a safety officer in a chemical plant, you test the air once a day for very small amounts of H_2S (hydrogen sulfide). You can tell how many of your three vats are out of adjustment and so producing the gas, but not which ones. The old vat is out of adjustment 5% of the time, the year-old vat is out 10% of the time, and the new vat is out 20% of the time. There is no connection among the three vats.

   a. What is the probability that exactly one vat is out of adjustment on a given day?

   b. This morning you detected the gas, enough to conclude that exactly one of the vats is out of adjustment. What is the probability that the new vat is at fault?

13. Five of the 23 people in your mechanics class are left-handed. A woman from the dean's office wants to interview one of the left-handed students about how well the left-handed desks in the room work.

   a. She talks to people as they leave the class, until one of them is left-handed. What is the probability she will have talked to more than six people?

   b. Furthermore, seven of the 28 people in your electronics class are left-handed. All you know is that the woman interviewed people in one of the two classes, but she tells you that it took her 4 interviews to find her left-hander. What is the probability it was the electronics class she was talking to?

14. You ship off your motorcycle to be sold at a used motorcycle fair. Unfortunately, you ship it at the last minute, on a standby basis. The shipper estimates a 35% chance that it will get there in time for the Saturday show, a 41% chance that it will arrive only in time for the Sunday show, and a 24% chance that it will arrive too late for the fair. Your experience with this fair is that there is a 28% chance that your motorcycle will sell on Saturday, if it has arrived. There is only a 15% chance that it will sell on Sunday, if it is there to be sold on Sunday.

   a. What is the probability that you will sell your motorcycle?

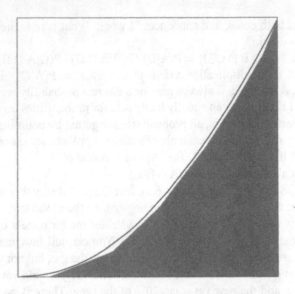

FIGURE 4.12. Exercise 15: Under a parabola

   b. You get word that your motorcycle was not sold. What is the probability
   that it arrived too late for the fair?

15. Let a random point be chosen uniformly on the unit square $(0, 1) \times (0, 1)$.
    What is the probability the point will land under the parabola $y = x^2$? (See
    Figure 4.12.)

16. Show that $\bigcup_{i=1}^{\infty} (1/(i + 1), 1/i] = (0, 1]$.

17. Prove that if $\{A_i\}$ is a countable collection of events in a σ-algebra, then $\bigcap_i A_i$
    is also an event.

18. Prove that in the Borel algebra on the real line, $[a, b]$ and $[a, b)$ are events.

19. Prove that in the Borel algebra on the real line, (a, ∞), $[a, \infty)$, $(-\infty, b]$, and
    $(-\infty, b)$ are events.

20. Prove that the entire plane is a Borel event. Prove that $[a, \infty) \times (-\infty, b]$ is
    a Borel event.

21. Let random outcomes be uniformly distributed (just as likely to hit anywhere)
    over the rectangle $(0, 3] \times (0, 2]$, with coordinates of the hit point (x, y) (see
    Figure 4.13). Consider any vertical strip A with $0 < a < x \le b < 3$ and
    any horizontal strip B with $0 < c < y \le d < 2$. Prove that the event of an
    outcome in A is independent of the event of an outcome in B.

4.11 Supplementary Exercises

22. List all the events in the smallest algebra of sets that contains the events
    $\{1, 2, 3\}$ and $\{2, 3, 4\}$.

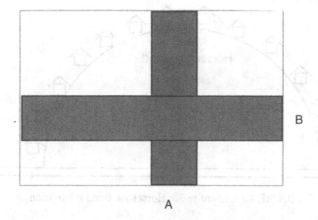

FIGURE 4.13. Exercise 21: Independent strips

23. Prove that for any events A and B (with a common condition),

$$P(A) + P(B - A) = P(B) + P(A - B).$$

24. You have a box with 8 nine-pin patch cords and 5 twelve-pin patch cords mixed up in it. You remove two patch cords at random from the box.

a. What is the probability that the two cords will have the same number of pins?

b. If, fortunately, your two cords do have the same number of pins, what is the probability that they are nine-pin cords?

25. A company makes three nut mixes in very similar cans: One is all peanuts, one is $\frac{1}{3}$ cashews and $\frac{2}{3}$ peanuts, and one is $\frac{2}{3}$ cashews and $\frac{1}{3}$ peanuts. A friend (who never looks at prices in the store) is equally likely to buy all three mixes. One evening you go to her house, sit down on the sofa, and take a nut from the can on the coffee table. It is a peanut.
What is the probability that the can is all peanuts?

26. Of middle-aged men who come to a clinic complaining of chest pain, 75% have heartburn, 20% have angina, and 5% have had a mild heart attack (the doctor records only the most important source of the pain. Other problems are too rare to be significant). It is then usual to take an EKG, which records heart activity. In 90% of heartburn cases, the EKG is normal. In 70% of angina cases, it is also normal. However, in mild heart-attack cases, only 20% of EKGs are within normal limits.

a. What is the probability that the next middle-aged male complaining of chest pain will have a normal EKG?

b. A 50-year-old man arrives at the clinic, reporting chest pain. His EKG is notably **abnormal**. What is the probability that he has had a mild heart attack?

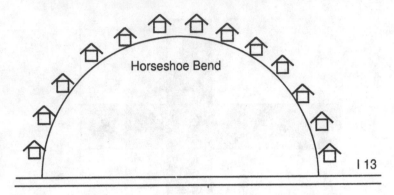

FIGURE 4.14. Exercise 30: Horseshoe Bend subdivision

27. The odds ratio is sometimes a useful way to write probabilities: If A and $\overline{A}$ are a partition of a general condition C, then define $O_C(A|B) = \frac{P(A|B \cap C)}{P(\overline{A}|B \cap C)}$. (As shorthand, we write $O_C(A|C) = O_C(A)$.)

   a. Write $P(\overline{A}|B \cap C)$ in terms of $P(A|B \cap C)$.
   b. The odds form of Bayes's theorem may be written $O_C(A|B) = O_C(A)K$, where K, a ratio of probabilities, is called the *Bayes factor* for the observation B. Derive a simple expression for K, using Bayes's theorem.

28. a. In Exercise 26, compute the odds ratio that a 50-year-old man complaining of chest pain has actually had a heart attack.
   b. Find the Bayes factor (Exercise 27) provided by the knowledge that this man has an abnormal EKG. Use it to compute the probability that he has had a heart attack. Verify that your answer is consistent with the answer to Exercise 26(b).

29. Let $\{B_i\}$ be a partition of C. Assume that for an event A that is a subset of C, you know all probabilities $P(A|B_i)$ and $P(B_i|A)$. Derive a formula for $P(A|C)$ that uses only these known probabilities.

30. You are shopping for a house. You read in the newspaper that a house is available in Horseshoe Bend, a subdivision of a great many houses spread uniformly along a semicircular street off a very noisy freeway (Figure 4.14). Obviously, the sites become more valuable as you move away from the noisy freeway. The semicircle has radius one kilometer.

   a. Find a formula for the probability that the house in the newspaper (which may be anywhere in the subdivision) is between a and b kilometers from the freeway ($0 \le a < b \le 1$) as the crow flies (in a straight line).
   b. Find the probability density for the distance of that house from the freeway.

CHAPTER 5

Discrete Random Variables I: The Hypergeometric Process

5.1 Introduction

You will have gathered from the first two chapters that the usual grist for the statistician's mill is data, in particular, numerical data (and often lots of it). Yet Chapters 3 and 4 wandered into the subject of probability, and even though many of the examples were from the practice of statistics, the connection may have been unclear. In this chapter we will study random experiments in which the outcomes are numbers. In other words, we will develop probability models to try to explain the variability in many sets of numerical scientific data. Quantitative outcomes to probabilistic experiments will be called *random variables*, a concept that pervades statistics. We will introduce some important families of interrelated random variables that have been found to be good descriptions of the outcomes of experiments. In this chapter we concentrate on families that arise in sampling from finite populations of subjects.

Of course, the interest in having numerical data is that we may construct useful arithmetic summaries. We will introduce the idea of the average value of a discrete random variable, called an *expectation*. Very often, too, the goal of an experiment will be to learn more about just which random variable best describes an experiment. We will begin to develop methods of *testing* and *estimation* designed to answer such questions.

Time to Review

Chapter 1, Section 7
Summing infinite series

5.2   Random Variables

5.2.1   Some Simple Examples

Definition. A **random variable** is a probability space whose outcomes are real numbers. Its **sample space** S is the collection of all possible outcomes of that random variable.

Example. (1) If you poll 100 randomly chosen voters to discover their presidential preference, one random variable of interest is the number who will say they support your candidate. The sample space is the set of integers between 0 and 100 inclusive.

(2) In studying a dangerous epidemic disease, doctors in an emergency room take the oral temperature of each patient who arrives. The temperature in degrees Celsius of the next patient is a random variable, and its sample space might conceivably be any real number higher than -273 (absolute zero).

(3) There are 7 bird's nests of the same species in a large tree. A biologist finds a hatchling on the ground at the base of the tree. How many nests will she have to check to find where the hatchling came from?

In other textbooks you may encounter a more sophisticated definition of random variable, in which the "sample space" is instead the set of all outcomes of an experiment, and the random variable is then a real-number-valued *function* defined on that set. These texts do this to be consistent with advanced graduate texts in mathematical probability. Since the distinction makes no difference in how we use the concept in this book (and very little difference in any case), we will use the simpler definition. We will use capital italic letters like X for random variables; we will think of the random variable as taking on a value as a result of the experiment, which justifies notation like $P(X = x | A)$, where x is one particular possible value that we are curious about.

In the first two experiments, we would have to know a lot more to be able to assign probabilities, but the third example is easy. Place W white marbles and a single black marble in a jar and shake well. Remove one marble at a time, without replacement, until you find the black marble. The number of white marbles you have removed is a random variable, and its sample space is $\{0, 1, 2, \ldots, W\}$. All the possible *hypergeometric processes* (see 3.3.3) are given by the case where the black marble comes first, the case where it comes second, and so on to the case where it comes last. It is reasonable to assume that these cases are equally likely, and there are $W + 1$ of them, so $P(X = x | x \in \{0, \ldots, W\}) = \frac{1}{W+1}$. This is an example of a uniform random variable:

Definition. A **(discrete) uniform** random variable is a random variable with finite sample space, each of whose outcomes is equally likely.

Example ((3) *cont.*). The hatchling problem is equivalent to a hypergeometric process with one black marble and 6 white marbles. The number of nests checked without locating the right one is a discrete uniform random variable as in the example above; the sample space is 0 to 6, and the probability of each value is $\frac{1}{7}$.

5.2.2 Discrete Random Variables

Of course, the various possible outcomes of an experiment need not be equally likely.

Example. A chain of 10 dry-cleaning stores has been robbed repeatedly, so the owner hires three security guards and hides them in three randomly chosen stores. If a robber tries to hold up a series of these stores, how many successes will he have before a security guard interrupts his career? Assuming that he has no idea where the guards are hidden, we calculate (let U mean Unguarded and G mean Guarded)

$$P(X = 0) = P(\text{first store guarded}) = P(G) = \frac{3}{10}.$$

$$P(X = 1) = P(\text{first store unguarded, second guarded} = P(UG)$$
$$= P(U) \cdot P(G|U) = \frac{7}{10} \cdot \frac{3}{9}.$$

$$P(X = 2) = P(UUG) = P(U) \cdot P(U|U) \cdot P(G|UU) = \frac{7}{10} \cdot \frac{6}{9} \cdot \frac{3}{8},$$

and so forth, until

$$P(X = 7) = \frac{7 \cdot 6 \cdot 5 \cdot 4 \cdot 3 \cdot 2 \cdot 1 \cdot 3}{10 \cdot 9 \cdot 8 \cdot 7 \cdot 6 \cdot 5 \cdot 4 \cdot 3}.$$

Again, there is an urn model for problems like this. In an urn with W white marbles and B black marbles, let X be the number of white marbles drawn without replacement before the first black marble is encountered. In our example, $W = 7$ and $B = 3$. Generally X is a random variable with sample space $\{0, \ldots, W\}$. The calculations above become

$$P(X = x) = \frac{W \cdot (W - 1) \cdot \cdots \cdot (W - x + 1) \cdot B}{(W + B) \cdots (W + B - x + 1) \cdot (W + B - x)}$$
$$= \frac{(W)_x}{(W + B)_{x+1}} B,$$

taking advantage of permutation notation.

With this random variable the probabilities of different numerical outcomes are not all the same, so it is an example of a discrete random variable

Definition. A **discrete** random variable is a discrete probability space (see 4.5.1) whose universe U is a set of real numbers (so that the sample space of the random variable S is equal to U).

Any discrete uniform random variable is also a discrete random variable. And in the example above of the number of white marbles found before the first black marble is encountered, we found that $U = \{0, \ldots, W\}$ and that $p_i = (W)_i/(W + B)_{i+1} B$. We think of p_i as a function of the corresponding value $x_i = i$ of the random variable.

Definition. The **probability mass function** (or **probability distribution function**) of a discrete random variable is $p(x) = P(X = x)$.

Therefore, $p(x_i) = p_i$.

It is sometimes convenient to organize the facts about a discrete random variable into a table:

x	0	1	2	3	4	5	6	7
$p(x)$	0.3	0.2333	0.175	0.125	0.0833	0.05	0.025	0.0083

This is the Table for the laundry-guarding problem. These tables are traditionally presented with the values of x in ascending order.

Proposition. (i) *For all $x \in S$, $p(x) \geq 0$; and*
(ii) $\sum_{x \in S} p(x) = 1$.

These assertions just restate the corresponding properties of discrete probability spaces.

5.2.3   The Negative Hypergeometric Family

Our next, more general, type of discrete random variable will turn out to be one of the most revealing, primarily because of its many ties to other variables.

Example. You have to get permission from your neighbors to build a fence around the back yard of your new house. There are 12 households, and 5 of them have a family member on the neighborhood council. You need to talk to a majority of those on the council, 3, in order to get permission. You have no idea where they live. What is the probability that you will have to visit only 4 houses to talk to that majority?

To model this problem, place W white marbles and B black marbles in an urn. Mix them up thoroughly and, then remove them one at a time without replacement until you have removed b black marbles (rather than just one, as in the preceding example). Then our random variable X will be the number of white marbles you have happened to remove along the way. In the example, call the houses with a council member black marbles and those without, white marbles. Therefore, $W = 7$, $B = 5$, and you must find $b = 3$ of them.

Definition. A **negative hypergeometric** (or **beta-binomial**) random variable $N(W, B, b)$ arises when all possible sequences of $W + B$ objects, W of the first kind and B of the second kind, are equally likely. The random variable X is the number of objects of the first kind that precede the bth object of the second kind in a given sequence.

Notice that we have described each of these variables with a notation $N(W, B, b)$ that gives each of the key quantities that determines how it arises. We call the negative hypergeometric variables a *family*, and the crucial numbers that tell you which specific one, W, B, and b, are called *parameters*. We already have two examples of this family: In the discrete uniform case when we were searching

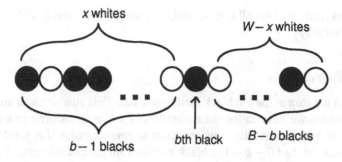

FIGURE 5.1. A negative hypergeometric urn experiment

for a single black marble, the number of white marbles found along the way was $N(W, 1, 1)$. When we were searching for the first black marble from W white ones and B black ones, the number of white marbles found was $N(W, B, 1)$. Any collection of related random variables whose members we single out by numerical indices will be a family.

The sample space of a negative hypergeometric random variable is obvious; no white marbles need precede our bth black marble, or all of them may. Therefore, S is the collection of integers in the range $0 \le X \le W$. Their probabilities may be computed by noticing that there are $\binom{W+B}{W}$ equally likely hypergeometric sequences (see 3.3.3). The ones that have X whites before the bth black may be counted by noting that among the first $X+b-1$ marbles we must distribute X white marbles; the $(b+X)$th marble must be black, and among the last $W+B-b-x$ marbles we must distribute $W-X$ white marbles. (See Figure 5.1.)

Therefore, we have established the following:

Proposition. *A negative hypergeometric* $N(W, B, b)$ *random variable has sample space* S *consisting of all integers in the range* $0 \le X \le W$ *and probability mass function*

$$P(X = x | W, B, b) = p(x) = \frac{\binom{x+b-1}{x}\binom{W+B-b-x}{W-x}}{\binom{W+B}{W}}.$$

You should verify that when we were looking for one black marble, the probability of each number of whites was $1/(W+1)$, by using this formula. Also verify that when we are looking for the first of B black marbles ($b = 1$), this big formula reduces to the simpler formula we derived for that case.

Example (*cont.*). In the quorum-search problem the question is, if the number of unsuccessful visits is negative hypergeometric, what is the probability of only $X = 1$ misses?

$$p(1) = \frac{\binom{3}{1}\binom{8}{6}}{\binom{12}{7}} = 7/66 = 0.106.$$

If the question is, how surprised should we be at so few unsuccessful visits, then we really want to know the probability of 1 or 0 misses: $p(0) + p(1) = 1/22 + 7/66 =$

0.152. This really was not all that surprising; if we get done that quickly, we were only a little lucky.

5.2.4   Symmetry

Notice that we stop at the bth black marble of a complete row of black and white marbles, which we have called the realization of a hypergeometric process (see 3.3.3). If we had laid out the same sequence in *reverse* order, that same marble would have been the $(B - b + 1)$st black marble from the end (the extra 1 appears because the stopping marble gets counted from either direction). In getting to it, we would have passed the other $W - X$ white marbles. But the probability of such a sequence is obviously exactly the same as the corresponding one in the original order (all sequences are equally likely). This lets us conclude a nice general fact:

Proposition (reversal symmetry).

$$P(x|N(W, B, b)] = P[W - x|N(W, B, B - b + 1)].$$

In the quorum problem, this is nothing more amazing than noticing that the probability of visiting one unnecessary house is exactly the same as of *not* visiting 6 unnecessary houses. This is an example of a *symmetry* in the family: Two probabilities from two family members can be demonstrated to be the same. This particular symmetry we will call *reversal symmetry*. If we are alert for these, they can help us avoid duplicate calculations. In fact, if there is an odd number of black marbles B, then $b = \frac{B+1}{2}$ is the middle black marble; then $b = B - b + 1$. We have

$$P\left[x|N\left(W, B, \frac{B+1}{2}\right)\right] = P\left[W - x|N\left(W, B, \frac{B+1}{2}\right)\right].$$

This is an example of a symmetry in a single random variable: The probabilities are the same as you look through the table from either end.

5.3   Hypergeometric Variables

5.3.1   The Hypergeometric Family

Looking at the hypergeometric process in a different way suggests another sort of random variable:

Example. Eight bottles of wine are submitted to two judges, who taste independently. Judge C picks the best three bottles, and Judge D picks the best four bottles. Since your bottle never does very well, you form the opinion that their choices are entirely capricious. If that is really so, what is the probability that their choices would have two bottles in common?

Imagine that Judge C surreptitiously puts a small white mark on the bottom of his winners that Judge D will not notice. If their judgments are indeed entirely capricious, then Judge D is picking his 4 at random and chanced to get two "white" bottles.

We can construct an urn model for this that will turn out to have many applications. Place W white marbles and B black marbles in an urn. Shake well and then reach in without looking and remove n marbles, without replacement. Then the unpredictable number X of white marbles you have removed is also a random variable. In our example, $W = 3$ (C's winners), $B = 5$ (C's losers), and $n = 4$ (D's winners); since two bottles with a white mark are the outcome we have asked about, $X = 2$.

Definition. A **hypergeometric** random variable with parameters $W + B$, W, and n is, given a set consisting of W elements of a first kind and B elements of a second kind, the number of elements of the first kind appearing in a randomly chosen subset of n elements, where every such subset is equally likely. We write $H(W + B, W, n)$.

How does this differ from the negative hypergeometric problem? In both cases, we remove the marbles from a jar in unpredictable order, stopping at some point to count white marbles. In the former case, we stop when we have found b black marbles. In the new, hypergeometric, case, we stop when we have removed a total of n marbles. We will see shortly a connection between their probabilities as well.

We need to determine the sample space of our new random variable. Obviously, $X \geq 0$. But notice also that if n is bigger than B, we may run out of black marbles, which places a higher minimum on the number of white marbles in our handful: $X \geq n - B$. In the same way, obviously $X \leq n$. But also there is a built-in limit to the number of white marbles in the handful, $X \leq W$. The sample space is the collection of integers that meets all four requirements.

The probability of a given outcome is easy to calculate, because we have done it before (see 3.4.1), the tea-tasting example. There are $\binom{W+B}{n}$ equally likely subsets. There are $\binom{W}{x}$ ways to get x white marbles and $\binom{B}{n-x}$ ways to choose the black marbles that make up the rest of your handful. We summarize these facts:

Proposition. *For a hypergeometric* $H(W + B, W, n)$ *random variable* X:

(i) *the sample space* S *is the set of integers that meet* $\max\{0, n - B\} \leq X \leq \min\{n, W\}$; *and*

(ii) *the probability mass function is* $P(X = x|H(W + B, W, n)) = p(x) = \binom{W}{x}\binom{B}{n-x}/\binom{W+B}{n}$.

The **max** function chooses the larger of the listed values (since X has to be bigger than both numbers); in the same way, **min** chooses the smaller.

Example (*cont.*). We can use this formula to solve the wine-judging problem with $W = 3$, $B = 5$, $n = 4$, and $x = 2$:

$$P(X = 2|H(8, 3, 4)) = \frac{\binom{3}{2}\binom{5}{2}}{\binom{8}{4}} = \frac{3}{7}.$$

If the two judges do choose two (or more) bottles in common, that is little evidence against your opinion that their choices are capricious. It would happen very often just by accident.

5.3.2   More Symmetries

Notice that of the $W + B - n$ marbles that get left behind in the jar, $W - X$ are white. But leaving marbles behind is just as good a way of selecting them as removing them is, as we noticed in some of our sampling problems. We express this as a formula:

Proposition. $P[x|H(W + B, W, n)] = P[W - x|H(W + B, W, W + B - n)]$,

which is a fundamental symmetry of the hypergeometric family; this is another instance of reversal symmetry. If the n marbles we remove are exactly half the marbles, then both sides describe the same random variable, which is therefore symmetric.

The hypergeometric family has a completely different sort of symmetry, as well. Our sampling process may be thought of as a cross-classification of the marbles: We are looking at all the possible ways of dividing the marbles into two groups, white and black. We are also at the same time classifying all the marbles into the two groups, sampled and unsampled:

	White	Black	total
Sampled	X	$n - x$	n
Unsampled	$W - X$	$B - n + X$	$W + B - n$
total	W	B	$W + B$

Notice that in this way of looking at it, we might just as well have picked out the ones to sample first, and which were to be painted white second. It is still the probability of the same table, in which we happen to have interchanged rows and columns, like taking the transpose of a matrix. We call this *transpose* symmetry and state it precisely:

Proposition (transpose symmetry).
$P[x|H(W + B, W, n)] = P[x|H(W + B, n, W)]$.

This corresponds to the obvious fact that in the wine-judging problem we could just as well have had judge D go first and mark his winners with white paint; the probability of what happened would still be the same, because the judges do not consult one another.

5.3.3   Fisher's Test for Independence.

We illustrated transpose symmetry with a two-by-two contingency table to display our results. You may remember from Chapter 1 that we were interested in models for the counts in such tables, and you are no doubt curious about any connection with hypergeometric experiments. Notice that the probability that any given marble appears in the sample, $n/(W + B)$, is the same whether the marble is black or white. Therefore, the hypergeometric experiment assumes *independence* of the two ways of classifying marbles: black–white and sampled–unsampled. If X were improbably large (or small), this would cast doubt on the appropriateness of the hypergeometric random variable, and therefore on the assumption of independence. This is essentially how we reasoned in the wine-tasting example.

Example. Mann in 1981 reported a survey in which incidents of a person threatening suicide by jumping from a tall building were recorded; it was noted whether or not the threat occurred during the summer months, and whether or not there was jeering or baiting of the subject by a crowd. A natural question was whether or not summer weather was associated with baiting behavior.

	Baiting	None	total
Summer	8	4	12
Other	2	7	9
total	10	11	21

We might reason as follows: If independence of the season and crowd behavior hold, then the results might have arisen by marking the 21 incidents as either summer or other, then choosing 10 of those incidents completely at random to have crowd baiting occur. Then X is the number of summer incidents at which baiting happened, and it is an $H(21, 12, 10)$ variable. To check how improbable our observation is, we compute the probability that there would have been 8 *or more* summer incidents with baiting. (We would have been even more surprised at the seasonal association if there had been 9 or 10 summer incidents):

$$P(X \geq 8) = p(8) + p(9) + p(10) = 0.0505 + 0.0056 + 0.0002 = .0563.$$

Our results were moderately improbable but could conceivably have arisen by accident. We take this as some evidence that independence does not hold and summer is associated with more baiting, but we would like a bigger survey in order to be sure.

This style of analysis of independence models for two-by-two contingency tables is called **Fisher's exact test**. Transpose symmetry promises us that it does not matter which we called the row classification and which we called the column. You may have noticed that we used a peculiar line of reasoning. Those statisticians whom we have called *frequentists* calculate the probabilities of various outcomes before they do an experiment; afterward, they compare those probabilities to what actually happened and come to conclusions. But in this example we calculated our probabilities using as parameters the marginal totals 21, 12, and 10, which of course we do not know until we do the experiment. It is as if we proceeded instead to do the experiment, then had an assistant tell us only the marginal totals. We calculate the probabilities of various complete outcomes, then look up the complete results and compare. Such a procedure is called *conditional inference*, because we calculate probabilities conditioned on partial information about the results. This is a bit controversial but is nevertheless plausible enough to be widely accepted. There was no difficulty with the wine-tasting experiment, because the marginal totals, the number of good bottles to be chosen by each judge, could be specified in advance.

5.3.4 Hypothesis Testing

Each of our examples of frequentist reasoning has followed a pattern. We start with a claim that might reasonably be made about how an experiment will work. This is conventionally called the *null hypothesis* about that experiment, independence of two ways of classifying is an important example. Then we look at the actual result and calculate the probability that the observed value, or some value casting even more doubt on the null hypothesis, would have happened. This probability is traditionally called the *p-value* for that hypothesis (in our suicide-baiting example it was 0.0563). If it is disturbingly small, so that we are uncomfortable calling our result an accident, we say that we *reject* the null hypothesis, and we report our experiment as evidence against it. In effect, the experimenter is saying that what happened was too much of a coincidence to be believed.

Scientists do not like to leave it up to the judgment of the individual experimenter whether to call a *p*-value disturbingly small. Conventions about when a probability is small have been adopted by the scientific community; the single most common one says that less than 0.05 will be generally accepted as fairly small. As a practical consequence, this means that about one in every twenty published sensible statistical experiments to test perfectly sound hypotheses will wrongly reject those hypotheses. But scientists know that they will sometimes be wrong and have decided to tolerate such error rates. The number 0.05 is called a *significance level*; if the *p*-value is less than that, we say that we *reject* the hypothesis at the 0.05 level of significance. If, as in our example, *p* is larger than 0.05, we simply say that we *fail to reject* the null hypothesis.

The value 0.05 is, of course, quite arbitrary. More stringent communities of scientists often demand significance levels of 0.01, or even 0.001. As we will see, this means that we need ever bigger experiments to have any hope of detecting deviations from hypotheses.

5.3.5 The Sign Test

Now we can do a probability-based test of a simple contingency-table model from Chapter 1. Can we test some of the models for measurements from the same place? Really satisfactory tests will have to wait quite a while, but it is possible to turn certain questions into questions about contingency tables. For example, if we have two levels of treatment and wish to decide whether they are really different, we may reason as follows: Split the sample into those above the *sample median* (see Exercise 1.17) of all measurements and those below the sample median. The result is a two-by-two contingency table.

Example. Exercise 2 from Chapter 1 quoted 24 DBH levels of psychotic and nonpsychotic patients collected by Sternberg. The sample median of the DBH levels is between 0.0200 and 0.0204, so we get counts as in the following table:

	below median	above median	total
psychotic	1	9	10
nonpsychotic	11	3	14
/bf total	12	12	

(If there is an odd number of observations, use any rule of thumb to split them unevenly.) Now, if there is no relationship between the two groups and the quantity being measured, we may imagine that the observations have been arbitrarily assigned to the above and below groups. Therefore, the random variable X is the number in level 1 who chanced to be assigned to the below-median group, and it is hypergeometric: $H(n, n_1, n/2)$. I am sure that you see where this is going: We do a Fisher's exact test for independence in our artificial 2-by-2 table. If independence fails, the measurements may be concluded to be different between the two levels.

Example (*cont.*).   Let our significance level be 0.05, and ask whether the number of psychotics with below-median DBH is surprisingly small: $P(X \leq 1) = p(0) + p(1) = 0.00138$. This is so improbable that we conclude that psychotics tend to have higher DBH than nonpsychotics.

This procedure is called a **sign test** for the difference of two groups of measurements (because traditionally it is carried out by writing a $(+)$ next to each above-median observation and a $(-)$ next to each below-median observation, as an aid to counting them). It is usually classified as a *rank* test, like those based on the Kruskal–Wallis statistic (see 2.5.5). This is because we could have done it by ranking the observations, then counting those above and below the middle rank.

The sign test has the advantage of other methods based on ranks that it is unaffected by peculiarities of the scale of measurement, such as miscalibration. It has, even more than the Kruskal–Wallis statistic, the disadvantage that it may waste a great deal of information. A student would not be very well informed who knew only that she scored above the middle of her class on an important exam.

5.4   The Cumulative Distribution Function

5.4.1   Some Properties

We often find ourselves computing not just the probability that we get a certain value, but that as in the quorum search example we get *at most* a certain value. Therefore, we have given this quantity a name.

Definition. The **cumulative distribution function** $F(x)$ of a random variable X is the probability that the variable will achieve at most the specified value x, that is, $F(x) = P(X \leq x)$.

Example. For a discrete random variable, the cumulative distribution function may be displayed as a third row in the table. Then it is a running (cumulative) total of the probabilities in the second row. In the example of searching for a quorum, we have the following table:

x	0	1	2	3	4	5	6	7
$p(x)$	0.0455	0.106	0.1591	0.1894	0.1894	0.1591	0.106	0.0455
$F(x)$	0.0455	0.1515	0.3106	0.5	0.6894	0.8485	0.9545	1.0

For example, the number $F(2) = 0.3106$ in the third column is just $0.0455 + 0.106 + 0.1591$, the sum of the probabilities of getting 0, 1, and 2.

Computer statistical programs often provide commands that calculate the cumulative distribution functions of important families of random variables. Notice that the same table or function will answer questions about the probability of *at least* some value:

$$P(\text{at least 8 incidents}) = P(X \geq 8) = 1 - P(X \leq 7) = 1 - F(7).$$

Thus it is particularly handy for computing p-values, since there we want the sum of the probabilities of our result and also more extreme results.

Example. In the hurricane problem, (see 4.5.2) $p(0) = \frac{1}{2}$, $p(1) = \frac{1}{4}$, $p(2) = \frac{1}{8}$, and so forth; so $F(0) = \frac{1}{2}$, $F(1) = \frac{3}{4}$, and $F(2) = \frac{7}{8}$. As an exercise, show that for any x in the sample space, $F(x) = 1 - 1/2^{x+1}$.

Example. In the N$(W, B, 1)$ cases, where we were looking for the first black marble, $F(x)$ is the probability that we get at most x white marbles. But that is the same as the probability that we do *not* get at first $x + 1$ or more white marbles in a row. The probability of $x + 1$ or more white marbles before the first black is just the probability that the first $x + 1$ marbles are all white, which is $(W)_{x+1}/(W + B)_{x+1}$, as you might remember from one of our first permutation problems. We conclude that for this class of random variables,

$$F(X) = 1 - \frac{(W)_{x+1}}{(W + B)_{x+1}}$$

As an exercise, compare this calculation to the running total in our table for the laundry problem.

5.4.2 Continuous Variables

In the last chapter we discussed probabilities of points on the real line; if such points have coordinate *numbers*, then we have a random variable. In this case, the cumulative distribution function $F(x) = P(X \leq x)$ is the probability of an outcome falling in the left half-line, which we required to be an event in the Borel algebra. In the calculator-generated random number example (see 4.2.1), $F(x) = P(X \leq x) = P(0 < X \leq x) = x - 0 = x$ when $0 < x < 1$. This random variable, whose outcomes are any numbers in an interval and not just a discrete set, is our first example of a *continuous* random variable. Another is the following:

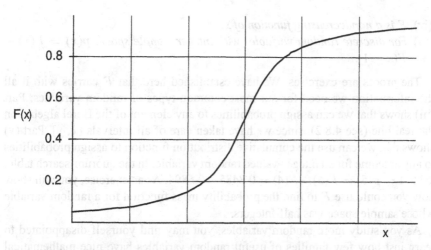

FIGURE 5.2. Cauchy cumulative distribution function ($m = 0, d = 1$)

Example. Let a random variable X be the coordinate of a bullet hole in the Great Wall of China problem in the last chapter (see 4.8.1). We found a formula for the probability that a hole would fall in any interval, so we can do the same for the half-infinite interval in the definition of the cumulative distribution function:

$F(x) = P(X \le x) = \frac{1}{2} + \frac{1}{\pi} \tan^{-1} \frac{x-m}{d}$, since as the point on the wall goes off to the left, to negative infinity, its arc tangent approaches $-\pi/2$. This function defines the Cauchy family of random variables, with parameters m and d (see Figure 5.2).

From the definition, we know that the height of this curve tells us the probability that X falls to the left of the point.

We pointed out that many problems of this type have *densities*, in this case,

$$F'(x) = f(x) = \frac{1}{\pi d(1 + \{(x-m)/d\}^2)}$$

is the density function for the Cauchy family.

From the last chapter (see 4.8.1), remember that $P(a < X \le b) = \int_a^b f(X)dX$, so the area under a piece of this curve gives us the probability that the variable will fall in that interval along the x-axis. In the preceding example, we had the relationship between the density and the cumulative distribution function $F(x) = \int_{-\infty}^x f(X)dX$. This is just the fundamental theorem of calculus, and so it holds quite generally for continuous random variables with densities.

We can make some general claims about cumulative distribution functions, which will hold both for discrete and for continuous random variables.

Proposition (properties of cumulative distribution functions).

(i) $\lim_{x \to \infty} F(x) = 1$.
(ii) $\lim_{x \to -\infty} F(x) = 0$.
(iii) $P(x < X \le y) = F(y) - F(x)$.

(iv) *F is a nondecreasing function of x.*
(v) *For discrete random variables with integer sample space,* $p(x) = F(x) - F(x - 1)$.

The proofs are exercises. We have established here that F carries with it all the information we need for our most common types of random variables: Part (iii) shows that we can assign probabilities to any element of the Borel algebra on the real line (see 4.8.2), since we have taken care of all intervals $(x, y]$. Part (v) shows that we can use the cumulative distribution function to assign probabilities to any outcome for an integer-valued random variable. In the quorum-search table, $0.1591 = p(5) = F(5) - F(4) = 0.8485 - 0.6894$. As an exercise, you will show how you could use F to find the probability mass function for a random variable whose sample space was half-integers.

As you study more random variables, you may find yourself disappointed to learn just how few families of useful random variables have nice mathematical expressions for their cumulative distribution functions, as several of our examples did. However, computer programs are widely available to compute a great many of these families of functions when we need them.

5.4.3   Symmetry and Duality

The cumulative distribution function will now allow us to find a useful connection between our deceptively similar families, hypergeometric and negative hypergeometric random variables. Such a connection between the probabilities in different families will be called a *duality*. Remember that the two families correspond to two criteria for stopping a search through a realization of a hypergeometric process (laying out a row of marbles on the table). Consider the statement that "at most x white marbles were found by the time the bth black marble was found"; this is exactly the same condition as "at most $b + x$ marbles were found by the time the bth black marble was found." But this is the same as "at least b black marbles were found in the first $b + x$ marbles," which is the same as "at most x white marbles were found in the first $b + x$ marbles." You may have to think about this for a while. The equations are as follows:

Theorem (positive–negative duality).

(i) $F[x|N(W, B, b)] = F[x|H(W + B, W, b + x)]$.
(ii) $F[x|H(W + B, W, n)] = F[x|N(W, B, n - x)]$.

Figure 5.3 shows how the theorem works: Any sequence of black and white marbles (bold path—'up' is a black marble, 'rightward' is a white marble) must cross the b(blacks) line and the $b + x$(total marbles) line on the same side of the x(whites) line. The second equation in the theorem just turns our sequence of equivalent statements around.

We need only have one set of tables or one computer program for the hypergeometric cumulative distribution function or only one for the negative hy-

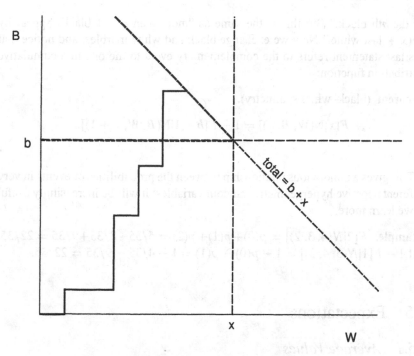

FIGURE 5.3. Positive–negative duality

pergeometric, not both. This is true even though the urn experiments, sample space, and probability mass functions are quite different.

Example. $F[2|N(4, 3, 2)] = p(0)+p(1)+p(2) = 5/35+8/35+9/35 = 22/35$. But $F[2|H(7, 4, 4)] = p(1) + p(2) = 4/35 + 18/35 = 22/35$.

There is one more important change of perspective we can apply to hypergeometric processes, which leads to useful symmetries in some of our families. What happens if we paint the black marbles white and the white marbles black? It is easy to see the effect of this black–white transformation on the hypergeometric family: We interchange W and B, and find ourselves counting the black marbles, the ones we did not count before, from our sample of n. Therefore,

$$P[x|H(W + B, W, n)) = P[n - x|H(W + B, B, n)].$$

However, you should convince yourself as an exercise that we could have figured this out by multiple applications of the reversal and transpose symmetries, so that we have learned nothing very new.

The black–white transformation has more interesting consequences for the negative hypergeometric family. Now the change of color interferes with our stopping rule, because we were using the number of black marbles to decide when to quit sampling. Instead, consider the cumulative distribution function. The event "at most x whites by the bth black" is identical to "the $(x + 1)$st white appears after the bth black." But this is the same as "more than $b - 1$ blacks appear by the$(x + 1)$st white." Now we exchange black and white marbles, and notice that this last statement refers to the complementary event to the one in a cumulative distribution function:

Theorem (black–white symmetry).

$$F[x|N(W, B, b)] = 1 - F[b - 1|N(B, W, x + 1)].$$

This gives a nonobvious relationship between the probabilities of events in very different negative hypergeometric random variables; it will be increasingly useful as we learn more.

Example. $F[2|N(4, 3, 2)] = p(0)+p(1)+p(2) = 5/35+8/35+9/35 = 22/35$. But $1 - F[1|N(3, 4, 3)] = 1 - p(0) - p(1) = 1 - 4/35 - 9/35 = 22/35$.

5.5   Expectations

5.5.1   Average Values

There must be some reason that we are interested in numerical outcomes for probabilistic experiments. Presumably, we want to be able to do various kinds of arithmetic in order to learn more about the data.

Example. I might randomly choose one of four treatments (with replacement) for each patient who enters a study. But these treatments have differing costs per week: $15, $28, $30, and $75. Therefore, the cost of continuing my experiment is affected by chance; it could be very expensive or relatively cheap. Intuitively, though, I believe that for a large number of patients, there is some sort of typical cost that I might reasonably expect. The weekly cost of a patient is an example of a discrete uniform random variable, with sample space {15, 28, 30, 75}. So we are seeking some sort of typical value for that random variable.

If I assigned treatments many times (with replacement), I would presumably get each one about equally often. My average costs per patient would then be just about the sample average of the possible prices for each treatment: $(15+28+30+75)/4 =$ $37. Therefore, it might be part of a sensible attitude in the long run to budget about $37 per patient per week.

Later in the course we will learn something about when such a policy is indeed sensible. But since it is at least plausible, we give it a name:

Definition. The **expectation** (or **expected value**) of a discrete uniform random variable is the average of the outcomes. If the variable is X, we write the expectation $E(X)$.

Since each outcome is equally likely, a simple average reflects the cost of an assignment. In the special case of a negative hypergeometric random variable in which we were chasing the only black ball in the urn, all outcomes $\{0, \ldots, W\}$ were equally likely. Therefore,

$$E(X) = \frac{0 + 1 + 2 + \cdots + W}{W + 1} = \frac{\binom{W+1}{2}}{W + 1} = \frac{W}{2}.$$

If we are searching for the one bad apple in a barrel of 10 apples, we will have to check an average of 4.5 good apples to find it. Notice that the expected value, which is a fraction, need not be a possible value, which must be an integer.

5.5.2 Discrete Random Variables

This idea of expectation of a random quantity promises to be useful enough that we would like to apply it to more cases than just the equally likely one. In the general negative hypergeometric case, we still have a discrete random variable, but all the different outcomes $\{0, \ldots, W\}$ are no longer equally likely, so our definition fails to apply directly. But we remember that the variable was just a number of white marbles up to a certain point in each of the $\binom{W+B}{W}$ equally likely sequences that realize the process. So we can use the definition to compute

$$E(X) = \frac{\sum_{\text{all sequences}}(\text{number of whites by } b\text{th black})}{\binom{W+B}{W}}.$$

Now group together in the numerator the sequences in which we drew a given number x of white marbles:

$$E(X) = \frac{\sum_{x=0}^{W} \sum_{\text{sequences with } x \text{ whites by } b\text{th black}} x}{\binom{W+B}{W}}$$

$$= \frac{\sum_{x=0}^{W} x \cdot (\text{number of sequences with } x \text{ whites})}{\binom{W+B}{W}}.$$

We have already computed the number of these sequences, so

$$E(X) = \frac{\sum_{x=0}^{W} x \cdot \binom{x+b-1}{x}\binom{W+B-x-b}{W-x}}{\binom{W+B}{W}}$$

$$= \sum_{x=0}^{W} x \frac{\binom{x+b-1}{x}\binom{W+B-x-b}{W-x}}{\binom{W+B}{W}} = \sum_{x=0}^{W} x \cdot p(x).$$

using our formula for the probability mass function. This last expression for the expectation is easy to interpret: To find the expected value of a discrete random variable, take a *weighted average* of its possible outcomes with the weights proportional to how probable that outcome is. The more likely a result, the more influence it will have on the expectation. We would like to apply the formula generally, letting $E(X) = \sum_i x_i p(x_i)$ for any discrete random variable; we will do essentially that.

If the probability mass function is given by a table, we can compute the expectation by attaching a product row and summing it. In the quorum search problem, we have the following table

x	0	1	2	3	4	5	6	7	
$p(x)$	0.0455	0.106	0.1591	0.1894	0.1894	0.1591	0.106	0.0455	total
$x\, p(x)$	0	0.106	0.3182	0.5682	0.7576	0.7955	0.636	0.3185	3.5

We come to the plausible conclusion that you must visit an average of 3.5 unnecessary houses to find the 3 people you need.

We must quibble a bit: If our discrete random variable has an infinite (but countable) sample space (remember the hurricane count (see 4.5.1)?), then to get the expectation we have to sum an infinite series. We can often do that; but if the outcomes include both positive and negative values, you may remember from calculus that sometimes the sum depends on the *order* in which you sum the terms. But if we think of the expectation as the average of a great many repetitions of the random variable, we see that we are in effect summing our series in random, unpredictable, order. You will see an example of this phenomenon in your exercises. This is unsatisfactory, so we will require that it never happen.

Definition. A sum $\sum_i a_i$ is said to be **absolutely convergent** if the positive and negative terms may be summed separately; in that case $\sum_i a_i = \sum_{a_i < 0} a_i + \sum_{a_i \geq 0} a_i$.

It should be obvious from the definition that if a series is absolutely convergent, then it does not matter in what order you add the terms; you always get the same answer. In that case, we can forget our quibbles and use our nice formula.

Definition. For a discrete random variable, $E(X) = \sum_i x_i p(x_i)$ whenever the series is absolutely convergent.

5.5.3 The Method of Indicators

Such calculations can get somewhat laborious as the number of discrete outcomes grows; we would like simpler expectation formulas, like the one we got in the search for a single black marble. One other case is almost as easy: when there is a single white marble, and we draw until we find the bth of B black marbles. Then our negative hypergeometric random variable, the number of white marbles found, can take on only two values: 1 if we find the marble, 0 if we do not. There are exactly $B+1$ equally likely realizations, according to where the white marble is. In b of those cases (just before the first black, just before the second, ..., just before the bth) we will find the white marble; in the other cases, we will not. Therefore,

$$E(X) = \frac{1 \cdot b + 0 \cdot (B + 1 - b)}{B + 1} = 0 \cdot (1 - \frac{b}{B + 1}) + 1 \cdot \frac{b}{B + 1} = \frac{b}{B + 1}.$$

Proposition. *For X an* $N(1, B, b)$ *random variable,* $E(X) = b/(B + 1)$.

Such a random variable with sample space only 0 and 1 is called a *Bernoulli*(p) variable, where the parameter p is the probability of getting a 1, $p = b/(B + 1)$. Its expectation gives us the clue we need to find the expectation of a negative hypergeometric random variable for any number of white marbles W. Remember that deciding when to stop is entirely determined by the black marbles; we ignore the white marbles until we have to count them at the end. Imagine that the white marbles are numbered, $i = 1, \ldots, W$; you ask W friends to help you by each keeping track of a different one of the white marbles. After you have removed b black marbles, you ask each friend to tell you "how many" of his white marbles have been removed along the way; he will tell you either 0 or 1. If he was looking for the marble numbered i, then his answer (0 or 1) we might call X_i. You add the numbers from each of your friends to get $X = X_1 + X_2 + \cdots + X_W$, the total of white marbles removed. For example, the result $1 + 0 + 0 + 1 + 1 + 0 + 0 = 3$ says that the white marbles labeled 1, 4, and 5 appeared, and 2, 3, 6, and 7 did not, during the draw.

Each friend need pay no attention to any white marble except the one with his number on it. Therefore, each of them is observing an $N(1, B, b)$ random variable X_i; the last proposition says that $E(X_i) = b/(B + 1)$. Each sequence a friend observes corresponds to an equal number of equally likely sequences from the original game (imagine the ways the other white balls may be scattered through his sequence). Therefore, the expectation is just the sum of the expectations for each friend. There are W friends, so we come to this conclusion:

Proposition. *For X a negative hypergeometric* $N(W, B, b)$ *random variable,* $E(X) = Wb/(B + 1)$.

For example, in the quorum search problem with $W = 7$, $B = 5$, and $b = 3$, we verify that indeed $E(X) = 7 \times 3/6 = 3.5$. You should check that this general formula matches each of the other examples and special cases we have studied. This method, in which we split a random variable up into simple, and usually equivalent, random variables $X = \sum_i X_i$ (often, but not always, each X_i is Bernoulli) and then reason that $E(X) = \sum_i E(X_i)$, is called the *method of indicators*. We shall give a general justification in a later chapter. As an exercise, you might use this approach to find a simple formula for the expectation of a hypergeometric random variable.

5.6   Estimation and Confidence Bounds

5.6.1   Estimation

In our applications so far, we have assumed that we had a random variable that was a sensible model for some real experiment. This allowed us to compute the probabilities of various outcomes. Then, if we were using the frequentist style of reasoning, we could check whether the actual numerical outcomes were surprisingly unlikely; if they were, we had reason to doubt that the model (or at least the claimed value of some parameter) really was appropriate.

As useful as this is, it has a disturbingly negative flavor; the only thing we seem to be able to do is doubt some claim. In this section we will look at a common class of problems, called *estimation* problems, in which we want to actually learn the unknown value of a parameter in some family of random variables. But estimating a parameter value will raise a harder question: How accurate is our estimate? With a bit of ingenuity, we will come up with a way to address this question using frequentist hypothesis testing to get a partial solution, called a *confidence bound*.

Example. An ichthyologist (who studies fish) tags 12 adult trout and returns them to their lake. After a brief period to let the tagged fish recover and spread through the lake, a fisherman sets out to fish the lake, and after catching 40 trout, hooks the first tagged one. Does the fisherman's experience tell the ichthyologist anything useful about the total trout population?

We start by imagining that the fisherman's experience is something like an $N(W, 12, 1)$ random variable, where he has observed $X = 40$, and W, the untagged trout population, is unknown. What would be a plausible estimate of W? A naive rule of thumb would be to guess that X is something close to its average value; and we know $E(X) = \frac{W}{B+1}$. Then just solve the equation $X \approx E(X) = \frac{W}{B+1}$ for W to get $\hat{W} = 40 \times 13 = 520$ untagged trout.

Notice that we have carried over the hat notation from when we were estimating parameters of a structural model for data, such as a regression line. This rule-of-thumb estimate, which matches a random variable to its expected value, will play the role that standard estimates played in Chapter 1. Much later in the book you will see sounder general principles for estimating parameters of families of random variables. In the meantime, the matching technique, called the *method of moments*, will be seen to work satisfactorily for a number of our favorite families. At the moment, of course, we have no idea how good an estimate of W it is.

5.6.2   Compatibility with the Data

Can we say something more useful about the true value of W? First of all, we know that it is at least 40, for obvious reasons. But we of course cannot place any corresponding upper bound. There could have been a million trout out there, and the fisherman was just lucky to catch a tagged one so soon.

Backing off a little from demanding such hard-edged knowledge (as statisticians must always do), is it not true that some values of W make us seem rather absurdly lucky? Let us try to see which values of W are implausibly large. We will proceed, using the frequentist style of reasoning, to ask when our observed value of X is improbably *small*, for a given large value of W. Of course, we would then find smaller values than the observed X even more unlikely, and so must include them in the probability to be calculated. Fortunately, this probability model has a simple expression for its cumulative distribution function:

$$P(X = 0, 1, \cdots, 40 | W) = F(40) = 1 - \frac{(W)_{41}}{(W + 12)_{41}}.$$

For example, if $W = 7000$, our p-value is 0.068. From the section on hypothesis testing, even though this probability is a bit small, we fail to reject this population size at the popular 0.05 significance level. Now, $W = 14{,}000$ has p-value 0.035; so we may reject this larger value. Using the 0.05 level, we tend to disbelieve a trout population of 14,000 but will tolerate the suggestion of 7000.

Still, our conclusions seem more than a bit weak. Our doubts range over thousands of different values of W. Worse, if we changed our significance level to, for example, 0.01, our examples of values of W that we would barely reject or barely accept would be in a very different range (exercise).

Now try for information about the compatibility of the fisherman's experiment with *small* trout populations W. We need to know for which values of W an X of 40 fish (or more) is improbably *large*:

$$P(X = 40, 41, \cdots | W) = 1 - F(39) = \frac{(W)_{40}}{(W + 12)_{40}}.$$

Trying out $W = 200$, we get a p-value of .0754—a little unlikely to find 40 or so untagged fish, but not to the traditional significance level. Now try $W = 150$: The p-value is 0.0289, and we are ready to reject the hypothesis that there are so few trout. Our information is much sharper; a substantial change in plausibility for a moderate change in parameter value. After some more calculation, we narrow it down to exactly when we start rejecting the proposed population size: For $W = 175$ we compute $p = 0.0503$, and for $W = 174$ we get $p = 0.0494$.

Finally, we are prepared to say something really useful to the ichthyologist: If we use the 0.05 significance level, then we find our experimental result consistent with any $W \geq 175$ untagged trout and inconsistent with any smaller values. (Actually, We have ducked one issue: We have not checked our statement for every possible value of W. In an exercise you will remedy that oversight.) This limit changes with significance level, but not nearly so radically as before. This could be of real scientific usefulness in monitoring the trout population. Certainly it is a clear improvement over our crude estimate, of unknown, and apparently very low, accuracy that there are 520 untagged trout.

5.6.3   Lower Confidence Bounds

The method we have devised is so widely useful that we give it a name.

Definition. Let a random variable X be from a known family, one of whose parameters θ is unknown. Assume that for all values $\theta < \theta_L$ we find that the observed value of X leads us to reject θ at the significance level α and that for all $\theta \geq \theta_L$ we fail to reject θ. Then we say that θ_L is a $(1 - \alpha)\%$ **lower confidence bound** for θ (or that $\theta \geq \theta_L$ is true **with** $100(1 - \alpha)\%$ **confidence**).

For our example, we discovered that 175 was a 95% lower confidence bound for the true number of untagged trout. As an easy exercise, you will write down the definition for an *upper* confidence bound θ_U.

I hope you are convinced from our example of the potential usefulness of confidence bounds. Unfortunately, our justification for them, based on which hypotheses we would reject or fail to reject, is subtle and rather hard to explain to scientists. People keep trying to say simpler things, like "the probability is 0.95 that $\theta \geq \theta_L$." Is something like that true? Remember that the probabilities in frequentist hypothesis testing are computed before the experiment is done. Afterward, of course, we know the exact value of X. So before the experiment, we imagine that there is a true value for θ (W in our example). The probability that the observed value of X will lead us to reject the (true) hypothesis that the parameter is θ is then (at most) the significance level α. But those are exactly the cases when we will, after the experiment, choose a θ_L for which $\theta < \theta_L$. Therefore, the lower confidence bound will later happen to be false, when it says $\theta \geq \theta_L$, with probability at most α. The fact that we do not know θ is irrelevant. We state this formally:

Proposition. *The probability that a $100(1 - \alpha)\%$ lower (or upper) confidence bound $\theta \geq \theta_L$ (or $\theta \leq \theta_U$) computed in a future experiment will be true is at least $1 - \alpha$.*

The practical implication of this result for me is that as a consulting statistician who will compute many 95% confidence bounds during the rest of my career, at least 95% of my claims should turn out to be correct.

5.7   Summary

Many different probability models will be needed to describe the enormous variety of kinds of data generated by the many profoundly different experiments one could perform. We began right away to group *random variables*, probability spaces with numerical outcomes, into *families*, the members of which may be distinguished by numerical "addresses," called *parameters*. In particular, we began to explore an especially rich family of random variables called the *negative hypergeometric* family, which comes about when we realize a *hypergeometric process*. Its *probability mass function* is

$$P(X = x | W, B, b) = p(x) = \frac{\binom{x+b-1}{x}\binom{W+B-b-x}{W-x}}{\binom{W+B}{W}}$$

for the member of the family N(W, B, b) (2.3). This family will turn out later in the book to be related to an amazing number of the most useful random variables in statistics. Like many families, it possesses *symmetries*, or relationships between the probabilities of different members of the family (2.4). We then derived a related family, the *hypergeometric* family, with mass function

$$P[X = x | H(W + B, W, n)] = p(x) = \frac{\binom{W}{x}\binom{B}{n-x}}{\binom{W+B}{n}} \quad (3.1).$$

An important practical applications is to *Fisher's exact test* of independence in contingency tables (3.3). This test illustrates a more formal way of interpreting statistical experiments, called *hypothesis testing* (3.4). The *cumulative distribution function* of a random variable, $F(x) = P(X \le x)$, can be used to calculate the probability that a random variable will fall in any interval, $P(x < X \le y) = F(y) - F(x)$ (4.2). Furthermore, it lets us express a mathematical relationship between the negative hypergeometric and hypergeometric families, called a *duality* (4.3).

We defined the *expectation* (average value) of a discrete random variable by $E(X) = \sum_i x_i p(x_i)$ (5.2). Then the first example of an important technique for finding expectations, the *method of indicators*, was applied to some of our families (5.3). This suggested a simple method of *estimation* of unknown parameters, by matching the observed result to its expectation (6.1). To get stronger information about an unknown parameter, we turned around the logic of hypothesis tests to construct *confidence bounds* (6.3).

5.8   Exercises

1. An ESP researcher makes up a deck of cards on each of which is printed one of four geometrical figures, one of which is a square. There are two cards with each figure, for a total of 8 cards. He places them face down on a table in random order and asks a subject to turn over cards until the first square is uncovered. He will be impressed if the subject finds one quickly.

   a. You believe that the subject has no idea where the squares are. What is the probability the subject will find one for the first time when the second card is turned over?

   b. Let a random variable X be the number of cards without squares that are turned over in the course of the experiment. Construct the table of its probability mass function.

2. Write down all the realizations of a hypergeometric process with 5 white marbles and 3 black marbles. Put a check mark next to all those in which the first black marble appears before the third white. Does the probability of this happening match our formula?

3. On the first day of class, my roll tells me that I have 6 sophomores among my 20 students. I need to find out how much calculus the sophomores know, so I want to interview 2 of them in depth. I do not know which person is which yet, so I simply go through the class at random, asking them if they are sophomores, until I find my two. What is the probability that I will ask 5 nonsophomores along the way? What sort of random variable am I asking about, and what are my parameter values?

4. According to a contractor's records, three very similar varieties of tree, 8 live oaks, 6 Brazos oaks, and 5 shady oaks, were ordered for planting 20 years ago along a street of a new subdivision. Sure enough, all 19 trees are still thriving. However, a tree surgeon must treat the live oaks to prevent a new blight. Unfortunately, the varieties cannot be distinguished except by careful examination of several leaves. The surgeon plans to check the trees one at a time and treat the live oaks she finds. Her day's work will be complete when she has treated 5 trees. What is the probability that she will identify 4 Brazos oaks and 2 shady oaks along the way?

5. Only 85 out of the 100 integrated circuits in a shipment meet design specifications (but the customer doesn't know that). She picks 8 at random and tests them carefully. What is the probability that three *or more* of the circuits she tests will fail to meet specifications?

6. Verify the proposition about reversal symmetry of the negative hypergeometric family by writing down the formulas for the two probability mass functions and showing that they are equal.

7. In the same manner as in Exercise 6, verify reversal symmetry in the hypergeometric family.

8. Verify transpose symmetry in the hypergeometric family, using the formula for the mass function.

9. It is folk wisdom that beer consumption is *countercyclical*; that is, more is purchased in bad economic times than in good. To study one aspect of this conjecture, you interview 30 working-age adults and ask whether or not they are currently gainfully employed and whether or not they have drunk at least one bottle of beer in the last 24 hours. Your results:

	employed	no
beer	6	8
no	13	3

Carry out Fisher's exact test of the independence of beer consumption and employment. What do you conclude? Use the 0.05 significance level.

10. Thomas and Simmons in 1969 reported on the sputum histamine levels of a number of allergic and nonallergic people; here are some of their results, in

parts per thousand:

nonallergic: 4.7 5.2 6.6 18.9 27.3 29.1 32.4 34.3 35.4 41.7 45.5 48.0 48.1

allergic: 31.0 39.6 64.7 65.9 67.9 100.0 102.4 1112.0 1651.0

Test whether allergic people tend to have higher histamine levels, using the sign test. Interpret it with a 0.01 significance level.

11. Write down the table for the cumulative distribution function for the random variable X of Exercise 1. What is the probability that the subject will turn over no more than 4 wrong cards?

12. Use the formula for the cumulative distribution function of an $N(W, B, 1)$ random variable to verify the numerical results we got in

    a. the laundry-guarding example (see Section 2.2); and
    b. Exercise 11.

13. In the last chapter (see 4.8.1) we considered the probabilities for outcome falling in a vertical strip of a circular dart board. Let a continuous random variable X be the x-coordinate of the point at which a dart hits. Find the cumulative distribution function of X.

14. Prove properties (iii)–(v) of cumulative distribution functions (see Section 4.2).

15. A certain random variable arising from a geometrical outcome on the interval $(0, 2)$ has density function $f(x) = \frac{3}{4}x(2 - x)$.

    a. Find its cumulative distribution function.
    b. Compute $P(1.5 < X \le 2)$.

16. In a candy jar with 7 chocolates and 5 caramels, remove candy at random until you encounter the second caramel. Calculate the probability that you will have found no more than 3 chocolates. In the original jar, remove 5 pieces of candy. Calculate the probability you will have found no more than 3 chocolates.

17. Here is a partial table of the cumulative distribution function of a negative hypergeometric $N(25, 14, 8)$ random variable:

x	10	11	12	13	14
$F(x)$	0.24344	0.32521	0.41584	0.51124	0.60665

In a graduate class of 39 students, 14 were undergraduate students at Tech. I work down my alphabetic roll of the class until I find 8 who were undergraduates at Tech. What is the probability that I will have passed exactly 14 other students along the way?

18. Show that we could have verified the black–white symmetry in the hypergeometric family $p[x|H(B, W, n)] = p[n - x|H(W + B, B, n)]$ by checking that the mass functions are the same.

19. There are 12 women and 15 men in the introductory statistics class. The grader brings me their first test, sorted in descending order of score.

a. I glance quickly down the pile until I find the fourth man's paper. If the sexes did about equally well on the test, compute the probability that I would have seen no more than 4 women's papers.

b. On the other hand, I might glance down the pile until I see the 5th woman's paper. Compute the probability that I would have seen more than three men's papers by that point.

20. A very new and complex computer chip is known to have a high rate of small defects. The manufacturer admits this but will sell you a box of 40 chips cheap, with the guarantee that no more than 8 are defective. You need 10 perfect chips for your process control computer, so you carefully test through the batch until you have found them. Unfortunately, you find 5 bad ones along the way, which is disturbing.

    Giving the manufacturer the benefit of the doubt, assume that exactly 8 are bad. What is the probability that you would have found 5 **or more** of the bad ones while retrieving 10 good ones?

21. Here is a table of the cumulative distribution function ($F(x) = P(X \leq x)$) for a certain discrete random variable:

x	0	1	2	3	4
$F(X)$	0.0182	0.2153	0.4871	0.8865	1.0

    Calculate E(X).

22. Find E(X) for the random variable in Exercise 1, (a) from your table of the mass function, and (b) using the formula for the expectation of a negative hypergeometric random variable.

23. Find E(X):

    a. for the negative hypergeometric random variable in Exercise 17; and

    b. for the number of nonsophomores questioned in Exercise 3.

24. There are n delicious strawberries in a basket, but there are in addition 2 contaminated strawberries, which look and smell exactly the same as the others. However, anyone who bites into a contaminated strawberry will find that it tastes so awful that he or she will have no further appetite. A person comes along and begins eating strawberries, and will stop only on biting into a contaminated one. Let a random variable X be the number of **good** strawberries eaten.

    a. Give the range of possible values of X and find its probability mass function $p(x) = P(X = x)$. If $n = 12$, what is the probability that 7 good strawberries will be eaten?

    b. For any n good strawberries, compute E(X).

25. If we have a random variable with *finite* sample space, why is our formula for E(X) always absolutely convergent?

26. A rancher has scattered 8 black sheep among his large flock. As a new shepherd, you count the sheep returning to their pen from a day of grazing, and the first black sheep is the 20th sheep you see.

    a. Use the method of moments to estimate the total number of sheep in the flock.
    b. Find a lower 95% confidence bound on the number of sheep in the flock.

27. State a precise definition of the upper $100(1 - \alpha)\%$ confidence bound for a parameter of a random variable. Find an upper 99% confidence bound for the flock size in Exercise 26.

28. In experiments whose outcome X is an $N(W, B, b)$ random variable, compute a method-of-moments estimate:

    a. for W, if B and b are known;
    b. for B, if W and b are known; and
    c. for b, if W and B are known.

5.9 Supplementary Exercises

29. Show that for the hurricane example, with $p(x) = 1/2^{x+1}$ for $x = 0, 1, 2, \ldots$, the cumulative distribution function is $F(x) = 1 - 1/2^{x+1}$.

30. In Exercise 30 of Chapter 4, let a continuous random variable X be the distance of a randomly chosen house from the freeway. Find its cumulative distribution function.

31. Using the definition of a limit from advanced calculus, prove properties (i) and (ii) of cumulative distribution functions.

32. From a list of 39 potential earthquake sites around the world, a psychic claims she can identify those that will have a 6.0 Richter or greater earthquake in the next 5 years. She writes down those 14 sites she believes are in the greatest danger and seals them in an envelope. In fact, 20 of the sites have earthquakes. What is the probability that the psychic will have identified at least 8 of them correctly, purely by chance?
    Hint: Use the table in Exercise 17, and do very little arithmetic.

33. Show that we could have verified the black–white symmetry in the hypergeometric family $P[x|H(W + B, W, n)] = p[n - x|H(W + B, B, n)]$ by multiple applications of reversal and transpose symmetries.

34. Computing cumulative distribution functions for negative hypergeometric random variables can be time-consuming, but there is a useful shortcut:

    a. write down the formula for $p(0)$; then
    b. write down a simple formula for $r(x) = p(x)/p(x-1)$, canceling as many factors as you can.

This lets you recursively compute the cases $x = 1, 2, 3, \ldots$ by the formula $p(x) = r(x)p(x - 1)$.

35. Use the formula from Exercise 34 to reconstruct the table in Exercise 17.

36. Invent a recursive computational procedure for the probabilities of any hypergeometric random variable, similar to the one in Exercise 34. Redo the calculations in Exercise 9 using your simplified arithmetic.

37. Some calculus books say that a sum $\sum_i a_i$ is absolutely convergent if $\sum_i |a_i|$ exists (so that for an expectation, $\sum_i |x_i| p(x_i)$ has a finite sum). Prove that this definition is equivalent to ours.

38. Find a simple expression for the expectation of any hypergeometric random variable, using the method of indicators.

39. Use the result of Exercise 38

   a. to find the expected number of bad circuits located in Exercise 5; and
   b. to find the expected number of predicted earthquake sites in Exercise 32.

40. Of 40 engineering majors in an engineering stat class, 12 are mechanical engineers and 15 are industrial engineers. The instructor chooses 10 to represent the class in a stat contest.

   a. If major should have no effect on who is chosen, what is the probability that 3 mechanical engineers and 5 industrial engineers will be chosen for the contest?
   b. On average, how many mechanical engineers would you expect to be chosen for the contest?

41. Consider the first n positive integers $\{1, 2, 3, \ldots, n\}$. Choose m of these numbers at random without replacement and call their sum X. (For example, if from the first 5 integers you chose the three numbers 4, 1, and 3, then $X = 8$).

   a. What is E(X)?
      Hint: Use the method of indicators and the fact that $\sum_{i=1}^{r} i = r(r + 1)/2$ (see Exercise 3.23).
   b. Therefore, in the discussion of rank statistics (see 2.5.5), assume that ranks are unrelated to level of the treatment, and compute E(W_i) and E($\overline{R}_i$).

42. A jeweler has a set of 100 identically cut diamonds in a drawer. By accident, someone mixes up in the drawer an unknown number of excellent fake diamonds of the same size and cut. You set out to find the fake diamonds by careful inspection. After finding 13 real diamonds, you locate the first fake one. You want to decide what this tells you about how many fake diamonds there are.
   Hint: A reasonable probability model for the number of real diamonds found so far is N(W, B, 1). But which parameter is unknown?

   a. Find a method-of-moments estimator to estimate the number of fake diamonds.
   b. Construct a lower 95% confidence bound on the number of fake diamonds.

    c. Construct an upper 95% confidence bound on the number of fake diamonds. Which bound gives you more useful information?

43. a. Using the result of Exercise 38 and given the result X of an experiment which is H($W + B$, W, n), find method-of-moment estimates, in turn, for B, W, and n, if the other two parameters are known.

    b. For the census data from Chapter 1, Exercise 32, use (a) to estimate the total population of that census tract.

CHAPTER 6

Discrete Random Variables II: The Bernoulli Process

6.1 Introduction

In the last chapter we looked at several families of random variables that arise from the hypergeometric process. As the size of the urn (out of which we are imagining we draw marbles) grows, the calculations we have to do to find probabilities become complicated. In this chapter we will explore some simpler approximate calculations, which will work when the number of marbles removed, or the number of marbles being counted, is relatively small. The approximations will be interesting random variables in themselves, and we will discover thereby several new families and a new stochastic process, the *Bernoulli* process, out of which they arise naturally. We think of this as sampling from *infinite* populations. As the outcomes being counted grow rarer, a further simplification is possible, leading to the *Poisson* family. On the way, we learn a new method for evaluating certain expectations and use it to measure population variability. Then we find ways of constructing simultaneous upper and lower confidence bounds for unknown parameters in our families.

Time to Review

Chapter 2, Sections 2–4
Limits of sequences
Power series for the exponential function.

6.2   The Geometric and Negative Binomial Families

6.2.1   The Geometric Approximation

We noted in an earlier chapter that in our urn problems, if the number of marbles is very large, then an experiment that involves removing relatively few marbles will not deplete the total very much.

Example. Kim is looking for a job. A helpful hostess holds a party in which 30 prospective employers and 50 other employment seekers are invited. She hopes that while having fun, some people will also find jobs. Kim arrives, and knows no one; the guests are milling around in a large ballroom. What is the probability that the fourth person Kim talks to will be the first employer?

In this case approximation by a draw *with* replacement (i.e., assuming independence of the draws) may work satisfactorily. The urn model might be $W = 50$ and $B = 30$, and a negative hypergeometric random variable in which we are looking for the first black marble [$N(W, B, 1)$]. Then $x = 3$, and our calculation is

$$p(x) = \frac{(W)_x}{(W + B)_{x+1}} B = \frac{50 \cdot 49 \cdot 48 \cdot 30}{80 \cdot 79 \cdot 78 \cdot 77} = 0.092945.$$

The practical consequence of the fact that we are not depleting the total supply of marbles (prospective employees) very much is that we would expect $50 \cdot 49 \cdot 48$ to be pretty close to $50 \cdot 50 \cdot 50 = 50^3$, and $80 \cdot 79 \cdot 78 \cdot 77$ to be pretty close to $80 \cdot 80 \cdot 80 \cdot 80 = 80^4$. Trying this approximate calculation, we obtain $p(x) \approx \frac{50^3}{80^4} 30 = 0.09155$, which is indeed fairly close to the same answer. Notice that the calculation is exactly the one we would do if we were doing our draws with replacement, and so not depleting the jar at all.

When does this approximation work well? In the birthday inequality (see 3.5.3), we discovered that $(n)_k / n^k$ is close to 1 when $\binom{k}{2}$ is small compared to n; that is, we permute so few of the available objects that drawing with and without replacement is almost the same. To make our approximation work, we would need to have that $\binom{x}{2}$ is small compared to W and $\binom{x+1}{2}$ is small compared to $W + B$. But $\binom{x}{2}$ is smaller than $\binom{x+1}{2}$.

Proposition. *For an* $N(W, B, 1)$ *random variable* X, *if* $\binom{x+1}{2}$ *is small compared to* W *then* $p(x) \approx W^x / (W + B)^{x+1} B$.

In the party example, our approximation could be expected to work because $\binom{4}{2} = 6$ is small compared to 50.

It will be interesting to rewrite this as $p(x) \approx \left(\frac{W}{W+B}\right)^x \frac{B}{W+B}$. The quantity $\frac{W}{W+B}$ is just the probability that the first marble one draws is white; let us give it a name, p. Then $\frac{B}{W+B} = 1 - p$ is the probability that the first is black. Then we can rewrite $p(x) \approx p^x (1 - p)$. This formula has the nice property that we need to work with only one parameter, p, rather than two, W and B, to use it. Remember that the calculation was exact for draws with replacement, that is, for a sequence of independent experiments.

6.2.2   The Geometric Family

The calculation from the last section suggests that we have a family of random variables of interest in itself:

Definition. Consider a sequence of independent trials for which two outcomes are possible at each trial. The probability of one outcome, usually called a *success*, is p (and the probability of the other, a *failure*, is $1 - p$), where $0 < p < 1$. (These are, of course, *Bernoulli* trials, see 5.5.3) Then the number of successes X before the first failure is a **geometric** random variable. Since the sequence can continue indefinitely, its sample space is $\{0, 1, 2, \ldots\}$.

We compute the probability of X successes in a row followed by a failure, when each trial is independent:

Proposition. *For X geometric,* $p(x) = p^x(1 - p)$.

Example. In the hurricane example (see 4.5.2), let the number of hurricanes be geometric with $p = 0.5$. Then the formula gives us $p(x) = 2^{-x-1}$, as claimed. The same random variable is a model for tossing a fair coin until you get the first tail.

6.2.3   Negative Binomial Approximations

The approximation method from Section 6.2.1 can be used on a more general problem:

Example. I learn from an anonymous survey that of a sample of 100 people, 40 admit to having cheated on their income tax. I want to do an in-depth, follow-up confidential interview of five cheaters. What is the probability I will have to talk to exactly nine people among the sample to find them?

This is negative hypergeometric, with $B = 40$, $b = 5$, $W = 60$, and $x = 4$, so $p(4) = (\binom{8}{4}\binom{91}{56})/\binom{100}{60}$ from our big formula. The way of organizing the calculation that allows the most cancellation, and so leaves us with the fewest multiplications, is

$$p(4) = \binom{8}{4}\frac{\dfrac{91!}{56!35!}}{\dfrac{100!}{60!40!}} = \binom{8}{4}\frac{\dfrac{60!\,40!}{56!\,35!}}{\dfrac{100!}{91!}} = \binom{8}{4}\frac{(60)_4(40)_5}{(100)_9} = 0.0937.$$

It occurs to us, as with the last such calculation, that, for example, $(60)_4 = 60 \cdot 59 \cdot 58 \cdot 57$ should be fairly close to $60^4 = 60 \cdot 60 \cdot 60 \cdot 60$. Here there are two other permutations where such an approximation is plausible. Using the condition from the birthday problem, we check that $\binom{x}{2} = 6$ is small compared to 60 and $\binom{b}{2} = 10$ is fairly small compared to 40; so we compute

$$p(4) \approx \binom{8}{4}\frac{60^4 40^5}{100^9} = 0.0929.$$

Since the calculation was notably easier, this is attractively close.

Generally, what we did was to rewrite the probability mass function to cancel as many large factorials as possible:

$$\frac{\binom{x+b-1}{x}\binom{W+B-b-x}{W-x}}{\binom{W+B}{W}} = \binom{x+b-1}{x}\frac{(W)_x(B)_b}{(W+B)_{x+b}}.$$

We can say when replacing the permutations by powers will work satisfactorily:

Proposition. *For an* $N(W, B, b)$ *random variable, when* $\binom{x}{2}$ *is small compared to* W, *and* $\binom{b}{2}$ *is small compared to* B, *then* $p(x)$ *is close to* $\binom{x+b-1}{x}W^x B^b/(W+B)^{x+b}$.

We have not needed to put in a condition for the denominator approximation, $\binom{x+b}{2}$ small compared to $W+B$; you will check in an exercise that this follows from the conditions we did give. Once again, let the quantity $W/(W+B)$, the probability that the first marble drawn is white, be called p. Since $B/(W+B) = 1 - p$, our approximation formula can be rewritten:

$$p(x) \approx \binom{x+b-1}{x}\frac{W^x}{(W+B)^x}\frac{B^b}{(W+B)^b} = \binom{x+b-1}{x}p^x(1-p)^b.$$

6.2.4 Negative Binomial Variables

We derived the above approximation by assuming that we were drawing so few marbles out of so many that the difference between drawing with and without replacement was relatively unimportant. What would happen if we really had drawn with replacement, and so had true independence between draws? Then the probability of a given sequence with x whites and b blacks is $p^x(1-p)^b$, because we simply multiply the probabilities p of each of the x white marbles and the probabilities $1 - p$ of each of the b black marbles. If this sequence has arisen in a search for the bth black marbles then the number of such sequences is the number of ways we can distribute x white marbles among the previous $b + x - 1$ draws, or $\binom{x+b-1}{x}$. This is a whole new family of random variables:

Definition. A **negative binomial** random variable (with parameters k where $k = 1, 2, 3, \ldots$, and p where $0 < p < 1$), $NB(k, p)$, is the number of successes X before the kth failure in a sequence of independent trials with probability p of success at each trial.

Proposition. *A negative binomial* $NB(u, p)$ *random variable* X *has sample space all nonnegative integers* $(X = 0, 1, 2, \ldots)$ *and* $p(x) = \binom{x+k-1}{x}p^x(1-p)^b$.

The sample space is unbounded because when we draw with replacement there is no limit to the number of white marbles we may encounter.

Notice that the geometric random variable was just the special case of looking for only one failure, $NB(1, p)$. But now there are others of possible usefulness:

Example. Every time I turn on the reading lamp on my desk, there is a probability of 0.05 that the bulb will blow out. I have two spare bulbs, in addition to the one in the lamp. What is the probability that I will have to shop for bulbs after turning on my lamp for the 60th time? This might be negative binomial with $k = 3$, $p = 0.95$, and $x = 57$ (because the other three times a bulb blew). Then $p(57) = \binom{59}{57}(0.95)^{57}(0.05)^3 = 0.01149$.

6.2.5 Convergence in Distribution

There is another way of looking at our result that if the number of marbles of each kind is large compared to the number we are looking for, then a negative hypergeometric random variable is well approximated by a certain negative binomial random variable. Instead, imagine that we have a sequence of urn problems in which we are always searching for the same number b of black marbles, but the number of white and black marbles is getting larger and larger. Then the negative binomial approximation to the probability mass function $p(x)$ is getting better and better. But for any given value of x, the cumulative distribution function may be written $F(x) = \sum_{y=0}^{x} p(y)$, so that it is the sum of a fixed, finite number of terms $p(y)$. We conclude that our approximation to F is also getting better and better. This is an example of an important phenomenon.

Definition. Consider a sequence of random variables $\{X_i\}$ and an additional variable X, each with sample space the integers, and with cumulative distribution functions F_i and F, so that for each x in the sample space of X, $\lim_{i \to \infty} F_i(x) = F(x)$. Then we say that the sequence $\{X_i\}$ **converges in distribution** to X. We write $X_i \to X$.

Another way of putting it is that the sequence of random variables $\{X_i\}$ is *asymptotic* to X. The importance of convergence in distribution is usually for applications just like the one we have seen: If we have reason to believe that a complicated random variable is far along in such a sequence, and X has some simple properties, then we hope to find that our random variable approximately shares those simple properties.

Proposition. *Let the sequence of negative hypergeometric random variables* $N(W_i, B_i, b)$ *be such that* $B_i \to \infty$ *and* $\frac{W_i}{W_i + B_i} \to p$ *as i goes to infinity, where* $0 < p < 1$. *Then the sequence converges in distribution to a negative binomial* $NB(b, p)$ *random variable.*

PROOF.   We must simply check that the approximation to $p(x)$ gets as good as we please for each x, because then the approximation to $F(x)$ gets as good as we please for each x, as we noticed earlier. But our condition for a good approximation requires that B_i be arbitrarily large compared to $\binom{b}{2}$; since b is fixed and the B's are going to infinity, that is certainly happening. Also, the W's must become large compared to $\binom{x}{2}$ for each fixed x; show as an exercise that this must happen because the W's approach a fixed proportion p of all the marbles, and the B's are getting numerous.   □

The relationship between the negative hypergeometric and negative binomial ought to give us more information. Perhaps we can see what happens to our formula for the expectation as the number of marbles grows:

$$\lim_{i \to \infty} E(X_i) = \lim_{i \to \infty} \frac{W_i b}{B_i + 1} = b \lim_{i \to \infty} \frac{W_i}{B_1 + 1} = b \lim_{i \to \infty} \frac{W_i}{B_i}$$

$$= b \lim_{i \to \infty} \frac{W_i/(W_i + B_i)}{B_i/(W_i + B_i)} = \frac{bp}{1 - p}$$

using standard facts about limits. We would like to say that if $X_i \to X$, then $E(X_i) \to E(X)$; therefore, for a negative binomial NB(k, p) random variable, $E(X) \stackrel{?}{=} kp/(1 - p)$. This last formula will turn out to be correct; but it is not always true that the expectation of the limit is the limit of the expectations (as you will check in an exercise). We will verify the formula in other ways, shortly. Meanwhile, notice that it predicts in the dice example that to get 3 sixes you will on average make $3 \cdot \frac{5/6}{1-5/6} = 15$ unsuccessful rolls, which is reasonable (5 nonsixes for each six). Of course, in practice you might make no bad rolls, or a million.

6.3 The Binomial Family and the Bernoulli Process

6.3.1 Binomial Approximations

Our urn problems can become painfully large in other, quite different, ways.

Example. A wealthy grandmother dies and leaves her estate to her 5 grandchildren, all of whom live in a small town (with 255 households). Unfortunately, none of them share her last name, and she did not give last names in her will. As executor, you will have to simply visit every house in town, until you find them. You decide to visit until you have crossed 100 homes *without* an heir off your list the first day (that is all the frustration you can stand). What is the probability you will find 2 heirs that day?

If you visit houses at random, this has an urn model with $W = 5$ successful marbles and $B = 250$ failures. We are doing a negative hypergeometric search with $b = 100$: Therefore,

$$p(2) = \frac{\binom{100+2-1}{2}\binom{250+5-100-2}{3}}{\binom{250+5}{5}} = \frac{\frac{101 \cdot 100}{2} \cdot \frac{153 \cdot 152 \cdot 151}{6}}{\frac{255 \cdot 254 \cdot 253 \cdot 252 \cdot 251}{120}} = 0.3422.$$

As unpleasant as this calculation was, we got a good deal of cancellation; the general situation is that when the number of black marbles B and the number of black marbles to be found b are large compared to the number of white marbles, then the most cancellation is gotten by organizing the negative hypergeometric calculation as

$$p(x) = \frac{\frac{(x+b-1)_x}{x!} \frac{(W+B-b-x)_{W-x}}{(W-x)!}}{\frac{(W+B)_W}{W!}} = \binom{W}{x} \frac{(x+b-1)_x(W+B-b-x)_{W-x}}{(W+B)_W}.$$

Since W is small compared to B and b, it is reasonable to presume that, for example, $255 \cdot 254 \cdot 253 \cdot 252 \cdot 251$ is close to 251^5. This is not quite the same approximation as the birthday-problem formula used at the beginning of this chapter: Notice that the approximation is on the low side of the exact value. Nevertheless, there is a similar bound to the error.

Proposition. $e^{1/(n+k-1)\binom{k}{2}} \le (n+k-1)_k/n^k \le e^{(1/n)\binom{k}{2}}$.

PROOF. Exercise. It precisely parallels the proof of the corresponding proposition in Chapter 3.5.3. In fact, if we had been imaginative enough to invent "negative" permutations (in which the products go up instead of down, as in some of the Urn Problem 4 (see Exercise 3.37) calculations, the two results could have been a single proposition. □

Applying this proposition to our rearrangement of the hypergeometric calculation, we get

$$p(x) = \binom{W}{x} \frac{(x+b-1)_x(W+B-b-x)_{W-x}}{(W+B)_W}$$

$$\approx \binom{W}{x} \frac{b^x(B-b+1)^{W-x}}{(B+1)^W}.$$

Proposition. Whenever $\binom{W}{2}$ is small compared to b and $B+1-b$, then

$$p(x) \approx \binom{W}{x} \frac{b^x(B+1-b)^{W-x}}{(B+1)^W}.$$

The approximation uses the last proposition and the fact that x and $W-x$ are no greater than W.

Example. In the problem with the heirs, $p(2) \approx \binom{5}{2} \frac{100^2(151)^3}{251^5} = 0.3456$, which is within about 1% of the correct answer.

6.3.2 Binomial Random Variables

Inspired by our earlier work, we simplify the expression by letting $b/(B+1) = p$, the probability that a single white marble will be selected: $p(x) \approx \binom{W}{x} p^x (1-p)^{W-x}$. As before, we would like to interpret this approximation as a probability of interest in itself. White marbles are rare in these urns, and so they are usually widely scattered through the sequence of marbles we draw. If we imagine creating our sequence by sowing white marbles at random into the long sequence of black marbles, it seems plausible that these drops are almost independent of one another, because earlier white marbles are so few as to have little effect on the next drop. This suggests the following definition.

Definition. In a sequence of n (= 1, 2, 3, ...) independent trials with probability p ($0 < p < 1$) of success at each trial, then X, the number of successes, is a **binomial** [$B(n, p)$] random variable.

We imagine a success to be a white marble that was dropped before the bth black marble, out of a total of W white marbles introduced. Each sequence of the desired number of successes and failures has probability $p^x(1 - p)^{n-x}$, because each trial is independent of the others, and so we just multiply. The number of sequences of n trials with x successes among them is, of course, just $\binom{n}{x}$.

Proposition. *The sample space of a binomial* $B(n, p)$ *random variable is* $\{0, 1, \ldots, n\}$, *and* $p(x) = \binom{n}{x}p^x(1 - p)^{n-x}$.

Compare this to our result about approximating negative hypergeometric random variables:

Proposition. *Let* X_i *be a sequence of* $N(W, B_i, b_i)$ *random variables such that* $B_i \to \infty$ *and* $b_i/(B_i + 1) \to p$ *where* $0 < p < 1$. *Then the sequence converges in distribution to a* $B(W, p)$ *random variable.*

Of course, this new family is not just an approximating device.

Example. A certain lung disease in newborns is fatal in 70% of cases. A new treatment has been proposed, but you doubt that it will improve the survival rate. Ten randomly chosen patients are to be given the new treatment. What is the probability that exactly 2 will die?

If you are right, then the number of survivors will be a binomial $B(10, 0.7)$ random variable, since presumably the newborn's chances are independent of one another. Then $p(2) = \binom{10}{2}(0.7)^2(0.3)^8 = 0.00145$, which is about one time in 700. Even if we add in the even rarer possibilities of 1 or 0 deaths, getting a p-value well under 0.01, this is so unusual that if it happens that way, you should rethink your skepticism about the new treatment.

We can calculate the limit of the expectations in the sequence above:

$$\lim_{i \to \infty} E(X_i) = \lim_{i \to \infty} \frac{Wb_i}{B_i + 1} = W \lim_{i \to \infty} \frac{b_i}{B_i + 1} = Wp,$$

which leads us to the conjecture that for a binomial $B(n, p)$ random variable, $E(X) = np$. Intuitively, the expected number of successes is the number of trials times the proportion of successes. This conjecture will turn out to be correct, later in the chapter. In our example, we would expect 7 patients to die, on average.

The binomial random variable has a symmetry to it that follows from the reversal symmetry in a hypergeometric variable: counting the white marbles that are drawn *after* the bth black marble. The probability that a single white marble will fall in that range is, of course, $\frac{B+1-b}{(B+1)} = 1 - p$, which we now think of as the probability of a failure. But after n independent trials, we conclude that

$P[x|B(n, p)] = P[n - x|B(n, 1 - p)]$; this just says that if we observe that a certain number of experiments are successes, the rest must be failures. Interestingly, a negative binomial family has no reversal symmetry, because the sequence of trials has no necessary end point.

6.3.3 Bernoulli Processes

It is natural to wonder whether there is a connection between negative binomial and binomial probabilities. The mass functions are obviously not the same. The relationship can be explained as follows:

Definition. A **Bernoulli(p) process** is a sequence of independent Bernoulli trials, with probability of success $p(0 < p < 1)$ at each trial, thought of as continuing indefinitely. A realization of such a process is a particular sequence of successes and failures.

For example, FSFFSFSSSFSFFSSFSFS is a segment of the realization of such a trial. Notice that the probability that a segment of this length will look like this is just $p^{10}(1 - p)^9$. We see that a negative binomial random variable is just the number of successes before the kth failure in a Bernoulli process. Furthermore, a binomial random variable is the number of successes in the first n trials of a Bernoulli process. Thus, the two are related just as negative hypergeometric and hypergeometric variables are related—two corresponding stopping rules in the same sort of stochastic process (see 5.4.3).

This tells us that we can use precisely the same reasoning as before to connect the cumulative distribution functions of the two random variables: "At most x successes precede the kth failure" is equivalent to "at most x successes are in the first $x + k$ trials." Therefore, we have a corresponding equality:

Proposition (positive–negative duality).

(i) $F[x|NB(k, p)] = F[x|B(x + k, p)]$.
(ii) $F[x|B(n, p)] = F[x|NB(n - x, p)]$.

Bernoulli processes, of course, have their own black–white transformation: We interchange those outcomes we call successes and those we call failures. The probability of success then becomes $1 - p$. In a binomial experiment, we are counting what used to be failures after n trials, which is, of course, all those that were not successes—we have simply rediscovered reversal symmetry. In a negative binomial experiment something more complicated happens, since we have changed the stopping rule. As in the negative hypergeometric case, we reason that "at most x successes by the kth failure" is equivalent to "more than $k - 1$ failures appear by the $(x + 1)$st success." Now interchange success and failure to get an important symmetry:

Proposition (black–white symmetry).
$F[x|NB(k, p)] = 1 - F[k - 1|NB(x + 1, 1 - p)]$.

6.4   The Poisson Family

6.4.1   Poisson Approximation to Binomial Probabilities

We invented negative binomial and binomial random variables to approximate certain urn problems that, though involving many marbles, in practice required us to count relatively few marbles. This does not mean that these new families are useful only in problems involving small counts.

Example. A manufacturer of integrated circuit chips says that the probability that one of his chips will be bad is no more than 2%. You will periodically test 100 chips, chosen at random, and you will complain to the manufacturer if you discover 6 or more bad chips. What is the probability that from a given experiment you will complain in error? The number of bad chips in a test batch might be a B(100, 0.02) random variable.

$$P(X \geq 6) = P(X > 5) = 1 - P(X \leq 5) = 1 - p(5) - p(4) - \cdots - p(0),$$

where $p(5) = \frac{100 \cdot 99 \cdot 98 \cdot 97 \cdot 96}{5 \cdot 4 \cdot 3 \cdot 2 \cdot 1} .02^5 .98^{95} = .035347$, and so forth. This is a longish, but not impractical, hand calculation. We conclude that the total probability of rejecting a batch is 0.01548; so we will not be sounding the alarm in error very often.

This calculation reminds me of cases where we could do simple approximations in earlier sections. When n is large compared to x, we would presumably organize the binomial calculation as $p(x) = \frac{(n)_x}{x!} p^x (1 - p)^{n-x}$. But we now know that if $\binom{x}{2}$ is small compared to n, then $(n)_x$ is well approximated by n^x. In this case, $p(x) \approx \frac{(np)^x}{x!}(1 - p)^{n-x}$.

Since n is large and x is small, we are presumably interested in cases where p is small; therefore, the quantity np is not too large compared to n. This leaves the exponent, $n - x$, the only irritatingly large part of this expression. Let us see whether we can simplify that as well: First factor it into a large and a smaller piece $(1 - p)^n / (1 - p)^x$. Remembering that $1 - p \leq e^{-p}$ (see Exercise 3.24), we have that $(1 - p)^n \leq e^{-np}$ using the basic multiplicative property of exponents. In the quality control problem, this means that $0.98^{100} = 0.1326 \leq e^{-100 \cdot 0.02} = 0.1353$. It seems that the exponential upper bound is fairly close; perhaps we may use it as the desired approximation? To do so we need to find out how close it is in general, which means that we need a lower bound. This will require a bit of ingenuity: $\frac{1}{1-p} = 1 + \frac{p}{1-p} \leq e^{p/(1-p)}$. But then

$$(1 - p)^n = \left(\frac{1}{1 - p}\right)^{-n} = \left(1 + \frac{p}{1 - p}\right)^{-n} \geq e^{-np/(1-p)}.$$

How close is this to the upper bound? A little algebra establishes that since $\frac{1}{1-p} = 1 + \frac{p}{1-p}$, we have $e^{-np/(1-p)} = e^{-np - np^2/(1-p)}$.

Proposition. (i) $e^{-np} e^{-np^2/1-p} \leq (1 - p)^n \leq e^{-np}$.

(ii) *If $np^2/(1 - p)$ is close to zero, then $(1 - p)^n / e^{-np}$ is close to one.*

The second fact follows because in that case the second exponential in (i) is close to 1. Furthermore, since $\binom{x}{2}$ is small compared to n, we have for our remaining piece $(1 - p)^x \approx e^{-xp} \approx 1$. We have now assembled the facts necessary to state a very useful approximation to a binomial random variable:

Theorem (Poisson approximation to the binomial). *For a binomial* $B(n, p)$ *random variable such that* $np^2/(1 - p)$ *is small, then if* $\binom{x}{2}$ *is small compared to* n, *we have* $p(x) \approx \frac{(np)^x}{x!}e^{-np}$.

Example. In the quality control problem with $n = 100$ and $p = 0.02$, we note that $\frac{100 \cdot (0.02)^2}{1 - 0.02} = 0.0408$ is much smaller than 1, and $\binom{5}{2}$ is small compared to 100. Then we feel free to try $p(5) \approx \frac{(2)^5}{5!}e^{-2}$, and so forth for 4, 3, ... The probability of rejecting a batch turns out to be approximately 0.0166, which is reasonably close to the exact answer, 0.01548.

Our approximation to the probability mass function is attractively simple, particularly so since the parameters of the binomial always just appear as the product np; this is the quantity we have claimed will turn out to be the expectation of a binomial. It is common to write this $\lambda = np$ (Greek letter lambda), so that our approximation looks like $p(x) \approx \frac{\lambda^x}{x!}e^{-\lambda}$.

6.4.2 Approximation to the Negative Binomial

Such a simple result deserves to be used in other problems, and justice triumphs. The same formula is useful in approximating certain *negative* binomial probabilities. The idea will be that if x is small enough and k is large enough, then in

$$p(x) = \binom{x + k - 1}{x} p^x(1 - p)^k = \frac{(x + k - 1)_x}{x!} p^x(1 - p)^k$$

we may sometimes be able to replace $(x + k - 1)_x$ with k^x. In a similar way to the binomial case, for p small we may sometimes be able to say that $(1 - p)^k$ is close to $e^{-kp/(1-p)}$. Notice that we contrived the exponent to match what we conjecture to be the expectation.

Theorem (Poisson approximation to the negative binomial). *For a negative binomial* NB(k, p) *random variable such that* $kp^2/(1 - p)$ *is small, then if* $\binom{x}{2}$ *is small compared to* k, *let* $\lambda = kp/(1 - p)$. *Then* $p(x) \approx \frac{\lambda^x}{x!}e^{-\lambda}$.

PROOF. Exercise. The argument parallels the previous one, with slightly more work required to arrange that the parameter equal the expectation.   ⊓

Example. The rare XXY configuration of the sex chromosomes occurs in about 1.5% of all human males. You require a sample of 400 men who do not possess this arrangement, so you test a random sequence of men until you have enough without this configuration. What is the probability of 3 or fewer XXY subjects that you must discard from your sample?

The negative binomial model is reasonable here, with $k = 400$ and $p = 0.015$. Then we calculate $p(3) = \binom{402}{3}0.015^3 0.985^{400} = 0.08591$; and so forth for 2, 1, 0 to get 0.1452. We suspect that the Poisson approximation might be appropriate, since $kp^2/(1 - p) = 0.09137$ and $\binom{x}{2}/k = 0.0075$ are fairly small. We have $\lambda = \frac{kp}{1-p} = 6.0914$, and so

$$p(3) \approx \frac{6.0914^3}{3!}e^{-6.0914} = 0.08522.$$

The total approximate probability is 0.1432, which is quite close to the exact calculation.

6.4.3 Poisson Random Variables

When we found useful approximations to probability mass functions earlier in the chapter, the new formulas turned out to be exact for certain new families of random variables. Our luck will hold, but unfortunately, our new family cannot be realized by some simple probability process that can be modeled exactly by draws from an urn, or rolling dice, or some such experiment. We shall have to wait to develop the tools to define this *Poisson process*; in the meantime, we have a probability mass function $p(x) = \lambda^x/x!e^{-\lambda}$, which may give us the probabilities we need. We note that for $x \geq 0$, the probabilities are positive. Furthermore,

$$\sum_{x=0}^{\infty} \frac{\lambda^x}{x!}e^{-\lambda} = e^{-\lambda}\sum_{x=0}^{\infty} \frac{\lambda^x}{x!} = e^{-\lambda}e^{\lambda} = e^0 = 1$$

by a standard infinite series you learned in calculus. Therefore, our probabilities sum to 1, and we have the information required to define a discrete random variable.

Definition. A **Poisson** random variable X with parameter $\lambda \geq 0$ has sample space $X = 0, 1, 2, \ldots$ and probability mass function $p(x) = (\lambda^x/x!)e^{-\lambda}$.

We gather clues from its applications so far as to how this family might be useful. In both the negative binomial and binomial cases, it approximately described a situation in which we counted successes in independent Bernoulli trials when the probability of success was very small, but the number of failures, or trials, was rather large. Generally, we will think of using Poisson random variables as models when we are counting rare, independent events. We may interpret λ, since it is np in the binomial case, as a measure of the average rate at which the rare events are happening.

Example. The lightning rod on the top of a certain skyscraper is hit by bolts of lightning at an average rate of about 3 times per year, based on many years of experience. What is the probability that it will be hit 6 or more times next year? Since these strikes are rare occurrences, and presumably independent when looked at over long time intervals, we presume that the number of hits is a Poisson variable with $\lambda = 3$. Then

$$P(X \geq 6 | \lambda = 3) = 1 - P(X \leq 5) = 1 - p(5) - p(4) - \cdots - p(0);$$

we calculate $p(5) = \frac{3^5}{5!}e^{-3} = 0.10082$. After calculating all 6 probabilities, our answer is then 0.0839.

We could have pretended that there were 1000 chances for lightning to strike in a year, with a probability of 0.003 that each would happen; then we would use the Poisson approximation to a binomial variable, with the same λ as before, and get the same answer. But we have no idea how many times lightning almost struck; so we use the Poisson model directly.

Our approximation results may be interpreted as limits.

Theorem (Poisson limits in a Bernoulli process). (i) *Given a sequence of negative binomial random variables* $\{X_i\}$ *distributed* $\mathrm{NB}(k_i, p_i)$, *where* $p_i \to 0$ *and* $k_i p_i / (1 - p_i) \to \lambda > 0$, *then the sequence converges in distribution to a Poisson*(λ) *random variable.*

(ii) *Given a sequence of binomial random variables* $\{X_i\}$ *distributed* $\mathrm{B}(n_i, p_i)$, *where* $p_i \to 0$ *and* $n_i p_i \to \lambda > 0$, *then the sequence converges in distribution to a Poisson*(λ) *random variable.*

We can get some idea of the expected value of a Poisson random variable by looking at the behavior of similar binomials: $\lim_{i \to \infty} E(X_i) \overset{?}{=} \lim_{i \to \infty} n_i p_i = \lambda$. After two speculative uses of limits, we conjecture that the expectation of a Poisson random variable simply equals λ; we will shortly verify that this is correct. Notice that we were taking advantage of this guess in the lightning problem: We would estimate the rate of strikes per year by finding the sample average number over many years.

Poisson random variables are so simple that they have no symmetries at all. Nevertheless, or perhaps because of this, we will find them enormously useful from now on.

6.5 More About Expectation

We have speculated about the expectations of some of our limiting families, using somewhat dubious limit arguments to get plausible-sounding results. Let us tackle these problems more directly from the probability mass functions.

Let X be Poisson(λ); then if the expectation exists, we would have

$$E(X) = \sum_X Xp(X) = \sum_{X=0}^{\infty} X\frac{\lambda^X}{X!}e^{-\lambda}.$$

The first term in this sum is zero, and in all others the X cancels the first factor of $X!$:

$$E(X) = \sum_{X=1}^{\infty} \frac{\lambda^X}{(X-1)!} e^{-\lambda}.$$

Except that X starts at one instead of zero, this reminds us of a sum of Poisson probabilities; so substitute $Y = X - 1$:

$$E(X) = \sum_{Y=0}^{\infty} \frac{\lambda^{1+Y}}{Y!} e^{-\lambda} = \lambda \sum_{Y=0}^{\infty} \frac{\lambda^Y}{Y!} e^{-\lambda}.$$

But the sum is just the total of all the probabilities of possible values for a Poisson(λ) random variable, which is, of course, 1. So $E(X) = \lambda$, as conjectured.

This technique, rearranging the expectation formula so that the hard part is a sum of all probabilities and so equal to 1, appears everywhere in statistics. We will call it the **inductive** method.

You may have noticed that when we used summation notation in our expectation formulas, we let the index of summation be written capital X or Y, as if the index were a random variable. It turns that the index of summation behaves just like a random variable in such formulas; we do not know its value yet, but it must be one from the list. This convention will be particularly helpful later, when our random variables are no longer discrete.

The same approach gives us the expectation of a binomial B(n, p) random variable:

$$E(X) = \sum_{X=0}^{n} X \frac{n!}{X!(n-X)!} p^X (1-p)^{n-X} = \sum_{X=1}^{n} \frac{n!}{(X-1)!(n-X)!} p^X (1-p)^{n-X}.$$

Once again it seems reasonable to substitute $Y = X - 1$:

$$E(X) = \sum_{Y=0}^{n-1} \frac{n!}{Y!(n-1-P)!} p^{1+Y} (1-p)^{n-1-Y}$$

$$= np \sum_{Y=0}^{n-1} \frac{(n-1)!}{Y!(n-1-Y)!} p^Y (1-p)^{n-1-Y}.$$

Now the part under the summation is the collection of all probabilities for a B($n - 1$, p) random variable, which sum to one; so as we hoped, $E(X) = np$. The sort of change from n to $n - 1$ often happens in this method and is why we chose to call it the inductive method, since it may remind you of proofs by induction in mathematics.

Proposition. *For X following the law*

(i) *If X is Poisson(λ), then* $E(X) = \lambda$.
(ii) *If X is B(n, p), then* $E(X) = np$.
(iii) *If X is NB(k, p), then* $E(X) = kp/(1 - p)$.

The proof of (iii) is an exercise, using the same inductive principle of rearranging the sum so that the hard part equals 1. You might also try some harder calculations, using this technique to verify our expressions for the hypergeometric and negative hypergeometric expectations (see 5.5.3).

Example. Approximately 10% of Americans are left-handed. You need 20 left-handers for a study of the relationship between left-handedness and left-footedness. How many people will you have to interview, on average, to get your 20?

Strictly, interviews are not independent: Since we do not interview anybody twice, we are really selecting without replacement. In practice, the number of Americans is so huge compared to the number we are interviewing that it might as well be with replacement. We pretend that interviews are independent, and then the number of righties interviewed is negative binomial. In this way, we do not even have to figure out how many Americans are eligible for the study; just the probability 0.1 of a success. The expectation is then $20 \cdot \frac{0.9}{1-0.9} = 180$ right-handers to be interviewed, for a total of 200 interviews.

Example. Generate a discrete random variable by the following procedure: (1) Use a calculator or a computer to generate a real-valued random number X uniformly on the interval from 0 to 1; (2) calculate $Y = 1/X$; and (3) write down Z, the largest whole number no bigger than Y. Then Z has sample space $1, 2, 3, \ldots$. For example, my calculator gets $X = 0.2289823$; then $Y = 4.36715$, and $Z = 4$. Now

$$F(z) = P(Z \le z) = 1 - P(Z \ge z + 1) = 1 - P(Y \ge z + 1)$$

$$= 1 - P(X \le 1/(z + 1)) = 1 - 1/(z + 1).$$

We use our rule for extracting the probability mass function $p(x) = F(x) - F(x-1)$ to conclude that $p(z) = 1 - 1/(z + 1) - (1 - 1/z) = 1/(z(z + 1))$. For example, $p(4) = 1/20$.

Now let us find the expectation of Z:

$$E(Z) = \sum_{Z=1}^{\infty} Z \frac{1}{Z(Z + 1)} = \sum_{Z=1}^{\infty} \frac{1}{Z + 1} = \frac{1}{2} + \frac{1}{3} + \frac{1}{4} + \cdots.$$

In case you do not remember how to sum this famous series (called the *harmonic* series) from calculus, let us see whether we can approximate the answer. Our approach will be to partition the sample space into a convenient collection of events: $C_1 = \{1\}$, $C_2 = \{2, 3\}$, $C_3\{4, 5, 6, 7\}$, and generally $C_i = \{2^{i-1} \le X < 2^i\}$. This is a useful partition because $P(C_i) = F(2^i - 1) - F(2^{i-1} - 1) = 2^{-i}$. Instead of multiplying each outcome by its probability and summing, we will find a *lower* limit for the expectation, by multiplying the probability of each element of the partition by the *smallest* value of its constituent outcomes:

$$E(X) = \sum_{X} X p(X) = \sum_{i} \left[\sum_{X \in C_i} X p(X) \right] \ge \sum_{i} \left[\min_{X \in C_i} X \sum_{X \in C_i} p(X) \right]$$

$$= \sum_i \min_{X \in C_i} XP(C_i).$$

In our problem, $\min_{X \in C_i} X = 2^{i-1}$, so our lower limit is

$$\sum_i \min_{X \in C_i} XP(C_i) = \sum_{i=1}^{\infty} 2^{i-1} 2^{-i} = \sum_{i=1}^{\infty} \frac{1}{2} = \frac{1}{2} + \frac{1}{2} + \cdots,$$

which is, of course, *infinite*. Since a lower bound on our expectation is infinite, we can only conclude that the expectation of our random variable is infinite. Some simple random variables do not possess a finite expectation.

What practical meaning does the lack of a finite expectation for the results of an experiment have? If you repeated, for example, a binomial experiment a great many times and averaged your results, you would find that with high probability, the answer would be close to our expected value np (as we will check later). But if you repeated the calculator experiment many times and took an average, the result would be highly variable, no matter how many times you repeated it. I generated 1000 independent copies of this random variable; my average was 7.80. I generated a second set of 1000 values; this time the average was 18.01. It showed no sign of settling down to some single value.

6.6   Mean Squared Error and Variance

6.6.1   Expectations of Functions

Random variables often represent efforts to measure some important number when there is random "noise" that keeps us from doing so accurately. For example, if 80% of the voters in a country favor some policy (though we do not know this), we might try to find this out by interviewing 100 people picked at random about their opinion. The result is unpredictable, but a reasonable model is that the number interviewed will be a binomial $B(100, 0.8)$ random variable. In our hearts, we believe that the "true" result of our experiment ought to be 80 in favor, so that the percentage is representative of the country as a whole.

In (5.6.1), we used the observed value of a random variable to get a method-of-moments estimate of a parameter in a family. We were seeing a parameter μ as an unknown, ideal value for which X is an erratic reflection. How good is X as a measure of μ? Statisticians use any of a number of standards of closeness of a random variable to some fixed value, but the single most useful one was popularized by the French mathematical astronomer Legendre about 1805. He proposed that the average value of the squared difference, $(X - \mu)^2$, was particularly easy to work with as a measure of how far X was, on the whole, from the ideal value. Clearly, this was inspired by the *sample* mean squared error from least-squares theory (see 2.2.2). For random variables, expectation embodies our idea of the average, but we apparently have to move beyond our basic idea of the expectation of X to the concept of the expectation of some function, call it $g(X)$. If our random variable were discrete uniform, then the expectation should still be a simple average, but now of the values of g, that is, $E[g(X)] = \frac{1}{n} \sum_{i=1}^{n} g(x_i)$ if there are n equally likely values. We should apply our weighted-average technique for the case of general discrete variables:

Definition. Let X be a discrete random variable and g a real-valued function defined on the sample space of X. Then $E[g(X)] = \sum_X g(X)p(X)$ whenever this sum is absolutely convergent.

Definition. The **mean squared error** of a random variable X with respect to a constant μ is $E[(X - \mu)^2]$.

Example. Consider a $B(3, 0.8)$ random variable. If we choose as its ideal value $\mu = 2$, then the mean squared error calculation would go as follows:

X	$p(X)$	$(X-2)^2$	$(X-2)^2 p(X)$
0	0.008	4	0.032
1	0.096	1	0.096
2	0.384	0	0
3	0.512	1	0.512
		total	0.64

We need to learn a bit more about the expectation of a function.

Theorem (expectation is a linear operator). *For X a discrete random variable:*

(i) *If a is constant, then $E(a) = a$.*
(ii) $E[ag(X)] = aE[g(X)]$ *whenever the second expectation exists.*
(iii) $E[g(X)+h(X)] = E[g(X)]+E[h(X)]$ *whenever the right-hand expectations exist.*

PROOF.   (i) $E[a] = \sum_x ap(X) = a \sum_x p(X) = a \cdot 1 = a$.
(ii) $E[ag(X)] = \sum_x ag(X)p(X) = a \sum_x g(X)p(X) = aE[g(X)]$.
(iii) $E[g(X) + h(X)] = \sum_x [g(X) + h(X)]p(X) = \sum_x g(X)p(X) + \sum_x h(X)p(X) = E[g(X)] + E[h(X)]$.   □

One important case of linearity is that $E(X + a) = E(X) + a$, applying (iii) and then (i) above. If there is a fixed cost every time we perform an experiment, the average cost is just that fixed cost, plus the average of the part of the cost that varies by chance.

We squared the distance from the reference point when defining a mean squared error in order that the result be a positive, or at least not a negative, number, to match our idea of a distance. Clearly, the average of positive numbers should be positive; and by staring at the definition we see that this is true for expectations:

Proposition (expectation is a positive operator).   (iv) *For $g(x) \geq 0$, $E[g(X)] \geq 0$.*

This must be, because all the terms in the sum are at least zero. An operator that is a linear operator and also meets this proposition is called a *positive linear operator*.

6.6.2 Variance

We will use these facts about expectations to extract some information about mean squared errors. An obvious limitation of mean squared errors as measures of the variability of a random variable is that they depend on your choice of ideal reference point, μ. As we did with samples (see 2.4.2), we look for a *minimum* possible value of the mean squared error. This would be a plausible measure of the uncertainty, or variability, inherent in that experiment. In this case, we make the following definition:

Definition. The **variance** of a random variable X is the minimum value among all possible mean squared errors with different centers μ. It is written Var(X).

Obviously, this was inspired by the *sample* variance. Let us assume that X has a variance, and that there is a number μ such that Var$(X) = E[(X - \mu)^2]$. Let us try to learn something about μ. First consider any other reference point v. Then by definition, Var$(X) = E[(X - \mu)^2] \le E[(X - v)^2]$. Now add and subtract μ inside the square on the right-hand side of the inequality:

$$E[(X - v)^2] = E[(X - \mu + \mu - v)^2] = E[(X-\mu)^2 + 2(\mu - nu)(X - \mu) + (\mu - v)^2].$$

Now we use the linearity properties of the expectation established earlier to get

$$E[(X - V)^2 = E[(X - \mu)^2] + 2(\mu - v)E(X - \mu) + (\mu - v)^2.$$

Comparing this to the equality above, we discover that for any value of v, we must have

$$2(\mu - v)E(X - \mu) + (\mu - v)^2 \ge 0.$$

What about μ would make this so? The second term is no problem, but it looks as if the first term could be of either sign and any size. However, if $E(X - \mu) = 0$, then the inequality is certainly always true, and this happens when $\mu = E(X)$. We have concluded that the minimum value of the mean squared error, which we now call the variance, measures deviations from the expected value. To summarize:

Proposition. *Let $\mu = E(X)$. Then*

(i) *for any number v, $E[(X - v)^2] = E[(X - \mu)^2] + (\mu - v)^2$ so long as the first expectation exists for some v. As a consequence,*

(ii) *Var$(X) = E[(X - \mu)^2]$ (since the previous equation shows that it must be the minimum value of the mean squared error), and*

(iii) *Var$(X) = E[X^2] - E(X)^2$ (by letting $v = 0$).*

We will call (iii) the short formula, since it often shortens our calculations.

Example. In the B$(3, 0.8)$ case above, $\mu = E(X) = 2.4$. We compute $E(X^2) = 6.24$; therefore, Var$(X) = 6.24 - (2.4)^2 = 0.48$ (see Figure 6.1).

It is worth noticing that Var$(a) = E(a^2) - (a)^2 = a^2 - a^2 = 0$. That is, a quantity that does not vary has no variance. Also,

$$Var(X + a) = E\{[(X + a) - E(X + a)]^2\} = E\{[X - E(X)]^2\} = Var(X),$$

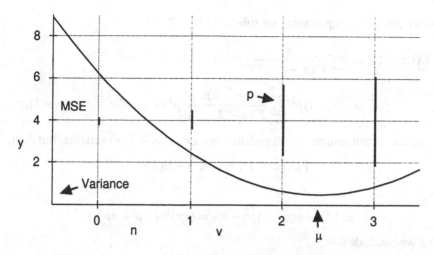

FIGURE 6.1. Mean squared error and variance

since the a's cancel. That is, adding or subtracting a constant amount to a random variable has no effect on its variability, as we would have hoped. Furthermore,

$$\text{Var}(aX) = \text{E}(a^2X^2) - \text{E}(aX)^2 = a^2\text{E}(X^2) - [a\text{E}(X)]^2 = a^2[\text{E}(X^2) - \text{E}(X)^2]$$
$$= a^2\text{Var}(X),$$

a somewhat less intuitive fact, to which we will return. These are important enough to summarize:

Proposition (properties of the variance).

(i) $\text{Var}(a) = 0$.
(ii) $\text{Var}(X + a) = \text{Var}(X)$.
(iii) $\text{Var}(aX) = a^2\text{Var}(X)$.

6.6.3 Variances of Some Families

We hope to find general formulas for the variance of whole families, for example, the binomial. Let X be $B(n, p)$. Try the inductive method. We might use the short formula, for which we need to calculate

$$E(X^2) = \sum_{X=0}^{n} X^2 \frac{n!}{X!(n-X)!} p^X(1-p)^{n-X}.$$

Unfortunately, only one of the X's cancels, and we are left with a bit of a mess. After a small flash of ingenuity, we calculate instead

$$E[X(X-1)] = \sum_{X=0}^{n} X(X-1) \frac{n!}{X!(n-X)!} p^X(1-p)^{n-X}$$

$$= \sum_{X=2}^{n} \frac{n!}{(X-2)!(n-X)!} p^X(1-p)^{n-X}.$$

As we did for the expectation, we substitute $Y = X - 2$:

$$E[X(X-1)] = \sum_{Y=0}^{n-2} \frac{n!}{Y!(n-2-Y)!} p^{2+Y}(1-p)^{n-2-Y}$$

$$= n(n-1)p^2 \sum_{Y=0}^{n-2} \frac{(n-2)!}{Y!(n-2-Y)!} p^Y (1-p)^{n-2-Y} = n(n-1)p^2,$$

since the second sum covers all probabilities for a $B(n-2, p)$ variable. But then,

$$E[X(X-1)] = E(X^2) - E(X),$$

so

$$E(X^2) = n(n-1)p^2 + np = (np)^2 + np - np^2,$$

and we conclude that

$$Var(X) = E[X^2] - E(X)^2 = np - np^2 = np(1-p).$$

Proposition.

(i) *If X is* $B(n, p)$, *then* $Var(X) = np(1-p)$.
(ii) *If X is Poisson(λ), then* $Var(X) = \lambda$,
(iii) *If X is* $NB(k, p)$, *then* $Var(X) = kp/(1-p)^2$.

Parts (ii) and (iii) are exercises, which should be done by the same method. It is possible to find the variance of hypergeometric and negative hypergeometric random variables by the same technique, though we will develop another, perhaps simpler, method shortly.

Though mean squared error and variance are very important concepts, they have little intuitive meaning to most of us as measures of the uncertainty in a random variable. For one thing, they are in units of the square of the original measurement. If the random variable is in dollars, its variance is in dollars-squared, whatever that means. We therefore find it useful to have the following definition:

Definition. The square root of a mean squared error is called a **root-mean-square** (rms) error. The square root of the variance is called the **standard deviation**, often denoted by σ_X.

This definition explains the common convention of denoting a variance by σ^2. Note that this is like calling the sample variance s^2 and the sample standard deviation s. From the corresponding fact about the variance, we discover by taking square roots that $\sigma_{aX} = |a|\sigma_X$. This means that the standard deviation is a measure of variability in the same units as X.

Example. If I toss a fair coin 100 times, I presume that the number of heads observed is $B(100, 0.5)$. The expected number of heads is, of course, $np = 50$, and the variance is $np(1-p) = 25$. This has little flavor, but the standard deviation is 5 heads. We might think of that as a typical deviation about the expectation, so that 45 heads would not be unusually small, and 55 would not be unusually large.

6.7   Bernoulli Parameter Estimation

6.7.1   Estimating Binomial p

The families of random variables in this chapter of course become more interesting when we want to learn the values of unknown parameters.

Example. You are a pollster and are hired by a candidate for governor to find what proportion of the likely voters in a large state would currently favor her for governor. You sample 200 voters, randomly selected from the pool of likely voters, and 107 favor her. What can you say about her actual statewide support?

First, we will assume that we have drawn few enough voters that we may safely pretend that we are sampling with replacement (see the exercises for the sorts of conditions we must meet). So a plausible model for our experiment is that 107 turned out to be a value from a B(200, p) random variable, where the unknown p is the probability that a random voter favors our candidate. The value of p is the most important question we are likely to be asked. As in the last chapter, we might as well let a standard estimate be the one suggested by matching expectation to observed value: $X \approx E(X) = np$, so $\hat{p} = X/n$. We will see sounder reasons why this is a good idea in later chapters. Meanwhile, we note without astonishment that it matches the standard estimate, the sample proportion, from Chapter 1 (see 1.7.1).

In our example, we estimate that a voter will favor your candidate with probability $\hat{p} = 0.535$. The next important question is, How close to the truth is this likely to be? It is, of course, itself a random variable, so

$$\text{Var}(\hat{p}) = \text{Var}\left(\frac{X}{n}\right) = \frac{1}{n^2}\text{Var}(X) = \frac{np(1-p)}{n^2} = \frac{p(1-p)}{n},$$

from what we have learned about variances. Then the standard deviation is $\sigma_{\hat{p}} = \sqrt{\frac{p(1-p)}{n}}$. Incidentally, the standard deviation of an estimate of a model parameter is often called its *standard error*. In (2.4.2), we mentioned that a rule of thumb for capturing much of the range of variation of a data set was a 2-s interval, which deviated up and down by two sample standard deviations from the sample mean. For random variables, particularly those that estimate quantities of interest, we define a corresponding 2-σ *interval*; in this case that would be

$$p - 2\sqrt{\frac{p(1-p)}{n}} \le \hat{p} \le p + 2\sqrt{\frac{p(1-p)}{n}}.$$

In later chapters we will learn something about how probable it is that the estimate falls in this range.

Of course, what we have written down is of little use, because once we do the poll, $\hat{p}$ is known, but p is still quite unknown. It would be more interesting to move things across the inequalities to get the mathematically identical statement

$$\hat{p} - 2\sqrt{\frac{p(1-p)}{n}} \le p \le \hat{p} + 2\sqrt{\frac{p(1-p)}{n}}.$$

Now the quantity we want to know is between limits, we hope with high probability. You are laughing at me, naturally, because you think that I have forgotten that the unknown p is still in those square-root terms. That is a problem, but since what we are doing is rough anyway, we do something crude but plausible: Replace these p's with their estimated value $\hat{p}$, to get the practically useful estimated 2-σ interval

$$\hat{p} - 2\sqrt{\frac{\hat{p}(1-\hat{p})}{n}} \le p \le \hat{p} + 2\sqrt{\frac{\hat{p}(1-\hat{p})}{n}}.$$

Example (*cont.*). The probability of a vote for your candidate has the 2-σ interval $0.4645 \le p \le 0.6055$.

That somewhat arbitrary trick of replacing the standard error by its rule-of-thumb estimate has one reassuring property: Although p and $\hat{p}$ are unlikely to be equal, it happens that the function $\sqrt{p(1-p)}$ changes rather slowly so long as we stay away from 0 and 1 (see Figure 6.2).

Therefore, it usually does not hurt much to replace the standard deviation by its estimate. This helps to explain why the experience of statisticians with this interval has been generally pleasant, despite its several arbitrary features.

6.7.2   *Confidence Bounds for Binomial p*

We learned in the last chapter how we could go beyond rules of thumb, and make definite probabilistic statements about the value of an unknown parameter, by constructing confidence bounds (see 5.6.2). Of course, we can do exactly the same

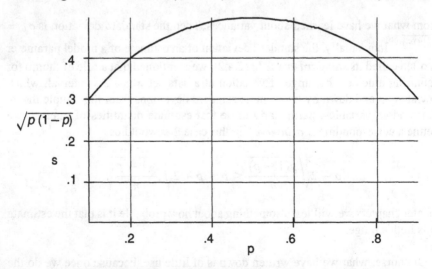

FIGURE 6.2. Binomial standard deviation as a function of p

thing for the p parameter in a binomial distribution. The one problem here is that earlier we used the formula for the cumulative distribution function that we fortunately had in that case. The binomial cumulative is a messy sum with no closed form. In the exercises, you will develop a simplified way to compute it; but even so, the author wrote a little computer program to aid him in doing the calculations in this section.

To get a lower, say, 95% confidence bound for p in our polling example, we want to find at what value the result of $X = 107$ favorable voters becomes improbable (at the 5% significance level). For us to decide that p is implausibly small, we will have to decide that X was improbably large; that is,

$$P(X \geq 107|B(200, p)) = 1 - F(106) = \sum_{X=107}^{200} \binom{200}{X} p^X (1 - p)^{200-X}$$

gets a small p-value. After a number of time-consuming calculations, I home in on a value that is barely compatible with the data:

p	$P(X \geq 107)$
0.5	0.179002
0.45	0.009668
0.48	0.068677
0.47	0.038404
0.475	0.051810
0.474	0.0488675
0.4744	0.050028

If I kept going, I could get as close to 0.05 as I pleased, but this will do. As a result of my poll, I believe that the proportion of voters favoring my candidate is $p \geq 0.4744$, with 95% confidence. I remember that what this really means is that before I took the poll, the probability was 95% that whatever lower confidence bound I set would be a correct inequality.

In this problem, an *upper* confidence bound turns out to be similarly useful. I look for the value of p at which counts of $X \leq 107$ become implausibly *small*, to conclude after many calculations that a 95% upper confidence bound would be $p \leq 0.5948$.

6.7.3 Confidence Intervals

I am sure that you are tempted to combine our two inequalities, to say $0.4744 \leq p \leq 0.5948$; this should tell us to what accuracy we have learned our degree of political support, with high probability. (It also looks a bit similar to the 2-σ interval from the last section.) But we need to be careful: Just what is the probability that this double inequality is correct? Turn the problem around, and ask the probability that such an interval would be wrong. Then for the (unknown) true value of p, either $X \geq 107$ has a low probability, or $X \leq 107$ has a low probability. These cannot both be the case, so long as $\alpha < 0.5$ defines a low probability, because the total of the two probabilities is at least 1. Therefore, either the first or the second inequality is false, but not both; the two events are mutually exclusive. We conclude from our addition rule that the probability that our interval above is false is $0.05 + 0.05 = 0.10$; it is therefore true with probability 0.90. We are ready to make a new definition:

Definition. Let a random variable X be observed from a family with unknown parameter θ. Let X lead us to reject θ as too small at a significance level α_L for exactly the values $\theta < \theta_L$; and for exactly the values $\theta > \theta_U$, let X lead us to reject θ as too large at the significance level α_U; and $\alpha_U + \alpha_L = \alpha$. Then we say that $\theta_L \le \theta \le \theta_U$ is a $(1 - \alpha) \times 100\%$ **confidence interval** for θ.

It seems that the interval above is only a 90% confidence interval for p.

Notice that to get a conventional 95% confidence interval for p, we must find lower and upper confidence bounds whose p-values sum to 0.05. There are obviously an infinite number of ways to do this. If we wish to be evenhanded about high and low misses, there are still several possibilities. Perhaps the best way is to reason that since we want to pin down the true value as precisely as possible, we should choose the *shortest* confidence interval such that for the two significance levels we have $\alpha_U + \alpha_L = \alpha$. This was not often done in practice, before computers were universal, because the computations may be a bit laborious.

The most popular way of constructing confidence intervals is simply to let $\alpha_U = \alpha_L = \alpha/2$. In the example, I proceed just as I did in the last section to find 97.5% upper and lower confidence bounds, and I conclude that $0.4633 \le p \le 0.6056$ is a 95% confidence interval. Notice that it is amazingly close to the 2-σ interval of the last section. It will turn out in a later chapter that this is not a coincidence; the 2-σ rule-of-thumb was invented to be an approximation to a 95% confidence interval in many important cases.

6.7.4   Two-Sided Hypothesis Tests

We have now seen a case in which we were simultaneously interested in the probability that a random variable might be surprisingly high and that it might be surprisingly low. This also happens sometimes in hypothesis testing.

Example. According to standard genetic theory, since brown eyes are dominant over blue, exactly 25% of the offspring of couples, both brown-eyed and heterozygotic for blue eyes, should turn out to be blue-eyed. You have a simple genetic test for heterozygoticity in this case. You will find brown-eyed couples who pass your test and continue the experiment until you have found 30 blue-eyed offspring of such couples. Naturally, you expect to find about 90 brown-eyed offspring along the way; if you get many more or many fewer than this, something has most likely gone wrong with either your experimental procedure or your genetic theory. It would be very interesting to discover when things indeed have gone wrong.

A reasonable model here is that the count of brown-eyed offspring should be NB(30, 0.75). We will set up a hypothesis test, with this as the null hypothesis. But we will reject it, at significance level, say, $\alpha = 0.01$, if the count of brown-eyed children is either surprisingly large (so 0.75 is an unrealistically low probability), or if the count is surprisingly low (so we will suspect that 0.75 is too high). To make sure that we will at most 1% of the time make a claim that Mendel was wrong (if he is indeed right), we follow the simple approach of the last section, allowing a probability of $\alpha/2 = 0.005$ that we will get too high a count, and the same probability that the count will be too low. We call this a *two-sided* hypothesis test.

After some laborious calculations with the aid of my computer, I find that $P(X \geq 146) = 0.00494$ and $P(X \leq 47) = 0.00477$ are the least extreme values I may use. Therefore, I decide that if I observe at least 146 brown-eyed offspring in the course of my experiment, or if I observe at most 47, I will decide to reject the null hypothesis at the 0.01 level of significance. People who do this frequentist style of reasoning call those conditions for rejection the *critical region* of the experiment.

If I am the research assistant who actually carries out the experiment and I observe 130 brown-eyed children, I use the negative binomial probabilities under the null hypothesis to discover that since this count seems a bit large, $P(X \geq 130) = 0.02726$. But if I know that my boss will be wanting to use a two-sided critical region, I must admit that he would have been willing to reject the null hypothesis for *small* values that had similarly low probabilities, too. So I *double* the probability I calculated, to include these hypothetical low values; my *p*-value is 0.0545. With far less work than in the previous paragraph, I know that he will fail to reject his null hypothesis at the 0.01 level and in fact will (barely) do the same if his preferred level was 0.05. This convenience is why computer statistics packages usually report a *p*-value; you can then compare it to whatever significance level you had in mind.

6.8 The Poisson Limit of the Negative Hypergeometric Family*

We diagram some things we have learned about limiting distributions in Figure 6.3.

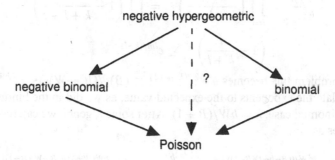

FIGURE 6.3. Poisson limit of negative hypergeometric variables

We have approximated negative hypergeometric probabilities in two very different ways; but under certain similar-sounding conditions we can approximate each of these cases by Poisson probabilities. The dotted arrow asks, can we then sometimes approximate negative hypergeometric probabilities directly by Poisson probabilities?

We proceed as before to look for simplification, when x is small compared to W and b, and these are small compared to B:

$$p(x) = \frac{(x+b-1)_x(W)_x(B)_b}{x!(B+W)_{x+b}} = \frac{(x+b-1)_x(W)_x(B)_b}{x!(B+W-b)_x(B+W)_b},$$

where at the second equality the permutation in the denominator was factored into two pieces in order to isolate all the terms that involve x. But our two permutation inequalities tell us that if $\binom{x}{2}$ is small compared to b and W, then

$$p(x) \approx \frac{[bW/(B+W-b)]^x(B)_b}{x!(B+W)_b}.$$

The last two permutations could be approximated using results we already know, but only at the cost of unnecessarily strong conditions ($\binom{b}{2}$ small compared to B). Instead, we work a little harder:

Proposition.

$$e^{lm/(k+l)} \le \frac{(k+l)_m}{(k)_m} \le e^{lm/(k-m+1)}.$$

PROOF.

$$\frac{(k+l)_m}{(k)_m} = \prod_{i=0}^{m-1} \frac{k+l-i}{k-i} = \prod_{i=0}^{m-1}\left(1 + \frac{l}{k-i}\right)$$

$$\le \left(1 + \frac{l}{k-m+1}\right)^m \le e^{lm/(k-m+1)},$$

where the second inequality works because we replaced each term by the largest term in the product. Similarly,

$$\frac{(k+l)_m}{(k)_m} = \prod_{i=0}^{m-1}\left(\frac{k-i}{k+l-i}\right)^{-1} = \prod_{i=0}^{m-1}\left(1 - \frac{l}{k+l-i}\right)^{-1}$$

$$\ge \left(1 - \frac{l}{k+l}\right)^{-m} \ge e^{lm/(k+l)}. \qquad \square$$

In our problem this becomes $e^{-bW/(B-b+1)} \le (B)_b/(B+W)_b \le e^{-bW/(B+W)}$. We can relate the exponents to the expected value, as we did in the binomial and negative binomial case: $\lambda = bW/(B+1)$. After some algebra, we can rewrite our inequalities as

$$e^{-\lambda}e^{-b^2W/((B-b+1)(B+1))} \le \frac{(B)_b}{(B+W)_b} \le e^{-\lambda}e^{bW(W-1)/((B+W)(B+1))}.$$

To guarantee that the complicated exponents are small, we need only know that λ^2/b and λ^2/W are small, since $B + W$ and $B - b + 1$ are not far from $B + 1$.

All we need now is to check that the expression to the xth power may be replaced by $\lambda = bW/(B + 1)$. But to compare denominators,

$$\frac{(B + W - b)^x}{(B + 1)^x} = \left(1 + \frac{W - b - 1}{B + 1}\right)^x \approx 1,$$

so long as $x\lambda$ is small compared to b and W. We summarize our conclusions:

Proposition (Poisson approximation to the negative hypergeometric).

  (i) *Let a random variable be* $N(W, B, b)$*; then letting* $\lambda = bW/(B+1)$*, we have* $p(x) \approx \lambda^x/x!e^{-\lambda}$ *whenever* $\binom{x}{2}$ *and* λ^2 *are small compared to* b *and* W.
 (ii) *A sequence of random variables* $N(W_i, B_i, b_i)$ *such that* $W_i \to \infty$, $b_i \to \infty$, *and* $0 < \lambda = \lim_{i\to\infty} b_i W_i/(B_i + 1)$ *will converge in distribution to a* Poisson(λ) *random variable.*

Example. A manufacturer sells batches of 1000 capacitors and promises that no more than 30 are defective. Give him the benefit of the doubt, and assume that exactly 30 are bad. You need 50 good capacitors, so you test through a batch until you have found 50 good ones. What is the probability that you will find 3 or more bad ones along the way?

A reasonable model for the number of bad ones is $N(30, 970, 50)$. You, of course, calculate $P(X \geq 3) = 1 - p(0) - p(1) - p(2) = 0.2025$. after many multiplications and divisions. But this seems a reasonable candidate for a Poisson approximation, since $\lambda^2 = 2.386$ is much smaller than either 30 or 50. Using $\lambda = 1.5448$, we find a Poisson $P(X \geq 3) = 0.2013$ with much less work.

As an exercise, you should find conditions under which a Poisson random variable is a satisfactory approximation to a hypergeometric random variable.

6.9   Summary

We found some simple approximate calculations for negative hypergeometric probabilities, which corresponded to experiments in a *Bernoulli* process, independent experiments that either succeed (with probability p) or fail (3.3). The first Bernoulli-based family we studied was the *geometric family*, the count of successes before the first failure, $p(x) = p^x(1 - p)$ (2.2). This generalizes to the *negative binomial* family, which was the number of successes before a certain number k of failures have happened, with mass function $p(x) = \binom{x+k-1}{x}p^x(1 - p)^k$ (2.4). The *binomial* family, on the other hand, counted successes in a fixed number n of trials. Its mass function is $p(x) = \binom{n}{x}p^x(1 - p)^{n-x}$ (3.2). In either case, if successes have very low probability, their number may be approximated by the *Poisson* family, where for average number of successes λ, we had $p(x) = \frac{\lambda^x}{x!}e^{-\lambda}$ (4.3).

We learned to evaluate expectations in families like these by the *inductive* method (5). Then we studied expectations of functions of random variables, including the *variance* $\sigma^2 = \mathrm{Var}(X) = \mathrm{E}[(X - \mu)^2]$, where $\mu = \mathrm{E}(X)$ (6.2). We were led to a simple estimate for an unknown binomial parameter $\hat{p} = X/n$, and to a 2-σ *interval*,

$$\hat{p} - 2\sqrt{\frac{\hat{p}(1 - \hat{p})}{n}} \leq p \leq \hat{p} + 2\sqrt{\frac{\hat{p}(1 - \hat{p})}{n}},$$

as a rough way of describing how accurately we know p (7.1). More careful analysis led to a *confidence interval* for our binomial parameter (7.3). We then developed *two-sided* hypothesis tests for cases in which we are interested in surprisingly large as well as surprisingly small values of our statistics at the same time (7.4). Finally, to show off how much we have learned about approximation to probabilities, we found conditions under which there are direct Poisson approximations to the negative hypergeometric family (8).

6.10   Exercises

1. In Exercise 19 of Chapter 5, there were 12 women and 15 men in a statistics class who took a test.

   a. What is the probability that the highest-scoring woman scored fourth highest overall?

   b. Recompute your answer using the geometric approximation. Was the geometric approximation appropriate here? Is the answer close?

2. I am going to roll a balanced die until I get three sixes. What is the probability I will have rolled the die exactly 12 times?

3. **a.** Derive a closed formula (no summation symbols or ...s) for the cumulative distribution function $F(x) = \mathrm{P}(X \leq x)$ of a geometric(p) random variable.

   b. The probability of snake eyes on rolling a pair of dice is $1/36$. I can keep rolling until I roll snake eyes. What is the probability that I will roll no more than 25 times?

4. You have invested in an oil exploration company that drills six oil wells a year. You estimate that the probability of striking oil is about 0.2 at each well. Of course, you want to be there when the first well strikes oil. Unfortunately, you will leave the country on sabbatical for one year, starting one year from now (and returning two years from now). What is the probability that you will be in the country for the first strike?

5. I am told at the beginning of a mushroom-hunter's guide that 28 of the 96 species described are good to eat, but the guide is not organized that way. I want to learn about edible mushrooms, so I decide that on my first day of study, I will read about species at random until I have read articles about 3 edible species.

a. What is the probability that I will have read about at most 4 inedible species?

b. Now redo the calculation, using the negative binomial approximation. Is that approximation plausible here? How close is your result to the exact answer?

6. The owner of a stable of racing cars knows that there is a 14% chance that the car she enters will be wrecked in a race. She will have to stop entering races for a while to rebuild her cars after she has wrecked three of them.

a. What is the probability that she will have entered cars in 10 races at the time she has to stop?

b. After 11 races, she finds that she has had two cars wrecked. What is the probability that she will still be entering cars in races after a total of 16 races?

7. I need to hire 5 new programmers for my software development group. In my experience, approximately 30% of applicants will be satisfactory, and any satisfactory applicants whom I interview I will hire immediately. What is the probability that I will hire my fifth programmer after 12 *or fewer* interviews?

8. Assume that $B_i \to \infty$ and $\frac{W_i}{W_i + B_i} \to p$ as i goes to infinity, where $0 < p < 1$. Show that $W_i \to \infty$.

9. Prove that $e^{\frac{1}{n+k-1}\binom{k}{2}} \leq \frac{(n+k-1)_k}{n^k} \leq e^{\frac{1}{n}\binom{k}{2}}$.

10. Your Halloween bag holds 30 chocolates and 3 caramels, thoroughly mixed. You eat them one at a time (over several days, of course) until you have eaten 20 chocolates.

a. What is the probability that you have eaten 2 or more caramels?

b. Redo this problem using an appropriate approximate technique.

11. Approximately 20% of job candidates turn out to be skilled in the use of a certain spreadsheet program, but you do not know in advance which ones will be. You interview 5 candidates picked at random for the job. Let X be the number interviewed who are skilled in using the spreadsheet.

a. What is the probability that all your candidates will be skilled with the spreadsheet (that $X = 5$)?

b. What is the probability that at least one candidate will be skilled with the spreadsheet (that $X \geq 1$)?

12. You and a friend flip a fair coin every week; heads he buys you a lottery ticket, tails you buy him one. The lottery has a chance of 20% of paying off. What is the chance you will win exactly one lottery payoff in the next six weeks?

13. There are 160 people on the voting rolls of a small town. A jury is selected by picking 12 different voters at random. In the next year, 10 juries will be selected; all voters are eligible to be on every jury, whether or not they have served previously. You are a voter in this town. What is the probability that you will serve on exactly two juries in the next year?

14. Show reversal symmetry for the binomial family by comparing the probability mass functions.

15. What is the probability that when you roll a die 12 times, you will get more than 2 aces (one pip)? Now roll a die until you *fail* to get an ace 10 times. What is the probability that you will get more than two aces along the way?

16. The probability that a baby will be a boy is 0.54. A family will keep having children until they have 2 boys. What is the probability that they will have no more than 3 girls? Another family will keep having babies until they have four girls. What is the probability that they will have more than one boy?

17. To study a rare, large species of starfish, you will make a series of dives during one day's work, during each of which you will try to bring up a starfish. Your chance of success on a given dive is about 15%. You imagine the success of each dive to be independent of the others.

    a. If each day you dive until you get a starfish, what is the probability, on a given day, that you require 4 or more dives?
    b. In the next week of work (6 days), what is the probability that on exactly 3 days you will require 4 or more dives to get your starfish?

18. 96% of students usually pass the introductory statistics final exam. Assume that they all have the same chance and perform independently of one another.

    a. What is the probability that 78 or more in a class of 80 will pass?
    b. Using a good approximate technique to simplify the calculation, redo (a). Compare your two answers.

19. Approximately 0.8% of oysters unexpectedly contain a jewelry-quality natural pearl. You have to provide 1000 oysters from an oyster bed to a restaurant, but if you find a pearl, you will keep the pearl and throw away the oyster. What is the probability that you will find 5 or fewer pearls? Calculate the answer by an exact calculation of an appropriate model and by a good approximate calculation.

20. 98% of clover plants have three leaves; the rest have four leaves. You search a field until you find 3 four-leaf clover plants.

    a. What is the probability that you will find at least 150 three-leaf clover plants along the way?
    b. Redo the calculation in (a) using a good approximate method. Why do you expect it to work well?

21. A certain fire station gets an average of five alarms per day. Assume that each of the very many different possible causes of alarms are independent of one another. The chief considers it a busy day if the station gets three or more alarms.

    a. What is the probability that a given day will be busy?
    b. In a seven-day week, what is the probability that no more than five days will be busy?

22. Use the inductive method to derive $E(X)$ for the NB(k, p) random variable.
23. For the random variable of Exercise 21 in Chapter 5,

    a. compute the mean squared error of X with respect to $\mu = 3$;
    b. compute Var(X) and σ_X.

24. Use the inductive method to find Var(X) when X is

    a. Poisson(λ).
    b. NB(k, p).

25. Find $E(1/(X + 1))$:

    a. for X a negative binomial NB(k, p) random variable.
    b. for X a Poisson(λ) random variable.

    Your expressions should have no summation signs or (...) in them.
26. Let X be a geometric(p) random variable.

    a. Find $E(2^x)$.
    b. In a certain gambling game, you roll a die (six sides) repeatedly until you **fail** to get a five. You start with \$1, and you double the amount of money you have each time you get a five. On the average, how much money will you have when the game is over?
    c. A bacterium divides into two exactly one minute after an experiment starts, the two bacteria each divide exactly one minute later, and so forth, with all bacteria dividing at each minute. You will use a random number generator immediately after each minute has passed to decide whether or not to look in the microscope. The probability that you will look each time is 0.4, and each decision is independent of the others. On the average, how many bacteria will you see the first time you look?

27. **a.** Find a method-of-moments estimate for the probability of success p for an NB(k, p) random variable.
    b. You are constructing a mailing list for the Citizens Party in your precinct. You visit voters at random until you have found 100 Citizens Party voters. On the way, you encounter 141 voters for other parties. Estimate the proportion of Citizens Party voters, and construct a 2-σ interval for your estimate.
28. A manufacturer of brake drums claims that only a very small percentage of their products are delivered with cracks. You maintain a large truck fleet and discover that the 75th drum you buy from them is cracked (though no previous one was).

    a. Find a 99% upper confidence bound for the probability that a given brake drum is cracked.
    b. Construct a 95% confidence interval for the probability that a given brake drum is cracked.

29. You are evaluating the balance of a die for use in gambling by counting the number of times a one comes up. You will use a two-sided test, at the $\alpha = 0.05$ significance level. Out of 300 rolls, one comes up 39 times. Do you reject the hypothesis that it is a balanced die?

6.11 Supplementary Exercises

30. To study the life spans of two species of mosquito, you introduce 400 newly hatched members of species A and 150 of species B into a terrarium. A colleague believes that species B lives longer, but you suspect that they are about the same. If you waited until even the few Methuselahs among them died, the experiment might take a long time, so you decide to stop when 390 of species A have died.

    a. If the two species are equivalent, what is the probability that at most 145 of species B will be dead?
    b. Do a good approximate recalculation of this probability, using only the proportions of the species in the terrarium, and not their total numbers.

    Hint: Counting living specimens is just as good as counting dead specimens.
31. If $\binom{x}{2}$ is small compared to W, and $\binom{b}{2}$ is small compared to B, then what can you say about the size of $\binom{x+b}{2}$ compared to $W + B$?
32. **The expectation of the limit may not be the limit of the expectation.** Define a random variable X_n with the probability mass function $p(0) = (n-1)/(n+1)$ and $p(i) = 2/(n(n+1))$ for $i = 1, \ldots, n$. Compute $E(X_n)$. Now find the random variable X that is the limit in distribution of the X_n as n goes to infinity. Compute $E(X)$. What do you conclude?
33. There are known to be 200 adult black bears living in a certain section of forest. You capture 10 of them at random and implant a miniature data recorder under the skin of the neck. A month later, you set out to find some of your recorders. If you stumble across one of your bears, it is easy to retrieve the recorder but a bear without a recorder will be very difficult to catch and check. Therefore, you assign yourself the task this week of checking bears at random until you have found 80 who do not have recorders.

    a. What is the probability that you will find exactly 3 bears who do have recorders.
    b. Recompute (a) using a plausible approximation. Is your approximation justified here?
34. Consider a hypergeometric $H(W + B, W, n)$ random variable in which W and B are very large compared to n. Find a simple approximation to $p(x)$ that uses the proportion of white marbles $p = W/(W + B)$ instead of W and B. Does this approximation look familiar?
35. I am interested in a hypergeometric random variable $H(W + B, W, n)$, in which the total number $W + B$ of marbles is large, the total number n that I

remove from the jar is large, and the number that remain in the jar after the draw $W + B - n$ is large, but the number of white marbles W, and therefore also X, is much smaller. Derive an approximate formula for the probability mass function $p(x)$ in which you do not mention B or n, just the proportion of marbles (which of course equals $n/(W + B)$) to be removed from the jar. Under what conditions would you expect your formula to work?

36. The registrar tells you that there are 8 National Merit Scholars among the 200 students in a freshman chemistry class. On the Friday before the UVa game, only 140 students show up for class.

   a. If the scholarship students behave pretty much like everybody else, what is the probability that 5 of them are in class on that Friday?
   b. Now use Exercise 35 to solve the problem approximately, and compare this to your exact answer.

37. In a small town with 114 registered voters, 39 are registered as Democrats. A polltaker interviews 10 voters chosen at random. (a) What is the probability that more than three will be Democrats? (b) Is the approximation in Exercise 34 plausible here? Calculate it and compare.

38. Derive a formula for the probability that if X is B(n, p), then X is an **even** number. **Hint:** Expand $[p - (1 - p)]^n$ and $[p + (1 - p)]^n$, using the binomial theorem from high-school algebra.

39. You need 100 perfect ball bearings for a particularly delicate application. In your experience, your vendor provides ball bearings that are perfect 97% of the time, so you purchase 105 bearings.

   a. What is the probability that you will get enough perfect bearings?
   b. Redo (a) using an appropriate approximate method. How close is your approximation?

40. Prove the theorem of the Poisson approximation to the negative binomial.
41. 10% of people in America are left-handed. In order to evaluate a new trackball designed for right-handers, I interview Americans at random until I have found 100 right-handed people for my study.
   You want to study how the trackball should be modified for left-handers; so you will work with the left-handed people that I encounter while finding my sample of 100 right-handed people.

   a. What is the probability that I will find fewer than 5 left-handers?
   b. Since there are relatively few left-handers, a simplified approximate calculation may be appropriate here. Use it to calculate an approximate probability that I will find fewer than five left-handers.

   Your answer will be quite a bit less accurate than most of our approximate calculations have been. Explain this fact.

42. a. For a B(n, p) random variable, find a simple expression for $p(x)/p(x-1)$, and use it to invent a recursive method for computing $p(x)$, starting with $p(0) = (1 - p)^n$.

b. Derive a similar procedure for computing $NB(k, p)$ and $Poisson(\lambda)$ probability mass functions.

43. Use the inductive method to check our expression for $E(X)$ for the $H(W + B, W, n)$ and for the $N(W, B, b)$ random variables.

44. Modify the computer generated random variable Z in Section 5 as follows: let $W = Z$ when Z is *odd*, and let $W = -Z$ when Z is *even*.

a. Compute the usual expression for $E(W)$, $\sum_{\text{all } w} W p(W)$. (You may need the help of your calculus book.) In particular, it has a finite sum.

b. Show that $\sum_{\text{all } w} W p(W)$ is not absolutely convergent (see 5.5.2). Therefore, W has no expectation. (You might then find it entertaining to generate a large number of values of W, and notice that indeed its average never seems to settle down anywhere.)

45. The *logarithmic* random variable X with parameter p has probability mass function $p(x) = p^x/(x \log[1/(1 - p)])$ for $X = 1, 2, 3, \ldots$.

a. Show that this really is a random variable (**Hint:** You may have to look up a fact in a good calculus book.)

b. Find closed formulas (no summations or omitted terms) for the expectation and variance of X.

46. The third central moment of a random variable X is given by $E[(X - \mu)^3]$, where $E(X) = \mu$. Let X be a binomial $B(n, p)$ random variable. Compute the third central moment of X.

47. In the last year, 57 cases of a very rare cancer were reported at a major cancer center. Assuming that these cases appear annually following a $Poisson(\lambda)$ law, construct a 98% confidence interval for λ. Compare it to a 2-σ interval.

48. Find conditions under which a Poisson random variable is a satisfactory approximation to a hypergeometric random variable.

49. Of 1000 entering freshmen at a small university, 28 have used heroin at least once.

a. In a confidential survey of a random sample of 50 freshmen, what is the probability that at least 3 will have used heroin?

b. Redo (a) approximately using the method of Exercise 48. Was that method appropriate here?

CHAPTER 7

Random Vectors and Random Samples

7.1 Introduction

Statistical experiments usually involve more than one measurement. We have already discussed replications under the same conditions, which we carry out so that we can allow for random error. More than that, though, we need to look at the several different aspects of each experimental subject that we may consider important. For instance, in a diet experiment we should record the heights as well as the weights of the participants, in order to put each weight in perspective. A poll would record the numbers of supporters of several different candidates. An ornithological survey might record all three coordinates for the location of a certain kind of bird's nest (east–west, north–south, height above the ground).

The several distinct numbers acquired during an experiment are called a *random vector*, because we think of the various values as coordinates in an abstract multidimensional space, whether or not they actually represent positions. We will develop tools for studying the interdependence of the different coordinates of a random vector. One important special case, where the various numbers represent attempts to measure the same thing in repeated, independent experiments, is called a *random sample*. This idea will allow us to treat sample means as random variables in themselves. We will then explore how sample means get ever closer to the true expectation as a sample grows. Finally, we will look at how an uncertain parameter and a random variable that depends on it give information about each other.

Time to Review

Chapter 2, Sections 3, 4, 7
Chapter 4, Section 8
Chapter 5, Section 4
Multiple integrals

7.2   Discrete Random Vectors

7.2.1   Multinomial Random Vectors

Experiments often measure several numbers at a time; for instance, the weather report from a certain time and place might include the temperature, humidity, barometric pressure, and wind velocity. These are, unfortunately, not very predictable in advance, so we might treat them as random quantities. Furthermore, it is wasteful just to report the separate measurements as if they were different experiments. For instance, the humidity has very different meaning in different seasons, since the capacity of air to hold water vapor rises with temperature. Therefore, we keep our random numbers together and interpret the different quantities in light of each other.

Definition. A **random vector X** is a probability space whose outcomes are vectors: ordered k-tuples of real numbers.

Example. A pollster wants to know whether the voters of a state favor candidates Smith or Jones (or neither) in the race for governor. Unbeknownst to the pollster, 40% favor Smith, 50% favor Jones, and 10% favor neither. If she collected a simple random sample of voters to interview, small enough that she could pretend it was with replacement and each interview was independent, how accurate would her sample proportions be?

The answer lies in a family of random vectors that generalizes the binomial family:

Definition. Consider a sequence of identical, mutually independent random experiments in which two or more outcomes are possible. Let the probabilities of the outcomes, numbered 1, 2, 3, ..., k, be $p_1, p_2, \ldots, p_k$, where $p_i > 0$ and $\sum_{i=1}^{k} p_i = 1$. If we perform n such experiments, let X_i be the number of experiments in which the ith outcome was observed. The random vector $\mathbf{X} = (X_i)^T$ is called a **multinomial vector**, $M(n, \mathbf{p})$.

In our example, if the pollster samples 100 voters, the result might be something like 43 for Smith, 53 for Jones, and 4 for neither. Thus $\mathbf{X} = (43, 53, 4)^T$ is a value of a multinomial $M(100, 0.4, 0.5, 0.1)$ random vector.

The first important fact we notice about such vectors is that the counts in the categories must sum to the total number of trials (each subject gets counted exactly once). That is, $\sum_{i=1}^{k} X_i = n$. This means that we can always solve for the count in some category, such as $X_k = n - \sum_{i=1}^{k-1} X_i$. In our example, that "Other" category is presumably of less immediate interest, so if X is the count of Smith voters, and Y the count of Jones voters, then we can quickly find the count of Other voters $100 - X - Y$. Therefore, three-category multinomial vectors (called *trinomial*, of course) may be thought of as vectors $(X, Y)^T$ in two-dimensional space. Generally, then, multinomial vectors live in a $(k - 1)$-dimensional vector space.

Binomial random variables, you might notice, are really a special case of multinomial vectors, with $k = 2$. Furthermore, X, the count of Successes, is a one-dimensional vector $(2 - 1)$, and the count in its Other category, Failures, we well know to be $n - X$. Then $p = p_1$ and $1 - p = p_2$.

7.2.2 Marginal and Conditional Distributions

Imagine that the pollster was hired by the Smith organization, so that X, the number of Smith voters, is itself a random variable of interest. If we ignore the distinction between Jones and Other voters, then our subjects have been split into the Smith voters and people who will not vote for Smith. We conclude that X by itself is a binomial $B(100, 0.4)$ random variable. More generally, any multinomial coordinate X_i, thought of in isolation, is a $B(n, p_i)$ random variable. We have a name for this thinking:

Definition. The probability space determined by the values of a single coordinate X_i of a random vector **X** is called a **marginal** random variable.

Proposition. *If* **X** *is* $M(n, \mathbf{p})$*, then* X_i *is marginally* $B(n, p_i)$.

Now we want to understand the connections among different random coordinates. First, what is the probability that the whole vector takes on a fixed value? For example, how probable was it that $X = 43$ and $Y = 53$? Using the multiplicative rule for the probability that two things both happen, $P(X = 43$ and $Y = 53) = P(X = 43)P(Y = 53 \mid X = 43)$ (where the common condition that we omitted was that our vector was $M(100, 0.4, 0.5, 0.1)$). We get the first probability from knowing that X is binomial. As for Y, once we know that $X = 43$, we can simply discard all the Smith voters and think of ourselves as interviewing the 57 voters who do not favor Smith. We are asking the probability that 53 of the 57 are, independently, Jones voters. But the probability that any one of these is a Jones voter is $P(\text{Jones} \mid \text{not Smith}) = \frac{0.5}{(1-0.4)} = \frac{5}{6}$. So the *conditional* random variable is again binomial, but now with $n = 57$ and $p = \frac{5}{6}$. We are able to compute the probability of the complete poll results by multiplying two binomial probabilities together:

$$P(X = 43, Y = 53) = \left[\binom{100}{43}0.4^{43}0.6^{57}\right]\left[\binom{57}{53}\left(\frac{5}{6}\right)^{53}\left(\frac{1}{6}\right)^{4}\right]$$
$$= 0.066729 \cdot 0.019382 = 0.0012933.$$

Let us look for the general formula for such trinomial probabilities. First, we need some notation:

Definition. A random vector X is discrete if its sample space is countable. Its probability mass function is the real-valued function $p(x) = P(X = x)$ defined on its sample space.

Obviously, $p(x) \geq 0$ and $\sum_X p(X) = 1$. This sum is really a multiple summation over several coordinates, which I have written more compactly as a sum over all values of a vector. In our example $p(43, 53) = 0.0012933$. The sample space of a multinomial random variable is obviously a finite set of possible vectors of nonnegative integers (each coordinate is an integer between 0 and n), so it is countable. We write the marginal probability mass function $p_{X_i}(x_i) = P(X_i = x_i)$. For the trinomial case, we can write one of the probability mass functions for a conditional random variable $p_{Y|X}(y|x) = P(Y = y \mid X = x)$. For more than two coordinates, you can see that there are a great many possible marginal and conditional distributions, depending on which coordinates you know and which ones you do not care about.

For a trinomial $M(n, p, q, 1 - p - q)$ vector $(X, Y)^T$, we reasoned that X is binomial $B(n, p)$ and that the conditional random variable $Y \mid x$ must be $B(n - x, q/(1 - p))$. Therefore,

$$p(x, y)$$
$$= p_X(x) p_{X|Y}(y|x)$$
$$= \frac{n!}{x!(n-x)!} p^x (1-p)^{n-x} \frac{(n-x)!}{y!(n-x-y)!} \left(\frac{q}{1-p}\right)^y \left(\frac{1-p-q}{1-p}\right)^{n-x-y}.$$

There are a two nice cancellations, after which we regroup to get

$$p(x, y) = \frac{n!}{x! y! (n-x-y)!} p^x q^y (1-p-q)^{n-x-y}.$$

This form is quite suggestive: The first part is a (surprise) multinomial symbol (see 3.3.4); the second contains the probability of each outcome to the power of the number of times it happens. We generalize to get the following:

Proposition. *A multinomial* $M(n, \mathbf{p})$ *vector has probability mass function*

$$p(\mathbf{x}) = \binom{n}{\mathbf{x}} \prod_{i=1}^{k} p_i^{x_i}.$$

PROOF. Imagine a generalization of a Bernoulli process in which each independent trial can fall in any one of k categories. Consider a string of n trials. If there are $x_1, x_2, \ldots, x_k$ outcomes of each of the types, then the probability of that particular string is $\prod_{i=1}^{k} p_i^{x_i}$. But from (3.3.4) we have already counted the number of sequences that would lead to a given vector of counts; it was the multinomial symbol

$$\frac{n!}{x_1! x_2! \cdots x_k!} = \binom{n}{x_1 x_2 \cdots x_k} = \binom{n}{\mathbf{x}}.$$

We are done.   □

When the sample space is finite, we can put the mass function for a two-coordinate random variable (called *bivariate*) in a table.

Example.

$x \backslash y$	0	1	2	3	$p_X(x)$
0	0.008	0.060	0.150	0.125	0.343
1	0.036	0.180	0.225	0	0.441
2	0.054	0.135	0	0	0.189
3	0.027	0	0	0	0.027
$p_Y(y)$	0.125	0.375	0.375	0.125	1.000

Notice that marginal probabilities for X are obtained by taking row sums; also, marginal probabilities for Y are column sums. (We see why the probabilities of individual coordinates are called marginal; they appear in the margins of the table.) This is because, for example, to get the probability that $x = 1$, we add the probabilities for the cases where $y = 0$, 1, and 2. We can summarize this as $p_X(x) = \sum_Y p(x, Y)$. Generally, to find any marginal probability mass function, we sum over the probabilities for all possible values of the other coordinates. Also, the grand total in the lower right corner verifies that our mass function sums to 1.

To find conditional probability mass functions, we just use the formula for introducing a condition (see 4.4.3), which becomes, for example, $p_{X|Y}(x|y) = \frac{p(x,y)}{p_Y(y)}$. This is just finding what proportion a table entry is of its column total, as $p_{X|Y}(1|1) = \frac{p(1,1)}{p_Y(1)} = \frac{0.180}{0.375} = 0.48$.

Combining these last two expressions, we can write $p_X(x) = \sum_Y p_Y(Y)p_{X|Y}(x|Y)$. That is, the marginal probabilities for one variable may be computed as an appropriate weighted average of its probabilities conditional on the other variable. We have seen this before in another guise—it is the division into cases formula (see 4.6.2) for discrete random variables. We shall have important applications for this shortly.

Writing tables for discrete random vectors raises a technical question: If the sample space of your random vector is all *pairs* of nonnegative integers, such as (10,17), (of which there are an infinite number), is the event countable (so that we really have a discrete random vector)? We use the integers to do the counting, and surely there are many more pairs of integers than there are integers. However, it turns out the pairs are indeed countable, as you can see, for example, from the counting scheme

$$
\begin{array}{cccc}
(0,0) & \rightarrow & (0,1) & \quad (0,2) & \rightarrow & (0,3) \\
 & & \downarrow & \quad \uparrow & & \downarrow \\
(1,0) & \leftarrow & (1,1) & \quad (1,2) & & (1,3) \\
\downarrow & & & \quad \uparrow & & \downarrow \\
(2,0) & \rightarrow & (2,1) & \rightarrow & (2,2) & & (2,3) \\
 & & & & & & \downarrow \\
(3,0) & \leftarrow & (3,1) & \leftarrow & (3,2) & \leftarrow & (3,3) \\
\downarrow
\end{array}
$$

where $(0, 0)$ is the first outcome, $(0, 1)$ is the second, $(1, 1)$ is the third, $(1, 0)$ is the fourth, and so on. Every pair gets counted eventually. This is essentially the reasoning Georg Cantor used to establish that the collection of all rational numbers p/q is countable.

For random vectors with more coordinates, there are similar counting schemes, and it is generally true that a finite-dimensional random vector whose sample space for each coordinate is countable, is itself countable.

7.3   Geometry of Random Vectors

7.3.1   Random Coordinates

Several of our examples of geometrical probability had outcomes on multidimensional objects (such as dart boards); so the coordinates of these outcomes are examples of random vectors, but no longer discrete. The probability of an outcome landing in an event A, P(A), we now write $P(\mathbf{X} \in A)$. If we are lucky, we have a **multivariate density function** $f(\mathbf{X})$, which we may integrate to compute these probabilities: $P(\mathbf{X} \in A) = \iiint_A f(\mathbf{X}) \, d\mathbf{X}$ (if we happen to have three random coordinates). You can see that it is time to review *multiple integrals* from your calculus course, if this notation is unfamiliar.

Example. In Chapter 4 (see 4.2.1) we proposed a circular dart board D of radius 1 and darts thrown from far enough away that if they hit the board, they seemed equally likely to hit anywhere. Put the origin of a coordinate system at the center of the board; a dart hit then gives us a random vector $(X, Y)^T$. We concluded that if we had a region $A \subset D$ whose volume can be computed, then $P(\mathbf{X} \in A) = \frac{V(A)}{V(D)} = \frac{1}{\pi} V(A)$. Expressing this as an integral, $P[(X, Y)^T \in A] = \iint_A \frac{1}{\pi} dX \, dY$. So in this case the density is $f(X, Y) = \frac{1}{\pi}$ for $(X, Y)^T \in D$.

Generally, the Cartesian coordinates of a uniform geometric probability space over some region have a constant density.

When we investigated the probability of landing in a vertical strip, we reduced the problem to the random behavior of the x-coordinate. This, then, had what we now realize was a *marginal density* $f_X(x)$ on $(-1, 1)$. What might the conditional behavior of the y-coordinate be if we know the value of $X = x$? That information pins the location of the dart down to a vertical line segment (the dotted line in Figure 7.1):

Since originally the dart was believed to be equally likely to fall anywhere on the disk, now that its horizontal location is known, presumably it is equally likely to be anywhere on that segment. Therefore, its *conditional density* will be *constant* over the segment, which goes from $(x, -\sqrt{1 - x^2})^T$ to $(x, \sqrt{1 - x^2})^T$. Then the segment is $2\sqrt{1 - x^2}$ in length. The conditional density has to be the constant value that will integrate to 1 over its length, so $f_{Y|X}(y|x)2\sqrt{1 - x^2} = 1$. Therefore, $f_{Y|X}(y|x) = \frac{1}{2\sqrt{1-x^2}}$. Remember that despite its appearance, this

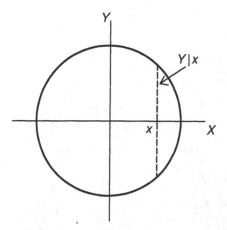

FIGURE 7.1. Vertical segment of a disk

function is constant over its sample space $(-\sqrt{1-x^2}, \sqrt{1-x^2})$; x is a known value that does not change, while Y is still random.

In the discrete case, the connection between the bivariate probability mass function, the marginal mass function, and the conditional mass function was just the multiplicative law for probabilities, $p_Y(y)p_{X|Y}(x|y) = p(x, y)$. Notice that in this continuous example $f_X(x)f_{Y|X}(y|x) = \frac{2}{\pi}\sqrt{1-x^2}\frac{1}{2\sqrt{1-x^2}} = \frac{1}{\pi} = f(x, y)$. To see that such a formula works all the time, it will be necessary to consider how to get from the multivariate density to the marginal and conditional densities.

First, ask yourself why you would want to know the marginal density of a continuous random variable? Presumably, to solve problems like "will the temperature be above freezing tomorrow morning (so that it will not kill my tomatoes)?" Humidity is an important weather fact, but the simple temperature number is most urgently needed at the moment. Generally, we want to compute things like $P(a < X \le b)$, ignoring Y (our vertical strip, again). Then we would use the marginal density to solve for the probability by $\int_a^b f_X(X)dX$. If unfortunately we only have the bivariate density handy, we have to compute instead the double integral $\int_{-\infty}^{\infty} \int_a^b f(X, Y)dXdY$. I hope you have finished your review of how to do this. You will have found that a famous fact, Fubini's theorem, says that if this integral makes sense, we may compute it by carrying out the two integrations, one at a time, in either order. So let us reverse the X and Y integrals to get $\int_a^b [\int_{-\infty}^{\infty} f(X, Y)dY]dX$. There is a subtlety here: The infinities in the limits stand for the limits in Y for each possible value of X, which is thought of as constant during the integration dY (they were $(-\sqrt{1-x^2}, \sqrt{1-x^2})$ in the example). Now compare this integral to the one of the marginal density above. We conclude that $f_X(x) = \int_{-\infty}^{\infty} f(x, Y)dY$. Generally, you can find a marginal density by integrating the multivariate density over all the possible values of all the other coordinates. You should check that this works in the dart board example. Now we can use this to define a conditional density $f_{Y|X}(y|x) = f(x, y)/f_X(x) = f(x, y)/(\int_{-\infty}^{\infty} f(x, Y)dY)$ for any x for which the marginal density is not zero, by analogy with the discrete case. You

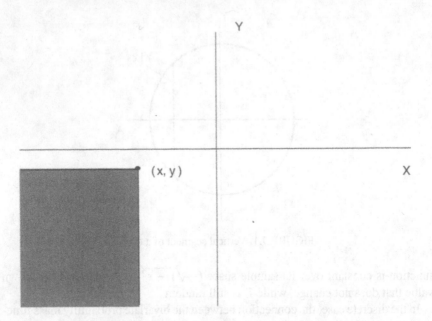

FIGURE 7.2. Cumulative distribution in a plane

should check as an exercise that this process really yields functions that can be densities.

7.3.2   Multivariate Cumulative Distribution Functions

The cumulative distribution function (see Chapter 5.4) was a useful tool for dealing with random variables; there is indeed a generalization for vectors.

Definition. The **cumulative distribution function of a random vector** is

$$F(x) = P(X_1 \leq x_1, X_2 \leq x_2, \ldots, X_k \leq x_k).$$

This awkward-looking quantity measures the probability that each random co-ordinate is at most the specified value. In the two-variable case, this amounts to the probability of the lower left-hand quadrant in a geometrical picture, Figure 7.2.

As you may remember from Chapter 4 (see 4.8.2), geometrical probabilities require us to be able to assign probabilities to all the events in a Borel algebra, which is built out of hyper-rectangles. The vector cumulative distribution function makes this possible; for example, in two variables,

$$P\{(a, b] \times (c, d] | (X, Y)\} = F(b, d) - F(b, c) - F(a, d) + F(a, c).$$

(See Figure 7.3) We took the probability of the large quadrant and subtracted off the lower right and upper left quadrants, which we did not need. But then we had subtracted the lower left quadrant twice, so we added it back in. As an exercise, you should find the corresponding formula for the probability of a three-dimensional box.

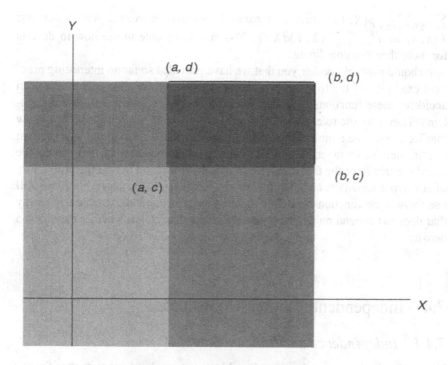

FIGURE 7.3. Probability of a rectangle

Example. Imagine a *square* dart board, with a coordinate system assigned so that the dart board is the set of coordinates $(0, 1) \times (0, 1)$. Then, if the player is so inept that the dart might land equally anywhere on the board, we see that for $0 < x < 1$ and $0 < y < 1$,

$$F(x, y) = P(0 < X \le x, 0 < Y \le y) = V\{0 < X \le x, 0 < Y \le y\} = xy,$$

since the total area is 1. As a somewhat more difficult exercise, you might find the cumulative distribution function for hits on our circular dart board.

It is easy to see what a marginal cumulative distribution function would be; for example, when we have two coordinates,

$$P(X \le x) = F_X(x) = \lim_{y \to \infty} P(X \le x, Y \le y) = \lim_{y \to \infty} F(x, y) = F(x, \infty),$$

where the last expression is a convenient but informal notation (infinity is not a number). With more than two coordinates, we can find the marginal cumulative distribution function for any one variable by simply placing an infinity symbol in the slot for each remaining variable.

Example. On our square dart board, infinity stands for the largest allowable value of a coordinate, 1. Therefore, $F_X(x) = x \cdot 1 = x$, as we might have expected.

For a bivariate discrete random vector, it is easy to see how to write the cumulative distribution function in terms of the probability mass function: $F(x, y) =$

$\sum_{X \leq x} \sum_{Y \leq y} p(X, Y)$. There is a parallel formula for vectors with a density: $F(x, y) = \int_{-\infty}^{y} \int_{-\infty}^{x} f(X, Y) dX dY$. You should be able to see how to do this for more than two coordinates.

It should probably bother you that we have provided so far no interesting practical examples of multivariate cumulative distribution functions. This is not an accident; these functions have very few direct applications to real-world problems. They play the role, rather, of a unifying mathematical device: If we know that we *can* define a multivariate cumulative distribution for a proposed random vector, then we know enough to study any possible behavior of that vector. We could see this from the fact that we could use the function to find the probability of any hyper-rectangle, and therefore of any Borel set. In the next section, we will use these same functions to define independence of random variables, in a way that does not depend on whether the vectors are discrete, or whether they have a density.

7.4   Independent Random Coordinates

7.4.1   Independence and Random Samples

Notice that in the square dart board problem, it turned out not to matter for our questions about the x coordinate, whether or not we knew something about the y coordinate. This sounds familiar.

Definition. X and Y are independent of one another whenever $F(x, y) = F_X(x) F_Y(y)$ for each $(x, y)^T$ in the sample space of our random vector.

This is because we may simply multiply the probabilities. Intuitively, two random variables are independent of one another when knowledge of one has no effect on our opinion about the other. The coordinates of hits on a square dart board are examples. As an exercise, notice that this is not true for circular dart boards. The concept is important, because it will result, when it applies, in great simplifications in our calculations.

Proposition. *For X, Y discrete and independent and for any pair of values in the sample space x and y, the events $X = x$ and $Y = y$ are independent; that is,*

$$p(x, y) = P(X = x, Y = y) = P(X = x)P(Y = y) = p_X(x)p_Y(y).$$

We will leave this for an exercise.

Statisticians often pursue independence when they design experiments. When a measurement is subject to much random error, we try to repeat it a number of times in hope that the truth will shine through the noise. For this technique to work well, each repetition of the experiment needs to be as similar as possible to the others, but not influenced by previous tries.

Definition. A random sample (or **independent identically distributed (i.i.d.)** sample) is a random vector such that the components each have the same marginal distribution F and they are mutually independent, so that $F(x) = \prod_i F(x_i)$.

Example. A particularly ambitious high-school senior takes the SAT test five times in quick succession, after taking an SAT practice short course. His total scores were 980, 1040, 990, 1080, and 1000. Test designers believe that there is little improvement due to practice; so we might imagine that these scores are a random sample attempting to measure the student's "true" SAT score. We will see much more of this concept later.

7.4.2   Sums of Random Vectors

Let X and Y be the discrete results of two independent experiments, for example, the costs of each. It is often natural to combine them to create a new variable $Z = X + Y$ (the total cost). What sort of random variable is Z?

In some particular cases, this is easy. Let X be binomial $B(n, p)$, and Y be $B(m, p)$. Then we can imagine that the first is the successes in n Bernoulli trials and that the second is the successes in the next m trials, all with probability p of success. This works because Bernoulli trials are always independent of each other. Then the total Z is the number of successes in $n + m$ trials and so is a $B(n + m, p)$ random variable. You should apply similar reasoning to find the behavior of the sum of a negative binomial $NB(k, p)$ and an independent $NB(l, p)$ variable.

In general, we would have to reason that $P(Z = z | Z = X + Y)$ is the sum over the probabilities of each pair of values of X and Y that sum to z. For example, if $z = 3$, we would have to add probabilities for the cases where $X = 0$ and $Y = 3$, where $X = 1$ and $Y = 2$, where $X = 2$ and $Y = 1$, and where $X = 3$ and $Y = 0$. We might write it $p(z) = \sum_X p(X, z - X)$, summing over the possible values of X, and the corresponding Y gotten by solving $X + Y = z$. If X and Y are independent, then we know that the probabilities factor:

$$p(x, y) = P(X = x \text{ and } Y = y) = P(X = x)P(Y = y) = p_X(x)p_Y(y),$$

and so $p(x, z - x) = p_X(x)p_Y(z - x)$.

For example, let X be Poisson(λ) and Y be independently Poisson(μ). Then

$$p(X, z - X) = \frac{\lambda^X}{X!}e^{-\lambda}\frac{\mu^{z-X}}{(z - X)!}e^{-\mu}.$$

Notice that X cannot exceed z, because Y cannot be negative. The two factorials remind us of the denominator of a combination, so we multiply and divide by $z!$ to get

$$p(X, z - X) = \frac{e^{-(\lambda+\mu)}}{z!}\frac{z!}{X!(z - X)!}\lambda^X \mu^{z-X}.$$

The second part reminds us of a binomial probability, if only λ and μ summed to 1. But we can force them to, by dividing by their sum: $\frac{\lambda}{\lambda+\mu} + \frac{\mu}{\lambda+\mu} = 1$. To do this in the probability formula we need to multiply and divide by $(\lambda + \mu)^z = (\lambda + \mu)^x (\lambda + \mu)^{z-x}$:

$$p(X, z - X) = \frac{(\lambda + \mu)^z e^{-(\lambda+\mu)}}{z!} \binom{z}{X} \left(\frac{\lambda}{\lambda + \mu}\right)^X \left(\frac{\mu}{\lambda + \mu}\right)^{z-X}.$$

We have managed to write the joint probability $p(z, x) = p(z)p(x|z)$, where the marginal distribution of Z is a Poisson$(\lambda + \mu)$ random variable, and the conditional distribution of X given $Z = z$ is $\mathrm{B}\left(z, \frac{\lambda}{\lambda+\mu}\right)$. We summarize

Proposition. *Let X be Poisson (λ) and Y be independently Poisson (μ). Then $X+Y$ is Poisson $(\lambda + \mu)$, and X conditioned on observing $X + Y = z$ is* $\mathrm{B}\left(z, \frac{\lambda}{\lambda+\mu}\right)$.

It is frustrating that this result is so similar to those for binomial and negative binomial probabilities yet requires a much more complicated argument. This will be remedied when we develop a probabilistic experiment out of which Poisson random variables arise naturally, a Poisson process, in a later chapter.

7.4.3 Convolutions

While studying the sums of independent Poisson vectors we found ourselves using a general argument about discrete vectors: When we are interested in the sum $Z = X + Y$, we may compute its probability mass function by summing over cases that can achieve the given value of the sum $p_z(z) = \sum_X p(X, z - X)$. In cases like ours in which X and Y are independent, we may factor to get $p_z(z) = \sum_X p_X(X) p_Y(z - X)$. Mathematicians have found this calculation so widely useful that they have immortalized it in the following definition.

Definition. Let f and g be functions defined on a countable set of real numbers. Then the **convolution** of f and g, written $f * g$, is a function defined by the formula $f * g(z) = \sum_x f(x)g(z - x)$ for any real z for which the formula makes sense.

We of course are interested in the case where f and g are probability mass functions, and we may state what we have learned as follows:

Proposition. *Let X and Y be independent discrete random variables. Then the probability mass function of $Z = X + Y$ is $p_Z = p_X * p_Y$.*

This is handy to know, because mathematicians have learned a great deal about convolutions; and now we can borrow from their results whenever we need to know about sums of random variables.

7.5   Expectations of Vectors

7.5.1   General Properties

Expectations of functions of discrete vectors work just as one would expect; the possibilities for functions have simply become richer.

Definition. Let $g(\mathbf{x})$ be a real-valued function defined on the sample space of a discrete random vector. Then the expectation of g is $E[g(\mathbf{X})] = \sum_{\mathbf{X}} g(\mathbf{X})p(\mathbf{X})$ whenever the sum is absolutely convergent.

Proposition. E *is a positive linear operator.*

The proof is identical to the one in the single-variable case (see 6.6.1). The interesting novelty is that we may not be concerned with all the coordinates. For example, in a poll, we might want to know the expected count for the one candidate who has hired us to do the poll. This means that the function g depends only on one coordinate. We compute

$$E[g(X_i)] = \sum_{\mathbf{X}} g(X_i)p(\mathbf{X}) = \sum_{x_i} g(x_i) \sum_{\substack{\text{all } \mathbf{X} \text{ with} \\ X_i = x_i}} p(\mathbf{X}) = \sum_{x_i} g(x_i)p_{X_i}(x_i),$$

which tells us that we may compute expectations having to do with single coordinates by ignoring the other coordinates and just using the marginal probabilities for that one.

Example. In the bivariate example given by a table in Section 2,

$$E(X) = 0 \cdot 0.343 + 1 \cdot 0.441 + 2 \cdot 0.189 + 3 \cdot 0.027 = 0.9.$$

In a multinomial experiment, the ith count is marginally binomial, so we know that its expectation is just np_i.

7.5.2   Conditional Expectations

Looking a little more closely at what we actually do to calculate an expectation in the case of two variables, we have to perform the double sum in some order. If we choose to sum over Y first with X held constant on each pass, then $E[g(X, Y)] = \sum_X [\sum_Y g(X, Y)p(X, Y)]$. But since X is constant during the inner sum, we can exploit our product rule $p(X, Y) = p_X(X)p_{Y|X}(Y \mid X)$ to factor out the marginal probability of X:

$$E[g(X, Y)] = \sum_X \left[\sum_Y g(X, Y)p_{Y|X}(Y \mid X) \right] p_X(X).$$

If you stare at the inner sum for a while, you will see that it looks like some sort of expectation by itself. For any fixed, known value of X, it is an expectation of g with respect to the conditional random behavior of Y:

Definition. For a discrete random vector with coordinates X, Y and a value x in the sample space of X, the **conditional expectation of Y given x** is

$$E_{Y|X}[g(X, Y) \mid x] = \sum_Y g(x, Y)p_{Y|X}(Y \mid x).$$

This has all the properties of a simple expectation, of course, because the conditional probability mass function is really just an ordinary mass function.

Example. If X, Y are trinomial $M(n, p, q, 1-p-q)$, then Y conditioned on $X = x$ turned out to be binomial $B(n-x, q/(1-p))$. But then the conditional expectation of Y is just the expectation of that binomial, $E_{Y|X}[Y \mid x] = (n - x)(q/(1 - p))$.

Now we can write the general expectation as

$$E[g(X, Y)] = \sum_X E_{Y|X}[g(X, Y) \mid X]p_X(X).$$

But now the sum over X looks like a (marginal) expectation.

Proposition. *For X, Y discrete,*

$$E[g(X, Y)] = E_X\{E_{Y|X}[g(X, Y) \mid X]\} = E_Y\{E_{X|Y}[g(X, Y) \mid Y]\}$$

whenever the first expectation exists.

We know that we can always do this because if the first expectation exists, then the double sum is absolutely convergent. But then we will get the same answer whatever the order of summation; and that leads to the other two expressions.

Example. For X, Y trinomial, $E(Y) = E_X[E_{Y|X}(Y|X)] = E_X[(n-X)(q/(1-p))]$. But X is marginally $B(n, p)$, so $E(Y) = (n - np)(q/(1 - p)) = nq$ after some cancellation (which we already knew by looking at the marginal distribution of Y).

7.5.3 Regression

If we manage to observe one coordinate X of a random vector, but not Y, we might be interested in predicting what Y will be. A plausible prediction would be its conditional average given $X = x$, $E_{Y|X}[Y \mid x]$. This may remind you of *regression* from Chapter 1. Even more, it is analogous to *least-squares* regression from Chapter 2. To see this, we might reasonably ask what the best possible prediction of Y would be in the form of a function $\hat{Y} = g(x)$ if we know $X = x$. Let our criterion for the best be that we minimize its mean squared error $E_{Y|X}[(Y - g(X))^2 \mid x]$ over all possible functions $g(x)$. But the conditional expectation says that we may do this one value of x at a time. In Chapter 6 (see 6.6.2) we showed that the mean squared error of a random variable is smallest about its expected value. We conclude that the least-squares prediction of Y as a function of X is given by its conditional expectation, $\hat{Y} = g(x) = E_{Y|X}[Y \mid X]$. Therefore, this function is sometimes called the *regression of Y on X*.

The corresponding analysis of variance expression says that for any function $h(x)$,

$$E_{Y|X}\left\{[Y - h(x)]^2 \mid x\right\} = E_{Y|X}\left\{[Y - E_{Y|X}(Y \mid x)]^2 \mid x\right\} + [E_{Y|X}(Y \mid x) - h(x)]^2.$$

The first term on the right is just the variance of Y, once you know x. In obvious notation,

$$E_{Y|X}\left\{[Y - h(x)]^2 \mid x\right\} = \text{Var}_{Y|X}(Y|x) + [E_{Y|X}(Y \mid x) - h(x)]^2.$$

This last expression has an interesting consequence. A naive prediction of Y (that is, ignoring X) would of course be just its average value $E(Y)$. Substitute this for $h(x)$ in the expression above to get

$$E_{Y|X}\left\{[Y - E(Y)]^2 \mid x\right\} = \text{Var}_{Y|X}(Y|x) + [E_{Y|X}(Y \mid x) - E(Y)]^2.$$

This has all been done for particular known values of X. Looking at the overall process of prediction, we should take expectations of this for all possible values of X. The proposition in the previous section tells us that $E_X[E_{Y|X}(Y \mid X)] = E(Y)$. Therefore, the third term is squared deviation about an average. When we average it over X, we get

$$E_X\left\{[E_{Y|X}(Y \mid X) - E(Y)]^2\right\} = \text{Var}_X[E_{Y|X}(Y \mid X)].$$

Applying the same proposition to the first term, we obtain

$$E_X\left([E_{Y|X}\left\{[Y - E(Y)]^2 \mid X\right\}\right) = E\left\{[Y - E(Y)]^2\right\} = \text{Var}(Y).$$

We combine these into a wonderful fact:

Theorem (conditional decomposition of variance).

$$\text{Var}(Y) = E_X[\text{Var}_{Y|X}(Y \mid X)] + \text{Var}_X[E_{Y|X}(Y|X)].$$

Remember this as "compute a variance by taking the average variance over cases and adding the variance of the average by cases." In the trinomial case (see Section 2.2) $M(n, p, q, 1-p-q)$, the variance of Y is, of course, $nq(1-q)$. The conditional expectation of Y for $X = x$ is $(n - x)(q/(1 - p))$; the variance of this conditional expectation over all X's is then $np(1 - p)(q^2/(1 - p)^2) = np(q^2/(1 - p))$. For the first term, the conditional variance is $(n - x)(q(1 - p - q)/(1 - p)^2)$. Its expectation is $(n - np)(q(1 - p - q)/(1 - p)^2) = n(q(1 - p - q)/(1 - p))$. Adding our two terms, we obtain $n(q(1 - p - q)/(1 - p)) + np(q^2/(1 - p)) = (nq/(1 - p))(1 - p - q + pq) = nq(1 - q)$, as the theorem promised.

7.5.4 Linear Regression

Our regression function $g(x)$ may take a great variety of functional shapes (just as in Chapters 1 and 2 we touched on the possibility of polynomial regression models). Notice, though, that in the trinomial example the conditional expectation of Y turned out to be a linear function of X, so this suggests that linear regression between random variables may be particularly interesting here, too. Let us proceed as in Chapter 2 (see 2.6.1) to find the generally best predictor of Y of the form $\hat{Y} = \mu + [X - E(X)]b$. Notice that we make it a centered model by subtracting $E(X)$ from X, as opposed to the sample mean $\bar{x}$ in Chapter 1. Now we want to choose μ and b to minimize the mean squared error $E[(Y - \hat{Y})^2] = E(\{Y - \mu - [X - E(X)]b\}^2)$. You may want to review how we found the corresponding answer in Chapter 2 (see 2.6.1) and note the parallels.

First, assume we know b, and treat $Y - [X - E(X)]b$ as a single random variable. Then we want to find the value of μ that makes $E(\{Y - [X - E(X)]b - \mu\}^2)$ as small as possible. But from Chapter 6 (see 6.6.2) we know that the expected value does it:

$$\mu = E(Y - [X - E(X)]b) = E(Y) - E[X - E(X)]b = E(Y).$$

Centering the model at $E(X)$ allowed it to be simplified.

Now, to find the best b, we must minimize $E(\{[Y - E(Y)] - [X - E(X)]b\}^2)$. This is similar to the simple proportionality between vectors that we worked on in Chapter 2 (see 2.3.2), and we will solve it in a similar way. Because it will turn out to be very useful elsewhere, we will look at the more general problem of when any two functions g and h of a random vector $\mathbf{X}$ are roughly proportional to each other. This means that for some unknown b, $g(\mathbf{X}) \approx bh(\mathbf{X})$. To find a reasonable b, we solve $\min_b E\{[g(\mathbf{X}) - bh(\mathbf{X})]^2\}$. A solution would be the number b such that for any other possible constant of proportionality c, $E\{[g(\mathbf{X}) - bh(\mathbf{X})]^2\} \leq E\{[g(\mathbf{X}) - ch(\mathbf{X})]^2\}$. Replacing c by $b + c - b$, expanding and rearranging terms in much the same way as when we were finding the variance, we get

$$2(b - c)E\{h(\mathbf{X})[g(\mathbf{X}) - bh(\mathbf{X})]\} + (b - c)^2 E[h(\mathbf{X})^2] \geq 0.$$

This will always be true if the first expectation is zero which happens when

$$E\{h(\mathbf{X})[g(\mathbf{X}) - bh(\mathbf{X})]\} = E[h(\mathbf{X})g(\mathbf{X})] - bE[h(\mathbf{X})^2] = 0.$$

This says that the best constant of proportionality is $b = E[h(\mathbf{X})g(\mathbf{X})]/E[h(\mathbf{X})^2]$ whenever the denominator is not zero.

By letting $Y - E(Y) = g(X)$ and $X - E(X) = h(X)$, we have solved the problem of finding the linear least-squares regression of Y on X, with coefficients $\mu = E(Y)$ and

$$b = \frac{E\{[X - E(X)][Y - E(Y)]\}}{E\{[X - E(X)]^2\}}.$$

The denominator is simply the variance of X, but the numerator we have never seen before. Since we are reviewing Chapter 2 as we go, we know what the corresponding quantity was called: the sample covariance (see 2.7.1).

Definition. The **covariance** of X and Y is given by

$$\text{Cov}(X, Y) = E\{[X - E(X)][Y - E(Y)]\}.$$

Now we can make the following assertion:

Proposition. *The least-squares linear regression of Y on X, $\hat{Y} = \mu + [X - E(X)]b$, is given by $\mu = E(Y)$ and $b = \text{Cov}(X, Y)/\text{Var}(X)$ whenever $\text{Var}(X) > 0$.*

7.5.5   Covariance

Notice that

$$E\{[X - E(X)][Y - E(Y)]\} = E(XY) - E[E(X)Y] - E[YE(X)] + E(X)E(Y)$$
$$= E(XY) - E(X)E(Y),$$

which is much like the short formula we got for the variance.

Example. In the bivariate example given by a table of p in Section 2, $E(XY)$ should be a sum of 25 terms; but in all but three cases, either X or Y or p is zero. Thus,

$$E(XY) = 1 \cdot 1 \cdot 0.18 + 1 \cdot 2 \cdot 0.255 + 2 \cdot 1 \cdot 0.135 = 0.9$$

From the marginal probabilities we found that $E(X) = 0.9$; similarly, $E(Y) = 1.5$. We conclude that $\mathrm{Cov}(X, Y) = 0.9 - (0.9)(1.5) = -0.45$. We compute further that $\mathrm{Var}(X) = 0.63$, and we have a regression equation $\hat{Y} = 1.5 - 0.71(X - 0.9)$.

Covariance measures the degree to which X and Y change linearly together.

Proposition (properties of the covariance).

(i) $\mathrm{Cov}(X, Y) = E(XY) - E(X)E(Y)$.
(ii) $\mathrm{Cov}(X, Y) = \mathrm{Cov}(Y, X)$.
(iii) $\mathrm{Cov}(X, X) = \mathrm{Var}(X)$.
(iv) $\mathrm{Cov}(a, X) = 0$.
(v) $\mathrm{Cov}(aX + bY, Z) = a\mathrm{Cov}(X, Z) + b\mathrm{Cov}(Y, Z)$.

The proofs of (ii)–(v) are easy but worthwhile exercises. You can get other interesting results by combining them. Parts (iv) and (v) together say that $\mathrm{Cov}(X + a, Y) = \mathrm{Cov}(X, y)$. Combining (ii) with either (iv) or (v) gives "right-hand" versions of those propositions.

Another important property can be seen by going back to the analysis of the regression of one function of $\mathbf{X}$ on another. By positivity of the expectation, we know that even at its minimum point, $E\{[g(\mathbf{X}) - bh(\mathbf{X})]^2\} \geq 0$. Using our best value for b, expanding and simplifying we get $E\{[g(\mathbf{X})]^2\} - \{E[g(\mathbf{X})h(\mathbf{X})]\}^2 / E\{[h(\mathbf{X})]^2\}$. Clearing the denominator, we get a very important fact:

Theorem (Cauchy–Schwarz inequality). $\{E[g(\mathbf{X})h(\mathbf{X})]\}^2 \leq E[g(\mathbf{X})^2]E[h(\mathbf{X})^2]$, *and the two sides are equal when g and h are proportional.*

If you stare at this result, and especially at the way we derived it, you will notice how closely it parallels the Schwarz inequality from Chapter 2 (see 2.3.5). The inequality is useful in many kinds of mathematics. Remembering that

$$\mathrm{Cov}(X, Y) = E\{[X - E(X)][Y - E(Y)]\},$$

our inequality says that

$$\mathrm{Cov}(X, Y)^2 \leq E\{[X - E(X)]^2\} E\{[Y - E(Y)^2]^2\} = \mathrm{Var}(X)\mathrm{Var}(Y)$$

in all cases.

We earlier found the regression of one trinomial on another,

$$E_{Y|X}[Y \mid x] = (n - x)\left(\frac{q}{1-p}\right).$$

Comparing this to our general linear regression formula with slope $b = \text{Cov}(X, Y)/\text{Var}(X) = -q/(1-p)$ and remembering that $\text{Var}(X)$ in this case is $np(1-p)$, we find that $\text{Cov}(X, Y) = -npq$. That this is negative reflects the unsurprising fact that the more observations get counted in one category, the fewer there tend to be in others. If we are looking at the covariance of two counts of a general multinomial, we can treat them as a trinomial, our two categories and an Other category combining all the remaining cases.

Proposition. *For a multinomial* $M(n, p_1, \ldots, p_k)$ *vector* $\mathbf{X}$, $\text{Cov}(X_i, X_j) = -np_i p_j$.

7.5.6   The Correlation Coefficient

By analogy with the sample correlation coefficient (see 2.7.1), there is a way to measure how strongly two variables are correlated, apart from the issue of how variable they are:

Definition. The **correlation coefficient** between random variables X and Y is $\rho_{XY} = \text{Cov}(X, Y)/\sigma_X \sigma_Y$.

Proposition. $-1 \leq \rho_{XY} \leq 1$

We check this by squaring the definition, applying the Cauchy–Schwarz inequality, and remembering that covariances may be either positive or negative.

Example. In a multinomial vector, $\rho_{X_i X_j} = -\sqrt{(p_i p_j)/((1 - p_i)(1 - p_j))}$. Notice that the number of trials n turns out to be quite irrelevant. This is a general phenomenon.

Proposition (properties of the correlation).

  (i) $\rho_{XY} = \rho_{YX}$.
 (ii) *If* $a > 0$, *then* $\rho_{aX,Y} = \rho_{XY}$. *If* $a < 0$, *then* $\rho_{aX,Y} = -\rho_{XY}$.
(iii) $\rho_{X+a,Y} = \rho_{XY}$.

Prove these for yourself from the corresponding properties of the variance and covariance. They tell us that the correlation coefficient reflects the tendency of two random variables to vary upward or downward together, without regard to their scale, or units, of measurement. We call such a quantity *dimensionless*. This suggests one reason why n did not appear in the multinomial correlation—it measures mainly the size of the experiment.

In Chapter 2 (see 2.7.1), we used correlation coefficients to write linear regression equations compactly. The same technique works here:

Definition. Let X be a random variable with $E(X) = \mu$ and $\text{Var}(X) = \sigma^2$. Then $Z = \frac{X-\mu}{\sigma}$ is called X **standardized**.

Proposition.

(i) $E(Z) = 0$.
(ii) $Var(Z) = 1$.
(iii) $\rho_{XY} = Cov(Z_X, Z_Y)$,

where of course Z_X, Z_Y are X and Y standardized. You should prove this proposition as an easy exercise. Now apply our linear regression equation:

Proposition. *The linear regression of Y on X may be written* $Z_{\hat{Y}} = \rho_{XY} Z_X$.

7.6   Linear Combinations of Random Variables

7.6.1   Expectations and Variances

We often find ourselves interested in *linear combinations* of the coordinates of a random vector, for example $aX + bY$, where a and b are constant.

Example. A salesman gets \$500 commission on each Corvette he sells, and \$400 on each Cadillac. The sales are unpredictable; call the daily number of Corvettes sold V, and of Cadillacs D. His daily earnings are then the random quantity $500V + 400D$.

Immediately from the fact that E is linear, we get that $E(aX + bY) = aE(X) + bE(Y)$. In our example, the salesman's expected daily earnings would be $500E(V) + 400E(D)$.

We might also be interested in the variance of a linear combination:

$$Var(aX + bY) = E\left\{[aX + bY - E(aX + bY)]^2\right\}$$
$$= E\left(\{a[X - E(X)] + b[Y - E(Y)]\}^2\right).$$

Expanding the square and applying the linearity of E, we get that this is equal to

$$a^2 E\left\{[X - E(X)]^2\right\} + 2abE\left\{[X - E(X)][Y - E(Y)]\right\} + b^2 E\left\{Y - E(Y)]^2\right\}.$$

Notice that we have here expressions for the variance of X and Y, and for their covariance.

We have discovered an important result:

Proposition.

(i) $E(aX + bY) = aE(X) + bE(Y)$.
(ii) $Var(aX + bY) = a^2 Var(X) + 2abCov(X, Y) + b^2 Var(Y)$.

In the special case where X and Y are trinomial,

$$Var(X + Y) = Var(X) + 2Cov(X, Y) + Var(Y).$$

But we know that $Var(X) = np(1 - p)$, $Var(Y) = nq(1 - q)$, and $Cov(X, Y) = -npq$; so

$$Var(X + Y) = np(1 - p) + nq(1 - q) - 2npq = n(p + q) - n(p + q)^2$$
$$= n(p + q)(1 - p - q).$$

Notice that $X + Y$ is just the total count not falling in the Other category, so it is $B(n, p + q)$. As it turns out, we should already have known the result of our variance calculation.

You should verify as an exercise that these results may be extended:

Proposition. *For a k-dimensional random vector* **X**,

(i) $E(\sum_{i=1}^{k} a_i X_i) = \sum_{i=1}^{k} a_i E(X_i)$.
(ii) $Var(\sum_{i=1}^{k} a_i X_i) = \sum_{i=1}^{k} a_i^2 Var(X_i) + 2 \sum_{i \leq i < j \leq k} a_i a_j Cov(X_i, X_j)$.

7.6.2   The Covariance Matrix

Our formula for the variance of a linear combination is fairly ugly. Matrix algebra will at least let us make the notation prettier. First of all, we can write $\sum_{i=1}^{k} a_i X_i = a^T X$.

Definition. Let $\mu = E(\mathbf{X})$ be the vector of expected values of the coordinates of **X**. Then the **covariance matrix** of **X**, $\Sigma = Var(\mathbf{X}) = E[(\mathbf{X} - \mu)(\mathbf{X} - \mu)^T]$.

Notice that the *outer square* of an n-dimensional vector, $\mathbf{v}\mathbf{v}^T$, is an $n \times n$ square matrix.

Proposition.

(i) *The diagonal elements* $\Sigma_{ii} = Var(X_i)$.
(ii) *For* $i \neq j$, $\Sigma_{ij} = Cov(X_i, X_j)$.
(iii) $Var(\mathbf{a}^T\mathbf{X}) = \mathbf{a}^T \Sigma \mathbf{a}$.

You should check (i) and (ii) by expanding the matrix product in the definition. Then check that (iii) is just a restatement of our formula for the variance of a linear combination.

Proposition.

(i) Σ *is a symmetric matrix; that is,* $\Sigma_{ij} = \Sigma_{ji}$ *(by one of the properties of the covariance).*
(ii) Σ *is a nonnegative definite matrix; that is, for any* **v**, $\mathbf{v}^T \Sigma \mathbf{v} \geq 0$.

(This is because (ii) is just the variance of the linear combination $\mathbf{v}^T\mathbf{X}$, and variances are always at least zero).

We shall have many uses for the matrix formulation later. Notice, though, that if the coordinates have zero covariance (they are said to be **uncorrelated**), the simplification is drastic even in the old notation:

Proposition. *If the coordinates of a vector* **X** *are pairwise uncorrelated, then*

$$Var\left(\sum_{i=1}^{k} a_i X_i\right) = \sum_{i=1}^{k} a_i^2 Var(X_i).$$

This is a promising formula, if only we had better than a qualitative idea of when variables might be uncorrelated.

7.6.3 Sums of Independent Variables

A lack of tendency to change together reminds me of probabilistic independence. Assume that X and Y are independent; we might ask ourselves to what extent we can compute $E[g(X, Y)]$ one coordinate at a time. If we can factor $g(X, Y) = g(X)h(Y)$, then

$$E[g(X)h(Y)] = \sum_X \sum_Y g(X)h(Y)p(X, Y) = \sum_X \sum_Y g(X)h(Y)p_X(X)p_Y(Y)$$

because of independence; and so factoring constants out of the inner sum, we obtain

$$\sum_X g(X)p_X(X) \sum_Y h(Y)p_Y(Y) = E[g(X)]E[h(Y)].$$

We summarize this as follows:

Proposition. *For X and Y independent,* $E[g(X)h(Y)] = E[g(X)]E[h(Y)]$.

But then $\text{Cov}(X, Y) = E(XY) - E(X)E(Y) = E(X)E(Y) - E(X) - E(Y) = 0$.

Proposition. *For X and Y independent,* $\text{Cov}(X, Y) = 0$.

This gets us the following weaker, but very useful, result:

Theorem (variance of independent sums). *If the coordinates of a vector $\mathbf{X}$ are pairwise independent, then*

$$\text{Var}\left(\sum_{i=1}^{k} a_i X_i\right) = \sum_{i=1}^{k} a_i^2 \text{Var}(X_i).$$

This beautiful and unexpected fact was one of the things that first convinced me that mathematical statistics was worth learning. I remember it by thinking of the case where all the a's are 1 and saying to myself, "With independence, the variance of a sum is the sum of the variances." Its uses are many, as we shall see.

Example. Your restaurant has a weekly profit that varies unpredictably, but the standard deviation is about $500. Over a year (52 weeks), how variable would your total profit be? It seems plausible that weeks should be independent of one another. The weekly variance is $500^2 = 250,000$; so over a year the variance would be 13,000,000 by our theorem. The standard deviation of your annual profit is $\sqrt{13,000,000} = \$3605.55$.

7.6.4 Statistical Properties of Sample Means and Variances

We have mentioned a particularly important sort of random vector, a random sample, in which we try to repeat an experimental measurement identically and independently a number of times, in order to try to see through the confusing effects of random noise. We then try to compute a summary measurement that we hope will be more accurate than any one measurement, for example, the ordinary average, or *sample mean*, written $\bar{X} = \frac{1}{n}\sum_{i=1}^{n} X_i$ when, in contrast to Chapters 1 and 2, we think of it as a random variable until we carry out the experiment. For example, our diligent college applicant who took the SAT five times has a sample mean score of 1018.

This points out a particularly easy case of the results of the last section, when we are interested in the simple sum of n random coordinates. Then our formulas reduce to $E\left(\sum_{i=1}^{n} X_j\right) = \sum_{i=1}^{n} E(X_i)$ (the expectation of the sum is the sum of the expectations) and the more complicated

$$\text{Var}\left(\sum_{i=1}^{k} X_i\right) = \sum_{i=1}^{k} \text{Var}(X_i) + 2 \sum_{i \leq i < j \leq k} \text{Cov}(X_i, X_j).$$

When we have pairwise independence, as in a random sample, we have seen that this reduces to $\text{Var}\left(\sum_{i=1}^{k} X_i\right) = \sum_{i=1}^{k} \text{Var}(X_i)$. When the marginal distributions of the coordinates are all the same, say that of a random variable X, then these simplify radically to $E\left(\sum_{i=1}^{n} X_i\right) = nE(X)$. When the joint distribution of each pair of coordinates is the same, then we get

$$\text{Var}\left(\sum_{i=1}^{k} X_i\right) = n\text{Var}(X) + 2\binom{n}{2}\text{Cov}(X, Y) = n\text{Var}(X) + n(n-1)\text{Cov}(X, Y),$$

since all the covariances are equal. We will see some lovely applications of this shortly. Of course, in the case of a random sample, where we have independence of the coordinates, this collapses again to $\text{Var}\left(\sum_{i=1}^{k} X_i\right) = n\text{Var}(X)$.

Now the sample mean divides the sum by n, so we get an important result.

Theorem (statistics of the sample mean).

(i) $E(\bar{X}) = E(X)$.
(ii) $\text{Var}(\bar{X}) = \frac{\text{Var}(X)}{n}$.
(iii) $\sigma_{\bar{X}} = \frac{\sigma_X}{\sqrt{n}}$.

You should finish proving these for yourself. This small result is among the most useful in all of statistics, for it tells us how much good replication—repeated experiments—can do us in the problem of measurement in the presence of noise. Our index of uncertainty, the standard deviation, gets steadily smaller as we increase the number of experiments. Unfortunately, the rate of improvement is only by the square root of n; so that for example, we must quadruple the amount of work we do in order to double the accuracy. You may hear $\sigma_{\bar{X}}$ called the standard *error* of the mean.

Example. The standard deviation of one person's total score on the SAT is about 50 points. Our student who averages his results on five tries is therefore measuring his performance with a standard deviation of $50/\sqrt{5} = 22.36$ points.

It is natural also to wonder what the statistical properties of the *sample variance* might be. For simplicity in notation, let $E(X) = \mu_X$. If we knew the expectation, then the obvious estimator of the true variance of X is $\hat{\sigma}_X^2 = \frac{1}{n}\sum_{i=1}^{n}(X_i - \mu_X)^2$.

Taking its expected value, we get $E(\hat{\sigma}_X^2) = \frac{1}{n} \sum_{i=1}^{n} E[(X_i - \mu_X)^2] = \frac{1}{n} \sum_{i=1}^{n} \sigma_X^2 = \sigma_X^2$ from the linearity of expectation. Whenever the average value of a statistic is equal to a parameter of interest, we call the statistic *unbiased* for that parameter.

Of course, this estimator is of little use in practice, because if we are trying to understand an unknown distribution by studying data, we are very unlikely to know μ_X. That is why we would presumably want to use the sample variance from Chapter 2 (see 2.4.2) to estimate the variance of X. We remember it as $s^2 = \frac{1}{n-1} \sum_{i=1}^{n} (X_i - \bar{X})^2$ but could compute it more generally by

$$s^2 = \frac{1}{n-1} \left[\sum_{i=1}^{n} (X_i - v)^2 - n(\bar{X} - v)^2 \right]$$

for any constant v. To find its expectation, you will not be surprised to hear that a convenient choice is $v = \mu_X$:

$$E(s^2) = \frac{1}{n-1} \left\{ \sum_{i=1}^{n} E\left[(X_i - \mu_X)^2\right] - nE\left[(\bar{X} - \mu_X)^2\right] \right\}$$

$$= \frac{1}{n-1} \left(n\sigma_X^2 - n\frac{\sigma_X^2}{n} \right) = \sigma_X^2.$$

Thus s^2 is also an unbiased estimate of the true variance of X. Now we see the most important reason to divide by $n - 1$ instead of n, so that on average we will be correct.

Proposition. *For any random variable X whose mean and variance μ_X and σ_X^2 exist, and random samples of size $n > 1$, $\hat{\sigma}_X^2$ and s^2 are unbiased estimates of σ_X^2.*

7.6.5   The Method of Indicators

Notice that the fact that the expectation of the sum is the sum of the expectations is a general justification for our use of the *method of indicators* in Chapter 5 (see 5.5.3). We broke a negative hypergeometric random variable into W equivalent pieces X_i, each telling us whether or not the ith white marble appeared before the bth black marble. We were able to calculate the expectation of that indicator, $b/(B+1)$. The sum of all W of the pieces then had expectation $Wb/(B+1)$. This method applies to a number of other problems. For example, in a binomial experiment let X_i be zero if the ith experiment is a failure, and one if it is a success. Then $X = \sum_{i=1}^{n} X_i$ is a Binomial(n, p) random variable. Now, $E(X_i) = 0 \cdot (1 - p) + 1 \cdot p = p$, so $E(X) = np$, as we learned before by a more complicated procedure.

We can use the same approach to calculate the variance of a binomial. Notice that the X_i are independent of one another, because they refer to different Bernoulli experiments:

$$\text{Var}(X_i) = E(X_i^2) - E(X_i)^2 = p - p^2.$$

(notice that $X_i^2 = X_i$, since the only values are 0 and 1), and so $\text{Var}(X) = np(1-p)$, since in this case the variance of a sum is the sum of the variances. As a slightly harder exercise, you should use the same technique to find the expectation and variance of a negative binomial random variable.

Calculating the variance of a negative hypergeometric variable is somewhat more difficult by the inductive method. Using indicators,

$$\text{Var}(X_i) = E(X_i^2) - E(X_i)^2 = \frac{b}{B+1} - \frac{b^2}{(B+1)^2} = \frac{b(B+1-b)}{(B+1)^2}.$$

Unfortunately, the X_i are by no means mutually independent. Intuitively, if one white marble falls before the bth black, it creates an additional slot into which the next white one might fall; therefore, we would expect them to be positively correlated. To calculate the covariance, pretend that only the ith and jth white marbles are present, so we have an N(2, B, b) variable:

$$E(X_i X_j) = P(\text{both before } b\text{th black}) = p(2) = \frac{\binom{b+1}{2}\binom{B-b}{0}}{\binom{B+2}{2}}$$

$$= \frac{b(b+1)}{(B+1)(B+2)},$$

$$\text{Cov}(X_i, X_j) = \frac{b(b+1)}{(B+1)(B+2)} - \frac{b^2}{(B+1)^2} = \frac{b(B-b+1)}{(B+1)^2(B+2)}.$$

Now we are ready to use our formula for variances of sums of identical variables from the beginning of this section:

$$\text{Var}(X) = \frac{Wb(B-b+1)}{(B+1)^2} + \frac{W(W-1)b(B-b+1)}{(B+1)^2(B+2)}.$$

Now simplify this:

Proposition. *If X is* N(W, B, b), *then*

$$\text{Var}(X) = (Wb(B-b+1)(W+B+1))/((B+1)^2(B+2)).$$

Example. 100 caribou are released into a wildlife preserve in which they had been extinct. Twenty-five of them have tiny data recorders implanted under the skin of the neck. After 6 months, scientists need to read 10 recorders, so they begin recapturing caribou. How many animals will need to be captured to get them?

This problem is negative hypergeometric, with $W = 75$, $B = 25$, $b = 10$, and X is the number of caribou captured without recorders. We know $E(X) = 750/26 = 28.85$, so they have to capture 39, on average. $\text{Var}(X) = \frac{75 \cdot 10 \cdot 16 \cdot 101}{(26)^2 27} = 66.40$, so that the standard deviation of the number captured is a little more than 8. A typical variation might be from 31 to 47 caribou captured.

This formula is impressively complicated, so let us try to interpret it. In the case where we used binomial approximation (see 6.3.1), we let $n = W$, and $p = \frac{b}{B+1}$. Then we can write $\text{Var}(X) = np(1-p)\left(1 + \frac{W-1}{B+2}\right)$. The final factor is called a *finite population correction*; it says that the binomial approximation compresses the variance by that factor. When the approximation is appropriate, of course W is small compared to B, and the correction is practically 1. As an exercise, you should show that the finite population correction to the variance when you try to apply the negative binomial approximation to negative hypergeometric random variables is roughly $1 - \frac{b+1}{B+2}$. Therefore, using this approximation inflates the variance (but only slightly in cases where the approximation is any good).

It should now be a straightforward exercise for you to find the variance of a hypergeometric random variable.

7.7 Convergence in Probability

7.7.1 Probabilistic Accuracy

In the last section we noticed that sample means had standard deviations (standard errors) that got smaller as the sample size grew; it seems reasonable to interpret this as saying that the sample mean became more accurate as an estimate of the expectation the more data we take. But does it really say that? We are going to come up with a more precise statement, in terms of probabilities, of what we really mean when we say that an estimator is "accurate." Of course, if an estimator were simply correct, this would not be a statistics course. So we say something weaker, like, "most of the time, the estimator is pretty accurate." To turn that into mathematics, let X_n be a sequence of random variables (statistics, presumably based on growing samples), and let μ be the "true" value that we wish the X_n's were equal to. Now let $d > 0$ be an error that for some purpose we are willing to tolerate. It is a reasonable question to ask how often the statistic is inside the error bound. That is, what is $P(|X_n - \mu| < d)$? And especially, does the probability of being this accurate get large as we go to bigger sample sizes? We use this idea to make a definition:

Definition. A sequence of random variables X_n is said to **converge in probability** to a constant μ if for any standard of accuracy $d > 0$, $\lim_{n \to \infty} P(|X_n - \mu| < d) = 1$.

So we could imagine a big enough experiment that would make us as sure as we could hope to be of meeting our standard of accuracy.

7.7.2 Markov's Inequality

Unfortunately, it is not at all clear how we would go about checking that some statistic converges in probability to the value we want. Our experience would suggest that those probabilities usually get more and more complicated to compute as the sample grows. So we must look for some indirect way, based on some qualitative summary of behavior (like the standard error), to check that we have convergence in probability.

There is a remarkably simple device for doing this. First turn the probability around, into the complementary one for exceeding the error bound; then express it as a sum: $P(|X - \mu| \geq d) = \sum_{|x_i - \mu| \geq d} p(x_i)$. Now notice that whenever

$|X - \mu| \geq d$, obviously $\frac{|X-\mu|}{d} \geq 1$. Multiplying each of our probabilities by this number that is at least 1, we get the inequality

$$P\big(|X - \mu| \geq d\big) \leq \sum_{|X-\mu| \geq d} \frac{|X - \mu|}{d} p(x_i).$$

Extending this sum over the whole sample space can only increase the right-hand side: $P\big(X - \mu| \geq d\big) \leq \sum_{\text{all } x_i} \frac{|X-\mu|}{d} p(x_i)$. Now the right-hand side is an expectation:

Proposition (Markov's inequality).   *For X a discrete random variable,*

$$P\left(|X - \mu| \geq d\right) \leq \frac{1}{d} E|X - \mu|.$$

As an exercise, you will compute some easy examples. Do not be misled into imagining that this is a useful inequality, helpful in calculating approximate probabilities. In almost every practical case it gets awful answers. Its main reason for being is that it immediately gives us a general truth:

Proposition. *Let X_n be a sequence of random variables with the property that for some constant μ, $\lim_{n\to\infty} E|X_n - \mu| = 0$. Then the X_n converge in probability to μ.*

This proposition holds because the right-hand side of Markov's inequality goes to zero, forcing the left side to zero as well. Therefore, its complement goes to 1.

This is a big improvement, because it connects an overall measure of accuracy, the expected absolute error, to convergence in probability. But it is no surprise that we have seen little of this measure; historically, it turned out to be hard to work with.

7.7.3   Convergence in Mean Squared Error

We would prefer to do everything in terms of our old friend, the mean squared error (MSE). But that is now easy:

$$(E|X - \mu|)^2 = [E(1 \cdot |X - \mu|)]^2 \leq E(1^2)E(|X - \mu|^2) = E[(X - \mu)^2]$$

by probably the easiest possible application of the Cauchy–Schwarz inequality (see Section 5.3). So if the MSE gets small, then we are sure that the expected absolute error gets small as well. We have finally figured out a widely applicable fact.

Theorem (convergence in MSE implies convergence in probability).   *Let X_n be a sequence of random variables with the property that for some constant μ, $\lim_{n\to\infty} E[(X_n - \mu)^2] = 0$. Then the X_n converge in probability to μ.*

This result will be easier to use than the one before it (we know much more about MSE), but you might remember that it says less. There are sequences of random variables that do not converge in MSE, but do converge in expected absolute error, as you will check in an exercise.

We are ready for our promised application. We found out in the last section that the variance of the sample mean, if there was one, decreased in proportion to the sample size.

Theorem (a law of large numbers). *If X has expectation μ and finite variance, then the sample means of random samples of size n, $\bar{X}_n$, converge in probability to μ.*

This goes a bit of the way toward justifying what scientists have always done: To get more accurate results in a noisy experiment, repeat the experiment as often as possible, then average.

Later in the book we will prove a variation of this theorem without having to assume that X has a finite variance. We might have guessed that something like this was so, because we started by studying the convergence of variables that had a finite absolute error (which means they need only have an expected value $E(X)$). Only then did we back off to weaker results about variables with finite variance, in order to make our math easier.

7.8   Bayesian Estimation and Inference

7.8.1   *Parameters in Models as Random Variables*

The frequentist style from Chapters 5 and 6 is not the only way of looking at problems of hypothesis testing and parameter estimation.

Example. A genetic crossbreeding experiment is believed to produce 25% seeds that are homozygotic for a lethal gene; it is believed that those seeds can never sprout. Further, it is impractical to count the seeds directly; the scientist can only count the sprouts that come up, and he believes that all seeds other than the homozygotic ones will sprout. He observes 81 sprouts. How many seeds were there originally?

It seems plausible to imagine that before the experiment, the number of sprouts would be expected to be a $B(n, 0.75)$ random variable, which was then observed to take on the value $X = 81$. The sample size n is unknown. As exercises, you should see what a method-of-moments estimate and a confidence interval tell you about n.

Instead, we will go back to the state of the experiment before the seeds sprouted. We do not know X, because we believe that it is a random variable; furthermore, we do not know n. Would it help us with our thinking to imagine that n is also a random variable, so that $(N, X)^T$ is a random vector?

Generally, imagine that before the experiment, we knew that there would be a discrete quantity X that we would measure and a discrete quantity θ that we cannot measure but would like to know. We believe that these quantities have some bivariate probability mass function $p(x, \theta)$. Once we have measured $X = x$, what do we know about θ? By the conditioning formula, we have that $p_{\Theta|X}(\theta|x) =$

$p(x,\theta)/p_X(x) = p(x,\theta)/(\sum_\Theta p(x,\Theta))$. We still do not know exactly the value of θ, but perhaps its conditional distribution will say something more about it than we knew before.

This leaves us with the problem of finding the bivariate mass function. Usually, we reason as follows: Thinking of the θ as the unknown parameter of a distribution for the random result X, its probability mass function is the other conditional $p_{X|\Theta}(x|\theta)$. In our example, we believed that X followed a binomial law with unknown parameter n. But then we imagine that before this random process determined X, another random process determined θ. Let this marginal random variable have mass function $p_\Theta(\theta)$; this is called the *prior* distribution of θ. Now the multiplicative rule gives us the bivariate mass function we needed, $p_\Theta(\theta)p_{X|\Theta}(x|\theta) = p(x,\theta)$. After the experiment is done, we calculate

$$p_{\Theta|X}(\theta|x) = \frac{p_\Theta(\theta)p_{X|\Theta}(x|\theta)}{\sum_\Theta p_\Theta(\Theta)p_{X|\Theta}(x|\Theta)}.$$

This conditional mass function for θ is called its *posterior* distribution. Notice that it is a version of Bayes's theorem, so that this style of reasoning, which uses experimental data as a bridge from the prior to the posterior distribution of an unknown parameter, is called *Bayesian inference*.

7.8.2 An Example of Bayesian Inference

We need to come up with a prior distribution for our number of seeds n in our genetics experiment. This is usually the hard part in a Bayesian analysis. Sometimes there will be a sound scientific basis for assuming a prior variability for the parameter, but very often, statisticians must just do the best they can to describe their uncertainty about its value in the form of a probability law. In our problem, let us say that before the experiment, the geneticist thought, on the basis of experience, that on average something like 100 seeds would have been formed. Let us declare that the prior number of seeds was a Poisson random variable with $\lambda = 100$, because this is a simple law we know quite a bit about. Then we multiply our Poisson and binomial mass functions to get a bivariate mass function:

$$p(n,x) = p_N(n)p_{X|N}(x|n) = \frac{\lambda^n}{n!}e^{-\lambda}\frac{n!}{x!(n-x)!}p^x(1-p)^{n-x}.$$

Bayes's theorem now requires us to divide this expression by its sum over all possible values of n, to arrive at a posterior mass function. As will often be the case, we can here avoid doing all that work. The variable part of the posterior is those terms in the bivariate mass function involving $n : \lambda^n(1-p)^{n-x}/(n-x)!$. Simplify it even further by factoring out the constant λ^x, to get $[\lambda(1-p)]^{n-x}/(n-x)!$. The mass function will be a constant multiple of this, which causes it to sum to 1 over all possible values of n. Now let the random variable instead be $Z = n - x$, the number of seeds that did *not* sprout. Then its posterior mass function is a multiple of $[\lambda(1-p)]^z/z!$. We conclude that Z is Poisson$[\lambda(1-p)]$ (because we have the variable part of its mass function, without the multiplicative constant $e^{-\lambda(1-p)}$).

This is intuitively plausible, since the parameter is just the average number of seeds times the proportion that do not sprout.

It is easy to find uses for the posterior random behavior of the unknown parameter. For example, a sensible estimate might minimize its mean squared error, and in an earlier section we learned that the expected value has this property. The estimate is then the *posterior mean*. In this problem, $\hat{n} = E(N|x) = E[x+Z] = x+\lambda(1-p)$. In the genetics example, if our scientist believed in advance that there would be an average of $\lambda = 100$ seeds, then after 81 sprouts came up he would estimate that $\hat{n} = 81 + 100 \times 0.25 = 106$ seeds had formed.

We also now know the posterior mean squared error, which is just the variance of the posterior distribution. Before the experiment, when the scientist thought there would be about 100 seeds, his standard deviation would be $\sqrt{100} = 10$, from what we know about Poisson variables. With the experiment behind him, he believes there were about 106 seeds. But now the standard deviation of that estimate is $\sqrt{\text{Var}(x + Z)} = \sqrt{\text{Var}(Z)} = 5$. The experiment has narrowed down its value quite a bit.

Bayesian thinking provides the analogue of a confidence interval, but it is somewhat easier to compute and to understand. The unknown parameter is now a random variable; so just find two values within which it falls with high probability:

Definition. A $100(1 - \alpha)\%$ **Bayes interval** for a parameter θ is a pair of numbers θ_L and θ_U and a posterior distribution for $\theta = \Theta$ conditional on experimental data x such that $P(\theta_L \le \Theta \le \theta_U | X = x) \ge 1 - \alpha$.

In the genetics experiment, since Z is Poisson(25), we discover that $P(Z \le 15) = 0.02229$ and $P(Z \ge 36) = 0.02245$; therefore, adding the known 81 sprouted seeds, $97 \le N \le 116$ is a 95% Bayes interval for n.

7.9   Summary

In this chapter we defined random vectors and the concepts of *marginal* and *conditional* distribution, whose mass functions in the discrete case are given by $p_X(x) = \sum_Y p(x, Y)$, and $p_{X|Y}(x|y) = p(x, y)/(p_Y(y))$ (2.2); we also defined *independence* of random variables (4.1). We then considered expectations of functions of random vectors (in the discrete case $E[g(\mathbf{X})] = \sum_{\mathbf{X}} g(\mathbf{X})p(\mathbf{X})$ (5.1)) and *conditional expectations* $E_{Y|X}[g(X, Y)|x] \sum_Y g(x, Y)p_{Y|X}(Y|x)$. These combine to give the useful formula $E[g(X, Y)] = E\{E_{Y|X}[g(X, Y)|X]\} = E_Y\{E_{X|Y}[g(X, Y)|Y]\}$ (5.3). This concept suggested the *regression* of one random coordinate on another. When such regression predictions are linear, this led to the ideas of *covariance* $\text{Cov}(X, Y) = E\{[X - E(X)][Y - E(Y)]\}$ (5.4) and *correlation* $\rho_{XY} = \text{Cov}(X, Y)/(\sigma_X \sigma_Y)$ of random variables (5.5). These tools allowed us to deal with *linear combinations* of random coordinates, in particular to their variance,

$$\text{Var}(aX + bY) = a^2\text{Var}(X) + 2ab\text{Cov}(X, Y) + b^2\text{Var}(Y). \quad (6.1).$$

This drastically simplifies in the case of independent observations to $\text{Var}(\sum_{i=1}^{k} a_i X_i) = \sum_{i=1}^{k} a_i^2 \text{Var}(X_i)$ (6.3). For example, we were able to study the uncertainty in a sample mean, including its *standard error* $\sigma_{\bar{X}} = \sigma_X / \sqrt{n}$ (6.4). At last, we have justified the *method of indicators* (6.5).

Our new information about the rate at which sample means converge to the expectation inspired the idea of *convergence in probability* (7.1) and a first example of a *law of large numbers* (7.3). Finally, we used the ideas of conditional and marginal distribution to demonstrate *Bayesian inference*, where we formalized our knowledge about an unknown parameter as its *posterior distribution* (in the discrete parameter case

$$p_{\Theta|X}(\theta|x) = \frac{p_\Theta(\theta)p_{X|\Theta}(x|\theta)}{\sum_\Theta p_\Theta(\Theta)(x|\Theta)}$$

after we have observed a sample of measurements whose probabilities depend on it (8.1).

7.10   Exercises

1. In a Mendelian crossing experiment, 25% of the third generation of white mice have genotype AA, 50% have genotype AB, and 25% have genotype BB. There are 40 mice born into the third generation.

   a. What is the probability that you will find 24 AB mice in your third generation?
   b. If you quickly discover that 9 are type BB, what is now the probability that 8 are of type AA?
   c. What is the probability that there will be 11 AA, 22 AB, and 7 BB in the third generation?

2. Here is the probability mass function $p(x, y)$ of a certain bivariate distribution:

		0	1	2	3	4
	0	0.06667	0.06667	0.04286	0.01905	0.00476
x	1	0.05000	0.08571	0.08571	0.05714	0.02143
	2	0.02143	0.05714	0.08571	0.08571	0.05000
	3	0.00476	0.01905	0.04286	0.06667	0.06667

   a. Compute $p_X(1) = P(X = 1)$.
   b. Compute $p_{Y|X}(2|1) = P(Y = 2|X = 1)$.
   c. Compute $E(X + 2Y)$.

3. Here is the probability mass function of a certain random vector (X, Y):

	y			
	0	1	2	3
0	0.027	0.108	0.144	0.064
x 1	0.081	0.216	0.144	0
2	0.081	0.108	0	0
3	0.027	0	0	0

a. If you know that $X = 1$, find the conditional probability mass function for Y.

b. Find the probability mass function for $Z = Y - X$.

c. What is $P(Y \geq X)$?

4. Let (X, Y) be trinomial $M(n, p, q, 1 - p - q)$. Start with the bivariate mass function $p(x, y)$ and work backwards to show that

a. X has marginally the mass function of $B(n, p)$; and

b. X has conditionally on $Y = y$ the mass function of $B(n - y, p/(1 - q))$.

5. A **negative** multinomial $NM(k, \mathbf{p})$ random vector, where $\mathbf{p} = (p_0, p_1, p_2, \ldots, p_l)$ are positive and sum to 1, is the vector of counts $\mathbf{X} = (X_1, X_2, \ldots, X_l)$ falling in categories 1 to l as a result of a sequence of independent experiments in which the p's give the probabilities of falling in the various categories. The novelty is that we stop when k experiments have fallen in the zeroth category.

a. Write down the probability mass function for a negative multinomial vector.

b. What is the marginal distribution of X_i? What is the conditional distribution of X_i given X_j?

6. We have 5 pea seeds homozygotic for smooth pod, 8 pea seeds homozygotic for wrinkled pod, and 12 heterozygotic pea seeds (these are nonoverlapping genetic categories). We pick 7 of these seeds at random for a cultivation experiment. Let the random vector (X, Y) be $X =$ number of seeds homozygotic for smooth pod chosen and $Y =$ number homozygotic for wrinkled pod chosen.

a. Compute $p(2, 3)$.

b. Compute the marginal probability $p_X(2)$.

c. Compute the probability that $Y = 3$ given that $X = 2$, $p_{Y|X}(3|2)$.

7. Consider a random vector (X, Y) with the following probability mass function:

	y		
	0	1	2
0	0.08	0.15	0.09
x 1	0.11	0.21	0.18
2	0.07	0.06	0.05

Compute $E(X|X + Y = z)$ for the special case $z = 2$.

8. Construct a table of the cumulative distribution function for the random vector of Exercise 7.

9. Let a random vector be the two rectangular coordinates of uniform (equally likely to be anywhere) hits on a circular dart board. Find the cumulative distribution function and show that the two coordinates are not independent.

10. For a random variable whose sample space consists of pairs of integers, find a formula that expresses the probability mass function $p(x, y)$ in terms of values of the cumulative distribution function.

11. Let X be NB(k, p) and Y be independently NB(l, p). Find the probability law for the variable $Z = X + Y$.

12. Let X be B(n, p) and Y be independently B(m, p). Derive the probability mass function for $Z = X + Y$ in a manner analogous to the method used in the Poisson case, using summations.

13. Prove properties (ii)–(v) of the covariance (see Section 5.5).

14. For the random vector of Exercise 2, compute Var(X), Var(Y), and Cov(X, Y).

15. For the random vector of Exercise 7, compute Var(X), Var(Y), and Cov(X, Y).

16. If Σ is the covariance matrix for $\mathbf{X}$, prove that (a) $\Sigma_{ii} = \text{Var}(X_i)$; (b) for $i \neq j$, $\Sigma_{ij} = \text{Cov}(X_i X_j)$; and (c) $\text{Var}(\mathbf{a}^T\mathbf{X}) = \mathbf{a}^T\Sigma\mathbf{a}$.

17. In a certain population, people's weights have mean 60 kg and standard deviation 12 kg; their heights have mean 160 cm and standard deviation 10 cm. The covariance of the two is 60. The Terrell Fat Index is (height − weight). (It tends to be large for thin people and small for fat people.) Write down the mean and standard deviation of the TFI.

18. Here is the probability mass function for the number of Corvettes (V) and Cadillacs (D) sold in one work day by a sales worker:

		d		
		0	1	2
v	0	0.03	0.11	0.16
	1	0.08	0.19	0.13
	2	0.14	0.09	0.07

The commission for selling a Corvette is \$500 and for selling a Cadillac is \$360. Find the expected value and standard deviation of the worker's daily commission.

19. Prove the three properties of the correlation (see Section 5.6).

20. For the random vector of Exercises 7 and 15, compute ρ_{XY}.

21. Derive the statistics of the sample mean.

22. I know that there are an average of 20 bullets that will not fire in each crate of cheap ammunition I sell, with a standard deviation of 6. A customer who buys in large quantities occasionally thoroughly tests a crate, to see whether I am maintaining my standards. If the customer counts the bad bullets in 12 crates a year and computes the sample mean of those 12 counts, what are the expected value, variance, and standard deviation of the sample mean he will compute next year?

23. Use the method of indicators to compute the expectation and variance of a negative binomial NB(k, p) random variable.

24. You run the computer maintenance facility at your company. Of the misbehaving computers you see, approximately 24% have primarily hard-drive problems, 38% have primarily display problems, 22% have primarily motherboard problems, and the rest have some other primary problem. One morning you arrive at work to find that 12 computers have arrived for repair.

    a. What is the probability that 5 have primarily a hard-drive problem, 2 have primarily display problems, 4 have primarily motherboard problems, and the other has something else?
    b. What is the probability that at least three have motherboard problems?

25. In the situation of Exercise 24, your average repair costs are as follows: $150 for hard drives, $275 for displays, $80 for motherboards, and $50 for other problems.

    a. On average, how much will it cost to fix the primary problem in those 12 computers?
    b. What is the standard deviation of the cost?

26. For the discrete uniform $\{0, \ldots, M\}$ random variable with M even, let the center $\mu = M/2$. For integer values of the error d, compute both sides of Markov's inequality. Check it for several values of d and M; note that it is usually very crude.

27. Define a sequence of random variables X_n for positive integers n, with mass functions
$$p(x) = \begin{cases} 1 - 1/n^2 & x = 0, \\ 1/n^2 & x = n. \end{cases}$$

    a. Show that the X_n converge in probability to $\mu = 0$.
    b. Show that the X_n converge in expected absolute error to $\mu = 0$.
    c. Show that the X_n do not converge in MSE to $\mu = 0$.

28. In the genetics problem of Section 8:

    a. Find a method-of-moments estimate of n.
    b. Find a 95% confidence interval for n.

29. In a survey of a wildlife refuge, you believe that in a systematic overflight in a small plane, you will have a 30% probability of seeing any particular adult brown bear, and the sightings are independent of one another. Your prior best guess of the total adult brown bear population is Poisson with a mean of 150. When you actually do the overflight, you see 48 bears.

    a. Using a Bayesian analysis, compute the mean and standard deviation of the posterior distribution of the total bear population.
    b. Find a 99% Bayes interval for the total adult brown bear population.

7.11   Supplementary Exercises

30. In a survey of galaxies, a sphere one million parsecs in radius is arbitrarily placed, and a right-angled coordinate system is defined with the origin at the center of the sphere and axes X, Y, and Z measured in units of a million parsecs. Since the sphere was arbitrarily located, the center of any galaxy that happens to fall inside this sphere may be thought of as a random vector uniformly distributed over the interior of the sphere.

    a. Find the marginal density for the X-coordinate of the center of an arbitrarily chosen galaxy inside the sphere.

    b. Find the marginal bivariate density of the coordinates (X, Z) of the galactic center (that is, ignoring Y).

    c. Find the conditional density of Y, given that $X = x$ (but ignoring Z).

31. Let $\mathbf{X}$ be a trivariate random vector. Find the formula, using cumulative distribution functions, for $P\{\mathbf{X} \in (a_1, b_1] \times (a_2, b_2] \times (a_3, b_3]\}$; that is, $\mathbf{X}$ is in a rectangular box parallel to the axes.

32. Using the results of Exercise 10, prove that for a random vector with sample space pairs of integers, if $F(x, y) = F_X(x)F_Y(y)$ for all (x, y), then $p(x, y) = p_X(x)p_Y(y)$ for all (x, y).

33. a. In the negative multinomial random variable of Exercise 5, find $\mathrm{Cov}(X_i, X_j)$.

    b. If (X, Y) is negative multinomial $\mathrm{NM}(k, 1-p-q, p, q)$, find an equation for the least-squares regression of Y on X.

34. Show that the finite population correction to the variance when using a negative binomial approximation for a negative hypergeometric random variable is roughly $1 - \frac{b+1}{B+2}$. **Hint:** Since in this case W and B should be large, let $p = \frac{W}{W+B+1}$ (instead of $\frac{W}{W+B}$ as we found convenient in (6.2.3)).

35. Find the variance of a hypergeometric $\mathrm{H}(W + B, W, n)$ random variable, using the method of indicators.

36. Find finite population corrections to the variance when binomial approximations to hypergeometric variables are used as in Exercises 6.34 and 6.35.

37. Sitting Bull's warriors have trapped General Custer's last 40 soldiers in a narrow valley. They are crowded so tightly together that any arrow aimed at them is sure to hit some soldier. However, the bowmen are standing at a safe distance, so that for all practical purposes any soldier is equally likely to be hit by any arrow.

One hundred arrows are released at the soldiers. What are the expectation and standard deviation of the number of soldiers who are still not hit by any arrow? **Hint:** Since the number of uninjured soldiers has a very complicated probability law, you might try the method of indicators.

38. Consider the collection of numbers $\{1, 2, \ldots, n\}$. Choose m of those numbers at random. Let X be the sum of the numbers you have chosen. We showed earlier (see Exercise 5.41) that $\mathrm{E}(X) = m\frac{n+1}{2}$. Find $\mathrm{Var}(X)$.

Hint: Let X be the sum of m variables X_i each of which is the value of the ith number chosen. At some point you may need to compute $\text{Cov}(X_i, X_j)$; one way to do this is to pretend temporarily that $m = n$, so that you are drawing all the numbers. In this special case, what is the variance of the total? Also, at some point you may need the results of Exercise 3.28.

39. Notice that Exercise 38 established the variance of a Wilcoxon rank sum W_i (see 2.5.5) under the hypothesis that ranks are unrelated to level of a treatment.

   a. Show that under this hypothesis, the expectation of the Kruskal–Wallis statistic is given by

   $$E(K) = \frac{12}{n(n+1)} \sum_{i=1}^{k} \frac{\text{Var}(W_i)}{n_i}.$$

   b. Therefore, $E(K) = k - 1$.

40. A couple has rather erratic income because of their jobs. He is a musician, who earns \$200 for each gig. Unfortunately, gigs arise quite unpredictably, though over the long run he averages 3 gigs per month. She is a mud wrestler, whose contract guarantees her exactly 8 matches per month. She has a 40% probability of winning any given match. When she wins, she earns \$300. What are the average and standard deviation of this couple's total income for one year (12 months)?

41. The *skewness* of a random variable is $k_1 = E[(X - \mu)^3]/\sigma_X^3$; the *kurtosis* is $k_2 = E[(X - \mu)^4]/\sigma_X^4$. Prove that $k_1^2 \leq k_2$. **Hint:** Try the Cauchy–Schwarz inequality.

42. Some statisticians would be unhappy with our use of a Poisson prior distribution to estimate a binomial sample size, because a Poisson distribution implies that we have too precise an opinion about what n should be. But we notice in Chapter 6 (see 6.6.3) that though the Poisson mean and variance are the same, the negative binomial has a larger variance than its mean; therefore, it is less precise.

   a. Derive the posterior distribution of binomial n, assuming that we know p, given that its prior distribution is $NB(k, q)$.

   b. In Exercise 29, the brown bear counting problem, let your prior for the brown bear population size be $NB(150, 0.5)$ (so it has the same mean as before). Now after seeing 48 bears, what is the posterior mean population size?

   c. Construct a 99% Bayes interval for the population size.

CHAPTER 8

Maximum Likelihood Estimates for Discrete Models

8.1 Introduction

You will remember that in Chapter 1 we introduced a variety of models for summarizing experimental data, both for measurement data and for counted data. Then in Chapter 2 we discovered a powerful general principle for choosing the parameters in our models for measurement data, the principle of least squares. This had the added advantage that it told us immediately how closely reality matched our theory, because we could compute mean squared errors. You may have noticed that we have no comparable way of dealing with counted experimental data; we proposed only standard estimates, based on the sample proportions, to estimate some of our models for contingency tables. But for other models, such as the linear logistic regression model with more than two values of the independent variable, we had no idea how to choose the parameters. Furthermore, in all cases of counted data, we had no way to quantify the distance of our model from the results of the experiment.

Now we know a great deal more about counted data, because in Chapters 5 and 6 we developed a number of possible probability models under which our results might have arisen by chance. This chapter will propose a general method for establishing distance from models to data, the *likelihood* (essentially the probability that you would observe what you did, given the model). This gives us plausible estimates for the parameters: those that give the largest possible value of this likelihood. We call this the method of *maximum likelihood*. (Later, we will learn that it is even more general than the principle of least squares, because in a certain sense least squares is a special case of maximum likelihood).

Time to Review

Finding the maximum of a function
Partial and total derivatives
Chapter 1, Sections 7 and 8
Chapter 6

8.2   Poisson and Binomial Models

8.2.1   Posterior Probability of a Parameter Value

We might well believe that the Poisson(λ) model is a reasonable description of some observation: for example, the number of car crashes in a year at a certain dangerous intersection. But what is λ? We need some way of estimating this parameter. If we in fact observed x crashes last year, then consider two possibilities, λ and μ, for the mean parameter. If we cannot in advance make a preference, we might say that from our ignorant point of view the two are equally probable: $P(\lambda) = P(\mu) = 0.5$. This is just a (discrete) prior distribution on the Poisson parameter, of the sort we studied in Chapter 7 (see 7.8.1). In that case, we might ask how probable the two are *after* we carry out the survey and get x crashes: What are $P(\lambda|x)$ and $P(\mu|x)$, the posterior probabilities of the parameter? Bayes's theorem, for example, tells us that

$$P(\lambda|x) = \frac{P(x|\lambda)P(\lambda)}{P(x|\lambda)P(\lambda) + P(x|\mu)P(\mu)} = \frac{P(x|\lambda)}{P(x|\lambda) + P(x|\mu)}$$

after we cancel the 0.5's. Then we might decide that one of the two parameter values is the better estimate if its posterior probability is the larger. Obviously, that depends on the relative size of $P(x|\lambda) = (\lambda^x/x!)e^{-\lambda}$ and $P(x|\mu) = (\mu^x/x!)e^{-\mu}$. If, say, $P(x|\mu) > P(x|\lambda)$, then $P(\mu|x) > P(\lambda|x)$, and we would argue that we had evidence favoring the model with mean μ.

Example. Two traffic experts propose average annual rates of severe accidents at our corner. One says that there are 10 accidents on average; the other says that there are 20. When we look up the records for 1997, we discover that there were actually $x = 15$. It sounds like a tossup, so we apply our probability criterion: $P(15|10) = 0.03472$ and $P(15|20) = 0.05165$. Both are a tad implausible, but surprisingly, the evidence gives a bit of an edge to 20.

We have now turned our thinking around and are calculating what probabilities would have been if the parameters were known and the random experiment had not been done yet (when in fact, x is known and we are trying to guess the parameter). We need some new language:

Definition. The discrete **likelihood** of a parameter (or vector of parameters) θ, given the discrete data (vector) $\mathbf{x}$, is $L(\theta|\mathbf{x}) = P(\mathbf{X} = \mathbf{x}|\theta)$.

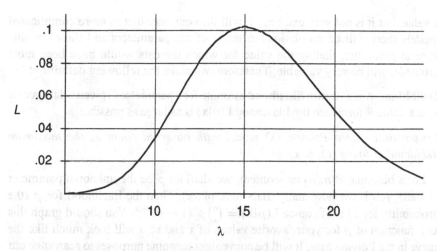

FIGURE 8.1. Poisson likelihood

The calculation in the example works for any finite number of possible parameter values: If we believe them equally likely to start with, then Bayes's theorem says that the likelihood measures which of them is most probable after the experiment. It would be interesting to graph the likelihood in our example as a function of possible values of λ; and we do this in Figure 8.1. This will be a very characteristic shape of likelihood curves.

In practice, the likelihood for even a good model may be rather small (there may be a great many reasonable possibilities for x), so we usually compare two likelihoods not by taking their difference, but by taking their ratio:

Definition. The **likelihood ratio** for comparing θ_1 to θ_2 is $R = L(\theta_1|\mathbf{x})/L(\theta_2|\mathbf{x})$.

In our traffic problem, the likelihood ratio for an average of 20 versus 10 accidents, when we have seen 15, is $0.05165/0.03472 = 1.4876$. Our results would happen about three times under the first model for each two times they would happen under the second.

8.2.2 Maximum Likelihood

We perhaps should try to find an estimate of λ by finding a value for which the likelihood of λ is largest over *all* possibilities. At what λ is our curve highest? Because the probability involves exponents, it will turn out that it is easier to find the maximum value of the *log*-likelihood $\log L(\lambda|x) = -\log(x!) + x\log\lambda - \lambda$. Since x is fixed and the best value of λ is unknown, we differentiate with respect to λ (using partial derivative notation) and set the result equal to zero: $[\partial \log L(\lambda|x)]/(\partial\lambda) = (x/\lambda) - 1 = 0$. Solving, we find that $\lambda = x$. We check that the second derivative is $[\partial^2 \log L(\lambda|x)]/(\partial\lambda^2) = -(x/\lambda^2)$, which is always negative. We recall from calculus that this value is indeed the λ of maximum probability (if there were any events to count). Therefore, our best guess for the Poisson mean parameter λ is just the observed count x of Poisson events. It is reassuring that it is so plausible a value, but it is not very exciting. It will turn out later that in more complicated models there will be no obvious estimate of the parameters and therefore this general procedure, finding the value for which the data would have been most probable, will be very valuable. Therefore, we make the following definition:

Definition. A **maximum likelihood** estimate for a parameter θ, given a data vector $\mathbf{x}$, is a value $\hat{\theta}$ for which the likelihood $L(\theta|\mathbf{x})$ is as large as possible.

Proposition. *For a Poisson* (λ) *model with observed count* x, *the maximum likelihood estimate is* $\hat{\lambda} = x$.

For a binomial $B(n, p)$ experiment, we shall let p be the unknown parameter (usually you know how many trials took place). Then the likelihood for p (the probability for x) is of course $L(p|x) = \binom{n}{x} p^x (1 - p)^{n-x}$. You should graph this as a function of p for your favorite values of x and n; it will look much like the curve in the Poisson case. It will be convenient for some purposes to rearrange our likelihood as

$$L(p|x) = \binom{n}{x} \left(\frac{p}{1 - p} \right)^x (1 - p)^n.$$

Once again, there are exponents, so we will want to take logarithms to make the maximum easier to find. We do this so often that we may as well have some notation: the *log-likelihood* is $l(\mathbf{x}|\theta) = \log L(\mathbf{x}|\theta)$. In the binomial case, this is

$$l(p|x) = \log \binom{n}{x} + x \log \frac{p}{1 - p} + n \log(1 - p).$$

Our rearrangement has broken it into three terms: one involving only the data, one involving both the data and the parameter, and the third involving only the parameter. You will notice that the log-likelihood for the Poisson problem broke up in the same way. Also, the middle term involves the *logit*, which was important in Chapter 1 (see 1.7.3).

To find a maximum likelihood estimate for p, we will differentiate l with p as the variable and set this derivative equal to zero. Remembering that $\log \frac{p}{1-p} = \log p - \log(1-p)$, we obtain $[\partial l(p|x)]/(\partial p) = (x/p) + x/(1-p) - n/(1-p) = 0$. You should take the second derivative to check that it is in fact the maximum. Adding the first two terms, we obtain $x/(p(1 - p)) = n/(1 - p)$; multiply both sides by $p(1 - p)/n$, and we have the maximum likelihood estimate $\hat{p} = x/n$. Reassuringly, this is the sample proportion that was our standard estimate for the multinomial proportions models (see 1.7.1).

Proposition.

   (i) *For* $B(n, p)$ *data* x, *the maximum likelihood estimate is* $\hat{p} = x/n$;
   (ii) *For* $NB(k, p)$ *data* x, *the maximum likelihood estimate is* $\hat{p} = x/(x + k)$.

You should derive (ii) as an exercise. Notice that the negative binomial estimate is still the sample proportion of successes, even though our stopping rule was different.

We justified the method of maximum likelihood by imagining that at the beginning all possible estimates were equally likely. If we believe the parameter to have more complicated prior probabilities (instead of just discrete uniform ones), then we would still use the likelihood in Bayes's theorem but might come to different conclusions about which values were most probable after the experiment. This is a sort of Bayesian estimation that uses the posterior *mode* (most probable value) instead of the posterior mean that we used in (7.8.2).

8.3   The Likelihood Ratio and the G-Squared Statistic

8.3.1   Ratio of the Maximum Likelihood to a Hypothetical Likelihood

Now that we have an estimate of the parameter from the data, we have a natural measure for how close a proposed value of the parameter is to that closest value. We simply take the likelihood ratio of the probability at the maximum to the probability at the proposed value: $R(\theta) = \frac{L(\hat{\theta})}{L(\theta)}$. Notice that always $R(\theta) \geq 1$, because the numerator is the largest possible value of L.

Example. A referee flips a purportedly fair coin 100 times and it lands heads 55 times. Should we be surprised by the apparent preference for heads? Using a binomial $B(100, p)$ model, the claim that the coin is fair says that $p = 0.5$, while the maximum likelihood estimate is $\hat{p} = 0.55$, we find a likelihood ratio $R(0.5) = \left[\binom{100}{55} 0.55^{55} 0.45^{45} \right] / \left[\binom{100}{55} 0.5^{55} 0.5^{45} \right] = 1.65$. So the observed value is only $\frac{5}{3}$ as likely at maximum as at the fair value. We seem to have little reason to believe the coin to be unfair.

If we plot $R(p)$, we get a curve of much the same shape as we did above for the Poisson likelihood as a function of λ (except, of course, upside down). We have noticed that the calculus is easier for log-likelihoods, which inspires us to try to understand the curve better by plotting its logarithm, $\log R(p) = x \log \frac{\hat{p}}{p} + (n - x) \log \frac{1-\hat{p}}{1-p}$ (solid curve in Figure 8.2). This sort of shape should now look familiar: It is very like a parabola (dotted curve). This is appealing, because we would like to use this as a distance measure, and SSE was parabolic as a function of parameters when we were doing least-squares fitting.

To compute the matching exact parabola, notice that the minimum value, zero, is at $\hat{p}$, and of course, the first derivative is zero there (because it is a minimum). The second derivative, with our computed value for $\hat{p}$ substituted in, is $n/(\hat{p}(1-\hat{p}))$. The parabola that almost matches our curve is then $(n(p-\hat{p})^2)/(2\hat{p}(1-\hat{p}))$ (the 2 appears when you differentiate the square). Now we can take exponentials to get rid of the logarithm, $e^{n(p-\hat{p})^2/(2\hat{p}(1-\hat{p}))} \approx L(\hat{p}|x)/L(p|x)$; and solve for the approximate shape of the binomial likelihood curve $L(p|x) \approx L(\hat{p}|x)e^{-n(p-\hat{p})^2/(2\hat{p}(1-\hat{p}))}$. This is an equation for the famous *normal curve*, which appears everywhere in statistics. As an exercise, you should derive the approximate normal curve for the Poisson likelihood.

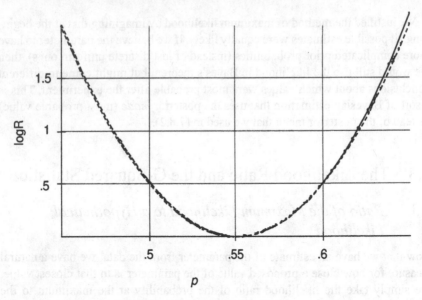

FIGURE 8.2. Log-likelihood ratio

8.3.2   G-Squared

We are ready to define the analog of the SSE for the distance from a model to the data as measured by likelihood:

Definition. The **likelihood ratio chi-squared** statistic is

$$G^2(\theta) = 2 \log \frac{L(\hat{\theta}|\mathbf{x})}{L(\theta|\mathbf{x})} = 2l(\hat{\theta}|\mathbf{x}) - 2l(\theta|\mathbf{x}).$$

The factor of 2 has the effect of canceling the 2 that appeared in the denominator in our parabolic approximation above. We will shortly see historical reasons for calling it G-squared. For now, it is reassuring that since the likelihood ratio is at least 1, our new statistic is always at least zero, as we would expect for a square.

In the binomial case,

$$G^2(p) = 2x \log \frac{\hat{p}}{p} + 2(n - x) \log \frac{1 - \hat{p}}{1 - p} = 2x \log \frac{x}{np} + 2(n - x) \log \frac{n - x}{n(1 - p)}.$$

In the coin flipping example, we find that $G^2(0.5) = 1.002$.

When we started, we assumed that we knew the parameter in the model; in this case G-squared is a measure of how far away the data varied by chance from its ideal value. If it is too large, of course, we begin to think that something went wrong, either in our experiment or in our assumption about the value of the parameter. In our parabolic approximation to a binomial likelihood ratio, let us assume that the sample proportion $\hat{p}$ is a reasonably accurate estimate of the true value p, at least good enough to estimate the denominator $p(1 - p)$. Then our approximate G-squared is given by $(n(p - \hat{p})^2)/(\hat{p}(1 - \hat{p})) \approx (n(\hat{p} - p)^2)/(p(1 - p))$ by adjusting the denominator. But since $\hat{p} = X/n$, we get that $E(\hat{p}) = E(X)/n = np/n = p$ from the expectation of a binomial. Similarly, $\text{Var}(\hat{p}) = p(1 - p)/n$. Combining these two, we find that $E[(n(\hat{p} - p)^2)/(p(1 - p))] = 1$. So a typical value of the binomial G-squared is something like 1. In our coin-tossing example, 55 heads turns out to be a thoroughly typical deviation from middle of fair-coin behavior.

If you try to calculate the expected value of G-squared exactly, it may bother you that our discrete models each have a finite, but usually tiny, probability that some category (e.g., either successes or failures) has exactly zero counts. But log(0) is negatively infinite. However, what you should really be calculating in those cases is $0 \log(0)$; to see what that should be, find $\lim_{x \to 0} x \log(x)$ by L'Hospital's rule (exercise). Your answer will be zero; and this causes no problem with the existence of the expectation.

8.4   G-Squared and Chi-Squared

8.4.1   Chi-Squared

Let us stare more carefully at the approximation to the binomial G-squared. Notice first that $\frac{1}{p(1-p)} = \frac{1}{p} + \frac{1}{1-p}$, so

$$\frac{n(\hat{p} - p)^2}{p(1 - p)} = \frac{n(\hat{p} - p)^2}{p} + \frac{n(\hat{p} - p)^2}{1 - p} = \frac{n(\hat{p} - p)^2}{p} + \frac{n[(1 - \hat{p}) - (1 - p)]^2}{1 - p},$$

where in the second term we rearranged the numerator to have $(1 - p)$'s to match the denominator. Now multiply numerator and denominator by n, and pull the n inside the square:

$$= \frac{(n\hat{p} - np)^2}{np} + \frac{[n(1 - \hat{p}) - n(1 - p)]^2}{n(1 - p)}.$$

Let $\hat{p} = X/n$ so that

$$= \frac{(X - np)^2}{np} + \frac{[n - X - n(1 - p)]^2}{n(1 - p)}.$$

We can interpret this as two terms, one each for the success and failure categories. In each category, from the observed count we subtract its expectation and then square. Finally, we divide by its expectation. This is a sort of weighted, squared Euclidean distance between theory and observation in vectors of cell counts. It is promising that our new measure of distance is roughly parallel to the sum of squares from least-squares theory. Generally, we have the following situation:

Definition. Given an experiment with k cells, E_i the expected count in the ith cell under some model, and observed count O_i in that cell, then the (Pearson's) **chi-squared** statistic for measuring the **goodness of fit** of that model is $\chi^2 = \sum_{i=1}^{k}(O_i - E_i)^2/E_i$.

(Do you recall this from the Introduction?) This measure of distance dates from the turn of the century and is perhaps the first important example of a test statistic. The approximation to G-squared discussed above is the chi-squared statistic for fit to a $B(n, p)$ model.

8.4.2 Comparing the Two Statistics

We will now worry about just when chi-squared is a good approximation to G-squared.

The likelihood ratio statistic for a Poisson(λ) experiment with observed count x is $G^2 = 2\log[x^x e^{-x}]/(\lambda^x e^{-\lambda}) = 2\left[x\log\frac{x}{\lambda} - (x - \lambda)\right]$. By judicious addition and subtraction, express G^2 in terms of $x - \lambda$:

$$G^2 = 2\left\{[\lambda + (x - \lambda)]\log\left(1 + \frac{x - \lambda}{\lambda}\right) - (x - \lambda)\right\}.$$

Now, by factoring out λ we can express everything in terms of the relative error $r = \frac{x-\lambda}{\lambda}$: $G^2 = 2\lambda\left[\left(1 + \frac{x-\lambda}{\lambda}\right)\log\left(1 + \frac{x-\lambda}{\lambda}\right) - \frac{x-\lambda}{\lambda}\right]$.

We want to establish how nearly the part in brackets, $(1 + r)\log(1 + r) - r$, is a parabola with minimum value 0 at 0. To do this, we will come up with a lemma much like the basic inequality for the logarithm in Chapter 3 (see 3.5.1). First notice that our expression is simpler than it looks. Take its derivative to get

$$[(1 + r)\log(1 + r) - r]' = \log(1 + r).$$

Therefore, we can express it as an integral:

$$(1 + r)\log(1 + r) - r = \int_0^r \log(1 + s)\,ds = \int_0^r \int_0^s \frac{dt}{1 + t}\,ds,$$

since the logarithm itself can be expressed as the inner integral. As we have done earlier, break up $1/(1 + t) = 1 - t/(1 + t)$, so that $\int_0^r \int_0^s (dt)/(1 + t)\,ds = \int_0^r \int_0^s 1\,dt\,ds - \int_0^r \int_0^s (t\,dt)/(1 + t)\,ds$. The first double integral immediately can be solved as $r^2/2$; we have our parabola.

The second double integral is the error in our approximation, so our remaining work will be to get some idea of how big it is. First, consider the case $r > 0$; then $1/(1 + t) \leq 1$, and $\int_0^r \int_0^s (t\,dt)/(1 + t)\,ds \leq \int_0^r \int_0^s t\,dt\,ds = r^3/6$. Furthermore, it is also true that $1/(1 + t) \geq 1/(1 + r)$. Then

$$\int_0^r \int_0^s (t\,dt)/(1 + t)\,ds \geq \int_0^r \int_0^s t/(1 + r)\,dt\,ds = r^3/(6(1 + r)).$$

Therefore,

$$-\frac{r^3}{6(1 + r)} \geq [(1 + r)\log(1 + r) - r] - \frac{r^2}{2} \geq -\frac{r^3}{6}.$$

On the other hand, if $r < 0$, we have to reverse the limits of both integrals, leaving the sign unaffected. We get exactly the same interval. We summarize our result:

Theorem (quadratic approximation to the log-likelihood). *For any $r > -1$, the difference between $(1 + r)\log(1 + r) - r$ and $r^2/2$ is between $-r^3/6$ and $-r^3/(6(1 + r))$.*

This says that the *relative* error in the approximation of $(1 + r)\log(1 + r) - r$ by $r^2/2$ is small if $r/3$ and $r/(3(1 + r))$ are both small in size. Recalling the definition of r, this says that $(x - \lambda)/(3\lambda)$ and $(x - \lambda)/(3x)$ are close to zero; informally, the approximation works if x and λ are both fairly good relative approximations to each other.

8.4.3   Multicell Poisson Models

If we have a contingency table with cells $i = 1, \ldots, k$, cell counts x_i, and a model in which the cells are independent Poisson variables with means λ_i, then the likelihood ratio is given by

$$R(\lambda) = \frac{\prod_{i=1}^{k} L(\hat{\lambda}_i | x_i)}{\prod_{i=1}^{k} L(\lambda_i | x_i)} = \prod_{i=1}^{k} \frac{L(\hat{\lambda}_i | x_i)}{L(\lambda_i | x_i)} = \prod_{i=1}^{k} R(\lambda_i).$$

But then $G^2 = 2 \log R(\lambda) = \sum_{i=1}^{k} 2 \log R(\lambda_i)$.

On the other hand, the chi-squared statistic happens to have a simple interpretation. We imagine that we *standardize* the count in each cell: $z_i = (x_i - E(x_i))/\sigma_{x_i} = (x_i - \lambda_i)/\sqrt{\lambda_i}$, each of which has expectation 0 and variance 1. Now notice that the sum of squares of the z_i is chi-squared: $\chi^2 = \sum_{i=1}^{k} z_i^2 = \sum_{i=1}^{k} (x_i - \lambda_i)^2 / \lambda_i$. Both G-squared and chi-squared are sums of cellwise distance measures. Use the theorem above to compare them cell by cell:

Theorem (equivalence of G-squared and chi-squared).   *In an independent Poisson model for a contingency table,* $G^2 \approx \chi^2$ *when all* $(O_i - E_i)/(3E_i)$ *and* $(O_i - E_i)/(3O_i)$ *are close to zero.*

Example. Historical records indicate that Louisiana, Mississippi, and Alabama have an average of 25, 42, and 27 documented tornadoes per year. Last year, there were 31, 45, and 35. Was this a surprising result? We assume independence of the states (questionable, but we do not know what else to do) and compute $G^2 = 1.3369 + 0.2094 + 2.1658 = 3.7120$. Also, $\chi^2 = 1.44 + 0.2143 + 2.3704 = 4.0247$. The two statistics differ by less than 10%. This is consistent with our theorem, as the largest of the error bounds, for Alabama, is 0.0988. Since the expected value of chi-squared under the Poisson model was 3 (adding one for each state), we had an unlucky, but not really surprising, year.

8.4.4   Multinomial Models

If you remember Chapter 1, you are probably thinking that the previous theorem is uninteresting, because most of our models for contingency tables were based on multinomial proportions. This presumably means that we had some sort of multinomial sampling design, not independent Poisson. Fortunately, this difference will not matter. For the multinomial case, all the factorials cancel out in the likelihood ratio, and we get

$$G^2 = 2 \log \frac{\prod_{i=1}^{k} \hat{p}_i^{x_i}}{\prod_{i=1}^{k} p_i^{x_i}} = 2 \sum_{i=1}^{k} x_i \log \frac{\hat{p}_i}{p_i} = 2 \sum_{i=1}^{k} x_i \log \frac{x_i}{np_i},$$

where we used the standard multinomial proportions estimate for $\hat{p}_i$. (You will check as an exercise that these are the maximum likelihood estimates.) Since $E(X_i) = np_i$, this looks remarkably like the G-squared for the Poisson case, except for a missing $x - \lambda$ term. But we will sneakily introduce that term: Remember that in a multinomial distribution $\sum_{i=1}^{k} p_i = 1$. Then $\sum_{i=1}^{k} np_i = n = \sum_{i=1}^{k} x_i$. So

$$G^2 = 2 \left(\sum_{i=1}^{k} x_i \log \frac{x_i}{np_i} - \sum_{i=1}^{k} x_i + \sum_{i=1}^{k} np_i \right) = 2 \sum_{i=1}^{k} \left[x_i \log \frac{x_i}{np_i} - (x_i - np_i) \right]$$

by subtracting and adding n. Now it exactly matches the Poisson case, and the theorem of the equivalence of G-squared and chi-squared applies here, too.

Example. In 1982, Wolf reported rolling a die 20,000 times, with the results

Face	1	2	3	4	5	6
Frequency	3407	3631	3176	2916	3448	3422

The obvious question to ask is, was the die fair? That is, is the result consistent with a multinomial probability of $\frac{1}{6}$ for each cell and therefore a cell expectation of 3333.33? We compute $G^2 = 95.80$ and $\chi^2 = 99.63$ (our relative error bound was 0.048, so this is about as close as expected.) In any case, these are amazingly large. I think that I would like to use this die in a game with a sucker.

Of course, we ducked the issue of just what a typical value was in the example. In an independent Poisson model, the expectation of chi-squared was just $E(\sum_{i=1}^{k} z_i^2) = \sum_{i=1}^{k} 1 = k$. Notice that this is the number of *degrees of freedom* in this model. Wonderfully enough, this is often true. In the multinomial case,

$$E(\chi^2) = E \left[\sum_{i=1}^{k} \frac{(X_i - np_i)^2}{np_i} \right] = \sum_{i=1}^{k} \frac{np_i(1 - p_i)}{np_i} = \sum_{i=1}^{k} 1 - \sum_{i=1}^{k} p_i = k - 1,$$

since the marginal distribution of each coordinate is binomial, and each numerator is a variance.

Proposition. *In the multinomial proportions model, the chi-squared statistic for the deviation of the sample proportion from the true probability has expectation* $k - 1$.

This is, of course, its degrees of freedom, because we have imposed the single constraint on our estimates that the sample proportions must sum to 1, as the true values do. Since it is almost the same, we will consider this to be a typical value for G-squared as well.

8.5   Maximum Likelihood Fitting for Loglinear Models

8.5.1   Conditions for a Maximum

Does the method of maximum likelihood help us estimate the parameters of more complicated models for contingency table experiments? Yes, and we shall illustrate this for the first interesting model, an independence model for a rectangular table with predictions $\hat{x}_{ij} = np_{i\bullet}p_{\bullet j}$. We could estimate the p's in this model directly, without much difficulty and with unsurprising results. But it will be much more revealing about fitting other models if instead we fit it in centered loglinear form, $\log \hat{x}_{ij} = \mu + b_i + c_j$, where the sum of all the b's and the sum of all the c's are zero (see 1.7.4).

Now for multinomial sampling in any *two-way* rectangular contingency table, the log-likelihood that we must maximize is $\log C + \sum_{i=1}^{k} \sum_{j=1}^{l} x_{ij} \log p_{ij} = \log C + \sum_{i=1}^{k} \sum_{j=1}^{l} x_{ij} \log(\hat{x}_{ij}/n)$, where C is the big multinomial symbol. But C does not depend on the unknown parameters and so is irrelevant to the maximization. Furthermore, since $\log(\hat{x}_{ij}/n) = \log \hat{x}_{ij} - \log n$, we can break off a double sum involving n that also involves only data, and so does not need to be calculated. To summarize, solving the maximum likelihood problem involves making only the simple expression $\sum_{i=1}^{k} \sum_{j=1}^{l} x_{ij} \log \hat{x}_{ij}$ as large as possible.

But we must be careful. We can make this expression grow forever by letting all the predictions $\hat{x}_{ij}$ get bigger and bigger. The problem is that we know in advance that $\sum_{i=1}^{k} \sum_{j=1}^{l} p_{ij} = 1$, so necessarily $n = \sum_{i=1}^{k} \sum_{j=1}^{l} np_{ij} = \sum_{i=1}^{k} \sum_{j=1}^{l} \hat{x}_{ij}$. We say that we have to do the maximization with the *constraint* that all the predicted counts must add up to n.

You may not yet have studied in your math classes how to maximize functions that have constraints, so we will use a trick similar to one used in the last section to make a multinomial G-squared look more like one for a Poisson problem. We just subtract the *constant* $\sum_{i=1}^{k} \sum_{j=1}^{l} \hat{x}_{ij} (= n)$ from the quantity to be made large (which will not affect the parameter estimates that make it largest), to get finally that we want to maximize

$$\sum_{i=1}^{k} \sum_{j=1}^{l} x_{ij} \log \hat{x}_{ij} - \sum_{i=1}^{k} \sum_{j=1}^{l} \hat{x}_{ij} = \sum_{i=1}^{k} \sum_{j=1}^{l} (x_{ij} \log \hat{x}_{ij} - \hat{x}_{ij}).$$

You should check that this is exactly the quantity we would want to maximize if it were a Poisson experiment; in any contingency table problem we will call this the **core** of the likelihood. We will have to maximize it and then check that indeed the solution meets our constraint.

Now we are ready to try to estimate our centered independence model. Replacing the predictions, we get $\sum_{i=1}^{k} \sum_{j=1}^{l} x_{ij}(\mu + b_i c_j) - \sum_{i=1}^{k} \sum_{j=1}^{l} e^{\mu + b_i + c_j}$. The first term becomes $\mu \sum_{i=1}^{k} \sum_{j=1}^{l} x_{ij} + \sum_{i=1}^{k} b_i \sum_{j=1}^{l} x_{ij} + \sum_{j=1}^{l} c_j \sum_{i=1}^{k} x_{ij}$. Using our notation for marginal totals, the core becomes $\mu n + \sum_{i=1}^{k} b_i x_{i\bullet} + \sum_{j=1}^{l} c_j x_{\bullet j} - \sum_{i=1}^{k} \sum_{j=1}^{l} e^{\mu + b_i + c_j}$. Notice an intriguing fact: The *only data* we will use in this estimation problem are the marginal totals that correspond to the parameters we have in the model. We have row adjustments b_i, so we need the row totals $x_{i\bullet}$, and so forth. The x_{ij} themselves are not needed, except when we sum them up to get marginal totals. These totals $x_{i\bullet}$ and $x_{\bullet j}$ are called **sufficient statistics**, which is generally what we call those functions of the data that we turn out to need in maximum likelihood estimation problems.

We are ready to maximize. Differentiate the core with respect to the b's to get $0 = \frac{\partial l}{\partial b_i} = x_{i\bullet} - \sum_{j=1}^{l} e^{\mu+b_i+c_j} = x_{i\bullet} - \sum_{j=1}^{l} \hat{x}_{ij} = x_{i\bullet} - \hat{x}_{i\bullet}$ in an obvious notation. We get a set of conditions for a solution $x_{i\bullet} = \hat{x}_{i\bullet}$. Similarly, by differentiating with respect to the c's, we require $x_{\bullet j} = \hat{x}_{\bullet j}$. First notice that we have indeed forced our constraint to hold, because necessarily the sum of the predicted counts equals the sum of the actual counts, which is n. Furthermore, we have presumably solved our estimation problem, because we have $k + l - 1$ distinct parameters to estimate (see 1.7.4), and by a similar counting procedure you should check that we have $k + l - 1$ independent marginal conditions to meet. Presumably, with a little arithmetic we are finished. Notice that this way of deriving a set of conditions, one for each parameter we need, would work for any loglinear model for a contingency table based on multinomial or Poisson sampling:

Theorem (maximum likelihood estimates for loglinear models).   *The maximum likelihood estimates for a loglinear model for any multiway rectangular contingency table obtained by multinomial, product-multinomial, or Poisson sampling may be obtained by requiring that the predicted marginal totals equal the actual marginal totals corresponding to each parameter in the model.*

You will check the claim about product-multinomial models in an exercise.

8.5.2   *Proportional Fitting*

We learned how to get standard estimates of the independence model in Chapter 1, and it would now be easy to check using our theorem that this is also the maximum likelihood estimate. Instead, we will find the maximum likelihood fit of the model directly, by a simple method that will work for many more problems. The idea is that we will construct the table of expectations by starting with a very simple table and forcing its marginal totals to be correct (as required by the theorem) one at a time. To demonstrate the process, recall the movie opinion survey from Chapter 1 (see 1.7.1):

	Male	Female	total
Like	51	83	134
Dislike	42	24	66
total	93	107	200

We start with a proposed table where all the coefficients are zero (since $\log 1 = 0$):

	Male	Female	total
Like	1	1	2
Dislike	1	1	2
total	2	2	4

The independence model says that we must adjust it to match the row totals, 134 and 66. The obvious way is to split these totals up for each row in proportion to what we have in the proposed table; so 134 is divided evenly between the males and the females, and similarly for allocating the 66 in the second row:

	Male	Female	total
Like	67	67	134
Dislike	33	33	66
total	100	100	200

Now we force the column totals to be 93 and 107 in the same way: Split the 93 up 67/100 to the likes (= 62.3) and 33/100 to the dislikes; similarly for the 107 females. We obtain

	Male	Female	total
Like	62.3	71.7	134
Dislike	30.7	35.3	66
total	93	107	200

which is identical to the "Expected" table we got another way in Chapter 1. Our measures of fit come straight from the original and final tables:

$$G^2 = 2\left(51\log\frac{51}{62.3} + 83\log\frac{83}{71.7} + 42\log\frac{42}{30.7} + 24\log\frac{24}{35.3}\right) = 11.69.$$

Since there are 4 degrees of freedom in the saturated model and 3 in the independence model we have fitted, if follows that this G-squared has one degree of freedom. Earlier results suggest that if the independence model is valid, we should expect this statistic to be about 1. As it is much larger, we seem to have evidence against the independence of gender and taste.

To extract coefficient estimates, we can now look at how the predictions change from cell to cell: For example, to find the male adjustment, we just find half the change to female $b_M = (\log 62.3 - \log 71.7)/2 = -0.07$.

To show how generally useful this method is, we write out formally what it says to do. At any given step, call the proposed expectations $\hat{x}_{ij}^{(0)}$. Now adjust these to give the right row totals $x_{i\bullet}$, in proportion to how large the entries were before, to give a modified expectation $\hat{x}_{ij}^{(1)} = \left(\hat{x}_{ij}^{(0)}/\hat{x}_{i\bullet}^{(0)}\right)x_{i\bullet}$. You should check as an easy exercise that we were successful, that $\sum_{j=1}^{J}\hat{x}_{ij}^{(1)} = x_{i\bullet}$. Then we do it again for columns, $\hat{x}_{ij}^{(2)} = \left(\hat{x}_{ij}^{(1)}/\hat{x}_{\bullet j}^{(1)}\right)x_{\bullet j}$, and in fact for all the indices corresponding to marginal totals we are required to match, in a multiway contingency table. This is called the method of **proportional fitting**. You will apply it to other models as exercises.

8.5.3 Iterative Proportional Fitting*

Unfortunately, the procedure of the last section does not work as expected for all models. A much harder problem would be a three-way contingency table like that in Exercise 1.35:

	Rural	Urban
Male	23	43
Female	27	52

Smokers

	Rural	Urban
Male	43	135
Female	32	118

Nonsmokers

You will show in exercises that proportional fitting will estimate the expectations for various possible models for this experiment. The most complicated model that is not saturated, though, is one with all possible associations of two factors, except that we assume no three-way association. This says that gender and location are indeed associated, as are gender and smoking, and location and smoking. But these associations are the same from level to level, so that for example, the relative odds for men and women smoking is the same whether they live in an urban or a rural setting. The loglinear model is

$$\log \hat{x}_{MRS} = \mu + b_M + c_R + d_S + e_{MR} + f_{MS} + g_{RS},$$

missing only the h_{MRS} to be completely saturated. From the theorem, we see that we need to match marginal totals that sum over each of the three variables in turn:

		Rural	Urban
(sum over	Male	66	178
smoking habit)	Female	59	170

		Smoker	Nonsmoker
(sum over	Male	66	178
residence)	Female	79	150

		Smoker	Nonsmoker
(sum over	Rural	50	75
gender)	Urban	95	253

corresponding to the three kinds of two-way association (for example, g_{RS} is the term that says we have to match $x_{\bullet RS} = 50$). Notice we do not need the sum corresponding to, for example, c_R, which is $x_{\bullet R\bullet} = 125$; because it is the sum of 66 and 59, which we already know we have to match.

We start with a table of ones and match each set of four totals in turn by proportional fitting to get an expected table (which you should do). But before we get excited, double check to see that we have indeed matched our marginal totals. Of course, the third table, the last one matched, is correct if we did our arithmetic correctly. But the other two are

	Rural	Urban
Male	64.968	179.032
Female	60.032	168.968

	Smoker	Nonsmoker
Male	66.194	179.032
Female	78.806	150.194

They are wrong. Proportional fitting does not solve this estimation problem.

Before we give up in despair, notice something slightly reassuring. The numbers in the second table are off by only 0.2; for that matter, those in the first table are off by only a little more than 1. We have approximately fitted the model. With a flash of ingenuity, we do the cycle of three proportional fittings of our tables of marginals again, but this time we start with the approximate expectations we just finished calculating. After much more arithmetic, we get a new table of expected counts, from which we can calculate our three tables of marginals. We again have the correct third table, but this time the first two are

	Rural	Urban
Male	65.954	178.046
Female	59.046	169.954

	Smoker	Nonsmoker
Male	66.003	177.997
Female	78.997	150.003

Now the second table is very close to what it is supposed to be, and even the first table is within 0.1 person. Knowing that we are "on a roll," we apply proportional fitting over and over again until the marginals tables match the truth to as high an accuracy as we want. This process usually works very fast (especially if you are using a computer). This technique for maximum likelihood estimation is called **iterative** proportional fitting. We will convince you that it always works, shortly.

After two more cycles, I am happy with the accuracy, and my table of expected counts looks like

	Rural	Urban
Male	23.679	42.321
Female	26.321	52.679

Smokers

	Rural	Urban
Male	42.321	135.679
Female	32.679	117.321

Nonsmokers

As exercises, you will estimate some of the coefficients in the loglinear model. The observed and expected counts are so close together that you will not be surprised that $G^2 = 0.088$. This is small compared to the one extra degree of freedom for the saturated model, so we conclude that our survey provided no evidence for three-way association.

8.5.4   Why Does It Work?*

The essential reason that iterative proportional fitting always leads to maximum likelihood estimates is that every time we force the expected table to match a marginal total, the likelihood increases. To see why this is so, remember that we modified the estimated expectations by the formula $\hat{x}_{ij}^{(1)} = (\hat{x}_{ij}^{(0)}/\hat{x}_{i\bullet}^{(0)})x_{i\bullet}$ to force the totals $\hat{x}_{i\bullet}^{(1)} = x_{i\bullet}$. This stands for a completely general step, in which j indexes the cells that get summed to create the total indexed by i. The core of the likelihood for the modified estimates $\hat{x}_{ij}^{(1)}$ is then $\sum_{i=1}^{k}\sum_{j=1}^{l}(x_{ij}\log\hat{x}_{ij}^{(1)} - \hat{x}_{ij}^{(1)}) = \sum_{i=1}^{k}\sum_{j=1}^{l}(x_{ij}\log(\hat{x}_{ij}^{(0)}/\hat{x}_{i\bullet}^{(0)})x_{i\bullet} - (\hat{x}_{ij}^{(0)}/\hat{x}_{i\bullet}^{(0)})x_{i\bullet})$. Now split the logarithm into two pieces to get

$$\sum_{i=1}^{k}\sum_{j=1}^{l}\left(x_{ij}\log\hat{x}_{ij}^{(0)} - \hat{x}_{ij}^{(0)}\right) + \sum_{i=1}^{k}\sum_{j=1}^{l}\left[x_{ij}\log\frac{x_{i\bullet}}{\hat{x}_{i\bullet}^{(0)}} - \left(\frac{\hat{x}_{ij}^{(0)}}{\hat{x}_{i\bullet}^{(0)}}x_{i\bullet} - \hat{x}_{ij}^{(0)}\right)\right].$$

Notice that we have subtracted and added $\hat{x}_{ij}^{(0)}$ in order to make the first sum the core of the likelihood under that previous set of estimates. Now sum the second part over j to get

$$\sum_{i=1}^{k}\left[x_{i\bullet}\log\frac{x_{i\bullet}}{\hat{x}_{i\bullet}^{(0)}} - \left(\frac{\hat{x}_{i\bullet}^{(0)}}{\hat{x}_{i\bullet}^{(0)}}x_{i\bullet} - \hat{x}_{i\bullet}^{(0)}\right)\right] = \sum_{i=1}^{k}\left[x_{i\bullet}\log\frac{x_{i\bullet}}{\hat{x}_{i\bullet}^{(0)}} - \left(x_{i\bullet} - \hat{x}_{i\bullet}^{(0)}\right)\right].$$

This should look familiar: It is one-half of the G-squared for how well a multicell Poisson model using our previous estimates would fit the collection of marginal totals indexed by i. Now, this is not to say we have a Poisson model (we may or may not); it is only to note that it is a G-squared, which is *guaranteed to be greater than zero* unless we had already matched the marginal totals at the previous step. So we have added a positive amount to the core of our likelihood under the estimates $\hat{x}_{ij}^{(0)}$. Therefore, iterative proportional fitting always increases the value of the log-likelihood function, so long as there are marginals not yet perfectly matched. That function is bounded above by the maximum likelihood, so a basic fact about limits from calculus says that it will converge. Since it must always improve by a positive amount governed by the imperfection of the matching, it cannot stop short; therefore, it converges to the maximum likelihood estimate.

Actually, we went too quickly over an important issue. If we had instead fitted a model with even higher association terms, we would still get expectations with the right marginal totals. To see this, imagine a model with a c_j term whose maximum likelihood estimates therefore match $x_{\bullet j\bullet}$. Now imagine the more complicated model that also has, for example, the g_{jk} association term. Its maximum likelihood estimates match the marginal $x_{\bullet jk}$, but by summing over all the levels of k, they match the $x_{\bullet j\bullet}$ marginal totals as well. So, how do we know that iterative proportional fitting has not accidentally estimated the wrong, more elaborate, model? Well, we started with expectations that were all ones; so $\log\hat{x}_{ijk}^{(0)} = 0$. You will show in an exercise that iterative proportional fitting never changes the zero values of those missing higher-order association terms. Therefore, iterative proportional fitting always gives us the maximum likelihood estimates for our loglinear model.

8.6 Decomposing G-Squared*

8.6.1 Relative G-Squared

Our emphasis on the G-squared statistic, instead of its close relative, chi-squared, for evaluating how well a model fits may surprise you. After all, chi-squared is easier to compute, and its expectation equals its degrees of freedom in important cases. Incidentally, it also behaves more reasonably in cases of poor fit.

Remember, though, that the measure of model fit we used in ANOVA and regression models in Chapter 2, the sum of squares, had a wonderful property: It could be decomposed, using generalizations of the Pythagorean theorem, into additive pieces that measured the influence of the various factors. Oddly enough, even though the chi-squared statistic looks like a sum of squares, it has no such decomposition. But G-squared does break up naturally into similar easy-to-interpret pieces. When you see why, you may be disappointed: The reason it decomposes is even more elementary than the Pythagorean theorem.

To illustrate, consider a three-way contingency table experiment. A complete independence model would include the simple terms for each of the three factors, which we shall call A, B, and C; that is, its loglinear model is $\log \hat{x}_{ijk} = \mu + b_i + c_j + d_k$. Let us write its G-squared as $G^2(A, B, C)$. If we suspect that some association might be present, for example between A and B (we will call it AB), we estimate a new model with the additional term e_{ij}. Call the new fit statistic $G^2(AB, C)$. (Notice that b_i and c_j are still in the model. Our compact notation presumes that they are present, because their association is.) Since we have allowed for a more complicated model, we might expect that this would be a smaller number—the fit is tighter.

We may in turn introduce the two other two-way associations, f_{ik} and g_{jk}, to get successively smaller statistics $G^2(AB, AC)$ and $G^2(AB, AC, BC)$. (As an exercise, write out the complete loglinear models that these refer to.) If we then add a final term h_{ijk} corresponding to the three-way association ABC, the model is now saturated; the cell expectations equal the cell counts, and G-squared is zero.

Recall that $G^2(A, B, C)$ is twice the logarithm of the likelihood ratio comparing that model to the saturated model, $L(ABC)/L(A, B, C)$. By a series of multiplications and divisions by the same amount, we can introduce all the other likelihoods that came up in our analysis:

$$\frac{L(ABC)}{L(A, B, C)} = \frac{L(AB, C)}{L(A, B, C)} \frac{L(AB, AC)}{L(AB, C)} \frac{L(AB, AC, BC)}{L(AB, AC)} \frac{L(ABC)}{L(AB, AC, BC)}.$$

The last of the four ratios is the likelihood ratio for the model discussed in the last section.

But notice that each of the four ratios is at least 1: The model in the numerator has one additional term over the model in the denominator, and all the terms are estimated by maximizing this likelihood. It is as if the denominator were estimated by arbitrarily restricting the extra term to be zero. Any time we restrict a search for the best value to a smaller neighborhood, our maximum will not be as good (the best pizza in town cannot be better than the best pizza in the state). Therefore, each numerator is at least as large as its denominator, and each ratio is at least one.

Now take twice the logarithm of both sides, and the additivity of logs separates the ratios:

$$2\log\frac{L(ABC)}{L(A,B,C)} = 2\log\frac{L(AB,C)}{L(A,B,C)} + 2\log\frac{L(AB,AC)}{L(AB,C)}$$
$$+ 2\log\frac{L(AB,AC,BC)}{L(AB,AC)} + 2\log\frac{L(ABC)}{L(AB,AC,BC)}.$$

We have decomposed our G-squared into four terms, the last of which is another G-squared. We will define the other terms as *relative* G-squared; they clearly measure the improvement in the fit from adding terms to the model. For example, write $2\log\frac{L(AB,AC)}{L(AB,C)} = G^2(AB,C|AB,AC)$. We interpret it as a measure of how well the model with only AB association fits compared to the improvement we would get if we included AC association. Its degrees of freedom are simply the extra degrees of freedom associated with the AC term, $(l-1)(p-1)$. Now we write our decomposition:

$$G^2(A,B,C) = G^2(A,B,C|AB,C) + G^2(AB,C|AB,AC)$$
$$+ G^2(AB,AC|AB,AC,BC) + G^2(AB,AC,BC).$$

This is the promised expression that corresponds to our decomposition of the sum of squares from least-squares theory. The connection with each of our earlier G-squared terms is obvious. For example,

$$G^2(AB,C) = G^2(AB,C|AB,AC) + G^2(AB,AC|AB,AC,BC)$$
$$+ G^2(AB,AC,BC).$$

Or we could work backwards and write things like

$$G^2(AB,C) - G^2(AB,AC) = G^2(AB,C|AB,AC).$$

This is exactly what we meant when we said that relative G-squared measures improvement in fit.

8.6.2 An ANOVA-like Table

Notice that the decomposition depends on the order in which we add terms. In practice, we add terms in descending order of how interesting they are to us or because we see from the data that they are important. Of course, you can also try several different orders of decomposition, in hope that they will tell you something interesting about the results of the survey.

Example. In the smoking survey, we might start with extremely simple loglinear models; if there is only a μ term, we are guessing that every cell is equally likely. If we introduce a term for smoking, the comparison is then asking whether or not there are equal numbers of smokers and nonsmokers. In this particular survey, we are not interested in such questions; we will start with the independence model, since we mainly care about associations between our classifications. Calling the three factors Smoking, Gender, and Location, we compute $G^2(S, G, L) = 10.25$. There are 4 degrees of freedom in the independence model, so there are 8 cells − 4 cells = 4 degrees of freedom for this statistic. We have suggested that a typical value of G-squared is the number of degrees of freedom; the actual value is enough larger to suggest strongly that the three factors are not, in fact, independent.

Staring at the data, we suspect that some of this association is between smoking and location—many of our nonsmokers live in cities. You estimated a model with an SL association introduced, and got (I hope) $G^2(G, SL) = 3.49$. There is an additional degree of freedom in this model, so we compute the relative term

$$G^2(S, G, L|SL) = G^2(S, G, L) - G^2(G, SL) = 10.25 - 3.49 = 6.76.$$

This is a strikingly large improvement for one degree of freedom; very likely, there is some association between where our subjects live and whether they smoke. On the other hand, the measure of fit for the new model, 3.49, is not impressive in light of the remaining three degrees of freedom. That single association may be all we have evidence for.

For completeness, let us add in one other apparent association, between gender and smoking. You have estimated this model, getting $G^2(SG, SL) = 0.36$, on 2 degrees of freedom. Our survey has found no evidence for any further association than this. On the other hand, $G^2(G, SL|SG) = G^2(G, SL) - G^2(SG, SL) = 3.49 - 0.36 = 3.13$ with one degree of freedom, suggests that we have found modest evidence that there is also a slight tendency for women to smoke more than men.

We already estimated a no-three-way-association model in an earlier section, and so the effect of the GL interaction is

$$G^2(SG, SL|GL) = G^2(SG, SL) - G^2(SG, SL, GL) = 0.36 - 0.09 = 0.27,$$

also negligible. Let us assemble these in an ANOVA-like table:

source	degrees of freedom	G-squared
S, G, L		
SL	1	6.76
SG	1	3.13
GL	1	0.27
SGL	1	0.09
total	4	10.25

You may wonder why we do not divide G-squared by its degrees of freedom, as with mean squares, so that it may be compared to 1. There is no good reason; it is simply not the reigning convention.

8.7   Estimating Logistic Regression Models

8.7.1   Likelihoods for General Bernoulli Experiments

In Chapter 1, Section 8, we did not find a convincing way to estimate the parameters in logistic regression models, except in simple cases where we could interpolate the cell logits. By now you will not be surprised to hear that the most widely used method for doing this is maximum likelihood. Very generally, in logistic regression we have an experiment in which we perform an independent sequence of Bernoulli trials; the result of each is either a "success" or a "failure". The probability of success is p_i for the ith trial; we try to estimate this so we can predict our chances of success in future trials. The likelihood of our results is then $\prod_{\text{successes } i} p_i \prod_{\text{failures } i}(1 - p_i)$, by independence. In Chapter 1 we were able to estimate some simple models by interpolating cells in a contingency table; then if the categories are $j = 1, \ldots, k$, the likelihood becomes $\prod_{j=1}^{k} p_j^{x_j}(1 - p_j)^{n_j - x_j}$, where p_j is the probability of success in that category, and x_j is the number of successes out of n_j trials. If we are interpolating and so can estimate each p separately, we see that we are just maximizing the cores of k binomial likelihoods, and the estimates are the sample proportions, as expected. Generally, any logistic regression model that came out of a contingency table has maximum likelihood at the standard estimates we got in Chapter 1 (see 1.8.1).

In the simplest case, with one numerical covariate with two values, the linear logistic regression model $l = \log\left(\frac{p_j}{1-p_j}\right) = \mu + b(x_j - \bar{x})$ corresponded to a saturated model fit to a two-by-two table. We noticed that an independence model was uninteresting, because then the conditional probability of success at each level of the independent variable was the same, and gave us no predictive value. But then the independence model corresponds to fitting a logistic model $l = \log\left(\frac{p_j}{1-p_j}\right) = \mu$. The slope b is assumed to be zero. Then the G-squared on 1 degree of freedom for testing independence is exactly the test that the slope is zero, as opposed to the saturated and interpolating alternative that it is not. Generally, our tests are exactly the same as the corresponding tests in contingency tables.

8.7.2   General Logistic Regression

Of course, maximum likelihood becomes particularly interesting when we apply it to problems that we do not know how to do otherwise.

Example. In 1991, Manly reported the mandible lengths in millimeters and by gender of 20 golden jackals:

length	105	106	107	107	107	108	110	110	110	111
gender	F	F	M	F	F	F	F	M	F	F
length	111	111	111	112	113	114	114	116	117	120
gender	M	F	F	M	M	M	M	M	M	M

There is a tendency for male mandibles to be longer. If we found a jackal mandible, could we predict whether it will turn out to be female?

We can no longer interpolate categories; we have almost as many lengths as subjects. But a linear logistic model for the probability of being female is plausible: $l = \log\left(\frac{p}{1-p}\right) = \mu + b(x - \bar{x})$, where p is the probability and x is the mandible length. We solve to find $p = e^{\mu+b(x-\bar{x})}/(1 + e^{\mu+b(x-\bar{x})})$; then the likelihood for all our successes and failures is

$$L = \prod_{\text{successes } i} e^{\mu+b(x_i \bar{x})}/(1 + e^{\mu+b(x_i-\bar{x})}) \prod_{\text{failures } i} 1/(1 + e^{\mu+b(x_i-\bar{x})}).$$

The log-likelihood is

$$l(\mu, b) = \sum_{\text{successes } i} [\mu + b(x_i - \bar{x})] - \sum_{\text{all } i} \log\left[1 + e^{\mu+b(x_i-\bar{x})}\right].$$

To find a criterion for a maximum value, we use calculus: Differentiate with respect to the unknown parameters μ and b, using partial differentation, and set each equal to zero.

$$0 = \frac{\partial l(\mu, b)}{\partial \mu} = \sum_{\text{successes } i} 1 - \sum_{\text{all } i} \frac{e^{\mu+b(x_i-\bar{x})}}{1 + e^{\mu+b(x_i-\bar{x})}},$$

$$0 = \frac{\partial l(\mu, b)}{\partial b} = \sum_{\text{successes } i} (x_i - \bar{x}) - \sum_{\text{all } i}(x_i - \bar{x})\frac{e^{\mu+b(x_i-\bar{x})}}{1 + e^{\mu+b(x_i-\bar{x})}}.$$

Recalling our expression for p, these may be rewritten as

$$0 = \sum_{\text{successes } i} 1 - \sum_{\text{all } i} p_i$$

and

$$0 = \sum_{\text{successes } i} (x_i - \bar{x}) - \sum_{\text{all } i}(x_i - \bar{x})p_i.$$

Our equations are simple, but it is hard to see what is going on. With a little ingenuity, think of the dependent variable, success or failure, as having the numerical value 1 or 0. It is then sort of an empirical probability corresponding to a cell with only one observation in it; we therefore call it $\hat{p}_i$. After a little rearrangement, our equations become

$$\sum_{\text{all } i}(\hat{p}_i - p_i) = 0 \text{ and } \sum_{\text{all } i}(x_i - \bar{x})(\hat{p}_i - p_i) = 0.$$

If you think of $\hat{p}_i - p_i$ as a residual, suddenly we have the normal equations from least-squares theory. (The first equation just says that the average estimate is just the average of the 1's and 0's). Are we finished? No; as lovely as these are, you must remember that the quantities we want are μ and b, and p is a *nonlinear* function of them. They cannot usually be solved for algebraically.

In small problems like our example, we may simply compute a number of values of the log-likelihood and graph the result (a computer math program helps here).

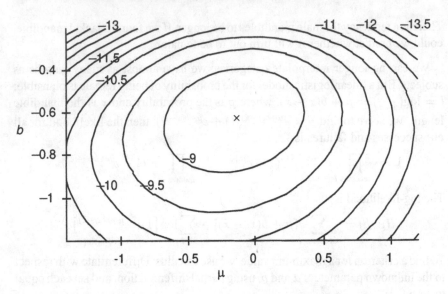

FIGURE 8.3. Log likelihood for a logistic regression model

Then we search for the maximum value over the range of our graph. (See Figure 8.3.) This is a contour plot, where all parameter pairs with the same likelihood are on a curve. This is therefore the picture of a likelihood "hill," with the top of the hill, the maximum of the likelihood, somewhere in the middle of the inner loop.

By focusing the search near the maximum, we find the maximum likelihood estimates $\hat{\mu} = -0.1508$ and $\hat{b} = -0.6085$, with a log likelihood -8.6294 there. Since $\bar{x} = 111$, we get a prediction equation $\hat{l} = -0.1058 - 0.6085(x - 111)$. If you should find an adult golden jackal mandible that is 109 millimeters long, we would predict a logit for it being female of 1.066; that gives a probability that it is female of 0.744.

We may ask, how sure are we that mandible length helps you identify gender at all? If we assume that $b = 0$, we are simply assuming a constant probability for each sex, estimated by the sample proportion $p = 10/20 = 0.5$. The log likelihood for that prediction is $10 \log 0.5 + 10 \log 0.5 = -13.863$. Taking the log-likelihood ratio for comparing the two classes of models, we get $G^2 = 2(-8.629 - -13.863) = 10.468$, on one degree of freedom. This is good evidence for the reality of a slope: longer mandibles suggest a male jaw.

8.8   Newton's Method for Maximizing Likelihoods

8.8.1   Linear Approximation to a Root

When you studied calculus, you may have learned a method attributed to Isaac Newton, of solving a nonlinear equation of the form $g(x) = 0$ for the variable x. The idea was that if you had a reasonably good first guess $x^{(0)}$, then the function may be almost a straight line between $x^{(0)}$ and the true value x. So we need to know

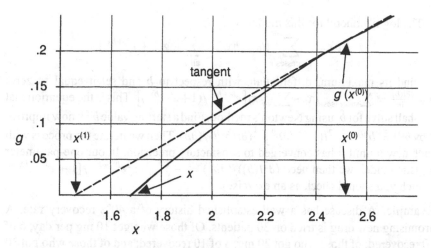

FIGURE 8.4. Newton's method

what straight line looks like. Calculus suggests that we find the *tangent* line to the curve at the point $x^{(0)}$ and then guess that the *secant* that takes us directly from there to the true solution is approximately the same as the tangent (Figure 8.4). That is, $\left(g(x)-g(x^{(0)})\right)/\left(x-x^{(0)}\right) \approx g'(x^{(0)})$. But since $g(x) = 0$, we find that $x-x^{(0)} \cong -g'(x^{(0)})^{-1}g(x^{(0)})$. We use this equation to calculate an improved guess to the solution $x^{(1)} = x^{(0)} - g'(x^{(0)})^{-1}g(x^{(0)})$; if the first guess was good enough and g is not too curved, this will be much better. We then use the new guess to calculate a third approximation $x^{(2)}$ and repeat until we have the solution to sufficient accuracy.

8.8.2 Dose–Response with Historical Controls

We will apply Newton's method to a maximum likelihood estimate of a logistic regression model with one parameter. Most interesting models have more than one parameter; we will return to that problem later. However, one reasonable model, a linear dose–response model with historical controls, has only a single parameter to estimate. This comes about when there is no standard drug available to treat some serious disease. So when a new drug comes out of the lab, with promising results on rats and on a handful of patients, doctors are eager to try it out on *all* their patients. They cite medical ethics when they refuse to include a control group of patients who get a dose of zero in the study, even though almost any statistician would agree that it would make for a much better experiment.

Our second choice would be to introduce recent experience with the disease into our study. We would assume that these *historical controls* (victims of the disease who did not get the drug because it had yet to be invented) had a certain probability of recovery, which we know accurately because there were a large number of them. Let that historical probability of getting well be p_0; then its logit is $\log \frac{p_0}{1-p_0} = l_0$. Our model will assume that the logit for recovery changes proportionally with the dose of the new drug, so $\hat{l} = l_0 + bx$, where x is the dose and b is the unknown slope parameter.

The log likelihood for this model is

$$l(b) = \sum_{\text{successes } i} [l_0 + bx_i] - \sum_{\text{all } i} \log[1 + e^{l_0 + bx_i}].$$

To find its maximum, differentiate with respect to b and set it equal to zero: $0 = \frac{\partial l(b)}{\partial b} = \sum_{\text{successes } i} x_i - \sum_{\text{all } i} x_i [e^{l_0 + bx_i}/(1 + e^{l_0 + bx_i})]$. This is the equation that we shall solve for $\hat{b}$, using Newton's method. Find a starting value $b^{(0)}$; now improve it by $b^{(1)} = b^{(0)} - [\partial^2 l(b^{(0)})/\partial b^2]^{-1}(\partial l(b^{(0)})/(\partial b))$. Then we iterate the process with each new b until it has converged to satisfactory precision. In our one-parameter logistic model, we then need $(\partial^2 l(b))/(\partial b^2) = -\sum_{\text{all } i} x_i^2 e^{l^{(0)} + bx_i}/[1 + e^{l^{(0)} + bx_i}]^2$ (which you should check as an exercise).

Example. A disease has a well-established history of a 40% recovery rate. A promising new drug is tried on 30 patients. Of those who got 10 mg per day, 6 of 10 recovered; of those who got 20 mg, 8 of 10 recovered; and of those who got 30 mg, 9 of 10 recovered. We will try the dose-response model $\hat{l} = l_0 + bx$, where x is the daily dose, and the zero-dose historical-control logit is $l_0 = \log\left(\frac{0.4}{1-0.4}\right) = -0.4055$. We will estimate the slope b by maximum likelihood. Let the starting value be $b^{(0)} = 0$. You should check my computation that $b^{(1)} = 0.0744$; then $b^{(2)} = 0.0859$, and $b^{(3)} = 0.0870$. The changes after that are negligible. Predicted rates of cure are 61.4% at 10 mg , 79.2% at 20 mg, and 90.1% at 30 mg—very close to the observed rates. The G-squared statistic, comparing the fit of a constant recovery rate of 0.4 to the one fitted by our model, has one degree of freedom and equals 19.32. It seems very likely that there is indeed a positive slope to this model. Within this range, the more of the drug, the better the chance of recovery.

8.8.3   Several Parameters*

In the more common models with several parameters, we can use a more sophisticated version of Newton's method. The condition for a maximum becomes $0 = (\partial l(\mathbf{b}))/(\partial \mathbf{b})$, a vector equation that has one coordinate equation for each of the k parameters being estimated. Then the second derivatives form a k-by-k matrix $(\partial^2 l(\mathbf{b}))/(\partial b_i \partial b_j)$. The approximation of the log-likelihood by a tangent plane at $\mathbf{b}^{(0)}$ is then $-(\partial^2 l(\mathbf{b}^{(0)}))/(\partial b_i^{(0)} \partial b_j^{(0)})(\mathbf{b}^{(1)} - \mathbf{b}^{(0)}) = \partial l(\mathbf{b}^{(0)})/\partial \mathbf{b}$ (you should review partial and total derivatives at this time). To solve for the improved vector of guesses $\mathbf{b}^{(1)}$ just requires you to solve a system of k equations in k unknowns. Then you iteratively compute new $\mathbf{b}$'s from old ones until they stop changing within the accuracy you are seeking. You will get a chance to try this in an exercise.

8.9   Summary

The likelihood of a parameter value θ once we have made a (discrete) observation is $L(\theta|\mathbf{x}) = P(\mathbf{X} = \mathbf{x}|\theta)$. We were able to compare the closeness to the data of two discrete models by taking the *likelihood ratio* $R = \frac{L(\theta_1|\mathbf{x})}{L(\theta_2|\mathbf{x})}$

(2.1). We then called the parameter value that made the likelihood greatest for a given observation its *maximum likelihood* estimate $\hat{\theta}$ (2.2). Comparing this likelihood to that for a hypothesized value θ gives us the G-*squared* statistic $G^2(\theta) = 2\log\frac{L(\hat{\theta}|x)}{L(\theta|x)} = 2l(\hat{\theta}|x) - 2l(\theta|x)$ (3.2). We found that for many discrete models this is almost a weighted squared distance measure, the *chi-squared* statistic $\chi^2 = \sum_{i=1}^{k}\frac{(O_i - E_i)^2}{E_i}$, where E_i are counts expected under the hypothesis and O_i are the counts actually observed (4.1). We found the maximum likelihood estimates for certain contingency table models (which use *sufficient statistics*, certain marginal total counts) (5.1). A general procedure for computing these estimates, *iterative proportional fitting*, was then derived (5.3). We evaluated our models using our distance measures; in particular, G-squared may be decomposed much like the sum of squares, to provide an ANOVA-like summary table (6.2). Finally we discovered that maximum likelihood may also be used to estimate logistic regression models (7.2). *Newton's method* for finding the roots of equations allowed us to compute the parameter estimates (8.2).

8.10 Exercises

1. A natural gas pipeline had 30 significant leaks last year. The operating company claims that the annual average is only 20. What is the relative likelihood of a true mean of 20 compared to a true mean of 30? Graph the likelihood of this observation.

2. A manufacturer admits to a 10% rate of defective compact digital discs. Of 120 disks you have bought in the last two years, 17 have been defective. What is the maximum likelihood estimate of the true rate of defectives? What is the likelihood ratio comparing that rate to the manufacturer's claim?

3. **a.** Derive the formula for the maximum likelihood estimate of p in an NB(k, p) model.

   b. You survey students until you find 10 who are left-handed. On the way, you notice that you have surveyed 87 right-handed students. What do you estimate is the population probability that a student is right-handed?

4. You perform a negative hypergeometric experiment with result x distributed N(W, B, b).

   a. If W is unknown, what is its maximum likelihood estimate?
   b. If instead B is unknown, what is its maximum likelihood estimate?

5. You perform a hypergeometric experiment with result X distributed H($W + B, W, n$).

   a. If W is unknown, what is its maximum likelihood estimate?
   b. If instead B is unknown, what is its maximum likelihood estimate?

6. Derive the maximum likelihood estimates for the vector of probabilities **p** in the multinomial random vector with k categories.

7. Use L'Hospital's rule to calculate $\lim_{x \to 0} x \log(x)$.

8. Compute the G-squared and chi-squared statistics for the claims in Exercises 1 and 2.

9. Use maximum likelihood to estimate the p's in the multinomial independence model for a rectangular table $\hat{x}_{ij} = np_{i \bullet} p_{\bullet j}$.

10. In Exercise 12 of Chapter 1 (status versus philosophy):

   a. Evaluate G-squared for the independence model. Compare it to the degrees of freedom. Conclusions?

   b. Compute chi-squared for the independence model. Check the criteria for a good match to G-squared. Are they consistent with the actual comparison?

11. In Exercise 30 of Chapter 1 (sex distribution in various cities)

   a. Evaluate G-squared for the independence model. Compare it to the degrees of freedom. Conclusions?

   b. Compute chi-squared for the independence model. Check the criteria for a good match to G-squared. Are they consistent with the actual comparison?

12. Show that a single-stage calculation in proportional fitting $\hat{x}_{ij}^{(1)} = \left(\hat{x}_{ij}^{(0)} / \hat{x}_{i \bullet}^{(0)} \right) x_{i \bullet}$ indeed enforces the correct row totals $\sum_{j=1}^{J} \hat{x}_{ij}^{(1)} = x_{i \bullet}$.

13. Estimate the independence model in Exercise 11 by proportional fitting.

14. Estimate the complete independence model in the smoking–gender–location survey (see Section 5.3) by proportional fitting. Compute G-squared, comparing it to the saturated model.

15. For the prediction of gender using mandible length, I proposed the linear logistic equation $\hat{l} = -0.1508 - 0.6085(x - 111)$. Show that these predictions meet the normal equations for maximum likelihood estimation.

16. The picture illustrating a step of Newton's method in Section 8 refers to the following problem: What Poisson mean λ would I need to have so that half the time the count was 0 or 1? That is, solve the equation $F(1) = 0.5$. This becomes $(1 + \lambda)e^{-\lambda} = 0.5$, or $g(\lambda) = 0.5 - (1 + \lambda)e^{-\lambda} = 0$. Let a starting guess be $\lambda^{(0)} = 2.5$, as in Figure 8.4. Compute several improved guesses using Newton's method until it stops changing to three significant figures. Compare your answer to Figure 8.4.

17. For the historical controls model in Section 8.2, verify that

$$\frac{\partial^2 l(b)}{\partial b^2} = -\sum_{\text{all } i} x_i^2 \frac{e^{l^{(0)} + bx_i}}{\left[1 + e^{l^{(0)} + bx_i} \right]^2}.$$

18. I purchased a balanced die, which I therefore assume has probability $\frac{1}{6}$ of coming up "six." But I want to try to "load" it so it will come up six more often. I inject 10 mg of lead into the opposite face, then roll it 60 times. I get 12 sixes. With 20 mg of lead, I get six in 21 of 60 rolls; and with 30 mg of lead, I get 23 sixes out of 60 rolls.

Let us guess that a linear logistic model $l(x) = l_0 + xb$ should work, with l the logit for coming up six, $x =$ mg of lead injected, and l_0 the logit of the balanced-die probability of $\frac{1}{6}$.

a. Estimate b by the method of maximum likelihood, using Newton's method.
b. Compute the G-squared for how well this fits the data, and compare it to the G-squared for a constant probability of $\frac{1}{6}$. What do you conclude?
c. Use your model to estimate the probability that a six will come up if you injected 25 mg of lead into the opposite face.

8.11   Supplementary Exercises

19. The method of maximum likelihood suggests yet another way to get an interval that reflects the uncertainty in a parameter estimate. The interval includes all values of the parameter that are at least $1/k$ times as likely as is the maximum likelihood estimate. This is called a *likelihood interval*.

   a. For the data of Exercise 2, find a $k = 7$ likelihood interval for possible values of the binomial probability of a defective disk.
   b. Find a 95% confidence interval for the binomial probability. (You will see the reason for the similarity of the two intervals in a later chapter.)

20. Let an observation x be Poisson (λ) with λ unknown. Derive the normal curve approximation to the likelihood $L(\lambda|x)$. Graph the true versus the approximate log-likelihood curves for the data of Exercise 1.

21. A very common way to survey a population is *stratified* sampling. For example, you may know the population proportion of some relevant groupings: gender, race, age. Then a simple random sample might, by accident, misrepresent one of those groups; if so, any conclusions on other issues could be distorted. Instead, sample within your groups, determining in advance how many of each you will take. Number your stratification variable $i = 1, \ldots, k$, and interview n_i in the ith stratum. Observe that each subject falls into categories $j = 1, \ldots, l$; then say that x_{ij} subjects from the ith stratum fell in the jth category.

   a. The usual model for this design would be the *product-multinomial* model: x_{ij} for $j = 1, \ldots, l$ are Multinomial(n_i, p_{ij} $j = 1, \ldots, l$), where $\sum_{j=1}^{l} p_{ij} = 1$ for each i. What is the core of the likelihood for this model? What are the maximum likelihood estimates of the p_{ij}?
   b. The *row homogeneity* model says that the stratification is irrelevant and the probabilities are the same in each row: $p_{ij} = p_j$. Find the maximum likelihood estimates for the p_j. How many degrees of freedom does it have?

22. In a precinct that is about 60% Democratic and 40% Republican, you locate 120 Democrats and 80 Republicans, and ask them whether they favor a new state lottery. You find

	For	Against	No Opinion
Democrats	73	27	20
Republicans	27	25	8

a. Find the parameter estimates and cell expectations for a row homogeneity model.

b. Find the G-squared statistic that compares this model to the (saturated) product-multinomial model. Compare it to the degrees of freedom, and comment.

23. In the smoking, gender, location survey (see Section 5.3):

a. Estimate the model with only SL interaction, by proportional fitting. Compute G-squared, comparing it to the saturated model.

b. Estimate the model with SL and SG interaction, by proportional fitting. Compute G-squared comparing, it to the saturated model.

24. In section 5.3 we estimated the $\hat{x}$'s from equations of the form $\log \hat{x}_{MRS} = \mu + b_M + c_R + d_S + e_{MR} + f_{MS} + g_{RS}$. Use the calculation method from (1.7.5) to find a numerical value for each of the seven parameters.

25. We want to show that proportional fitting never adds higher-order terms to your model. We will do it for the simplest case, association in a two-by-two table. Say that your current table has association $\rho = \left(x_{11}^{(0)} x_{22}^{(0)}\right) / \left(x_{12}^{(0)} x_{21}^{(0)}\right)$. Now, show that in the course of fitting an independence model, if you force either set of marginals to hold, after an iteration you still obtain $\rho = \left(x_{11}^{(1)} x_{22}^{(1)}\right) / \left(x_{12}^{(1)} x_{21}^{(1)}\right)$. Therefore, if your starting table has no higher-order terms ($\rho = 1$), then neither will your final table.

26. In 1987 Freeman reported a survey linking survival of infants to age one year to prematurity, mother's age, and whether she smoked:

		Premature		Full Term	
		Dead	Alive	Dead	Alive
Younger	No	50	315	24	4012
	Smokes	9	40	6	459
Older	No	41	147	14	1594
	Smokes	4	11	1	124

a. Other studies have suggested the plausibility of each of the six two-way associations here. Write down the loglinear model that has all those two-way associations (but no three-way associations). Interpret each of those associations in words.

b. Write down the marginal totals that are the sufficient statistics for this model.

c. Compute the predicted counts in the model, to within 0.1 person, by iterative proportional fitting.

d. Compute the G-squared, comparing this model to the saturated model. How many degrees of freedom does it have? What do you conclude about the model?

27. Use Newton's method to maximize the likelihood in the linear mandible-length model (Section 7.2), by simultaneously solving $0 = (\partial l(\mu, b))/(\partial \mu)$ and $0 = (\partial l(\mu, b))/(\partial b)$. You will need the matrix

$$\begin{pmatrix} \dfrac{\partial^2 l(\mu, b)}{\partial \mu^2} & \dfrac{\partial^2 l(\mu, b)}{\partial \mu \partial b} \\ \dfrac{\partial^2 l(\mu, b)}{\partial \mu \partial b} & \dfrac{\partial^2 l(\mu, b)}{\partial b^2} \end{pmatrix}$$

to construct your linear system of two equations in two unknowns. Let $\mu^{(0)} = 0$ (for an average mandible we guess equal likelihood that it is male or female), and $b^{(0)} = 0$ (maybe mandible length does not matter). Do several iterations, until your estimates stabilize; compare them to my graphical estimates.

CHAPTER 9

Continuous Random Variables I: The Gamma and Beta Families

9.1 Introduction

Many statistical applications seem not to be about discrete random variables, taking on values only from a manageable list. Rather, we see random quantities that might include any number in whole intervals, perhaps because they are measurements of time, weight, length, and so forth. These are instances of *continuous* random variables. We shall find ourselves using new mathematical techniques, often from calculus, to study them.

We shall start by inventing a class of experiments ruled by chance, called a *Poisson process*, out of which Poisson variables arise naturally. In addition, an important family of random variables with continuous values, described by its probability density, appears in a Poisson process. We will go on to investigate another chance process, the *Dirichlet* process, which is related to binomial random variables. Here, too, important continuous variables arise. Finally, we will study relationships between these processes; and inferences in them.

Time to Review

Chapter 4, Section 8
Chapter 5, Section 4.3
Chapter 6, Sections 3–4

9.2   The Uniform Case

9.2.1   Spatial Probabilities

We have already considered a class of continuous random variables: the coordinates of random points in geometrical probability problems. For example, where does a dart hit along the horizontal axis of some rectangular target? We suggested when first introducing the idea of a random variable that the cumulative distribution function $F(x) = P(X \leq x)$ should carry all the information we need to describe its random behavior (see 5.4.2). This is so because the *sigma algebra* (see 4.8.2) for geometrical probabilities in one dimension was built out of intervals like $(a, b]$, which just says that we need to know the probability of the variable falling in such intervals. But $P\{X\varepsilon(a, b]\} = F(b) - F(a)$ (you should remind yourself why this is so).

For example, mark off the horizontal axis of that rectangular target as $(0, 1]$ and imagine, if you can, that I am so inept at darts that every point is as likely a hit as every other point. Then, if $0 \leq a < b \leq 1$, we get $P(X\varepsilon(a, b]) = b - a$, since the longer an interval is, the more likely I am to hit it. This suggests what the cumulative distribution function should be: $F(x) = x$ on $(0, 1]$. This particular random variable is called a Uniform $(0, 1)$ random variable, because, like discrete uniform variables, it does not prefer any outcome to any other. Interestingly enough, it is the random variable you will usually get (approximately) when you hit a button called "random" or something like it on a calculator or invoke a random number generating function in a computing system (see also 4.2.1). We will see later why this simple example is so useful.

9.2.2   Continuous Variables

If you graph the cumulative distribution function $F(x) = x$ (a straight $45°$ segment), you should notice an important difference between it and the one for all our discrete random variables: It is a continuous function. If we try to graph the discrete case, F has to "jump" up by an amount p_i at each of our list of values x_i, creating a graph with many breaks. But since no single value from the infinity of possible coordinates of a geometrical outcome has substantial probability, we cannot have jumps anywhere; and in fact, the curve is continuous. We will let this be the characterizing feature of continuous random variables:

Definition. A **continuous** random variable is one whose cumulative distribution function is a continuous function on its sample space.

So the way to see whether it is continuous is to check: Can you graph the cumulative distribution without lifting your pencil from the paper? This has a peculiar consequence. What is the probability of a given point, say a? It should be quite small, since there are uncountably many possible points in the sample space. We do not have a probability mass function yet, only probabilities of intervals, so let us sneak up on it with smaller and smaller intervals that in limit contain only the one point:

$$P(X = a) = \lim_{\delta \to 0} P\{X\varepsilon(a-\delta, a]\} = \lim_{\delta \to 0}[F(a) - F(a-\delta)] = F(a) - \lim_{\delta \to 0} F(a-\delta).$$

Now, the formal definition of a continuous function in calculus (which you should review) says that the limit of its values as we approach a point is just the function evaluated at that point; this is just being precise about not lifting the pencil as we draw the graph. So $\lim_{\delta \to 0} F(a - \delta) = F(a)$ because we have assumed that F is continuous. Thus $P(X = a) = F(a) - F(a) = 0$. We conclude that the probability of any given outcome is not merely tiny, it is exactly zero.

Proposition. *If X is a continuous random variable, then for any number a in its sample space,* $P(X = a) = 0$.

In other textbooks, this property is used to define continuous random variables. Notice its peculiar effect on our intuition: certainly, very many of these values are possible outcomes—the dart really might hit there. Therefore a "probability of zero" does not mean the same thing as "impossible" (see 4.2.2). Looking at the complementary event, a "probability of one" does not mean the same thing as "certain." We shall have to think further to find a reasonable interpretation of probability zero. Imagine that somebody offered you a really wonderful wager, that she will pay you a million dollars if some perfectly possible event happens, for example, if a uniform random number comes out equal to $\pi/4 = 0.785398163\ldots$. This once-in-a-lifetime deal will only cost you a penny. Should you take it? Calculating the positive part of the expectation: $\$1,000,000 p(\pi/4) = \$1,000,000 \times 0 = \$0$, which means that on average you have nothing to gain from the deal. Therefore, zero probability means something like "never bet on it"; and probability one means "always worth betting on."

We can see another difficulty with continuous random variables: The probability mass function is always zero, and so is of no interest. But this function was usually the simplest way to describe discrete probabilities, and was needed to calculate expectations. We shall have to find other methods to accomplish these things, shortly.

9.3   The Poisson Process

9.3.1   How Would It Look?

We proposed applying Poisson random variables to counting rare, independent events (see 6.4.3). For example, we might wish to talk about the number of babies born in a given month in a town of several thousand people (counting twins once, to preserve independence.). This might well have something like a Poisson pattern of probabilities, with mean equal to some long-term average of monthly births. But there is a slightly different way we could look at that same problem: the sequence of times and dates, continuing indefinitely, at which a baby is born.

These random quantities can be any real number, and so in some sense are continuous random variables, even though the number of babies in a period is discrete. The mathematical description of the whole scheme is contained in the following definition.

Definition. A **(standard) Poisson process** is a random sequence of real numbers $0 < t_1 < t_2 < t_3 < \cdots$ such that (1) in any interval $0 \le a < b$, the number of t's that fall in the interval $|\{t_i \mid a < t_i \le b\}|$ is a Poisson$(b - a)$ variable; and (2) the number of t's in any two nonoverlapping intervals is independent.

Any stochastic process consisting of a countable sequence of increasing real numbers like this is called a *point process*.

The t's in our example are just the times that babies are born. There are two peculiarities in how we measure time: It is always measured from a starting time we call 0 that represents the moment we start counting our events. Even more oddly, we have let our unit of time be the length of an interval in which an average of one event happens. This is why we call it a *standard* process. In our example, if an average of 5 babies are born each month, then we are using the peculiar unit of time of about $30/5 = 6$ days. Then a period of 72 days is 12 of our perverse units, because an average of 12 babies are born in such a period. This convention will simplify our algebra; and as you will see, it is very easy to translate back and forth in practical problems.

9.3.2   How to Construct a Poisson Process

Of course, we have no reason to believe that any such Poisson processes exist; we have reasoned from a qualitative example. As with Poisson variables themselves, though, we can construct the process as a limit of things we do know to exist. A Bernoulli process, a sequence of independent trials that either succeed or fail independently of one another (see 6.3.3), might be given a very low probability of success at each trial. Then successes are rare. To connect it to a Poisson process, imagine that the many failures are ticks of a rapid clock, so that failed trials count off the passage of time. If we asked how many successes had taken place in a certain amount of "time," we are really counting successes before a certain number of failures, so that successes are negative binomial. In a standard Poisson process, a unit of time has an average of one event; in a negative binomial NB(k, p) variable, there are an average of $kp/(1 - p)$ successes. To approximate a Poisson process by a Bernoulli process, we need to synchronize their clocks; we will choose k such that $kp/(1 - p) = 1$. (I, of course, use small values of p that let k be a large integer.) Then the length of time measured by a tick is $p/(1 - p) = 1/k$ standard units.

Now we describe our Bernoulli process as if it were a point process: Let s_i be the number of failures that precede the ith success. Then the numbers $0 \le s_1 \le s_2 \le s_3 \le \cdots$ tell us exactly what happened. For example, 2, 5, 7, ... says that the Bernoulli sequence started out FFSFFFSFFS.... . We translate these "ticks" into our time units by computing $t_i = s_i/k$; this step is an example of a very important

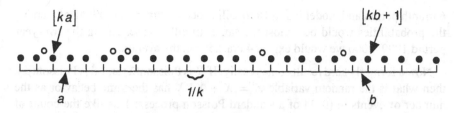

FIGURE 9.1. Part of a Bernoulli process

sort of transformation called *standardization* or *renormalization*. Now we want to know that for p small, our new description of the process $0 \leq t_1 \leq t_2 \leq t_3 \leq \cdots$ behaves approximately like a Poisson process.

Example. In a sequence of trials with $p = 1/101$, so that $k = 100$, I observed successes after 73, 208, 292, 428, 499, ... failures. We standardize these to get events at times 0.73, 2.08, 2.92, 4.28, 4.99, ...

Our second requirement is clearly met: The successes in two nonoverlapping time intervals are the successes in two separate stretches of a Bernoulli sequence, which are always independent of one another. It remains to check the probabilities for the number of t's in an interval, say $(a, b]$. This, of course, is the same as the number of s's in the interval $(ka, kb]$. They are in the sequence that begins with the next trial after failure number $\lfloor ka \rfloor$ (this is the *floor* function, the largest integer no bigger than that value; for example $\lfloor 3.14159 \rfloor = 3$) and ends with failure number $\lfloor kb + 1 \rfloor$.

(In Figure 9.1, a black marble corresponds to a failure and a white marble to a success.) Then our number of successes is negative binomial, NB($\lfloor kb + 1 \rfloor - \lfloor ka \rfloor, p$), because we are counting until that many more failures have happened. But in a series of such processes in which p gets small (and so k gets large), we know that this variable converges in distribution to a Poisson random variable with mean

$$\frac{(\lfloor kb + 1 \rfloor - \lfloor ka \rfloor)p}{1 - p} = \frac{\lfloor kb + 1 \rfloor - \lfloor ka \rfloor}{k} \approx b - a.$$

The last approximation holds because the two floor functions are within 1 of kb and ka, and k becomes large. We have shown the following fact:

Theorem (Poisson limit of Bernoulli processes). *Consider a series of Bernoulli processes in which $p \to 0$; standardize each sequence of successes by $t_i = s_i p/(1 - p)$. Then the processes realized by the sequences of t's converge in distribution to a standard Poisson process.*

Now we know that there is such a thing as a Poisson process, because we can construct simple experiments that behave as much like one as we please.

Example. If there are an average of 2 fatal commercial airline accidents per year, we might well model their times as a Poisson process (with standard time unit 6 months). We might almost as well note that there are about 100,000 safe flights per

6 month period and model it as a Bernoulli process with $p = 0.00001$ of a crash; the probabilities would be almost the same. In either case, during the two-year period 1999–2000 we would expect 4 crashes, on the average.

Now a formerly hard result comes easily: If X is Poisson(λ) and Y is Poisson(μ), then what is the random variable $Z = X + Y$? X has the same behavior as the number of events in $(0, \lambda]$ of a standard Poisson process; Y is like the count of events in the time interval $(\lambda, \lambda + \mu]$, and the two are independent. Therefore, Z is the count in $(0, \lambda + \mu]$ and is Poisson($\lambda + \mu$).

9.3.3 Spacings Between Events

Poisson counts are, of course, discrete; but now $T = t_1$, the first time something happens, is presumably a continuous random variable. If our first baby of 1998 was born on January 15 at 1:25 A.M., then $t_1 = 2.3432$ in standard time units (6 days each). In an approximating Bernoulli sequence, as k gets large, the possible values get closer and closer together. The number of successes before the first failure is s_1, with a probability $1 - p$ of success; therefore it is Geometric($1 - p$). At this point, we could calculate the cumulative distribution function of $T = s_1 p/(1 - p)$ by an easy limit argument; we will leave that to you as an exercise. Instead, we will use a slicker argument that will be useful later. Notice that we have invoked a black–white transformation on our Bernoulli process: What were successes are now failures. We found a black–white duality between certain negative binomial random variables; now we will use precisely that duality in a Poisson process to find the properties of T.

$$F(t) = P(T \leq t) = P(\text{first Poisson occurrence happens by time } t)$$
$$= P(\text{at least one happens by time } t) = 1 - P(\text{no occurrences by time } t)$$
$$= 1 - P[X = 0 \mid X \text{ is Poisson}(t)] = 1 - p(0) = 1 - e^{-t}.$$

This is important enough to be given a name:

Definition. A **(standard) negative exponential** random variable is the value of the first event in a (standard) Poisson process. Its sample space is all positive real numbers.

Proposition. (i) *A negative exponential random variable is continuous, with cumulative distribution function $F(t) = 1 - e^{-t}$ for all positive t.*

(ii) *Let U_i be a sequence of Geometric($1 - p_i$) random variables in which $p_i \to 0$. Then $T_i = \frac{U_i p_i}{1 - p_i}$ converge in distribution to a negative exponential variable.*

Example. Cosmic rays enter a cloud chamber and are recorded in an experiment on average once every five seconds. Separate events are independent of one another. What is the probability that the first cosmic ray will arrive within 12 seconds of the beginning of the experiment? This is presumably a Poisson process, with unit of time 5 seconds. We are asking about a length of time $12/5 = 2.4$ units. The problem is then asking $P(t \leq 2.4) = 1 - e^{-2.4} = 0.9093$.

Notice that it is still true that random variables converge in distribution (see 6.2.5) to a *continuous* distribution when their cumulative distribution functions have the right limit.

The starting time for a Poisson process seems pretty much arbitrary in each of our applications. This is no accident: Consider any positive time a; now Consider all the events that happen after that time, $a < t_i < t_{t+1} < \cdots$. Let $t'_j = t_{i+j-1} - a$, the amount of time after a until each later event; then $0 < t'_1 < t'_2 < \cdots$. It is easy to see that these form a new Poisson process: Intervals correspond to intervals of the same length in the original process, and if new intervals do not overlap, neither did the ones they derived from. Furthermore, anything that happened until time a is independent of the new things that happen after a. Thus resetting our clock to zero at any time leaves us with a clean slate; this is called the *memoryless* property of a Poisson process. It follows from the fact that any given stretch of Bernoulli trials is independent of the successes and failures that came before. This has an interesting consequence:

Proposition. *The intervals between successive events of a Poisson process $t_i - t_{i-1}$ are each negative exponential and are independent of each other and of t_1.*

This is because we can just imagine that we are starting the timer again as each event happens. If we had a source of independent negative exponential random variables V_i, we could have constructed a Poisson process by letting $t_1 = V_1, t_2 = t_1 + V_2, t_3 = t_2 + V_3$, and so forth.

9.3.4   Gamma Variables

Example. Negative exponential random variables are often used as models for time to failure of mechanical systems. Imagine that a space probe bound for Mars has on board a critical navigation computer and four identical backup computers that come on line as earlier ones fail. Let failures be a Poisson process that experiments suggest has a rate of failure of one per six months. It takes two years to reach Mars. What is the probability we will reach our destination before all five computers have failed?

We seem to have asked here about the fifth Poisson event rather than the first. Obviously, $T = t_i$ is also a continuous random variable for any positive integer i.

Definition. The time T to the αth event t_α in a standard Poisson process is a **Gamma(α)** random variable.

Thus a negative exponential variable is also Gamma(1). Our comment above tells us the following:

Proposition. (i) *If V_i are independent negative exponential variables, then $T = \sum_{i=1}^{\alpha} V_i$ is a Gamma(α) variable (since it is just the total of the waiting times until the αth event);*
   (ii) *if T is Gamma(α) and U is independently Gamma(β), then $V = T + U$ is Gamma($\alpha + \beta$).*

As interesting as these facts are, they do not tell us very much at this point about the probabilities of gamma random variables. Instead, use the black–white duality argument:

$$F(t) = P(T \leq t) = P(\alpha\text{th Poisson occurrence happens by time } t)$$
$$= P(\text{at least } \alpha \text{ occurrences happen by time } t) = P[X \geq \alpha | X \text{ is Poisson}(t)].$$

Theorem (gamma–Poisson duality).

$$F[t|\text{Gamma}(\alpha)] = 1 - F[\alpha - 1|\text{Poisson}(t)] = \sum_{i=\alpha}^{\infty} \frac{t^i}{i!} e^{-t}.$$

Example (*cont.*).   In the space probe problem, 2 years is 4.0 six-month periods, and we are concerned whether the fifth failure will happen after that time, so

$$P[t > 4.0|\text{Gamma}(5)] = 1 - F(4.0) = \sum_{X=0}^{5-1} \frac{4^X}{X!} e^{-4} = 0.6288.$$

This is not much of a safety margin.

We derived a Poisson process from a Bernoulli process with rare successes; but the failure count s_α before the αth success may be thought of as a Negative Binomial $(1 - p, \alpha)$ random variable, where p is now the small probability of success, by black–white duality. This just generalizes our observation about the first failure being a geometric random variable.

Theorem (gamma approximation to the negative binomial).   (i) *If* $\binom{\alpha}{2}$ *is small compared to* x, *and* $xp^2/(1 - p)$ *is small, then if* $t = xp/(1 - p)$, *we have*

$$F[x|\text{NB}(\alpha, 1 - p)] \approx F[t|\text{Gamma}(\alpha)].$$

(ii) *If* X_i *is* NB$(\alpha, 1 - p_i)]$ *and* $p_i \rightarrow 0$, *then* $T_i = X_i p/(1 - p)$ *converge in distribution to Gamma*(α).

You may check this by applying black–white duality to the negative binomial, then the Poisson approximation to the negative binomial, then the gamma–Poisson duality.

Example.   We have noted that 10% of the population is left-handed. There are three left-handed desks in a classroom, and the Equal Opportunity office requires that we start a new section as soon as a fourth left-hander enrolls in a course. What is the probability that we will have no more than thirty students in a given section? This is negative binomial with $p = 0.9$ and $k = 4$; we compute $F(27) = 0.3762$. Since lefties are relatively rare, try a Gamma(4) approximation with $t = \frac{27 \times 0.1}{0.9} = 3$. Then $F(3) = 0.3528$, which is not bad, given that p is not all that close to 1.

9.3.5   Poisson Process as the Limit of a Hypergeometric Process*

While we are here, we might as well use what we already know to construct a Poisson process from a hypergeometric process. The idea is that in an urn with a great many black marbles and relatively few but still numerous white marbles, we may treat the black marbles as the ticks of a clock (or perhaps better, as grains of sand falling through the neck of an hourglass). Then the white marbles are noteworthy events, and we can treat the "times" at which they occur as the t's in a roughly Poisson process. We already know that under certain conditions the random number of white marbles by the time we get a fixed count of black marbles is approximately a Poisson variable (see Chapter 6.8). We can specify exactly how the realization of the process has gone by simply letting s_i be the number of black marbles that have been removed by the time the ith white marble appears. (The sequence is now finite, only W numbers, but that is still a great many.) To standardize these counts we remember that the average of a negative hypergeometric variable is $bW/(B+1)$ white marbles by the bth black; therefore, the number of ticks of the clock, or grains of sand, that corresponds to one standard unit of time should be $b = (B+1)/W$. Now let $t_i = s_i/b$ convert our count into times at which our nearly Poisson events happen.

Theorem (Poisson limit of a hypergeometric process).   *In a series of hypergeometric processes in which $b = (B+1)/W \to \infty$ and $W \to \infty$, the sequences of numbers $t_i = s_i/b$ converge in distribution to a Poisson process.*

PROOF.   To check that the counts in a given time interval are Poisson, use the same procedure we used in the Bernoulli case and our result about the Poisson approximation to a hypergeometric variable. This time, though, independence of nonoverlapping intervals is not obvious, because the counts in different parts of a hypergeometric process are obviously not independent. Let X be the count in $(c, d]$ and Y be the count in nonoverlapping $(e, f]$. Then $p(x)$ is approximately Poisson$(d-c)$. Now, $p(x \mid y)$ is just a hypergeometric probability in which y white marbles and approximately $b(f - e)$ black marbles have been removed from the jar (because we know they appear at another time). These numbers are each small parts of the totals, as the urn grows. Thus, $p(x \mid y)$ is approximately Poisson with mean

$$\frac{b(d - c)(W - y)}{B + 1 - b(f - e)} = (d - c)\frac{1 - (y/W)}{1 - (f - e)/W} \approx (d - c),$$

since W gets large. Since the conditional probabilities converge to the same values as the unconditional, we have asymptotic independence of the intervals as the urns grow.

Since s_α is the count of black marbles before the αth white, by switching black and white marbles in the process, we see that it is an $N(B, W, \alpha)$ variable.   □

Theorem (gamma approximation to the negative hypergeometric).
(i) $F[x \mid N(B, W, \alpha)] \approx F[t \mid \text{Gamma}(\alpha)]$, *where we have standardized by* $t = xW/(B + 1)$, *when* $\binom{\alpha}{2}$ *and* t^2 *are small compared to* x *and* W.

   (ii) *If X is $N(B, W, \alpha)$ and we let $T = XW/(B + 1)$, then in a sequence of urns in which $W \to \infty$ and $(B+1)/W \to \infty$, T converges in distribution to Gamma(α).*

You should check this by imitating the proof that told us when we could make a gamma approximation to a negative binomial (see 3.4).

Example. Of the 100 people in a precinct who voted in the last election, only 10 voted for your candidate. You want to interview some people who voted for her, to find out what, if anything, your candidate did right. What is the probability that you will find one such person by your tenth interview of a voter? We compute $F[9 \mid N(90, 10, 1)] = 0.6695$. Your voters are fairly rare, and you are asking about only one of them, so we try $F[90/91 \mid \text{Gamma}(1)] = 0.6281$. Not too bad, and much easier arithmetic.

9.4   Probability Densities

9.4.1   Transforming Variables

As you will remember, we arbitrarily scaled time in a standard Poisson process so that the time interval in which an average of one event happens is of length one. This is hardly ever true, so we had to express time in these units before we could do any practical calculation. Instead, let a more general Poisson process have an average of one event in each time interval of length β. In the space-probe problem, for example, if β is 0.5 years, then all our calculations could use time in years. Now a Gamma(α, β) random variable will be the time to the αth event, when an average of one event happens in a period of length β. We have stretched our time measurements in proportion to β, so we may make the following definition.

Definition. If T is Gamma(α), then if $S = \beta T$ ($\beta > 0$), we call S a **Gamma(α, β)** random variable.

The probabilities are easy to calculate:

$$F[x \mid \text{Gamma}(\alpha, \beta)] = P(S \leq s) = P(\beta T \leq s) = P(T \leq s/\beta)$$
$$= F[s/\beta \mid \text{Gamma}(\alpha)].$$

Substituting, we get a formula:

Proposition. $F[s \mid \text{Gamma}(\alpha, \beta)] = \sum_{i=\alpha}^{\infty} (s^i/\beta^i i!) e^{s/\beta}$.

As impressive as this looks, it teaches us little; it simply points out the change of scale we knew we had to do anyway.

This was an example, though, of a very important operation on random variables, a *change of variables*. We have a variable X, and we want to use it to study a related variable Y. Let the connection be $X = g(Y)$ where g is a *nondecreasing* function on the sample space of Y. (That is, for any numbers $a \leq b$, it is always true that $g(a) \leq g(b)$.) In the case of the gamma family, this relationship is just $T = S/\beta$. Then it is easy to get the cumulative distribution function for Y:

$$F_Y(y) = P(Y \leq y) = P[g(Y) \leq g(y)] = P[X \leq g(y)] = F_X[g(y)],$$

where the second equality uses the fact that g is nondecreasing.

Proposition. (i) *If* $Y = g(X)$ *is a nondecreasing change of variables, then* $F_Y(y) = F_X[g(y)]$.
  (ii) *If* $Y = g(X)$ *is a nonincreasing change of variables, then* $F_Y(y) = 1 - F_X[g(y)]$.

Proof of the second part is an exercise.

Example. General negative exponential random variables (Gamma$(1, \beta)$) are sometimes studied on a log scale; that is, we work with $X = \log(T)$. Then $T = e^X$. Since $F_T(t) = 1 - e^{-t/\beta}$ on $T > 0$, then $F_X(x) = 1 - e^{-e^x/\beta}$, where X may be any real number. The variable X is an example of a **Fisher-Tippet** random variable.

This device of rescaling the time variable allows us to define a general (as opposed to a standard) Poisson process. We started by assuming that we measured time in units so convenient that an average of 1 event happened in each time period of length 1.0. But later we rescaled by $S = \beta T$. Now, for each time period of length β, in S-units, we still have an average of one Poisson event. Then for each time period of length 1.0 (in S-units), we must have a Poisson count with mean $1/\beta$. Let $\lambda = 1/\beta$, and make the following definition.

Definition. A **Poisson(λ) process** is a random sequence of real numbers $0 < t_1 < t_2 < t_3 < \cdots$ such that (1) in any interval $0 \le a < b$, the number of t's that fall in this interval, $|\{t_i \mid a < t_i \le b\}|$, is a Poisson$[\lambda(b - a)]$ variable; and (2) the number of t's in any two nonoverlapping intervals is independent.

For example, it is much more natural to let births in our small town be a Poisson(5) process, where the time unit is simply months.

9.4.2 Gamma Densities

You may be disturbed by the complexity of the cumulative distribution function for the gamma family; we have expressed it as an infinite sum. Of course, by taking complements we can reduce it to a finite sum, which still may require a lot of calculation if α is large. We remember that our major discrete families also lacked a simple expression for their cumulative distribution function; we tended to work instead with their relatively simple probability mass functions. Unfortunately, continuous random variables have no useful probability mass functions. As it happens, another function we have encountered in geometric problems, the *probability density* (see 5.4.2), plays somewhat the same role as did the probability mass function.

The trick was to *differentiate* the cumulative distribution function. In the Gamma(α) case,

$$F'(t) = \frac{d}{dt}\left(\sum_{i=\alpha}^{\infty} \frac{t^i}{i!} e^{-t}\right) = \sum_{i=\alpha}^{\infty} \frac{t^{i-1}}{(i-1)!} e^{-t} - \sum_{j=\alpha}^{\infty} \frac{t^j}{j!} e^{-t}$$

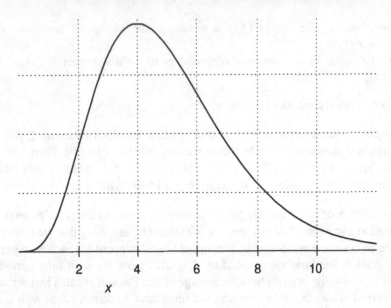

FIGURE 9.2. $F'[t|\text{gamma}(5)]$

by using the product rule for derivatives. Let $i = j + 1$ in the second sum, to get

$$F'(t) = \sum_{i=\alpha}^{\infty} \frac{t^{i-1}}{(i-1)!} e^{-t} - \sum_{i=\alpha+1}^{\infty} \frac{t^{i-1}}{(i-1)!} e^{-t}.$$

All terms but the first term in the first sum cancel, so $F'(t) = [t^{\alpha-1}/(\alpha-1)!]e^{-t}$, a remarkable simplification (Figure 9.2).

But can we use this expression to extract probability information about the random variable? Of course we can. Recall the fundamental theorem of calculus, which essentially says that integration undoes differentiation. We use it to express

$$P(a < X \leq b) = F(b) - F(a) = \int_a^b F'(X)\,dX.$$

For example, the probability that the fifth computer on our Mars craft will fail between 2 and 3 years out may be written $\int_{4.0}^{6.0} \frac{T^4}{4!} e^{-T}\,dT$ (Time is in units of 6 months). This corresponds to the area under a curve (see Figure 9.3).

This is a pleasingly compact expression, even though at the moment we would still have to calculate it with the sum formula. We will see by other examples that important cumulative distribution functions are very often simplified by differentiation.

9.4.3   General Properties

The above discussion motivates the following definition.

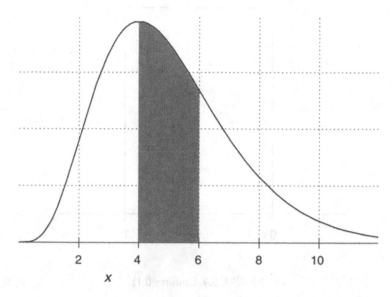

FIGURE 9.3. Area under a density curve

Definition. The **density** $f(x)$ of a random variable X is a nonnegative function defined on its sample space such that for $a < b$, $P(a < X \leq b) = \int_a^b f(X)dX$. A random variable that has a density is said to be **absolutely continuous**.

Proposition. (i) *On any interval on which the cumulative distribution function F is differentiable, $f(x) = F'(x)$.*
(ii) $F(x) = \int_{-\infty}^x f(X)\,dX$.
(iii) $\int_{-\infty}^\infty f(X)\,dX = 1$.

Check these as a very easy exercise. Notice that we use capital letters like X for the variable of integration when densities are involved, as we did with indices of summation in the discrete case (see Chapter 6.5). It will turn out shortly that variables of integration behave mathematically much like absolutely continuous random variables.

Example. (1) A Gamma(α, β) random variable has density

$$f(t) = \frac{t^{\alpha-1}}{\beta^\alpha(\alpha-1)!}e^{-t/\beta}$$

on $t > 0$.
(2) A Uniform(0, 1) random variable has density $f(x) = 1$ on (0, 1).
(3) The Fisher-Tippet variate described above has density

$$f(x) = \frac{1}{\beta}e^{x/\beta}e^{-e^{x/\beta}}$$

on the entire real line.
These are easy calculus exercises.

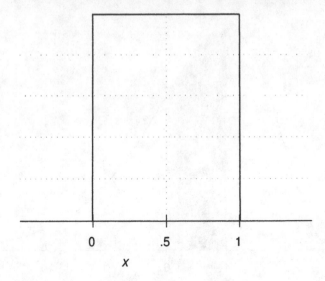

FIGURE 9.4. Uniform(0,1)

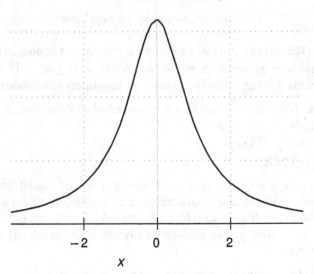

FIGURE 9.5. Cauchy(1,1)

9.4.4   Interpretation

Our densities have been reasonably simple, but we promised more for them; they should do some of the things for us that mass functions did. For one thing, mass functions told us immediately how relatively likely particular outcomes were. Look at some graphs of densities (Figures 9.4–9.7).

We have not shown numerical values on the axes, because we here wanted to show some of the qualitative variety of shapes that densities may have. Now, a Cauchy random variable is just one that follows the Cauchy law (see 4.8.1). What

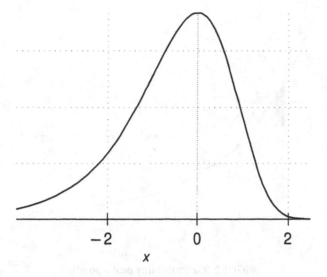

FIGURE 9.6. Fisher–Tippet

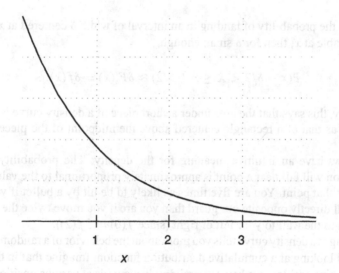

FIGURE 9.7. Negative exponential

does it mean, for example, that the Cauchy density is five times higher at 0 than at 2.0? Recall its appearance in the Great Wall of China problem, the probability of a bullet hitting at exactly certain points along the wall is, of course, zero. Instead, ask yourself about the probability of hitting *near* a point x; that is,

$$P(x - \delta/2 < X \le x + \delta/2) = \int_{x-\delta/2}^{x+\delta/2} f(X)\,dX = F(x + \delta/2) - F(x - \delta/2)$$

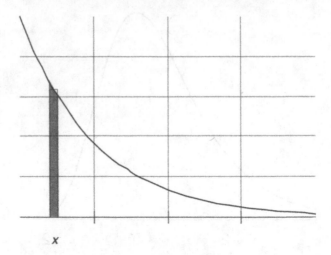

x

FIGURE 9.8. Probability near a point

measures the probability of landing in an interval of width δ centered at x. If F is differentiable at x, then for δ small enough,

$$P(x - \delta/2 < X \leq x + \delta/2) \approx \delta F'(x) = \delta f(x).$$

Pictorially, this says that the area under a short piece of a density curve is roughly the same as that of a rectangle centered above the midpoint of the piece (Figure 9.8).

We now have an intuitive meaning for the density: The probability that an observation will fall near a point is approximately proportional to the value of the density at that point. You are five times as likely to be hit by a bullet if you stand at the wall directly opposite our guard than you are if you move twice the distance from him to the wall to your left or right (since $f(0) = 5f(2)$).

Looking at a density curve tells you more about the behavior of a random variable than does looking at a cumulative distribution function. Imagine that in the Great Wall of China problem we have several drunken guards at various positions along the wall, firing at different rates. Then the density of a random bullet hole on the wall might be like that depicted in Figure 9.9.

We can conclude without knowing anything else that there seem to be three guards firing and that they are equally spaced along the wall. The middle guard seems to be firing more often than the other two (greater area). The one on the left is closer to the wall than the others—his bullet holes seem more narrowly concentrated. If we graph a large random sample of bullet holes on the same axis, we can see the relationship of the sample to the density (Figure 9.10).

The density is a measure of how concentrated observations are likely to be near a point, hence the name "density."

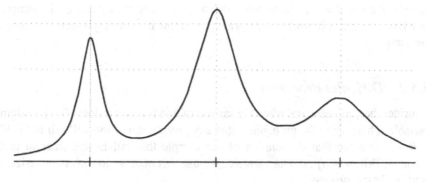

FIGURE 9.9. Complicated density

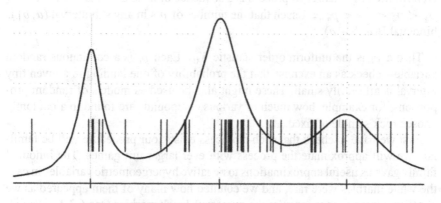

FIGURE 9.10. Sample from that density

9.5   The Beta Family

9.5.1   Order Statistics

Another important point process comes about from the *order statistics* of a random sample. Imagine that we have done a trial repeatedly and independently to get a random sample $X_1, \ldots, X_n$. One way to get a clear picture of the results is to sort the numbers; statisticians traditionally write them in ascending order, $X_{(1)} \leq X_{(2)} \leq \cdots \leq X_{(n)}$.

Definition. The ith value in ascending order from a random sample of n, $X_{(i)}$, is called the i**th order statistic** of that sample.

To illustrate, our guard who stands 5 meters from his wall might hit at -5.57, 2.59, -3.79, -8.99, and 0.90 meters from the point opposite him. We sort these to get $-8.99 < -5.57 < -3.79 < 0.90 < 2.59$; then, for example, the 4th order statistic is 0.90 meters. If his sergeant came along the next day and tried to guess from the bullet holes where his guard had been standing, a reasonable guess might be opposite the middle bullet hole, at -3.79 meters. Generally, if we have an odd sample size $n = 2m - 1$, the middle value $X_{(m)}$ is called the *median* of the sample (see Exercise 1.19). It is often used as a typical value in summary statements about the sample.

9.5.2 Dirichlet Processes

Consider the simplest case, when the sample of n is from a Uniform(0, 1) random variable. Then, since the probability that any given observation will fall in $(a, b]$ is $b - a$, we see that the number of our sample that fall in that interval is a binomial B$(n, b - a)$ random variable. We use this observation to characterize a new stochastic process:

Definition. A **Dirichlet**(n) process is a sequence of n random real numbers $0 < p_1 < p_2 < \cdots < p_n < 1$ such that the number of p's in any subinterval $(a, b]$ is binomial B$(n, b - a)$.

Then a p_i is the uniform order statistic $X_{(i)}$. Each p_i is a continuous random variable—check as an exercise that the probability of one landing in a given tiny interval is arbitrarily small. These might also be used as models of random proportions; for example, how much of various compounds are found in a randomly chosen rock sample of fixed size.

First we have to check that such a process exists; our procedure will be familiar: We will approximate the process with ever larger urn games. The binomial family gave us useful approximations to negative hypergeometric variables in case the white marbles were rare, and we counted how many of them appeared as we checked a large proportion p of the numerous black marbles (see 6.3.1). We will be interested in the random locations of the white marbles in a sequence drawn from the jar. Let the urn have n white marbles (which will remain fixed) and B black marbles (which will be allowed to grow later). Remove them all at random, and let b_i be the number of black marbles that appear before the ith white. Then $0 \le b_1 \le b_2 \le \cdots \le b_n \le B$ completely describes a realization of a hypergeometric process. Since we are interested in the proportion of the entire process these represent, standardize by $p_i = b_i/B$. Then $0 \le p_1 \le p_2 \le \cdots \le p_n \le 1$. We need to check that as B gets large this behaves more and more like a Dirichlet process.

To count the p's in an interval $(a, b]$, notice that it represents the stretch of draws from the one after the $\lfloor aB \rfloor$th black marble and ending with the $\lfloor bB + 1 \rfloor$th black (the floor function, again). The number of events, white marbles, in this stretch is therefore negative hypergeometric N$(n, B, \lfloor bB+1 \rfloor - \lfloor aB \rfloor)$. The possibility that some white marbles may have appeared earlier is irrelevant, because we of course do not know whether they did or not. This may be approximated by a binomial distribution with parameters n and $(\lfloor bB + 1 \rfloor - \lfloor aB \rfloor)/(B + 1)$ for n, b, and a fixed and B large enough (see 6.3.2).

Theorem (Dirichlet limit of hypergeometric processes). *Consider a sequence of hypergeometric processes with a fixed number n of white marbles and increasing numbers B of black marbles; describe a realization by $0 \le b_1 \le b_2 \le \cdots \le b_n \le$*

B where b_i is the number of black marbles that appear before the ith white. Let $p_i = b_i/B$. Then the $0 \le p_1 \le p_2 \le \cdots \le p_n \le 1$ converge in distribution to a Dirichlet(n) process.

Example. Five students from a small high school in far western Virginia are in the 1993 graduating class of 3871 students at Virginia Tech. Their class ranks are 73, 1298, 2525, 2682, and 3517. If these students are, in fact, a random selection from all the graduates, we might think of this as the realization of a hypergeometric process with five white marbles and 3866 black marbles. Since recruiters do not care about the total number of Tech graduates, we might more usefully express these as proportions of the way through the class: 0.0186, 0.3352, 0.6524, 0.6927, 0.9084. We conclude that one student was in the top 5% of his class and two were in the top half. Before we saw these results, we would have expected, for example, that the number of these five that would have made the top quarter would be Binomial(5, 0.25).

We can handle these random counts; the novelty here is that the p_i are continuous random variables, which have sample space $(0, 1)$. To discover their behavior, start with the simplest case, $n = 1$. The approximating hypergeometric process has a single white marble among many blacks; a partial search of this urn is binomial with probability of a success $b/(B + 1)$. Now apply the black–white transform to get an urn with only a single black marble in it among B whites. In our new notation, its random location is after $X = b_i$ white marbles; this is, of course, a discrete uniform random variable on $\{0, \ldots, B\}$. Standardizing, we get a random variable $P = p_1 = X/B$. The cumulative distribution function of the discrete uniform variable is $F(x) = x/B$; therefore, the cumulative distribution function of P is $F(p) = p$. As the number of marbles in the urn grows, of course, P may be arbitrarily close to any number between zero and one. We conclude that the single event in a Dirichlet(1) process has the same cumulative distribution function as a Uniform(0, 1) random variable; thus, it *is* a Uniform(0, 1) random variable. A random point on an interval is much like randomly placing a black marble in a long row of white marbles.

9.5.3 Beta Variables

Now we need to know the behavior of the real number p_i in a Dirichlet(n) process. Give it a name:

Definition. A **Beta**(α, β) random variable is the αth event in increasing order in a Dirichlet($\alpha + \beta - 1$) process.

That is, our p_i is a Beta($i, n + 1 - i$) random variable. We just noticed that a Beta(1, 1) variable is also Uniform(0, 1). You will not be surprised when we use black–white duality to find the cumulative distribution function of other beta variables. Let $P = p_i$ be the ith of n Dirichlet events, and thus a Beta($i, n + 1 - i$) variable:

$$F_P(p) = P(p_i \le p) = P(\text{at least } i \text{ events before } p) = P[X \text{ a } B(n, p) \text{ is at least } i].$$

But we know how to compute this:

Theorem (binomial–beta duality).

$$F[p|\text{Beta}(\alpha, \beta)] = 1 - F[\alpha - 1|\text{Binomial}(\alpha + \beta - 1, p)]$$

$$= \sum_{i=\alpha}^{\alpha+\beta-1} \binom{\alpha + \beta - 1}{i} p^i (1 - p)^{\alpha+\beta-i-1}.$$

The peculiar choice of parameters may be easier to remember if you notice that the event of interest is the αth from the beginning and the βth from the end of the interval. So that we will not count it twice, we subtract 1 to get the total number of events.

Example. Five children were arbitrarily assigned to a kindergarten reading group. What is the probability that the second-brightest child in the group is above-average among children in general? That second child's ability ranking, as far as we know, must be the 4th uniform order statistic of a sample of 5, a Beta(4, 2) variable, counting from the bottom. Our theorem says we can calculate the probability that he or she is above average by finding the probability that no more than 3 are below average (or that it is false that 4 or 5 are below average) when the probability of being below average is 0.5: $1 - 0.15625 - 0.03125 = 0.8125$.

Since a Dirichlet process is a limiting case of a hypergeometric process, it seems likely that under certain circumstances (black marbles rare?) beta probabilities would be useful approximations to negative hypergeometric probabilities.

Theorem (beta approximation to the negative hypergeometric). (i) *If $\binom{B}{2}$ is small compared to x and $W - x$, then*

$$F[x|\text{N}(W, B, b)] \approx F[x/W|\text{Beta}(b, B + 1 - b)].$$

(ii) *Let X_i be $\text{N}(W_i, B, b)$ and $P_i = X_i/W_i$. Then as $W_i \to \infty$, P_i converges in distribution to Beta$(b, B + 1 - b)$.*

By now you should find proving this familiar: apply black–white duality twice and the binomial approximation to the negative hypergeometric in between.

Example. I am in charge of maintenance for a large office building. A salesman wants to sell me a new, longer-lasting (but more expensive) brand of light bulb. I am skeptical of her claims about the new bulb, so I design an inexpensive experiment: I mix 7 of the new bulbs among 100 old-style bulbs that I install around the building this week. Then I make a note to check how many of the old ones have failed by the time the 4th new one burns out. If I am right to be skeptical, and the new are only about as reliable as the old, what is the probability that more than 50 old ones will have failed by that time?

There is a negative hypergeometric model for this: $P[X > 50|\text{N}(100, 7, 4)] \approx 0.4895$. But since $\binom{7}{2} = 21$ is fairly small compared to 50, we feel free to try the approximation $P[P > 50/100|\text{Beta}(4, 4)] \approx 0.5$, which is close.

Why did the beta probability come out so simple? Notice that beta variables possess a reversal symmetry: If P is Beta(α, β), then $Q = 1 - P$ is Beta(β, α). This is because if we have a Dirichlet(n) process with events $\{p_i\}$, then the process with events $p'_i = 1 - p_i$ (with order reversed) also meets the definition of a Dirichlet(n). But we see that if $\alpha = \beta$, Q and P must have the same distribution; then $P(P > 0.5) = P(P < 0.5) = 0.5$.

9.5.4   Beta Densities

The cumulative distribution function for the beta is complicated, but our experience with gamma variables gives us hope that its density is simpler:

$$f(p) = F'(p) = \left[\sum_{i=\alpha}^{\alpha+\beta-1} \binom{\alpha+\beta-1}{i} p^i (1-p)^{\alpha+\beta-i-1} \right]'$$

$$= \frac{(\alpha+\beta-1)!}{(\alpha-1)!(\beta-1)!} p^{\alpha-1} (1-p)^{\beta-1}$$

after many cancellations (which you should check).

Proposition. *The density of a Beta(α, β) variable is* $\frac{(\alpha+\beta-1)!}{(\alpha-1)!(\beta-1)!} p^{\alpha-1}(1-p)^{\beta-1}$.

(See Figures 9.11 and 12.)

Example. From a Uniform(0, 1) sample of 11, the median is the 6th counting from either end, $X_{(6)}$; it is therefore a Beta(6, 6) random variable. Its density is shown in Figure 9.13.

We often use the median as a clue to the location of the middle of a random variable; but we can see that even with as many as 11 samples it is quite variable. The probability is substantial that it will be as low as 0.3 or as high as 0.7.

Notice that α and β play equivalent roles in the density; we can express this as follows:

Proposition (reversal symmetry of the betas). *If P is a Beta(α, β) random variable, then $Q = 1 - P$ is a Beta(β, α) random variable.*

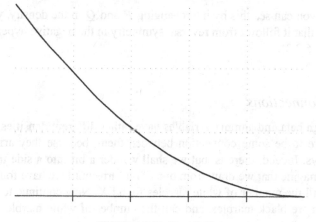

FIGURE 9.11. Beta(1,3)

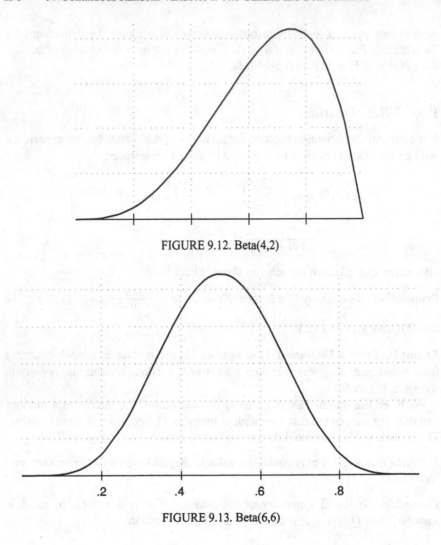

FIGURE 9.12. Beta(4,2)

FIGURE 9.13. Beta(6,6)

Though you can see this by interchanging P and Q in the density, you should also notice that it follows from reversal symmetry in the negative hypergeometric family (see 5.2.3).

9.5.5 Connections

Even though beta and gamma variables have rather different densities, we might expect there to be some connection between them, because they arise in such similar ways. Indeed, there is, but we shall wander a bit into a side trail to help us see it. Imagine that we draw from one of our urns until we have found b black marbles; call the number of white marbles found X. Now continue to draw until we find c more black marbles, and call the number of white marbles appearing along the way Y. Then X is $N(W, B, b)$, Y is $N(W - X, B - b, c)$, and the white total $X + Y$ is $N(W, B, b + c)$.

Imagine that you missed the drawing in the above experiment, and your friend could only remember that the grand total found was $X + Y = z$. Do you then know anything more about X, the number initially drawn? Surely you must; for one thing, its maximum possible value is now z instead of the maximum of W it could have been originally. In fact, it is easy to say precisely what you know, that is, what the conditional distribution of X given $X + Y = z$ is. You know that exactly $b + c$ black marbles and z white marbles were removed from the jar, and they could have been removed in any order at all, with each order equally likely. The total numbers of marbles in the original jar has become completely irrelevant. As far as you are concerned, they could have been drawing from a jar with only the $z + b + c$ marbles they actually chose in it until they found b black ones, at which point they wrote down the unpredictable number X of white ones found. This variable sounds familiar:

Proposition. *If X is $N(W, B, b)$ and Y is $N(W - X, B - b, c)$, then X conditioned on knowing that $X + Y = z$ is $N(z, b + c, b)$.*

This has an interesting extension to the case that the original urn is arbitrarily large but b and c are comparatively small. Let $p = W/(W + B)$, and let $\binom{b}{2}$ and $\binom{c}{2}$ be small compared to B. We learned long ago that X is then approximately negative binomial $NB(b, p)$ (see 6.2.3). But since we have not significantly reduced either the number of black or white marbles for moderate values of z, it is also true that Y converges in distribution to $NB(c, p)$, independent of X. The growth in the original urn does not affect our small, conditional urn. Thus, it is just negative hypergeometric.

Proposition. *If X is $NB(k, p)$ and Y is independently $NB(l, p)$, then X conditioned on knowing that $X + Y = z$ is $N(z, k + l, k)$.*

Actually, we could have reasoned this out directly by imagining that the original experiment was Bernoulli; we could then have simulated our ignorance about X by writing the observed totals of successes and failures on slips of paper and tossing them into an urn. Our draws from that urn are without replacement, because we are drawing against these fixed totals. Notice that for the first time, we have constructed a hypergeometric experiment from a Bernoulli experiment, rather than the other way around.

This simple, fairly intuitive, proposition will lead to two important results. First, consider what happens if p gets small while k and l are relatively large. At some point, X and Y are approximately Poisson (see 6.4.2) with (fixed) means $\lambda = kp/(1 - p)$ and $\mu = lp/(1 - p)$. Then we consider a fixed value of z as k and l get large. Our X conditioned on z is approximately binomial with z trials and probability $\frac{k}{k+l} = \frac{\lambda}{\lambda+\mu}$ of success at each trial (see 6.3.2).

Theorem (conditioning on a Poisson total).   *If X and Y are independently Poisson with means λ and μ, then X conditioned on knowing $X + Y = z$ is binomial, $B(z, \lambda/(\lambda + \mu))$.*

Notice that we already derived this from the probability mass function (see 7.4.2). This proposition and its generalization to more than two variables are important in applications. Consider two varieties of a rare disease, cases of which appear in a certain hospital as something like two Poisson processes with different time rates. If we collect all the z cases for a year, then the number of those cases that will turn out to be of one variety is binomial with z trials and probability the average proportion of that variety. Generally, we find the proposition useful when we are interested in studying relative numbers of observations of various types, and we consider the total number of observations (which is usually just the size of our study) scientifically irrelevant.

Returning to our result about negative binomials, we observe that the other extreme is when k and l stay fixed but p approaches 1. Then X and Y tend to be large; standardize them to approximate times in a Poisson process by letting $T = \frac{X(1-p)}{p}$ and $S = \frac{Y(1-p)}{p}$. Then in the limit, T is Gamma(k) and S is Gamma(l). Now assume that we know $z = X + Y$; then $U = X/z = T/(T + S)$ is the unknown proportion of the successes represented by the first count. It converges in distribution to a Beta(k, l). Something fascinating has happened: We found the distribution assuming that we knew the total count; but this total has canceled out. This is no longer a conditional result. We have discovered an important fact:

Theorem (beta is a gamma proportion).   *If T is Gamma(α) and S is independently Gamma(β), then $U = T/(T + S)$ is Beta(α, β), independent of $(T + S)$.*

This elegant result will find application somewhat later.

9.6   Inference About Gamma Variables

9.6.1   Hypothesis Tests and Parameter Estimates

The fact that sums of independent gamma variables with the same scale parameter β are also gamma will allow us to do a very good job of testing hypotheses about and making estimates of this scale parameter.

Example. A brand of electrical fuses burn out in what we believe to be a Poisson process, and the manufacturer asserts that it has time scale $\beta = 300$ days. Your experiences suggest to you that the fuses last, in fact, a shorter typical time. You decide that you will let the manufacturer's claim be the null hypothesis, and test whether it might be shorter, with a significance level of 0.05. To improve your experiment, you test 10 fuses until they burn out. Their life spans are 7, 226, 17, 88, 50, 24, 244, 214, 435, and 321 days.

Under the null hypothesis, the sum of these would be a gamma random variable with $\alpha = 10$ and $\beta = 300$; in fact, it is 1726. The p-value for the fuses being worse than we assume is then $P[T \leq 1726|\text{Gamma}(10, 300)]$, the probability that we would get a performance that poor or even poorer. Our sum formula for the cumulative distribution function in Section 3.4 gives us a probability of 0.06799 (which you should check). This is low, but not less than 0.05. We cannot publish a challenge to the claimed lifetime by our own standards. However, it looks poor enough to me that I might be tempted to test, say, 30 more fuses.

Since β is a sort of typical time between Poisson events (remember that it is 1 over λ, the expected Poisson count in unit time), then it seems reasonable to try to estimate its value from the data. The obvious statistic to try is the sample mean of the times to failure, $\bar{t} = \frac{1}{n} \sum_{i=1}^{n} t_i$. In our problem, $n = 10$, and $\bar{t} = 172.6$. But if we look at what we knew before the experiment and consider what happens by chance, this particular statistic $\bar{T}$ is just the Gamma(n, β) random variable we found above, but rescaled by a factor $1/n$. Therefore, we believed that $\bar{T}$ would be a Gamma(n, β/n) random variable.

What is $\bar{T}$ like as an estimate of β? In exercises, you will learn that the point at which a density is largest, and therefore the random variable occurs relatively most often, is called the mode. It is one way of seeing where our statistic is centered. You will find that the mode of our sample mean is $\frac{n-1}{n}\beta$. This says that as we take ever bigger samples, the mode gets ever closer to the correct value; in some sense, $\bar{T}$ is a plausible estimate of β. Of course, the estimate must be rather variable. We have already noted that if, as claimed, $\beta = 300$, then there is a probability of 0.068 that our estimate $\hat{\beta} = \bar{T}$ would be 172.6, or even less. A similar calculation concludes that the probability that $\hat{\beta}$ will be at least 500 is 0.031; it sounds like even this rather wrong answer still comes up occasionally. We will find later that the answer does get better as n gets larger, and also that, surprisingly, if β is really unknown, it is hard to do better than $\hat{\beta} = \bar{T}$.

9.6.2 Confidence Intervals

Of course, we know another way to pin down an unknown parameter: a confidence interval. Still pursuing our brand of electric fuses, we once again look at what values of β would fail to be rejected, using the 10 observed lifetimes. To get a 95% confidence interval, we will once again use the convention that we will tolerate a 0.025 probability that the data values are too large, and a 0.025 probability that they are too small. The improbability that the data are so **large** will of course establish a **lower** bound on β and vice versa (see 6.7.3). After many calculations searching for the right p's (with the aid of a little computer program to sum the series), I conclude that $101 \leq \beta \leq 360$ days is a 95% confidence interval for the average life of the fuses. It looks as though we might need to test many more fuses to get reasonably high accuracy.

Notice that since any independent gamma random variables with the same scale parameter sum to yet another gamma variable, we may do inferences on cases other than $\alpha = 1$.

Example. The lifetimes of mammals who live to adulthood and then die of natural causes are believed by some of my colleagues to follow roughly gamma laws with $\alpha = 5$, as if they had systems redundant enough to absorb 4 major breakdowns without dying. (I suppose that if they were cats, then α would be 9.) We observe in a species of desert mouse the following 20 lifetimes in days past maturity:

392	300	604	235	182
293	575	310	502	437
294	460	377	221	350
380	224	563	519	568

I would like to test the hypothesis, quoted in the standard reference book on desert mice, that their natural spans past maturity are no more than 300 days. I will use a significance level of 0.01 in my test. First I note that under the gamma assumption, life span would correspond to a time between system failures of $\beta = 60$ days, because there should be $\alpha = 5$ such failures in 300 days. Those 20 independent observations sum to 7742, which under our hypothesis is a Gamma$(20 \times 5 = 100, 60)$ random variable. Now a computer program is essential to discover that $P[T \geq 7742|\text{Gamma}(100, 60)] = 0.00353$. (In the next chapter we will discover a useful approximation to this calculation.) This is less than 0.01, so we reject the null hypothesis that the typical life is as short as 300 days (even though 6 of our mice did not live that long).

We also note that the sample mean life length was 387 days and that this is a Gamma$(100, \beta/100)$ random variable. To get an idea how this narrows down the plausible values, we will construct a confidence interval as before, this time a 99% interval. We split the probability of extreme values into 0.005 for each of high and low directions, and after many calculations conclude that $60.6 \leq \beta \leq 101.8$ is a 99% confidence interval. Translating this back into lifetimes, we are 99% confident that these mice live typically between 303 and 509 days past maturity, when we allow for 5 system failures.

As before, it is plausible in such problems to estimate the typical life span from the sample mean, $\alpha\hat{\beta} = \bar{T}$. Then $\hat{\beta} = \bar{T}/\alpha$. In the mammalian life example, it was believed that $\alpha = 5$, so that $\hat{\beta} = 76.35$ is our estimate of the number of days between successive system failures. Indeed, as in the previous section, this will turn out to be quite a sound method of estimation.

9.6.3 Inferences About the Shape Parameter

Of course, the problem could have been presented to us in quite a different form. If zoologists were confident that they knew, that $\beta = 60$ days was close to the correct gamma parameter for organisms of this type, then the question might be whether or not the hypothesis of $\alpha = 5$ tolerable major system failures was sound. We might proceed as above to estimate $\hat{\alpha}\beta = \bar{T}$, so that $\hat{\alpha} = \bar{T}/\beta$. In the zoology example, $\hat{\alpha} = 6.45$. Since at the moment we do not know how to interpret anything but an integer value of α, we might say that the most plausible values of α were 6 or 7. We will see in a later chapter that this is not a very sound method of estimation; however, people often do it anyway, because better methods are so much more complicated.

It is, however, still reasonable to do hypothesis tests. In fact, the earlier example amounted to testing the hypothesis $\alpha = 5$ and $\beta = 60$, which we rejected because we got a one-sided p-value of 0.00353. Then, because we considered $\beta = 60$ to be the more dubious assumption, we concluded that in fact $\beta > 60$. Now, though, we consider $\alpha = 5$ to be the more scientifically controversial; from the test we reject it, and decide that likely $\alpha > 5$.

If hypothesis tests work, then presumably we can use the same thinking to construct confidence intervals for α. After trying many values of α with my computer program, I obtain $P(T \geq 7742|\text{Gamma}(108, 60)) = 0.0248$ and $P[T \leq 7742|\text{Gamma}(153, 60)] = 0.0230$. From this we conclude that the 95% confidence interval on the shape parameter for a sum of 20 observations is from 109 to 152 inclusive. Dividing by 20, we interpret this as a confidence interval for the shape parameter for each lifetime of $5.45 \leq \alpha \leq 7.6$. Since we believe only in integer values, only 6 or 7 seems plausible from the data. You will be delighted to learn in the next chapter that fractional values of α may sometimes make sense, too.

If we have no firm belief in the value of either α or β, we need to estimate and test two parameters at once. We will leave that challenging problem for later.

9.7   Likelihood Ratio Tests

9.7.1   Alternative Hypotheses

When we construct a frequentist-type test of some hypothesis, which we will reject at some significance level α, it is natural to ask just how good our test is. For example, would it make more sense to ask whether the *median* of some sample is improbably large, instead of asking about the mean? To tackle this issue, we will have to ask ourselves just why the hypothesis might be rejected. That is, if the null hypothesis is not true, just what is true? This other possible truth about the world will be called an *alternative hypothesis*.

For example, instead of the typical lifetime of our desert mice (see Section 6.2) being 300 days, we might be trying to show that it is more like 400 days. To keep straight this distinction, we let the null hypothesis be denoted by H_0, and the alternative on which we are concentrating will be H_1. In a test of two alternative values of this population parameter, we will write

$$H_0 : \mu = \mu_o,$$
$$H_1 : \mu = \mu_1,$$

where the Greek letter μ is often used to stand for a typical value of a random variable. In our example, then, $\mu_0 = 300$ and $\mu_1 = 400$. Our hypothesis test is then supposed to decide between the two.

Now that we have two competing theories, we need to stop and ask why one is called null and the other alternative, instead of the other way around? As we will see, it does not matter mathematically, but it will be important to our thinking. We usually let the null hypothesis be a conservative position on the issue being studied. In science, it is based on the accepted laws (or at least the prevailing wisdom) in its field. In commerce, it might be the manufacturer's claim about the properties of his product. Then the alternative hypothesis is a challenge to that position. In science, the alternative hypothesis is the claim that something surprising is going on; perhaps this motivated the research in the first place. One does not receive a Nobel prize for finding that what everybody believed is in fact true. In business, the alternative might be to doubt that some product really meets its specification; if we decide that this is so, we might decide to change suppliers.

Since we are designing a frequentist experiment, we will reject the null hypothesis if the results are so extreme that they will only happen with a small probability called (confusingly enough) α if the null hypothesis is indeed true. In the mouse problem, we shall let this significance level be 0.01. But what do we mean by the data being extreme? Here, since we are investigating the possibility of *longer* lifetimes than is usually believed, an extreme result is a *large* average life in our sample. After many of the same sort of calculations we did before, we find that $P[\bar{T} > 374 | \text{Gamma}(100, 3)] = 0.01$. So we plan to reject the hypothesis of a typical life of 300 days if in fact that inequality holds (in our experiment, it did). We will call the region of the sample space in which we reject the null hypothesis, in this case $\bar{T} > 374$, the *rejection region*. Let us denote it by a capital letter, like R, since it is an event in our sample space. Then the key fact that determined R is that $P(R \mid \theta_0) = \alpha$, where θ_0 is the value of the parameter corresponding to the null hypothesis. We say in this case that the *size* of the test is the significance level α.

If the alternative hypothesis should be true, then we are interested in $P(R \mid \theta_1) = \rho$, the probability that we will (correctly) reject H_0 when it is in fact false. We call this probability the *power* of the test; the larger it is, the better. In the mouse lifetime example, we obtain $P[\bar{T} > 374 | \text{Gamma}(100, 4)] = 0.736$, using our alternative average of 400 days. Therefore, if our conjecture is right, almost 3 times out of 4 our experiment will reject the conventional wisdom. Note that if we construct a similar test with smaller significance level α, our ρ will decrease; the more demanding of reliability we are, the less powerful the test.

9.7.2   Most Powerful Tests

To see how good a test is, we will compare it to others, in particular to other tests of the same size. For example, we might simply count how many of our lifetimes are greater than 300 days. Under the null hypothesis, the probability of a single mouse outliving 300 days is 0.44. Therefore, the number of mice X exceeding 300 days should be binomial, $B(20, 0.44)$. Then $P(X \geq 15 | 300 \text{ days}) = 0.005$ is as

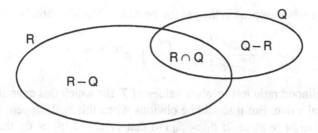

FIGURE 9.14. Two rejection regions

close as we can get to the same size (because the test is discrete). But its power when the typical life is really 400 days may be checked to be (exercise) 0.334. This competing test made perfect sense (and the computations were much easier), but it was a much less powerful way to explore our hypotheses in this experiment.

Let Q stand for the rejection region of any other test of the same size α as the test with rejection region R, so that $P(Q \mid \theta_0) = \alpha$ also. We can break each of these events up into two pieces that do not overlap, $R = (R - Q) \cup (R \cap Q)$ and $Q = (Q - R) \cup (R \cap Q)$. (See Figure 9.14.)

The fact that these tests are the same size says that $P(R-Q \mid \theta_0) = P(Q-R \mid \theta_0)$, because the rest of each rejection region is the same. Then if R is at least as powerful as Q, we know that $P(R - Q \mid \theta_1) \geq P(Q - R \mid \theta_1)$. If these two tests are not essentially the same, so their differences have positive probability, we may divide to get $P(R - Q \mid \theta_1)/P(R - Q \mid \theta_0) \geq P(Q - R \mid \theta_1)/P(Q - R \mid \theta_0)$.

This relationship gives us the crucial hint as to how we might construct a test that is at least as powerful as any other. We concentrate on the case where the observations are absolutely continuous. If we guarantee that for any observation $x \in R$ and any observation $y \notin R$ we have for their densities $f(x \mid \theta_1)/f(x \mid \theta_0) \geq f(y \mid \theta_1)/f(y \mid \theta_0)$, then when we integrate the densities over the events $R - Q$ and $Q - R$, we certainly get the probability inequality above, for any test Q of size α whatsoever. Therefore, we *define* our test by this inequality. In parallel with the discrete case (see 8.2.1), let the (absolutely continuous) *likelihood* of θ be $L(\theta \mid x) = f(x \mid \theta)$.

Definition. The **likelihood ratio test** of size α for the null hypothesis $\theta = \theta_0$ and the alternative hypothesis $\theta = \theta_1$ has the rejection region $R = \{x \mid \frac{L(\theta_1 \mid x)}{L(\theta_0 \mid x)} \geq C\}$, where the constant C is chosen so that $P(R \mid \theta_0) = \alpha$.

We have shown that this test is the best we can do, because the density inequality above certainly has to be true for our R; the x's are on one side of C, and the y's are on the other.

Theorem (Neyman–Pearson lemma). *A likelihood ratio test comparing two hypotheses is the most powerful test of its size for those hypotheses.*

The discrete case can be proved in the same way, using mass functions instead of densities (exercise).

Let us see what happens in the gamma problem: The likelihood ratio is

$$\frac{\frac{T^{\alpha-1}}{\beta_1^\alpha \Gamma(\alpha)}e^{-T/\beta_1}}{\frac{T^{\alpha-1}}{\beta_0^\alpha \Gamma(\alpha)}e^{-T/\beta_0}} = \left(\frac{\beta_0}{\beta_1}\right)^\alpha e^{T(1/\beta_0 - 1/\beta_1)}.$$

Then a likelihood ratio test involves values of T for which this quantity exceeds some critical value. But it should be obvious when this will happen, because T appears in only one place. If $\beta_1 > \beta_0$ (so that $1/\beta_0 - 1/\beta_1 > 0$), then our test looks like $R = \{T \geq C\}$; if $\beta_0 > \beta_1$, then it looks like $R = \{T \leq C\}$. Then C is determined by requiring that $P(R \mid \beta_0) = \alpha$. Something wonderful has happened in this case: The form of the test does not depend on our null hypothesis, and the test itself depends on the alternative hypothesis only as far as β_1 is above or below β_0. This makes tests easy to construct, because we do not have to construct new ones for each value of β_1; families that work this way are called *monotone likelihood ratio* families. You will see in exercises that a number of our favorite families have this property. Then we have simple most powerful tests for hypotheses like

$$H_0 : \theta = \theta_0,$$
$$H_1 : \theta > \theta_0,$$

or the reverse.

In our mouse life span example, you should check that the sample mean of the life spans is the only function of the data in our likelihood ratio, and it has a gamma distribution; so the test we constructed was indeed the most powerful test of size 0.01.

9.8   Summary

We defined a *Poisson process*, a model for independent events happening over time; then we constructed such a process as the limit of a Bernoulli process in which successes were very rare (3.2). The times at which the αth events happen in this process are continuous random variables of the *gamma* family, with cumulative distribution function $F(t) = \sum_{i=\alpha}^{\infty} t^i/(\beta^i i!)e^{-t/\beta}$, where β is the average time between events (4.1). This was a case in which *density* functions turned out to be much the simpler way to study a random variable: In the gamma family, $f(t) = F'(t) = t^{\alpha-1}/(\beta^\alpha(\alpha-1)!)e^{-t/\beta}$ (4.3).

Out of a hypergeometric process with black marbles very rare we constructed in the limit a *Dirichlet process* (5.2). The continuous locations of its events were a new sort of continuous random variable, from the *beta* family. It has density $f(p) = (\alpha + \beta - 1)!/((\alpha - 1)!(\beta - 1)!)p^{\alpha-1}(1 - p)^{\beta-1}$ (5.4). We then discovered a very useful relationship between the gamma and beta families: A beta variable is the proportion that one variable makes up of the sum of two independent gamma variables (5.5). From there we looked at the problem of estimation, testing, and confidence intervals in the gamma family (6). It turns out that for problems of this sort, a *likelihood ratio test* with *rejection region* $R = \{x \mid L(\theta_1 \mid x)/L(\theta_0 \mid x) \geq C\}$ is *most powerful* (7.2).

9.9 Exercises

1. A random variable X has the cumulative distribution function $[P(X \leq x)]$

$$F(x) = \begin{cases} 0 & x < -1, \\ \dfrac{3}{8} + \dfrac{x}{4} + \dfrac{x^2}{8} & -1 \leq x \leq 0, \\ \dfrac{5}{8} + \dfrac{x}{8} + \dfrac{x^2}{8} & 0 \leq x < 1, \\ 1 & x \geq 1. \end{cases}$$

   a. Compute $P(-\frac{1}{2} \leq X \leq \frac{1}{2})$.
   b. Compute $P(0 \leq X \leq \frac{1}{3})$.
   c. Is this a continuous random variable? Explain.

2. I flip a coin many times, and I get HHTHTTTHTHTTTHHTHHT.... This is a realization of a Bernoulli process. Rewrite it as an increasing sequence of s_i's, where these are the numbers of tails preceding the ith head.

3. Every day as you drive to work, you hit a pothole that has a 5% chance of blowing a tire. You have one spare tire.

   a. What is the probability that you will have to get a tire fixed within 50 work days?
   b. Do the calculation again using the gamma approximation. Compare.

4. You are on a long walk through a dense forest. There are a great many deer in this forest, but you see very few of them because of the density of the forest. Treating the sighting of each family of deer as an independent event, you predict from experience that you will see one family of deer on average for each 2000 meters you walk.
   What is the probability that at the point you see your fourth family of deer, you will have walked no more than 5000 meters?

5. In the theorem about hypergeometric processes converging to a Poisson process, check that the number of t's in a fixed interval indeed converges to the correct Poisson random variable.

6. Let X be a continuous random variable, and let $X = g(Y)$ be a **nonincreasing** function wherever X is defined (for any $a \leq b$, always $g(a) \geq g(b)$). Derive an expression for F_Y the cumulative distribution function of Y, in terms of F_X the cumulative distribution function of X, and the function g.

7. Prove the 3 basic properties of densities.

8. Compute the densities of the Uniform[0, 1] and Fisher–Tippet random variables.

9. The **mode** of an absolutely continuous random variable is the value of the variable at which its density is greatest. Since outcomes are most concentrated there, we sometimes use the mode as a typical value.

   The Weibull(a, b) family of random variables, often used in the study of the reliability of a system, has cumulative distribution function $F(x) = 1 - e^{-(x/b)^a}$ for $a, b > 0$, $x \geq 0$.

   a. Find the density of a Weibull(a, b) variable.
   b. Find the mode of a Weibull(a, b) random variable.
   c. For a Weibull$(2, 3)$ variable X, find $P(3 < X \leq 5)$.

10. Find the mode of a Gamma(α, β) random variable.

11. Let X be a Weibull(a, b) random variable (Exercise 9). Find the sample space and cumulative distribution function of $Y = \log(X)$.

12. A random variable X has density $f(x) = \frac{3}{(1+x)^4}$ on $X > 0$. Find its cumulative distribution function.

13. Show that the probability of a Dirichlet(n) event falling in an interval goes to zero as the length of the interval goes to zero.

14. A scheme for getting lower bids for contracts works as follows: All bids are sealed. Low bidder gets the contract, but they are paid the amount the *second*-lowest bidder had asked.

    Seven of the many hundreds of eligible bidders make bids. What is the probability that the amount paid will exceed the amounts 20% of the eligible bidders would have asked for?

15. Compute the density of a Beta(α, β) random variable.

16. By past experience, about 6 men and 2 women per month come into your clinic complaining of chest pain. You enroll the next 10 people to appear with chest pain into a study of a new drug. What is the probability that you find at least 3 women?

17. The year's 15th accident at a certain dangerous intersection occurs on the 305th day of that year. Assuming that accidents are a Poisson process, estimate the typical time between accidents β, and construct a 95% confidence interval on it.

18. In general, 40% of the management trainees at a large automotive firm are women. To check whether women are proportionally represented among those who then get low-level management assignments within two years, I am going to sample 40 assignees.

    a. Construct a likelihood ratio test (of size as close as possible to 5%) of the hypothesis that women are proportionally represented, against the alternative that only about 20% of those who get the assignments are female.
    b. What will be the power of your test?

19. In Exercise 18 you found a test of the binomial proportion p. In general, show whether or not the test depends on the exact value of the alternative p

or only on its direction from the null p. (If not, you have shown that this is a monotone likelihood ratio family.)

9.10 Supplementary Exercises

20. Write down the cumulative distribution function of a Geometric($1 - p$) random variable X. Now renormalize it (to expectation one) by substituting $T = Xp/(1 - p)$. Find the cumulative distribution function of T. Now find its limit as $p \to 0$; the result is the cumulative distribution function of the time to first failure of a Poisson process.
    Hint: We had to find a similar limit in Chapter 6, when we were deriving the Poisson approximation to the binomial distribution.

21. Prove the theorem of the gamma approximation to the negative binomial.

22. You buy water glasses in packages of 6. Occasionally, you will accidentally drop and break one, but they never wear out. You drop, on the average, 2 of them per year. What is the probability that a new package will last until you finish graduate school, in 2.5 years?

23. You have a huge mass of rubber bands in the back of your drawer, of which a small proportion are green. If you rapidly remove the rubber bands from your drawer, one at a time, you will find an average of 2 green rubber bands a minute.

    a. What is the probability that you will find 5 or more green rubber bands in a 2 minute period?
    b. Write down the probability density function for the time (in minutes) until you find the first green rubber band.

24. The network server to which you are connected goes down inexplicably on average every 40 hours. You decide that you will change servers in disgust after the sixteenth crash.
    What is the probability that you will still be with the server after 30 days?

25. Prove the theorem of the gamma approximation to the negative hypergeometric.

26. Prove the proposition about reversal symmetry in the beta family.

27. Prove the theorem about the beta approximation to the negative hypergeometric.

28. A random, anonymous survey of 1000 men in a large city discloses that 7 of them are HIV positive. You have the names of those surveyed, but not, of course, of those who were HIV positive. You decide that you must locate some of them for possible inclusion in an experimental drug treatment program; you will reinterview people one at a time until you find 5 who are HIV positive.

    a. What is the probability that you will have to interview no more than a total of 800 people to find them?

Hint: If you do this in the obvious way, you find yourself with a horrible sum to calculate. You might try using one of our symmetry results (black–white symmetry) to turn it into a very short sum.

b. Approximate the probability in (a) by instead looking at a continuous random variable that behaves like the proportion $P = X/993$ of your way through the healthy men in the survey. How close is your approximation?

29. You believe that two brands of electric fuses burn out from unpredictable power surges at about the rate of one every six months. You install them, one at a time, in a circuit. After burning out 2 fuses of the first brand, you switch to the second brand and burn out 3 more fuses. What is the probability that you will have had the first brand in place for more than half the time?

30. Our technique for constructing confidence bounds will also apply to Beta(α, β) random variables. A certain middle-school student is informed that she scored in the 94th percentile for her grade nationally on the SAT math test. Then she receives a prize for having had the second-highest score at her middle school. Assuming that the students who took the SAT at her middle school are typical of those who took it nationally in her grade, place a 95% confidence bound on the number of students at her middle school who took the test.

31. Prove the Neyman–Pearson lemma for discrete families of random variables.

32. **a.** Show that the negative hypergeometric N(W, B, b) family, where you want to test hypotheses about possible values of W, is a monotone likelihood ratio family.

    b. Show that the hypergeometric H($W + B$, W, n) family, where you want to test hypotheses about possible values of W, is a monotone likelihood ratio family.

CHAPTER 10

Continuous Random Variables II: Expectations and the Normal Family

10.1 Introduction

Expectation of a random variable is such a useful idea that in this chapter we apply it to all sorts of random variables, not just to discrete ones. Our goal will be to define the expectation in a general form. We will then apply it to our examples of absolutely continuous variables (those with densities) from the last chapter.

Continuing our program of discovering how random variables can have simple approximations even as their exact expressions get more complicated, we will find an important approximation for gamma variables when α is large, called the *normal* approximation. It will turn out to apply to Poisson probabilities as well. The method will turn out to be important enough that we will pause to derive a number of its properties. We will then exploit these two classes of approximations to get approximate confidence intervals for the Poisson parameter λ and the gamma parameter β.

Time to Review

Chapter 8, Sections 3–4
Intermediate value theorem
Integration by parts
Polar coordinates

10.2   Quantile Functions

10.2.1   Generating Discrete Variables

We noticed earlier that $E(X)$ made no sense for continuous random variables; we lacked a mass function $p(x)$ for the summation formula. Yet, our intuition demands that there should usually be some such quantity, corresponding to the idea of an average over many experimental measurements. We will indeed find a not-too-difficult way to achieve this goal.

Let me motivate our procedure with a practical problem: How do computer programs generate random numbers for simulation experiments? At the time this is being written, the method works something like this: The program has a way to generate, very rapidly, streams of floating point numbers that behave very much like independent Uniform(0, 1) variables. (Later, we will say a little about how this is done.) Then, the program transforms these numbers into the kind of random variables it needs. How does it find such a transformation? In practice, there are a great many ingenious tricks used to accomplish this. We will discuss only one method, which may not always be the best but is completely general: Any random variable at all can be generated this way.

Start by thinking about how, given a Uniform(0, 1) random variable U, you would obtain a Bernoulli variate X that is 1 with probability p and 0 with probability $1 - p$. One obvious way is to let X be 0 if $0 < U \leq 1 - p$, and let X be 1 if $1 - p < U < 1$. You should check that X has the right probabilities. It may occur to you that I could have selected the two values in reverse order, 1 first; but then X would not be an increasing function of U, which will be convenient later.

This simple device extends easily to any discrete random variable with a finite number of values: $x_1, \ldots, x_k$ with probabilities $p_1, \ldots, p_k$. If $0 < U \leq p_1$, we let $X = x_1$. If $p_1 < U \leq p_1 + p_2$, then $X = x_2$. If $p_1 + p_2 < U \leq p_1 + p_2 + p_3$ we set $X = x_3$. Continuing up the list of values, we finish with $p_1 + \cdots + p_{k-1} < U < p_1 + \cdots + p_k = 1$, giving $X = x_k$. Every value of X has been generated with the correct probability, and something has been done for every possible U.

This transformation can be thought of as a function, which we shall call $Q(u)$, the *quantile* function, whose graph is depicted in Figure 10.1.

10.2.2   Quantile Functions in General

We need to think of a way to define Q formally that will apply in as many cases as possible. Our procedure has, for each value of u, assigned it the smallest x whose cumulative probability is at least u (think about it).

Definition. The **quantile function** of a random variable X is

$$Q(u) = \min_x \{x : P(X \leq x) \geq u\} = \min_x \{x : F(x) \geq u\}$$

for $0 < u < 1$, where F is the cumulative distribution function of X.

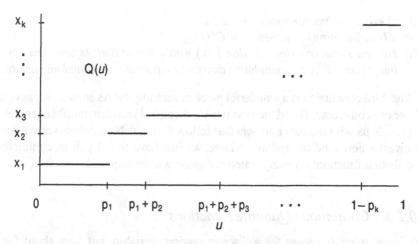

FIGURE 10.1. A discrete quantile function

You should translate this definition into words, to check that it does what we suggested it ought to do. We proposed this definition because in a very special circumstance, it turned a uniform variable into a random variable we wanted. Actually, it always does this.

First, notice that for any random variable X and for any $0 < u < 1$, Q has a value. To see this, we need to notice two things: that there is indeed a lower limit to the possible values of x in the set $\{x : F(x) \geq u\}$, and that there are always some members of this set, so we can find their minimum. Since F approaches zero as x gets small (review the properties of cumulative distribution functions, in (5.4.2)), there are always values of x that make F smaller than our $u > 0$; so we have a lower limit. Since F approaches 1 as x gets large and $u < 1$, there is also sure to be at least one x in our set $\{x : F(x) \geq u\}$. Therefore, we can always hope to find a minimum value of the set; and so Q always has a value.

Next we need to check that Q always transforms a uniform variable into one with the cumulative distribution function we want. That is, if U is a Uniform$(0, 1)$ random variable, then let $X = Q(U)$; now, can we be sure that $P(X \leq x) = F(x)$, as it should be?

$$P(X \leq x) = P[Q(U) \leq x] = P[\min_{y}\{y : F(y) \geq U\} \leq x].$$

But to say that the smallest y that satisfies an inequality (which is always true for larger y's, since F is nondecreasing) is no bigger than x is just to say that x satisfies that same inequality. Therefore,

$$P(X \leq x) = P[F(x) \geq U] = F(x).$$

The second equality is just the cumulative distribution function of Uniform$(0, 1)$ random variables. This is just what we wanted to know.

Theorem (the quantile transform). *If X is any random variable, then:*

(i) *Q exists for any random variable X.*

(ii) *if U is Uniform(0, 1), then X = Q(U).*

(iii) *For any nondecreasing function F(x) whose lower limit is zero and upper limit is one, F is the cumulative distribution function of a random variable.*

The third conclusion is a wonderful piece of serendipity. As soon as we have an F, we can construct a Q and then use our Uniform(0, 1) random number generator to provide us with random numbers that follow that law. Now, whenever we want to invent a new kind of random variable, we just have to tell you its cumulative distribution function; no complicated urn game will be required.

10.2.3   Continuous Quantile Functions

We know what Q means for a discrete random variable, but how about for a continuous one? If $x < y$ are in the sample space of X, so that

$$F(y) - F(x) = P(x < X \leq y) > 0,$$

then $F(x) < F(y)$. Thus, F is strictly increasing on its sample space. To see what this says about Q, observe that $Q[F(x)] = \min_y[y : F(y) \geq F(x)]$. But if x is in the sample space, where F is increasing, then the smallest y where F passes $F(x)$ is just x, so $Q[F(x)] = x$. I hope that this looks familiar: It is half of the requirement for Q to be the function *inverse* to F. We need further to investigate $F[Q(u)] = F\{\min_x[x : F(x) \geq u]\}$; since F is continuous, there is an x that solves the equation $F(x) = u$ (this is called the *intermediate value theorem* from calculus, which you should review). For the smallest such x, we still have $F[Q(u)] = u$. These two facts may be combined:

Proposition. *For a continuous random variable X, Q is the inverse function for F : Q = F⁻¹.*

Example. If T is negative exponential, then $F(t) = 1 - e^{-t}$. We find its inverse function by solving the equation $u = 1 - e^{-t}$ for t, to obtain

$$t = Q(u) = -\log(1 - u).$$

Thus, to generate a negative exponential random variable on a computer, we ask for a U that is Uniform(0, 1) and compute $T = Q(U) = -\log(1 - U)$.

If we have the happy situation that our random variable is absolutely continuous, then all we need to know about it is its density. We integrate the density f to get the cumulative F, as in the last chapter. Then we construct Q, and the random variable may be generated using Uniform(0, 1) variates.

Theorem (specifying a variable by its density).
Let $f(x) \geq 0$ have $\int_{-\infty}^{\infty} f(X)dX = 1$. Then f is the density of an absolutely continuous random variable.

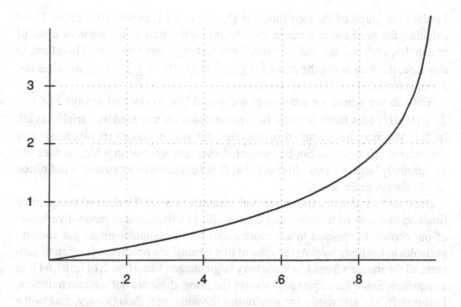

FIGURE 10.2. Negative exponential quantile function

10.2.4 Particular Quantiles

The quantile function also gives us a way to point out certain important features of a random variable. For example, $Q(0.5)$ is the smallest number such that the probability of not exceeding it is at least 0.5; that is, it is halfway through the possible outcomes. This is a bit like the value halfway through a sorted random sample, its median. Therefore, we call $Q(0.5)$ the median of the random variable (or the *population* median). For a negative exponential variable, $Q(0.5) = -\log(0.5) = 0.69315\ldots$. When we use negative exponential variables to predict the decay of radioactive particles, this number is called the half-life (in standard time units) of the particles, because half of them would be expected to decay by that time.

More generally, the quantile function points out values a certain part of the way through the values of a variable. Recall the *percentiles* from standardized tests that you took in grade school. The 90th percentile is $Q(0.9)$, just good enough to beat 90% of all test scores. The word quantile was coined by analogy with percentile.

10.3 Expectations in General

10.3.1 Expectation as the Integral of a Quantile Function

By now you are wondering what all this has to do with expectations. In the case of a discrete distribution with finite sample space, we know that $\mathrm{E}(X) = \sum_{i=1}^{k} x_i p_i$.

Look at the graph of the step function Q, Figure 10.1, in the last section: From calculus, the *area* under a curve, its integral, is the sum of the areas of a set of rectangles, each x_i high and p_i wide. Each rectangle has area $x_i p_i$. Therefore, in this case, the area under the curve for Q, $\int_0^1 Q(U)dU = \sum_{i=1}^k x_i p_i$, matches our formula for $E(X)$.

You can see where we are going; we would like to say that always $E(X) = \int_0^1 Q(U)dU$. This formula could be written down for any random variable at all. In fact, you may have learned in calculus that any *monotone* (nonincreasing or nondecreasing) function can be integrated over any interval in which it does not get infinitely large (you should check that Q is nondecreasing); so such a definition would always make sense.

But does this idea of expectation make intuitive sense? We derived the quantile function as a way of transforming Uniform(0, 1) variables into random variables of our choice. I promised to say more about how a computer might get uniform variables in the first place. At the time of this writing, the procedure is often to generate all the integers from 1 to some very large integer, like $M = 2,147,583,647$, in a sequence. Since the sequence is always the same, these are not random numbers. Fortunately, the arithmetic for generating the sequence, though easy, makes the ordering of the numbers in the sequence very peculiar and unexpected. Since we usually start the sequence at some arbitrary position, the numbers generated seem for quite a while random and independent, because they are unpredictable if you do not check the rather grubby arithmetic. Finally, the computer divides its integer by M, to get a number between 0 and 1. This is the *pseudo*-random number we pass on to the transformation formula.

Our random number generating procedure for X thus may give us M different values of $Q(U)$, where the U's are distributed at equal, tiny intervals over (0, 1). Furthermore, it gives all M values equally often because it is going through a complete sequence. Therefore, the average value of X generated is just the average of the M values of Q. From calculus, we suspect that an extremely good approximation of that average is just the average height of the Q curve, $\int_0^1 Q(U)dU$. Our proposed formula seems plausible.

What shall we do about the more general expectation $E[g(X)]$? Since this is just the average value of the quantity $g(X)$, it is just the result of averaging $g[Q(U)]$ for all values of U. Finally we are ready:

Definition. Let X be a random variable with quantile function Q. Then:

(i) $E(X) = \int_0^1 Q(U)\,dU$ whenever the integral exists.

(ii) $E[g(X)] = \int_0^1 g[Q(U)]\,du$ whenever that integral exists.

Proposition. *If X is discrete, it is still true that* $E(X) = \sum_i x_i p_i$ *and* $E[g(X)] = \sum_i g(x_i)p_i$.

Example. (1) If T is negative exponential, then $E[T] = \int_0^1 - \log(1 - U)dU$ must be evaluated. You may need to review your basic integration techniques. The one we will use is very handy in statistics: integration by parts: $\int u\,dv = uv - \int v\,du$. Let

$u = -\log(1 - U)$ and $dv = dU$. Then we compute $du = \frac{dU}{1-U}$ and $v = -(1 - U)$ (where we have chosen the additive constant in our antiderivative to cancel the other factor):

$$\int_0^1 -\log(1 - U)dU = \log(1 - U)(1 - U)|_0^1 - \int_0^1 -dU = 0 - 0 + 1.$$

We will leave it as an exercise to check that the first term was indeed zero: try L'Hospital's rule from calculus. Therefore, $E(T) = 1$. Notice that the *average* time to the first Poisson event is very different from the *median* time (which was about 0.69).

(2) If X is Cauchy(0, 1), then we found $F(x) = \frac{1}{2} + \frac{1}{\pi} \arctan(x)$. You should check that the quantile function is $Q(u) = \tan[\pi(u - \frac{1}{2})]$. Then

$$E(X) = \int_0^1 \tan\left[\pi\left(U - \frac{1}{2}\right)\right] dU.$$

You should remind yourself how to integrate tangents; you will get

$$E(X) = -\frac{1}{\pi} \log\cos\left[\pi\left(U - \frac{1}{2}\right)\right]\Big|_0^1 = -\infty - (-\infty).$$

Apparently, this Cauchy random variable has no expectation, since we cannot always cancel infinities. The practical implication is that the averages even of a great many repetitions settle down to no particular value.

It is time for some general facts about our new version of the expectation.

Theorem (expectation is a linear operator).

(i) *if a is constant, then* $E(a) = a$.
(ii) $E[ag(X)] = aE[g(X)]$ *whenever the second expectation exists.*
(iii) $E[g(X)+h(X)] = E[g(X)]+E[h(X)]$ *whenever the right-hand expectations exist.*

Proposition (expectation is a positive operator).   (iv) *For* $g(x) \geq 0$, $E[g(X)] \geq 0$.

As an easy but very important exercise, you should check both of these; use basic facts about integrals. These are exactly the same as some propositions we established several chapters back about our old definition of expectation (see 6.6.1); we say that expectation is always a positive linear operator. This is very important to us, because the definition and basic properties of the *variance* that we then worked out required us to know *only* these facts. You should review that section, because we are now going to assume that we know all about variances in general, and not just for discrete random variables.

Example. Let X be Uniform(0, 1); so $F(x) = x$, and $Q(u) = u$. Then $E(X) = \int_0^1 U dU = \frac{1}{2}$, to no one's surprise. But we established, using only the fact that expectations were linear, that $\text{Var}(X) = E(X^2) - E(X)^2$. Now $E(X^2) = \int_0^1 U^2 dU = \frac{1}{3}$; and $\text{Var}(X) = \frac{1}{3} - \left(\frac{1}{2}\right)^2 = \frac{1}{12}$, which was not obvious.

10.3.2 Markov's Inequality Revisited

Before we pursue the practical issues that come up when we want to compute expectations, let us notice that we can connect expectations with probabilities exactly as we did in Chapter 7.7. This follows because Markov's inequality still holds:

$$P(|X - \mu| \geq d) = \int_{|Q(u)-\mu| \geq d} 1 \, du$$

because for U Uniform$(0, 1)$ we know that $Q(U)$ has the same distribution as (any) X. Now we do the same trick as before; since $\frac{|Q(u)-\mu|}{d} \geq 1$ over the range of integration,

$$P(|X - \mu| \geq d) \leq \int_{|Q(u)-\mu| \geq d} \frac{|Q(u) - \mu|}{d} \, du \leq \frac{1}{d} \int_0^1 |Q(u) - \mu| \, du$$

when we extend the range.

Proposition (Markov's inequality). *For any random variable X for which the expectation exists, any constant d > 0, and any constant μ,*

$$P(|X - \mu| \geq d) \leq \frac{1}{d} E(|X - \mu|).$$

Then our easy consequences also work for all random variables: Convergence in expected error implies convergence in probability. Convergence in mean squared error implies convergence in expected error. (After all, the Cauchy–Schwarz inequality is still true, because it depends only on the fact that expectation is a positive linear operator.) For any variable with a variance, sample means converge in probability to their expectation as the sample size grows.

As you will soon see, we try to compute expectations directly from the definition only rarely. In most cases, writing down Q in a convenient form is hard to do. If F is already messy, you can imagine that finding its inverse is usually messier still. For example, you might try to write down Q for a gamma variable with $\alpha > 1$. In the next section we shall develop a much more practical technique for calculating expectations, which applies when the random variable has a density.

You may have had the following idea already: Since expectation seems still to work much as it did in the discrete case, why not use the method of indicators? We know that if $\{V_i\}$ are independently negative exponential, then $T = \sum_{i=1}^{\alpha} V_i$ is Gamma(α). It is plausible that

$$E(T) = \sum_{i=1}^{\alpha} E(V_i) = \alpha.$$

The answer is correct; and in fact, our reasoning is correct. Unfortunately, we really do not yet know what a multivariate expectation means in the continuous case. We shall see in the next chapter that it will have all the nice properties that we have hoped for.

10.4 Absolutely Continuous Expectation

10.4.1 Changing Variables in a Density

To compute expectations with the aid of densities, we shall need first to learn what effect change of variables has on a density. At this point, you should remind yourself what change of variables does to a cumulative distribution function (see 9.4.1).

Example. Remember that if T is Gamma(α), then $S = \beta T$ is Gamma(α, β). We discovered that $F(s) = \sum_{i=\alpha}^{\infty} s^i / (\beta^i i!) e^{-s/\beta}$. By the usual laborious differentiation and cancellation, we discovered that its density is $f(s) = s^{\alpha-1}/(\beta^\alpha(\alpha-1)!)e^{-s/\beta}$. That looks reasonable; but if you stare at it long enough, you may notice something peculiar: it is no longer quite a Gamma(α) density with s/β in place of t. There is one extra power of β in the denominator.

Actually, this should not surprise us. If, for example, β is greater than 1, it spreads out the values of the random variable by that factor. But every density must integrate to 1: The area under its graph does not change. Therefore, the wider, transformed density must shrink in height by the factor β to compensate (Figure 10.3).

We can easily work out the effects of a transformation more generally. Let X have density f_X and consider the transformation $X = g(Y)$, where g is nondecreasing. Then

$$f_Y(y) = \frac{d}{dy}F_Y(y) = \frac{d}{dy}F_X[g(y)] = f_X[g(y)]\frac{d}{dy}[g(y)],$$

where we have used the *chain rule* from calculus. You should do a similar calculation for a nonincreasing change of variables and combine them:

Theorem (change of variables in a density). *For a monotone (either nondecreasing or nonincreasing) change of variables $X = g(Y)$ that is differentiable*

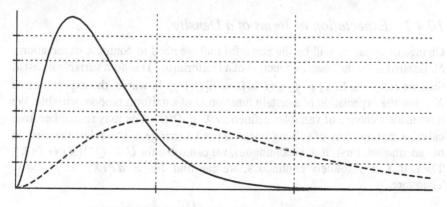

FIGURE 10.3. Gamma(3,2) and Gamma(3,5) Densities

at y,

$$f_Y(y) = f_X[g(y)] \left| \frac{d}{dy}[g(y)] \right|.$$

If you stare at it long enough, this complicated expression may look familiar, from calculus. When you try to integrate a function (say f_X) by the method of change of variables ($X = g(Y)$), the right hand side in the theorem is your new integrand. This is the long-promised reason why we prefer to write such variables of integration as if they were random variables, with capital letters. Absolutely continuous random variables transform exactly like variables of integration.

Example. If P is a Beta(α, β) random variable, define an F(α, β) (named after R. A. Fisher) random variable on $(0, \infty)$ by $Y = (P/\alpha)/((1 - P)/\beta)$. This peculiar formula comes about because P is α events into the interval, and $1 - P$ is β events from the other end. Numerator and denominator are both average spacing between Dirichlet events. Therefore, Y is in some sense centered at 1, and deviations from its typical behavior are easy to see. We invert the change of variables to get $p = (\alpha Y)/(\beta + \alpha Y)$. The derivative of this is $(\alpha\beta)/(\beta + \alpha Y)^2$. Since a beta density is $(\alpha + \beta - 1)!/((\alpha - 1)!(\beta - 1)!)p^{\alpha-1}(1 - p)^{\beta-1}$, we can substitute our value for P and multiply by the derivative to get

$$f(y) = (\alpha + \beta - 1)!/((\alpha - 1)!(\beta - 1)!)\alpha^\alpha \beta^\beta y^{\alpha-1}/(\beta + \alpha y)^{\alpha+\beta}$$

on $y > 0$. You should discover as an exercise that $1/Y$ is an F(β, α) random variable.

Proposition (location and scale changes in a density).
  (i) *If* $Y = X + m$, *then* $f_Y(y) = f_X(y - m)$.
  (ii) *If* $Y = dX$, *then* $f_Y(y) = \frac{1}{|d|} f_X\left(\frac{y}{d}\right)$.

The proofs are easy exercises. The second one shows us that what happened in the gamma example above always happens.

10.4.2 Expectation in Terms of a Density

Change of variables will be the powerful tool we need to compute expectations. Notice that since the quantile function of a Uniform(0, 1) random variable is just u, then we can write $E(X) = \int_0^1 Q(U)dU = E[Q(U)]$. In words, the expectation of X is just the expectation of a certain function Q of a uniform random variable. We have made a change of variables defined by $X = Q(U)$. You may remember from calculus that one way of solving integrals used a change of variables: the method of *substitution*. First, if X is continuous, we can solve for $U = Q^{-1}(X) = F(X)$. Then, if X is absolutely continuous, we can find $dU = dF(X) = f(X)dX$. Therefore,

$$E(X) = \int_0^1 Q(U)dU = \int_{Q(0)}^{Q(1)} Xf(X)dX.$$

$Q(0)$ is the lower bound of the sample space of X, and $Q(1)$ is its upper bound. We usually write these limits in explicitly, or if we leave them out, we mean "integrate over the entire sample space of X." We have found a fundamental fact:

Theorem (expectations of a density). *If X is absolutely continuous, then*

(i) $E(X) = \int Xf(X)dX$ *whenever the integral exists.*

(ii) $E[g(X)] = \int g(X)f(X)dX$ *whenever that integral exists.*

You should check that the second result is true for the same reason. Even though these expressions seem to be just one of a number of possible ways of evaluating our defining integral, they have turned out to be so extraordinarily useful that we usually try them first. In fact, many books more elementary than this one use them as the definition of expectation for absolutely continuous random variables.

Let T be a Gamma(α) variable. Our theorem says that

$$E(T) = \int_0^\infty T\frac{T^{\alpha-1}}{(\alpha-1)!}e^{-T}dT = \int_0^\infty \frac{T^{\alpha}}{(\alpha-1)!}e^{-T}dT.$$

The function we are integrating looks familiar: It is a Gamma($\alpha + 1$) density, except that the constant in the denominator is wrong. We can patch it using the fact that $\alpha! = \alpha(\alpha-1)!$; multiply and divide by α to get

$$E(T) = \alpha \int_0^\infty \frac{T^{\alpha}}{\alpha!}e^{-T}dT = \alpha \cdot 1 = \alpha,$$

since the integral of a density over the whole sample space is always 1. Our speculative calculation using indicators is borne out. This method of calculation should remind you of the *inductive* method, which we used repeatedly to calculate discrete expectations in some of our families; there we used the fact that mass functions sum to 1 (see Chapter 6, Section 5).

Proposition.

(i) *If T is Gamma(α), $E(T) = \alpha$.*

(ii) *If S is Gamma(α, b), $E(S) = \alpha\beta$.*

(iii) *If P is Beta(α, β), $\alpha/(\alpha + \beta)$.*

The proofs of (ii) and (iii) are exercises; in (ii), do not work very hard—use the definition of S and general properties of expectations.

To calculate the *variance* of T, we need

$$E(T^2) = \int_0^\infty T^2\frac{T^{\alpha-1}}{(\alpha-1)!}e^{-T}dT = \int_0^\infty \frac{T^{\alpha+1}}{(\alpha-1)!}e^{-T}dT$$

$$= \alpha(\alpha+1) \int_0^\infty \frac{T^{\alpha+1}}{(\alpha+1)!}e^{-T}dT.$$

Since the last integral is 1, we have $E(T^2) = \alpha(\alpha+1)$. We are ready to compute

$$Var(T) = E(T^2) - E(T)^2 = \alpha(\alpha+1) - \alpha^2 = \alpha.$$

Proposition.
(i) *For T a Gamma(α) variable,* Var(*T*) = α.
(ii) *For S a Gamma(α, β) variable,* Var(*S*) = αβ².
(iii) *For* P *a Beta(α, β) variable,* Var(P) = (αβ)/((α + β)²(α + β + 1)).

Part (ii) is an easy exercise, and (iii) is a little more fun.

Example. The fuse protecting an expensive circuit board blows out because of unpredictable power fluctuations and must be replaced. Past experience suggests that an average of one fuse will blow in five days. I bought a box of two dozen fuses; what can I say about how long the box will last?

Blown fuses might plausibly be modeled by a Poisson process, and so the life of the box in days is a Gamma(24, 5) variable. We can compute its cumulative distribution and its density precisely, but you should note that these are complicated. We will use expectations to summarize its properties. The average life of the box is 120 days; its variance is 600. As usual, this is hard to interpret, but the standard deviation is about 24.5 days. We would not be surprised if the box lasted only 95 days, nor if it lasted 145.

Example. An F(1, 1) variable has density $1/(1 + y)^2$. Therefore,

$$E(Y) = \int_0^\infty \frac{Y}{(1+Y)^2} dY = \int_0^\infty \left[\frac{1}{(1+Y)} - \frac{1}{(1+Y)^2} \right] dY,$$

where we have applied a partial fraction decomposition, which yields that

$$E(Y) = [\log(1 + Y) + 1/1 + Y]\big|_0^\infty = \infty.$$

No matter how often we do experiments that give this variable, averages will not settle down to some consistent value.

10.5   Normal Approximation to a Gamma Variable

10.5.1   Shape of a Gamma Density

Most of our families of random variables have arisen when we tried to approximate some other random variable in the case that certain parameters got large enough to make calculations unwieldy. You may have noticed that we have not finished. For example, in a negative hypergeometric random variable, what happens if W, b, and $B - b$ all get painfully large? Or in a binomial variable, what if n gets large but neither p nor $1 - p$ are small? Or with a gamma random variable, what if α gets large? It will turn out that we can do useful approximations in these cases. The miracle will be that the same technique, called *normal* approximation will solve each of these problems, and many more.

Some pictures of densities will suggest what happens to gamma random variables as α grows (see Figures 10.4–10.7).

These are far from being on the same scale (they would not have fit very well on one graph), but the increasing similarity of *shape* is striking. It should remind

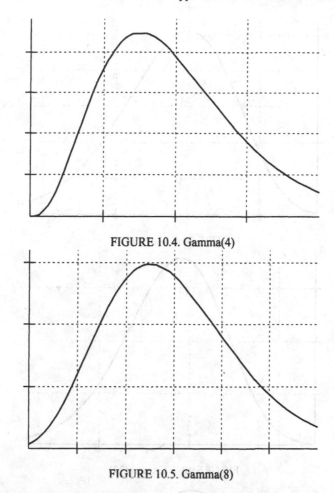

FIGURE 10.4. Gamma(4)

FIGURE 10.5. Gamma(8)

you of the shapes of certain likelihood functions in Chapter 8 (see 8.2.1). We will discover the mathematical reason for this pattern and use it to find useful approximations. Let me show you what happens when we transform each of these graphs so they fit on a single set of axes (see Figure 10.8).

We have matched center, curvature, and height of the four densities; you will see shortly how this was done. The common form is becoming clear—it is traditionally called "bell-shaped," for obvious reasons. What is the mathematical nature of this curve? Put these same graphs on a semilog scale, that is, let the vertical axis be logarithmic (Figure 10.9).

10.5.2   Quadratic Approximation to the Log-Density

Now we can guess what our approximate shape is: The curves look more and more like a parabola—the graph of a quadratic function. The same phenomenon arose when we looked at likelihoods in Chapter 8 (see 8.3.1). We will try to pin down a

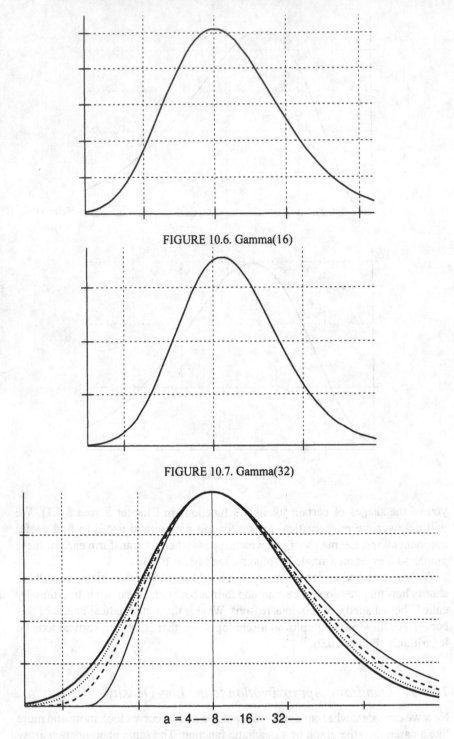

FIGURE 10.6. Gamma(16)

FIGURE 10.7. Gamma(32)

a = 4 — 8 --- 16 ⋯ 32 —

FIGURE 10.8. Rescaled gamma densities

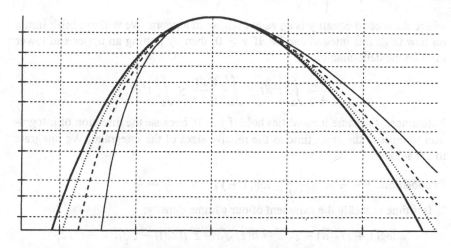

FIGURE 10.9. Rescaled gamma log-densities

conjecture: The logarithm of the density of a gamma random variable with α large is approximately quadratic (at least near its maximum value). First, so we will not always be subtracting 1, write $a = \alpha - 1$. Then put all t's in the exponent of the gamma density, to facilitate study of its logarithm:

$$f(t) = \frac{t^a}{a!}e^{-t} = \frac{1}{a!}e^{a\log t - t}.$$

We need to find the point at which the exponent of the density is maximal: $\frac{d}{dt}(a\log t - t) = (a/t) - 1 = 0$, so by elementary calculus the only possible maximum is at $t = a$. We use the second derivative to find the curvature there: $(d^2/dt^2)(a\log t - t) = -(a/t^2)$, which is negative. At the maximum, this curvature is $-1/a$. In order to compare many different densities, we will use a linear change of variables to *standardize*, so that the maximum is at zero and the second derivative of the logarithm there is -1. You should show as an exercise that the change of variables $z = (t - a)/(\sqrt{a})$ accomplishes this. Then

$$f(z) = \frac{\sqrt{a}}{a!}e^{a\log(a+z\sqrt{a})-a-z\sqrt{a}}.$$

We want to approximate the exponent by a second-degree polynomial in z. A polynomial approximation to a logarithm is easiest to find for $\log(1+y)$; with a little ingenuity we rearrange $\log(a+z\sqrt{a}) = \log[a(1+z/\sqrt{a})] = \log a + \log(1+z/\sqrt{a})$. Now collect constants and variable terms separately to get

$$f(z) = \frac{a^{a+1/2}e^{-a}}{a!}e^{a\log(1+z/\sqrt{a})-z\sqrt{a}}.$$

We need an approximation to the logarithm that is more accurate than the one in the birthday problem (see 3.5.1), but we will proceed similarly.

$$\log(1 + y) = \int_0^y \frac{dt}{1+t} = \int_0^y \left(1 - t + \frac{t^2}{1+t}\right) dt = y - \frac{y^2}{2} + \int_0^y \frac{t^2 dt}{1+t},$$

where the second equality is an easy exercise in algebra. We will establish limits on how large this integral can be. If $y \geq 0$, then by putting an upper and lower limit on the denominator,

$$\frac{1}{1+y} \int_0^y t^2 dt \leq \int_0^y \frac{t^2 dt}{1+t} \leq \int_0^y t^2 dt.$$

Futhermore, the same inequalities hold if $y < 0$, because the direction of integration reverses at the same time as the relative sizes of the integrand. We integrate to get a bound:

Proposition. $y - \frac{y^2}{2} + \frac{y^3}{3(1+y)} \leq \log(1+y) \leq y - \frac{y^2}{2} + \frac{y^3}{3}.$

It is time to tackle the exponent of our gamma density:

$$a \log(1 + z/\sqrt{a}) - z\sqrt{a} \approx a(z/\sqrt{a} - z^2/(2a)) - z\sqrt{a} = -z^2/2,$$

where the proposition tells us that the approximation works whenever $z^3/(3\sqrt{a})$ and $z^3/(3(\sqrt{a} + z))$ are small in size. When $|z|$ is small compared to $\sqrt{a}$, the two conditions say the same thing because the denominators are about the same size. (Using the definition of z, these two error estimates are $(x - a)^3/(3a^2)$ and $(x - a)^3/(3ax)$.) When we put back the definition of $a = \alpha - 1$, we get the complete approximation:

Proposition. *If T is Gamma(α), let $Z = (T - (\alpha - 1))/\sqrt{\alpha - 1}$. Then the density of Z is $f(z) \approx ((\alpha - 1)^{\alpha-1/2}e^{-(\alpha-1)})/(\alpha - 1)!e^{-z^2/2}$ whenever $\sqrt{\alpha - 1}$ is large, and in particular, large compared to $|z^3|/3$.*

This approximation answers our question about why the gamma densities have a family resemblance, but it is hard to imagine its practical use. The constant in front is as complicated to calculate as the original density. Notice, though, that the variable part involving z, $e^{-z^2/2}$, does not depend at all on the (large) parameter; this is as we would have hoped from past experience with asymptotic approximation. It would also be nice if that messy constant did not depend on α.

10.5.3 Standard Normal Density.

But wait, the constant *should not* depend on α. In the past, our approximations actually corresponded to random variables. If we insist that our approximation be a true density, its integral should be 1. But then since the variable part, the exponential, does not depend on α, integration determines the constant, and it in turn would not depend on α. Emboldened by these thoughts, we make a definition:

Definition. A **(standard) normal** (or **Gaussian**) random variable Z has sample space $(-\infty, \infty)$ and density $f(z) = ke^{-z^2/2}$, where k is a positive constant such that the integral of the density is 1.

The normal random variable will turn out to be perhaps the most useful continuous variable of all.

Notice that since the sample space of T is $(0, \infty)$, the sample space of Z above was $(-\sqrt{\alpha - 1}, \infty)$. For α large, the lower bound is a large negative number and is well outside the range in which we trust our approximation. Therefore, we have replaced it with negative infinity.

You might think that we could figure out k by elementary calculus, but you should verify that none of the standard methods apply. We shall have to use a trick that is not at all obvious. But first let us see what we can learn about Z without knowing k:

$$E(Z) = k \int_{-\infty}^{\infty} Z e^{-Z^2/2} \, dZ = -k e^{-Z^2/2} \big|_{-\infty}^{\infty} = -0 - -0,$$

$$\text{Var}(Z) = E(Z^2) = k \int_{-\infty}^{\infty} Z^2 e^{-Z^2/2} \, dZ.$$

The previous integral suggests integration by parts: $dv = Z e^{-Z^2/2} \, dZ$ and $u = Z$; then $v = -e^{-Z^2/2}$ and $du = dZ$.

$$\text{Var}(Z) = -k Z e^{-Z^2/2} \big|_{-\infty}^{\infty} + \int_{-\infty}^{\infty} k e^{-Z^2/2} \, dZ = 0 + 1.$$

You should check that the first term is zero using L'Hospital's rule. The second is just the integral of the normal density.

Proposition. *If Z is standard normal then:*

(i) *(reversal symmetry)* $f(z) = f(-z)$.
(ii) $E(Z) = 0$.
(iii) $\text{Var}(Z) = 1$.

We will not know the vertical scale until we compute k; but it is reassuring to see that we have indeed captured the qualitative shape of our gamma densities.

It is time to evaluate k by $\frac{1}{k} = \int_{-\infty}^{\infty} e^{-Z^2/2} dZ$. The trick will be to calculate instead its square,

$$\frac{1}{k^2} = \int_{-\infty}^{\infty} e^{-Z^2/2} dZ \int_{-\infty}^{\infty} e^{-W^2/2} dW.$$

We are going to pretend that the product of integrals is really a single bivariate integral over the (Z, W) plane, evaluated by *Fubini's theorem* from multivariable calculus (which you should review). Therefore

$$\frac{1}{h^2} = \int \int e^{-(Z^2 + W^2)/2} \, dZ \, dW.$$

When a function of two variables depends only on $Z^2 + W^2$, a bell should ring. Perhaps it is more natural to express it in *polar* coordinates: $r = Z^2 + W^2$ on $(0, \infty)$, and $\theta = \arctan(W/Z)$ on $[0, 2\pi)$. You should check that the *Jacobian* of this change of variables (time to review another fact from multivariable calculus) is $\frac{1}{2}$; in terms of elements of integration, we write this fact as $dZ\,dW = \frac{1}{2} dr\,d\theta$.

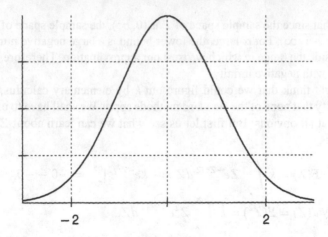

FIGURE 10.10. Standard normal density

Therefore our double integral above becomes

$$\int_0^{2\pi} d\theta \int_0^\infty \frac{1}{2} e^{-r/2} dr = 2\pi.$$

(The Jacobian method of changing variables will be reviewed in much more detail in a later chapter.) We conclude that $1/k^2 = 2\pi$.

Proposition. *A standard normal density is* $f(z) = \frac{1}{\sqrt{2\pi}} e^{-z^2/2}$.

10.5.4   Stirling's Formula

We needed the constant for calculation, but we can learn something else from it. This theorem and the previous theorem about the shape of a gamma distribution tell us that for large α, the transformed gamma density is approximately proportional to the normal density. But since all densities integrate to 1, the constant of proportionality must be approximately 1; that is, the constants in front are nearly the same: $1/\sqrt{2\pi} \approx ((\alpha - 1)^{\alpha - 1/2} e^{-(\alpha - 1)})/(\alpha - 1)!$. Solving for the factorial, we learn a marvelous fact.

Theorem (Stirling's formula).   *For n large,* $n! \approx \sqrt{2\pi} n^{n+1/2} e^{-n}$.

So an integer function, factorial, may be approximated by the sorts of functions one sees in calculus. In fact, for n very large, computers sometimes use extensions of Stirling's formula to save time. For example, $10! = 3,628,800$; and the approximation is $3,598,696$. It is already within 1% for n only 10.

10.5.5   Approximate Gamma Probabilities

We return to the problem of finding a useful normal approximation to the gamma family for α large. In the past, we have divided by a constant (standardized) in order to use gamma or beta approximations (see, for example, 9.3.4); one of the properties of this constant has been that the approximate variable had the same expected value as the original variable. Now we are using a two-part standardization in which we centered the variable at zero and then divided by a constant to make scales similar. Does this give the gamma variable the same expectation and scale (say, standard deviation) as the normal? Since the gamma expectation is α and the standard deviation is $\sqrt{\alpha}$, then $Z = \frac{T-\alpha}{\sqrt{\alpha}}$ has expectation 0 and standard deviation 1 (see 7.5.6), which matches the normal case. But this is not the same standardization as we used to prove the theorem (we had $\alpha - 1$ in place of α). However, for α large, the difference between the two is too small to matter (exercise).

Theorem (normal approximation to a gamma variable). *Let T be Gamma(α); and let $Z = \frac{T-\alpha}{\sqrt{\alpha}}$. Then:*

(i) *For $\sqrt{\alpha}$ large and $|z^3|/3$ small compared to $\sqrt{\alpha}$, $f(z)$ is approximately standard normal.*

(ii) *For a sequence of gamma variables for which $\alpha \to \infty$, Z converges in distribution to a standard normal variable.*

To see (ii), we must show that the cumulative distributions get close. These are obtained by integrating the density, whose approximation is known to be close for $|z|$ not too large. But the probability that $|z|$ is too large for the approximation to work becomes arbitrarily small as α grows, so the functions we are integrating are close over part of their range, and the integrals over the rest are too small to matter.

10.5.6 Computing Normal Probabilities

There is one last, unexpected, difficulty to overcome before we can use our approximation. Probabilities come from the cumulative distribution function, and we do not have a formula for the normal cumulative. The density is simple, so we will integrate it. But we have already noticed that this is hard to do. In fact, it cannot be expressed in terms of the usual functions from calculus. We shall have to find numerical methods to approximate it to the accuracy we need. One way would be to expand the density in a *power series* and then integrate the series term by term. From calculus, you should remind yourself,

$$e^x = 1 + x + \frac{x^2}{2} + \frac{x^3}{6} + \cdots = \sum_{i=0}^{\infty} \frac{x^i}{i!}.$$

Therefore,

$$e^{-z^2/2} = 1 - \frac{z^2}{2} + \frac{z^4}{8} - \cdots = \sum_{i=0}^{\infty} (-1)^i \frac{z^{2i}}{2^i i!}.$$

Now,

$$F_Z(z) - F_Z(0) = \int_0^z \frac{1}{\sqrt{2\pi}} e^{-z^2/2} \, dZ,$$

where $F_Z(0)$ is just the probability that the variable is negative; but the density is symmetric about zero, so it is equally likely to be positive or negative. Thus, $F_Z(0) = \frac{1}{2}$. Now use the series to integrate the density term by term:

Proposition. *The standard normal cumulative distribution is*

$$F_Z(z) = \frac{1}{2} + \frac{1}{\sqrt{2\pi}} \left(z - \frac{z^3}{6} + \frac{z^5}{40} - \cdots \right) = \frac{1}{2} + \frac{1}{\sqrt{2\pi}} \sum_{i=0}^{\infty} (-1)^i \frac{z^{2i+1}}{(2i+1)2^i i!}.$$

You should check that the series is absolutely convergent for any value of z.

Example. The probability that Z is at most 1 is given by $F_Z(1) \approx 0.841344$, the accuracy to which the series has settled after 6 terms (check the arithmetic for yourself). Since our series has alternating signs for $z > 0$, you should remember from calculus that as soon as the terms begin to shrink, we know that the sums of odd and even numbers of terms are upper and lower bounds of the correct answer. This is convenient in cases where it would take many terms for the series to converge; instead, take a few terms and see whether it is close enough for your purposes. For example, using 5 and 6 terms, we find that $0.9729 \leq F_Z(2) \leq 0.9922$.

10.5.7 Normal Tail Probabilities

As z becomes large—say, 3 or so—using this series becomes less satisfactory, for two reasons. We have already seen that it takes more and more terms to achieve a given number of significant figures. Furthermore, since the answer is close to 1, we are likely to be more interested in the small probability of exceeding z, $1 - F_Z(z)$, the *tail* probability. Now the interval in the example becomes $0.0078 \leq 1 - F_Z(2) \leq 0.0271$; we have hardly narrowed the answer down at all, after a good bit of calculation. Fortunately there is a simple way to get close: We write $P(Z > z) = \frac{1}{\sqrt{2\pi}} \int_z^{\infty} e^{-Z^2/2} dZ$. Limit ourselves to the case $z > 0$, and proceed to integrate this by (unexpected) parts: $dv = Ze^{-Z^2/2} dZ$ and $u = 1/Z$. Then $v = -e^{-Z^2/2}$ and $du = -dZ/Z^2$, and so

$$P(Z > z) = \frac{1}{\sqrt{2\pi}} \left(\frac{1}{z} e^{-z^2/2} - \int_z^{\infty} \frac{1}{Z^2} e^{-Z^2/2} dZ \right).$$

The integral is always positive (if we stay away from 0), so we learn something useful: $P(Z > z) < \frac{1}{\sqrt{2\pi}z} e^{-z^2/2}$. Integrate by parts again, using the same dv:

$$P(Z > z) = \frac{1}{2\pi} \left(\frac{1}{z} e^{-z^2/2} - \frac{1}{z^3} e^{-z^2/2} + \int_z^{\infty} \frac{3}{Z^4} e^{-Z^2/2} dZ \right).$$

Again the integral is positive, so we get an inequality in the other direction:

Proposition. *For Z standard normal and $z > 0$,*

$$\frac{1}{\sqrt{2\pi}} e^{-z^2/2}(1/z - 1/z^3) < P(Z > z) < \frac{1}{\sqrt{2\pi}z} e^{-z^2/2}.$$

For example, $0.02025 < P(Z > 2) < 0.02670$, which is much more precise than the previous bounds, after less work. The new proposition, in contrast to the old, gives more and more precise results as z gets large. As an exercise, you should continue our integrations by parts to get a series. We have not bothered to state it as a proposition, because it never converges. The successive pairs of bounds require larger and larger z's before they give us improved precision (such an object is called an *asymptotic* series).

You will find that you will not need to do calculations of the normal cumulative very often. It is such a useful random variable that tables of the function are widely available. Almost any package of statistics programs will have a function to calculate it, and statistical calculators should have a button to do it. You should find out which of these resources are available to you and learn to use them.

Example. Recall the fuse that burns out once in five days (see Section 4.2). We might ask what the probability is that our box of 24 will last no more than 100 days. If our unit is 5 days, then we are asking about 20 time units. An exact, rather painful calculation gets a probability of 0.2125. Since α is fairly large, we might try the normal approximation, the probability that a standard normal variable is at most $(20 - 24)/\sqrt{24} = -0.8165$. After an easier calculation we get 0.2071, which is quite close.

In this example, we worked the more general problem of normal approximation to a Gamma(α, β), by first translating time into standard units by dividing by β before computing Z. Combining the two, $Z = (T - \alpha\beta)/(\beta\sqrt{\alpha})$. But this is just subtracting the average and dividing by the standard deviation, as before. We could have done it all in one step.

10.6 Normal Approximation to a Poisson Variable

10.6.1 Dual Probabilities

We went to considerable trouble to find an approximation to gamma variables for large α; so you are probably hoping we will have other uses for the normal random variable. One is obvious: by gamma–Poisson duality (see 9.3.4), we must be calculating some Poisson probabilities already when we use the normal approximation. Let X be Poisson(λ), where we assume that λ is large. Then

$$F[x|\text{Poisson}(\lambda)] = 1 - F[\lambda|\text{Gamma}(x + 1)].$$

Standardizing $z = (\lambda - x - 1)/\sqrt{x + 1}$; we approximate our probability using $1 - F_Z(z)$. But by the symmetry of the normal, this is $F_Z(-z)$ (exercise). We can thus do a direct normal approximation to the Poisson by standardizing $z = (x - \lambda + 1)/\sqrt{x + 1}$ in the first place. We know that it works when $|x^3|/3$ is small compared to $\sqrt{x + 1}$.

Although this is satisfactory (it is, of course, exactly as accurate as the corresponding approximation to a gamma variable), it is not of the same form as the earlier method, which matched mean and standard deviation by a linear transformation. In other words, could we standardize by $z = \frac{x-\lambda}{\sqrt{\lambda}}$ instead? Using corresponding conditions that $\sqrt{\lambda}$ is large and $|z^3|/3$ is comparatively small, we will do a direct comparison of the two z's:

$$\frac{x+1-\lambda}{\sqrt{x+1}} - \frac{x+1-\lambda}{\sqrt{\lambda}} = (x+1-\lambda)\left(\frac{1}{\sqrt{x+1}} - \frac{1}{\sqrt{\lambda}}\right).$$

Now add and subtract λ in the first denominator:

$$= (x+1-\lambda)\left[\frac{1}{\sqrt{\lambda+(x+1-\lambda)}} - \frac{1}{\sqrt{\lambda}}\right]$$

$$= \frac{(x+1-\lambda)}{\sqrt{\lambda}}\left[\frac{1}{\sqrt{1+(x+1-\lambda)/\lambda}} - 1\right]$$

after factoring out $\sqrt{\lambda}$. Now we need a basic approximation to the square root, similar to approximations to logarithms from earlier chapters: Let a be a constant that is small compared to 1. Then $1 - \frac{1}{\sqrt{1+a}} = \frac{\sqrt{1+a}-1}{\sqrt{1+a}}$. Now use a standard algebraic trick for simplifying square roots that you may remember from high school:

$$\frac{\sqrt{1+a}-1}{\sqrt{1+a}}\frac{\sqrt{1+a}+1}{\sqrt{1+a}+1} = \frac{a}{1+a+\sqrt{1+a}} = \frac{a}{2+a+(\sqrt{1+a}-1)}.$$

Repeat the trick in the denominator to get

$$1 - 1/\sqrt{1+a} = a/(2+a+a/(\sqrt{1+a}+1)).$$

We can summarize this:

Proposition. *For $|a|$ small compared to 1, $1 - \frac{1}{\sqrt{1+a}} \approx \frac{a}{2}$.*

When this is applied to the comparison above, we obtain

$$(x+1-\lambda)/\sqrt{x+1} - (x+1-\lambda)/\sqrt{\lambda} \approx -(x+1-\lambda)^2/(2\lambda^{3/2}),$$

since if $|z^3|/3$ is small compared to λ large, so is $(x+1-\lambda)/\lambda$. Rearranging slightly, we find that $(x+1-\lambda)/\sqrt{x+1} - (x-\lambda)/\sqrt{\lambda} \approx 1/\sqrt{\lambda} - z^2/(2\sqrt{\lambda})$. The right-hand side is obviously small.

Theorem (normal approximation to the Poisson).

(i) *Let X be Poisson(λ) where $\sqrt{\lambda}$ is large, and let $z = (x-\lambda)/\sqrt{\lambda}$ where $|z^3|/3$ is small compared to $\sqrt{\lambda}$. Then*

$$F[x|\text{Poisson}(\lambda)] \approx F(z|\text{Normal}) = F_Z(z).$$

(ii) *Given a sequence of Poisson(λ) random variables X_i such that $\lambda_i \to \infty$, let $Z_i = (X_i - \lambda_i)/\sqrt{\lambda_i}$. Then the Z_i converge in distribution to a standard normal variable.*

To complete the proof, you need to check as an exercise that our new conditions on z and λ indeed imply those required by the normal approximation to the gamma.

10.6.2 Continuity Correction

Before we try out our new method, let me point out a peculiarity. We approximate $P(X \le x)$ by computing $z = \frac{x-\lambda}{\sqrt{\lambda}}$ and finding the normal $F_Z(z)$. We could also write $P(X \le x) = 1 - P(X > x) = 1 - P(X \ge x + 1)$, since Poisson variables take on only integer values. Then let $z^* = (x + 1 - \lambda)/\sqrt{\lambda}$ and compute $1 - [1 - F_Z(z^*)] = F_Z(z^*)$. We have two, different, normal approximations to the same probability. Which is better? There is no obvious way to decide.

To remove the ambiguity, let us improve the approximation in the previous section. Taking expectations over all possible values of the difference $1/\sqrt{\lambda} - z^2/(2\sqrt{\lambda})$, we get $1/\sqrt{\lambda} - 1/(2\sqrt{\lambda}) = 1/(2\sqrt{\lambda})$ (since $E(Z^2) = 1$). Then we have a better approximation $(x + 1 - \lambda)/\sqrt{x+1} \approx (x - \lambda)/\sqrt{\lambda} + 1/(2\sqrt{\lambda}) = (x + 1/2 - \lambda)/\sqrt{\lambda}$. Notice that this standardization is halfway between the two choices above: $z = (x + 1/2 - \lambda)/\sqrt{\lambda}$. This is called a *continuity correction*, because it adjusts for the incompatibility between the discrete Poisson and the continuous normal random variables.

Example. A certain large city park has an average of 64 mugging cases a month. The police department will be overloaded if in a given month the number exceeds 70. What is the probability of overload next month? Model the number as Poisson(64); after a long calculation, we find that the probability is 0.20616. Without the continuity correction we let $z = \frac{70-64}{8}$, and the normal approximation to our probability is 0.22663. With the continuity correction, $z = \frac{70.5-64}{8}$; the normal approximation is 0.20825, which is much better. The continuity correction often improves the result considerably with little extra trouble.

Notice that we have not found an approximation here for the mass function of a Poisson variable. Since we were using a continuous random variable, it was more natural to start with the cumulative. However, we can always calculate $p(x) = F(x) - F(x - 1)$ using the normal approximation. With the continuity correction, this becomes

$$p(x) \approx F_Z\left(\frac{x + 1/2 - \lambda}{\sqrt{\lambda}}\right) - F_Z\left(\frac{x - 1/2 - \lambda}{\sqrt{\lambda}}\right).$$

This should be easy to remember; we took the probability of all real numbers between $x - \frac{1}{2}$ and $x + \frac{1}{2}$ that round off to the integer x. For example, in the mugging problem, what is the probability of exactly 60 in a month? The exact answer is $p(60) = 0.04527$; its normal approximation is $0.33087 - 0.28689 = 0.04398$.

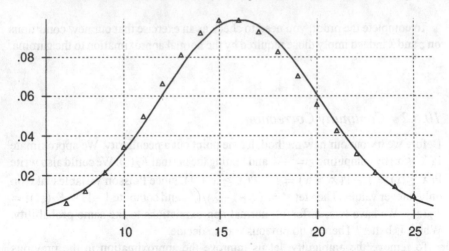

FIGURE 10.11. Poisson(16) mass function and its Normal(16, 4) approximation

10.7   Approximations to Confidence Intervals

10.7.1   The Normal Family

Given a standard normal random variable Z, we can create a new variable $X = \mu + \sigma Z$ that still has sample space all real numbers, but has expectation μ and variance σ^2. We have tailored the center and scale to our own tastes. This change of variables makes the normal into a two-parameter family:

Definition. If Z is standard normal, then $X = \mu + \sigma Z$ is called a **Normal**(μ, σ^2) random variable.

Proposition. *If X is Normal(μ, σ^2), then $f(x) = \frac{1}{\sqrt{2\pi}\sigma} e^{-(x-\mu)^2/2(\sigma^2)}$.*

You should prove this by carrying out the change of variables.

We use this idea to avoid having to standardize a variable before we compare it to the normal curve; we might say, for instance, that for large λ a Poisson random variable is asymptotically Normal(λ, λ). This gives us another way to look at the approximation to the mass function: $p(x) \approx F_X(x + \frac{1}{2}) - F_X(x - \frac{1}{2}) = \int_{x-1/2}^{x-1/2} f(X)dX$, where X is Normal(λ, λ). But for large $\sqrt{\lambda}$ the normal density changes fairly slowly over this interval of width 1. Pretend that it is approximately constant, and write $p(x) \approx \int_{x-1/2}^{x-1/2} f(X)dX \approx f(x)$. In our example above, $p(60) \cong f(60) = \frac{1}{\sqrt{2\pi 8}} e^{-(60-64)^2/2\cdot64} = 0.04401$, which is actually (barely) better than the earlier approximation, for much less work. (Figure 10.11)

In case we have a random variable X believed to be in the normal family, with σ known but with μ unknown, it is easy to construct a confidence interval for μ. For example, for a 95% confidence interval we discover after much computing with standard normal probabilities that $P(Z > 1.96) = 0.025$. Then by the symmetry of Z about zero, $P(|Z| > 1.96) = 0.05$, and so $P(-1.96 \le Z \le 1.96) = 0.95$. But we asserted above that $Z = \frac{X-\mu}{\sigma}$, and substituting, we obtain $P(-1.96 \le \frac{X-\mu}{\sigma} \le$

1.96) $= 0.95$. You should check for yourself that this is algebraically equivalent to $P(X - 1.96\sigma \leq \mu \leq X + 1.96\sigma) = 0.95$, which defines a 95% confidence interval for the normal mean.

Notice, by the way, that this is very close to a 2–σ interval (see 6.7.1) with 2 replaced by 1.96. In fact, this will be our promised justification for this conventional interval—if the observations are approximately normal, then a 2–σ interval is an approximate 95% confidence bound. As we learn about many more cases in which random variables are asymptotically normal, the popularity of this crude device will seem better justified.

10.7.2 Approximate Poisson Intervals

At the moment, though, our only applications for this are to approximate gamma and Poisson confidence intervals:

Example. There were 38 muggings in a certain public park last month. Let us assume that these were all independent incidents, and so this might be an instance of a Poisson(λ) random variable. The mean rate λ is unknown, but presumably fairly large (of the order of magnitude of 38). Let us try to construct an approximate 95% confidence interval for λ, using the normal approximation above.

We immediately get in trouble, because we are supposed to know $\sigma = \sqrt{\lambda}$, and of course, we do not. But the last time we ran into this problem, in the binomial case (see 6.7.1 again), we simply estimated the standard deviation from the available data. So now let $\hat{\lambda} = x$, and $\hat{\sigma} = \sqrt{\hat{\lambda}} = \sqrt{38}$. The square root changes slowly as X varies, so this involves less relative variation than does the variation in X that we are studying. We have an approximate confidence interval for the Poisson mean:

$$P(X - 1.96\sqrt{X} \leq \lambda \leq X + 1.96\sqrt{X}) \approx 0.95.$$

In our mugging example, the roughly 95% confidence limits on λ are $25.92 \leq \lambda \leq 50.08$ muggings per month, on average.

We can compare this to the exact 95% confidence interval for λ, corresponding to the method used in the binomial case (see 6.7.3 and Exercise 6.42). After a great deal of calculation using a handy computer program, I obtain $P(X \leq 38|\lambda = 52.16) = 0.025$ and $P(X \geq 38|\lambda = 26.89) = 0.025$. From this I conclude that an exact 95% confidence interval would be $26.89 \leq \lambda \leq 52.16$, which is fairly close to our normal approximate interval.

Perhaps a better, and certainly easier, way to look at the value of our approximate interval would be to see how often it would be violated. We get $P(X \geq 38|\lambda = 25.92) - 0.0153$ and $P(X \leq 38|\lambda = 50.08) - 0.0463$; combining the two tail probabilities, we find that our normal interval is a 93.8% confidence interval, instead of the 95% we advertised. This 0.938 is called a *coverage probability*. Approximate intervals would, of course, tend to get closer to exact as the X's that we observe in our problems gets larger, because the error estimates in the normal approximation to the Poisson get smaller.

10.7.3 Approximate Gamma Intervals

Example. We can reason in much the same way about gamma variables. From the last chapter, consider the observed lifetimes of 20 desert mice (see 9.6.2), which were assumed to be each Gamma($\alpha = 5, \beta$) days. The interesting new wrinkle here was that we had a sample of $n = 20$ observations rather than just a single one. We were, with considerable calculation, able to construct a 99% confidence interval for the average lifetime $\alpha\beta$ of the mice. If a good normal approximation to that interval could be found, it might save us quite a bit of trouble. The mean of the grand total of lifetimes, observed to be 7742, is $n\alpha\beta$, where $n = 20$ is the sample size. Its standard deviation is then $\sqrt{n\alpha}\beta$.

For a normal 99% confidence interval, we shall need a value of Z that is exceeded half of 1% of the time. Quantities like that are used so often that we have a notation for them: z_α shall stand for the solution of $P(Z > z_\alpha | Z$ standard normal$) = \alpha$. Thus $z_{0.025} = 1.96$. After much calculation (or more likely, getting a table or computer statistical package to extract it for us), we obtain $z_{0.005} = 2.576$. Generally, if X is Normal(μ, σ^2), then a $1 - \alpha$ confidence interval is given by $P(X - z_{\alpha/2}\sigma \leq \mu \leq X + z_{\alpha/2}\sigma) = 1 - \alpha$.

Example (*cont.*).   We find for the lifetime data that

$$P\left(\sum_{i=1}^{n} X_i - z_{\alpha/2}\sqrt{n\alpha}\beta \leq n\alpha\beta \leq \sum_{i=1}^{n} X_i + z_{\alpha/2}\sqrt{n\alpha}\beta\right) \approx 1 - \alpha$$

for $\sqrt{n\alpha}$ large. To put this in the form of a confidence interval for the mean lifetime of an individual mouse, divide all three sides by n to get

$$P\left(\bar{X} - z_{\alpha/2}\frac{\sqrt{\alpha}\beta}{\sqrt{n}} \leq \alpha\beta \leq \bar{X} + z_{\alpha/2}\frac{\sqrt{\alpha}\beta}{\sqrt{n}}\right) \approx 1 - \alpha.$$

We again use the obvious crude device for estimating the unknown standard deviation of each lifetime $\sqrt{\alpha}\beta$; we replace its β by the observed value that is supposed to have β as an average, $\hat{\beta} = \frac{7742}{20\times 5} = 77.42$. We finally have an approximate 99% confidence interval $387.1 - 99.7 \leq \alpha\beta \leq 387.1 + 99.7$, which is $287.4 \leq$ mean life ≤ 486.8.

This may be compared to the exact confidence interval from 303 to 509 days; or we may check that $P(X \geq 7742 | \alpha\beta = 287.4) = 0.00077$ and $P(X \leq 7742 | \alpha\beta = 486.8) = 0.01488$. Thus the coverage probability of our approximate 99% interval is 98.43%, which is not too bad.

Notice that for a value of α fixed by the models we are studying, the approximate normality depends on n, since we can estimate the quality of the approximation from the size of $\sqrt{n\alpha}$. That is, normal confidence intervals in this sort of problem are more accurate if we take larger independent samples. Restate the interval as

$$P\left(\bar{X} - z_{\alpha/2}\frac{\sigma}{\sqrt{n}} \leq \mu \leq \bar{X} + z_{\alpha/2}\frac{\sigma}{\sqrt{n}}\right) \approx 1 - \alpha,$$

where we actually have to go one step further and estimate σ from the data. This should remind you of our result (statistics of the sample mean) that the expected value of $\bar{X}$ is just the expectation μ of a single observation. Furthermore, the standard deviation of $\bar{X}$, called the standard error of the mean, is $\frac{\sigma}{\sqrt{n}}$, where σ is the standard deviation of a single observation (see 7.6.4). It will turn out to be very often true that sample means are approximately normal random variables for large enough n; theorems that say things like this are called *central limit theorems*. In such cases, approximate confidence intervals for the population mean may take the form above. We will see several other applications in later chapters.

10.8   Summary

The *quantile function* $Q(u)$ for any random variable is a way to generate one of its values by transforming a Uniform(0, 1) variable; for continuous variables it reduces to the inverse of the cumulative distribution function, $Q = F^{-1}$ (2.3). A completely general definition of an expectation for one random variable is then $E[g(X)] = \int_0^1 g[Q(U)]dU$, whether the variable is discrete or continuous (3.1). We went on to discover how densities may be used to find expectations for absolutely continuous random variables, $E[g(X)] = \int g(X)f(X)dX$ (4.2).

Continuing our program of finding good approximate calculation methods for when our favorite families become hard to work with, we discovered the *standard normal density* $f(z) = \frac{1}{\sqrt{2\pi}}e^{-z^2/2}$ (5.3). For gamma random variables with large α, we found that $Z = (T - \alpha\beta)/(\beta\sqrt{\alpha})$ has approximately standard normal probabilities (5.5). By duality, a normal approximation may be constructed for the Poisson family with large λ, $Z = \frac{X+1/2-\lambda}{\sqrt{\lambda}}$ (6.2). As usual, our asymptotic method amounts to a family of random variables of great interest in itself, the *normal* family, with densities $f(x) = \frac{1}{\sqrt{2\pi}\sigma}e^{-(x-\mu)^2/(2\sigma^2)}$ with mean μ and standard deviation σ (7.1). We studied some of its basic properties and investigated how to calculate its probabilities. Normal confidence intervals were particularly easy to construct, $P(X - z_{\alpha/2}\sigma \le \mu \le X + z_{\alpha/2}\sigma) = 1 - \alpha$, and so let us build simple approximate confidence intervals for gamma and Poisson parameters, and for the mean of any approximately normal random variable X from which we have a sample of n,

$$P\left(\bar{X} - z_{\alpha/2}\frac{\sigma}{\sqrt{n}} \le \mu \le \bar{X} + z_{\alpha/2}\frac{\sigma}{\sqrt{n}}\right) \approx 1 - \alpha \quad (7.3).$$

10.9   Exercises

1. Write down the quantile function for a binomial B(5, 0.4) random variable.
2. Write down the quantile function for a Beta(7, 1) random variable. Find its population median and its 25th percentile.

3. Using the quantile function, find a formula for the expectation and variance of Beta(α, 1) random variables.
4. Prove that expectation is a linear operator and is nonnegative for nonnegative g.
5. Derive a formula for how the density transforms when g is *nonincreasing*. Use it to verify the monotone change of variables formula.
6. Show that if Y is an F(α, β) random variable, then $1/Y$ is an F(β, α) random variable.
7. A logistic random variable X has cumulative distribution function

$$F(x) = 1/(1 + e^{-x}).$$

   a. Find the density of this logistic random variable.
   b. Compute the probability that this logistic random variable will fall between -1 and $+2$.
   c. Find the density of the random variable $Y = e^X$. Be sure to state its range.

8. Remember that a Gamma(α) random variable T has density $\frac{T^{\alpha-1}}{(\alpha-1)!}e^{-T}$ for $\alpha > 0$ on the sample space $T > 0$. Then the random variable $X = \log(T)$ is a (generalized) *Gumbel* density with parameter α. Find the density, sample space, and mode (point at which the density is greatest) of X.
9. Prove the proposition about location and scale changes in a density (see 4.1).
10. Find the expectations of (a) Gamma(α, β), and (b) Beta(α, β) random variables.
11. Find the variances of (a) Gamma(α, β), and (b) Beta(α, β) random variables.
12. Find the expectation, variance, and standard deviation of the random variable X in Exercise 12 of Chapter 9.
13. Let T be a Gamma(α, β) random variable. Then $S = 1/T$ is called an *Inverse Gamma*(α, β) random variable (appropriately enough).

   a. Find the sample space and density of S.
   b. Find the expectation and variance of S.

14. Remember that a Gamma(α, β) random variable may be thought of as the time until the αth unpredictable event when the average time between events is β. Let T be a Gamma(4, 3) random variable and S be a Gamma(6, 5) random variable. They are independent of one another. Now consider the random variables $X = 2T + S$ and $Y = 5T - 7S$. What are the expected values and variances of X and of Y, and what is the covariance of X and Y? (Assume that the results of Chapter 7 apply to continuous random variables. We will make sure of that in the next chapter.)
15. The exponent of a Gamma($a + 1$) density turned out to be $a \log(t) - t$. Make the change of variables $Z = \frac{T-a}{\sqrt{a}}$, and calculate the logarithm of the resulting density. Show that its maximum value is at $z = 0$, and that its second derivative there is -1.
16. Check that $\frac{1}{1+t} = 1 - t + \frac{t^2}{1+t}$.

17. Using the inequality for the natural logarithm in Section 5.2, find upper and lower bounds for log(1.5) and log(0.9). Compare them to the exact value.

18. Find using L'Hospital's rule $\lim_{z \to \infty} z e^{-z^2/2}$ and $\lim_{z \to \infty} z e^{-z^2/2}$.

19. If Z is standard normal, show that $F(-z) = 1 - F(z)$.

20. Calculate 25! exactly. Now calculate it approximately by Stirling's formula and compare.

21. **a.** Compute a formula for the probability that a fair coin tossed $2n$ times will get exactly half heads (n heads).

    b. Using Stirling's formula, find an approximation to (a) for large n that has no factorials in it. (You should simplify it where possible.)

    c. The lottery draws numbers from 1 to 40. Compute (exactly) the probability that half of the next 20 lottery numbers drawn (with replacement) are even. Now use (b) to recompute this probability approximately and compare.

22. For the standard normal random variable, calculate $F(-1.5)$ to within 0.0001. Write down an upper and lower bound that proves it is this accurate.

23. Carry out the standard normal approximation method for computing the answer to Exercise 24 of Chapter 9.

24. At a large allergy center, people with severe allergy to bee sting appear for treatment unpredictably, at an average rate of 5 per month. You need to find 20 patients for a research project.

    a. What is the exact probability that you will get your 20th patient within 5 months?

    b. Recalculate this probability using the normal approximation. Compare.

25. Use the proposition that $1 - \frac{1}{\sqrt{1+a}} \approx \frac{a}{2}$ to find approximate values for $\sqrt{2}$ and $\sqrt{0.85}$; compare them to the exact values.

26. As a seller of extremely expensive Swiss watches, you find that you average 28 sales in a normal month.

    a. How probable is it that you will sell fewer than 24 in the next normal month?

    b. Is a normal approximation appropriate here? Why or why not? Use one of the methods you have learned for calculating normal probabilities to prove that the normal approximation to the answer to (a) is less than 0.25

27. An amateur astronomer sees an average of 36 meteors in an hour of observation on an August night.

    a. What is the probability that in a given hour on an August night she will see at least 32?

    b. What is the probability she will see exactly 39?

    c. Recalculate (a) and (b) using normal approximations and compare.

28. In Exercise 17 from Chapter 9, construct the normal approximation to a 95% confidence interval on the average time between accidents β. What is the actual coverage probability of this interval?

10.10   Supplementary Exercises

29. Show that any quantile function is nondecreasing (if $a < b$, then $Q(a) \le Q(b)$).

30. A *congruential* pseudo-random number generator might work as follows: p is a positive integer called the *modulus*, m a positive integer called the *multiplier*, and x_0 a positive integer called the *seed*. Then we repeatedly calculate $x_i = mx_{i-1} \bmod p$, where *mod* means divide the first term by the second and take the remainder ($13 \bmod 5 = 3$). Now turn it into a Uniform(0, 1) pseudo-random number by letting $U_i = x_i/p$. With an intelligent choice of the modulus and the multiplier, the x's will run in an unpredictable order through the numbers from 1 to $p - 1$, and the U's will appear quite random.

    For example, let the modulus be 257 and the multiplier be 49. The seed is less important; let us use 180. Then x_1 is the remainder when we divide 257 into 180×49, or 82. Then $U_1 = 0.319$. Calculate U_2 through U_{11}.

31. Use the uniform pseudo-random numbers from Exercise 28 to generate

    a. 10 B(5, 0.4) pseudo-random numbers, using the quantile function from Exercise 1.
    b. 10 Beta(7, 1) pseudo-random variables, using the quantile function from Exercise 2.

32. Let T be a Gamma(α) random variable; let $Y = T^{1/3}$.

    a. Find the density of Y. (This is called a *symmetrizing* transform.)
    b. Graph the density of T and the density of Y for $\alpha = 10$ to see that the latter is more nearly symmetric about its maximum.

33. Let Z have the standard normal density

$$f(z) = \frac{1}{\sqrt{2\pi}} e^{-z^2/2}.$$

    Find the density of $Y = Z^3$.

34. Prove that $z = (t - \alpha)/\sqrt{\alpha}$ is close to $z^* = (t - (\alpha - 1))/\sqrt{\alpha - 1}$ when z is small compared to α.

35. Prove that our power series for the standard normal cumulative distribution is absolutely convergent for any value of z (you may have to review how such things are done from advanced calculus).

36. Apply our method of quadratic approximation to the log-density to establish conditions under which a Beta(α, α) is approximately normal.

37. Here is an accurate and fairly fast way to generate approximately standard normal random variables, using the results of Exercise 36: (1) generate 23 independent Uniform(0, 1) variates; (2) find their median P; and (3) compute $Z = 10 \times (P - 0.5)$.

    Find the exact value of $P(Z > 2)$. Now compare it to the standard normal probability for the same inequality.

38. Highly prized morel mushrooms grow unpredictably and one at a time in Midwestern woods in May (though there are a great many of them to be found). Even an expert hunter averages five minutes between finds.

   a. What is the probability that our expert will find her fourth mushroom within 16 minutes? What model are you using, and why is it plausible?

   b. She has promised a French restaurant that she will deliver twelve dozen (144) mushrooms. She begins hunting at 7:00 A.M. What is the probability that she will complete her promised shipment by the time it is dark, at 8:00 P.M? If you find the calculation time-consuming, you may use a good approximate method; if so, explain why your approximation is appropriate.

39. Remember that the density of a Normal(μ, σ^2) random variable X is $1/(\sqrt{2\pi}\sigma)e^{-(X-\mu)^2/(2\sigma^2)}$. Then the random variable $Y = e^X$ is called a **lognormal** random variable.

   a. What is the density of Y?

   b. What is E(Y)? (**Hint:** E(Y) = E(e^X).)

   c. What is Var(Y)?

40. Derive the general series (the *Laplace* series) for the tail probability of the standard normal distribution. Notice that it does not converge for $z = 4$.

41. a. Using the algebraic trick from (6.1), derive the *continued fraction*

$$\sqrt{1+\alpha} = 1 + \cfrac{a}{2 + \cfrac{a}{2 + \cfrac{a}{2 + \cfrac{a}{2 + \cdots}}}}.$$

   b. Show that stopping after odd and even numbers of fractions gets under- and overestimates of the limiting value, for $a > 0$.

   c. Use the continued fraction to calculate $\sqrt{1.2}$ to six significant figures.

42. In the lifetimes of mice data from (9.6.2), we also assumed that $\beta = 60$ was known, and we estimated α.

   a. Find a normal approximation to the 95% confidence interval for α.

   b. (Using a computer) find the coverage probability for the interval you constructed in (a).

CHAPTER 11

Continuous Random Vectors

11.1 Introduction

Here we develop some tools for dealing with several continuous random variables, simultaneously measured. First, we decide what we generally mean by an expectation in such a situation. Then, as an important special case, we explore how several uniform order statistics from the same sample might behave jointly. We go on to look at general properties of any multivariate density function. You will not be surprised to hear that a particularly important case will be the *multivariate normal* family. Sums of squares of independent standard normal vectors will lead us to consider gamma and beta random variables with *fractional* shape parameters. Then, our newfound expertise with more than one continuous variable will make for new Bayesian inference arguments, since now we can imagine absolutely continuous prior distributions on parameters that have a continuous range of possible values.

The normal approximation from Chapter 10 will turn out here to apply to an amazing variety of our favorite families of random variables, when certain parameters become painfully large. We will thereby establish why normal random variables are considered among the most important in all of statistics.

Time to Review

Determinants
Solving multiple integrals by change of variables
Chapter 7

11.2 Multivariate Expectations

11.2.1 Discrete Conditional Expectations

One tool we will surely need if we are to investigate continuous random vectors
is a general notion of what it means to take a multivariate expectation, such as
$E[g(X, Y)]$. For example, we will presumably be interested at some point in the
covariance of two such random measurements, $\mathrm{Cov}(X, Y) = E(XY) - E(X)E(Y)$.
There is a slick way to define such things, using the quantile function method of
the last chapter (see 10.3.1) (you might see whether you can think of it); but it is
not very helpful when we actually have to compute expectations. Instead, let us
start with a nice property of the discrete case:

$$E[g(X, Y)] = E_X\{E_{Y|X}[g(X, Y)|X]\}$$

(see 7.5.1). You should be careful to decode the complicated-looking notation here,
because this says something quite intuitively reasonable: To average a function over
a rectangular table (plane), first average it over each row (line) in which the first
coordinate is constant. Then average these averages over all such rows.

11.2.2 The General Case

Since it is so plausible and useful to do everything one coordinate at a time, we
will make it the definition in the general case, which includes continuous random
variables. From (10.2.1) we know what an expectation means for any random
variable, which tells us what a *marginal* expectation always is: $E_X[g(X)]$ is the
expectation of a function g of the variable X when we have no information about
any other variables. A *conditional* expectation $E_{X|Y}[g(X, y)|y]$ is no more than
the expectation of g as a function of the variable X when Y is known to be the
constant value y. In each case, there is only one random variable, so we may use
what we know about expectations to work with them.

One quibble may have occurred to you: In general, can we be sure that computing
using X first will get the same answer as using Y first? Surely, it should be so (it
is so in the discrete case), and we shall require it.

Definition. For any random vector (X, Y),

$$E[g(X, Y)] = E_X\{E_{Y|X}[g(X, Y)|X]\} = E_Y\{E_{X|Y}[g(X, Y)|Y]\}$$

whenever the two latter expressions exist and are equal.

The generalization to more than two coordinates is obvious (exercise). You
should do some simple examples as exercises, as well. You should check, as well,
that these expectations have the usual linearity and positivity properties.

One special case is particularly nice. Assume that X and Y are *independent*, and
we want to know $E[g(X)h(Y)]$. Notice that for each value of $Y = y$,

$$E_{X|Y}[g(X)h(y)|y] = h(y)E_{X|Y}[g(X)|y] = h(y)E_X[g(X)].$$

The first equality holds just because expectation is linear, and the second because independence just means that information about Y tells us nothing about X. Now apply the definition above: $E[g(X)h(Y)] = E_Y\{h(Y)E_X[g(X)]\} = E_X[g(X)]E_Y[h(y)]$. You may recall (see 7.6.3) that we showed in the special case of X and Y discrete that when X and Y are independent, they are also uncorrelated. Now we know that this is always true.

11.3   The Dirichlet Family

11.3.1   Two Order Statistics at Once

We shall now explore an important family of random vectors, making use of the above expressions. The coordinates of a random point, on a plane or in space, was our first example of a random vector (or, as we sometimes say, a multivariate random variable). Other, and often more interesting, examples arise when we measure several distinct numbers from a single experiment. Without realizing it, we have studied an important example already.

Example. Five friends finish the Boston Marathon. Since there are thousands of finishers, they rate themselves by the proportion of people they beat; so 0.74 means that 74% of the other finishers were slower. If the friends are typical contestants, we already know that the second-best result among them is approximately a Beta(4, 2) random variable (see 9.5.2), and the fourth-best is a Beta(2, 4) variate. But these two numbers are not independent. Say the fourth finished 0.56; then we know with certainty that the second finisher did better, so her finish must be greater than 0.56. Her conditional distribution can no longer be Beta(4, 2), which could be anything between 0 and 1.

Generally, let $U_{(1)} \leq U_{(2)} \leq \cdots \leq U_{(n)}$ be the *order statistics* from a Uniform(0, 1) random sample. Pick out $P = U_{(i)}$ and $Q = U_{(j)}$, where $i < j$, as the coordinates of the random vector of interest. We should be able to answer any questions we have using the cumulative distribution function $F(p, q) = P(P \leq p, Q \leq q)$. (see 7.3.2) We have

$$F(q, p) = P(i\text{th is at most } p, j\text{th is at most } q)$$

$$= P(\text{at least } i \text{ before } p \text{ and at least } j \text{ before } q)$$

$$= \sum_{Y=j}^{n} \sum_{X=i}^{Y} P(X \text{ before } p \text{ and } Y \text{ before } q)$$

$$= \sum_{Y=j}^{n} \sum_{X=i}^{Y} P\{X \text{ in } (0, p], Y - X \text{ in } (p, q], n - Y \text{ in } (q, 1)\}.$$

This is a multinomial probability with the probability of a point falling in each interval proportional to its length:

Proposition. *The ith and the jth Uniform(0, 1) order statistics P and Q from a random sample of n, $1 \leq i < j \leq n$, have the joint cumulative distribution*

$$F(p, q) = \sum_{Y=j}^{n} \sum_{X=i}^{Y} \frac{n!}{X!(Y-X)!(n-Y)!} p^X (q-p)^{Y-X} (1-q)^{n-Y},$$

where $0 < p < q < 1$.

This intimidating expression is necessary when you want actual probabilities. As an exercise (easier than it looks) derive the even more intimidating cumulative you get when you are interested in the joint distribution of more than two order statistics.

11.3.2 Joint Density of Two Order Statistics

Such quantities were at least easier to write when in the last chapter we used densities. We will get to that form using expectations, as in the last section. The connection is through something that should look a bit familiar.

Definition. The **indicator function** of a set A is $\chi_A(x) = \begin{cases} 1, & X \in A, \\ 0, & X \notin A. \end{cases}$

Then obviously $E(\chi_A) = P(A)$. (What is the connection with the method of indicators?) Now we can turn our cumulative function into an expectation: Denote the indicator function of $\{P \leq p, Q \leq q\}$ by χ_{pq}. Then $F(p, q) = E(\chi_{pq}) = E_P[E_{Q|P}(\chi_{pq}|P)]$.

To compute the inner expectation, let $P = p$ be fixed, and look at the random behavior of Q. If it has a density, then we will have $E_{Q|P}(\chi_{pq}|p) = P(Q \leq q|p) = \int_p^q f(Q|p)dQ$. We must have $n - i$ points scattered independently and uniformly over the interval $(p, 1)$, since the ith smallest is known to be at P. Therefore, Q, the jth order statistic, is the $(j - i)$th order statistic from these $n - i$ values. Let us transform Q into a random variable on $(0, 1)$ by $R = \frac{Q-p}{1-p}$. Now, R is a Uniform(0, 1) order statistic, so it is a Beta$(j - i, n + 1 - j)$ random variable, with density $\frac{(n-i)!}{(j-i-1)!(n-j)!} R^{j-i-1}(1-R)^{n-j}$. Change the variable back to Q (not forgetting the derivative of the transformation $1/(1-p)$), to get the density of Q conditioned on knowing P:

$$f(Q|P = p) = \frac{(n-1)!}{(j-i-1)!(n-j)!} \frac{(Q-p)^{j-i-1}(1-Q)^{n-j}}{(1-p)^{n-i}} \quad \text{on } (p, 1).$$

We already know the (marginal) distribution of P: Since it is an ith uniform order statistic, it is Beta$(i - 1, n - i)$. Then our double expectation expression gives us

$$F(p, q) = E_P[E_{Q|P}(\chi_{pq}|P)] = \int_0^p E_{Q|P}(\chi_{pq}|P)f(P)dP$$

$$= \int_0^p \left[\int_p^q f(Q|P)dQ \right] f(P)dP.$$

Now, Fubini's theorem from calculus (time to review) says that this integral of an integral is the same as an integral over a two-dimensional region:

$$F(p, q) = \int_0^p f(P) \int_p^q f(Q|P) dQ dP = \int_0^p \int_p^q f(P) f(Q|P) dQ dP.$$

Now we have a plausible candidate for the density of a random vector in two dimensions:

$$f(x, y) = f(x) f(y|x) = f(y) f(x|y).$$

In the beta example, we get

$$f(p, q)$$

$$= \frac{n!}{(i-1)!(n-i)!} p^{i-1}(1-p)^{n-i} \frac{(n-i)!}{(j-i-1)!(n-j)!} \frac{(q-p)^{j-i-1}(1-q)^{n-j}}{(1-p)^{n-i}}$$

$$= \frac{n!}{(i-1)!(j-i-1)!(n-j)!} p^{i-1}(q-p)^{j-i-1}(1-q)^{n-j} \text{ on } 0 < p < q < 1.$$

11.3.3 Joint Densities in General

Now, using the techniques of the last section, we can draw several conclusions. We begin with a definition.

Definition. The density of a random vector $\mathbf{X}$ is a function $f(\mathbf{X}) \geq 0$ such that for any event A, $P(A) = \int_A f(\mathbf{X}) d\mathbf{X}$. Any random vector with a density is said to be absolutely continuous.

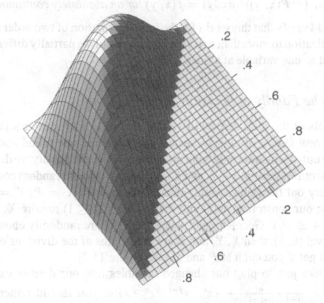

FIGURE 11.1. Joint density of 2nd and 4th of uniform sample of 5

Proposition. *For an absolutely continuous random vector* (X, Y), $f(x, y) = f(x)f(y|x) = f(y)f(x|y)$, *and analogously for more dimensions.*

Proposition.
For an absolutely continuous random vector $\mathbf{X}$, $E[g(\mathbf{X})] = \int g(\mathbf{X})f(\mathbf{X})d\mathbf{X}$.

Fubini's Theorem tells us that for absolutely continuous random vectors, we do not have to worry about the order of evaluation of that expectation. (In fact, we never do, but this fact we will leave for more mathematically advanced courses.)

While we are here, we can figure out the relationship between the cumulative distribution function and the density, corresponding to the ones we found before (see 9.4.3). One way comes straight from the definition:

$$F(x, y) = P(X \leq x \text{ and } Y \leq y) = \int_{-\infty}^{x} \int_{-\infty}^{y} f(X, Y)dYdX,$$

where we could have integrated, of course, in either order. As we did it, the first $-\infty$ stands for the lower limit of the x-coordinate of the sample space. But, be careful— the second $-\infty$ infinity is the lower bound of the y-coordinate, *conditional* on the current value of the x-coordinate. (Notice our example.)

Now go the other way, starting with the cumulative distribution function and trying to calculate the density, but this time using *partial* differentiation, first with respect to y: This letter does not appear in the x-limits, so calculus says we can differentiate under the integral sign: $(\partial F(x, y))/(\partial y) = \int_{-\infty}^{x} \frac{\partial}{\partial y} \int_{-\infty}^{y} f(X, Y)dYdX$. Then apply the fundamental theorem of calculus to the inner integral, to get $(\partial F(x, y))/(\partial y) = \int_{-\infty}^{x} f(X, y)dX$. We now differentiate with respect to x to get a simple formula:

Proposition. $(\partial^2 F(x, y))/(\partial x \partial y) = f(x, y)$ *for an absolutely continuous vector.*

You should verify that this works for the joint distribution of two order statistics. The generalization to more than two variables just involves partially differentiating with respect to one variable at a time.

11.3.4   The Family of Divisions of an Interval

Our result about the joint density of several uniform order statistics inspires us to define a new family of random vectors, not quite the one found above, but a simpler one out of which we may build order statistics. In the density we derived, the essential quantities were p, $q - p$, and $1 - p$; these will be our random coordinates. We will carry out the (multivariate) change of variables $X = P$, $Y = Q - P$, $Z = 1 - Q$; our constraints on Q and P ($0 \leq P \leq Q \leq 1$) require $X, Y, Z \geq 0$ and $X + Y + Z = 1$. Geometrically, since P and Q are randomly chosen points on the interval $(0, 1)$, then X, Y, and Z are the lengths of the divisions of the unit interval you get if you cut it at P and Q (see Figure 11.2).
We would like just to plug our changed variables into our density expression: $f(x, y, z) \stackrel{?}{=} \frac{n!}{(i-1)!(j-i-1)!(n-j)!}x^{i-1}y^{j-i-1}z^{n-j}$. But you should remember that previously a change of variables in the density required a factor, the derivative of

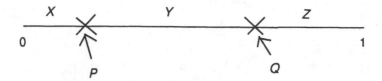

FIGURE 11.2. Random divisions of an interval

the transformation, that corresponded to a change of scale (see 10.4.1). We shall have to see what that means in this case.

11.4   Changing Variables in Random Vectors

11.4.1   Affine Multivariate Transformations

When we transformed a single variable, you will remember that we applied an increasing (or monotone) change of variables, looked at the simple way the cumulative distribution function changed, and differentiated to get the new density. You can try this in the vector case, as an exercise. But this approach has a serious limitation: You need a transformation that is monotone increasing or decreasing in each coordinate. By no means all, even very simple, transformations preserve the order of each coordinate. In the order statistic case, for example, knowing only that P and Q have increased, we just cannot say whether $Y = Q - P$ will have increased or not. Let us look for more general sorts of variable changes; we need ones that will, for example, take the original P and Q and change them to X, Y, and Z by a series of functions $X = P$, $Y = Q - P$, $Z = 1 - Q$. We will write it $\mathbf{X} = \mathbf{g}(\mathbf{Y})$. Notice that each term in this expression is a vector: $\mathbf{Y} = (Y_1, Y_2, \ldots, Y_k)^T$, $\mathbf{X} = (X_1, X_2, \ldots, X_k)^T$, and the equality relates the X's to a vector of functions $X_i = g_i(Y_1, Y_2, \ldots, Y_k)$.

This should remind you of a *multivariate change of variables* from advanced calculus (which you should review). As in the one-variable case, we need to adjust by a factor that reflects the fact that *volume* is measured differently in terms of $\mathbf{X}$ and of $\mathbf{Y}$. When you were first doing integrals, $d\mathbf{X}$ was thought of as the limit of the volumes of tiny little cubes

$$(x_1, x_1 + dx_1] \times (x_2, x_2 + dx_2] \times \cdots \times (x_k, x_k + dx_k].$$

You then multiplied that volume by a height $f(x)$ and summed over all the cubes that "tile" the region of integration, to approximate the integral. What volume in terms of $\mathbf{Y}$ does that correspond to?

We will concentrate for a while on the easy case that we need here, when the change of variables is *affine*. This says that it may be written $\mathbf{x} = \mathbf{g}(\mathbf{y}) = \mathbf{a} + \mathbf{G}\mathbf{y}$; where $\mathbf{G}$ is a square matrix. (We are assuming that the vectors $\mathbf{X}$ and $\mathbf{Y}$ are of the same dimension.) In the order-statistic problem we are working on, $\mathbf{a} = \mathbf{0}$ and

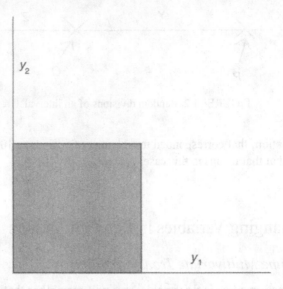

FIGURE 11.3. Cube I

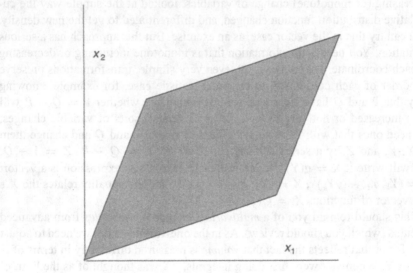

FIGURE 11.4. Parallelogram $G(I)$

$G = \begin{pmatrix} 1 & 0 \\ 1 & 1 \end{pmatrix}$. (You should verify that indeed $G \begin{pmatrix} X \\ Y \end{pmatrix} = \begin{pmatrix} X \\ X+Y \end{pmatrix} = \begin{pmatrix} P \\ Q \end{pmatrix}$;

we do not need Z, which is determined by the other variables.)

What effect does this change of variables have on the volumes of those little cubes? Notice that an affine function changes a cube into a multidimensional version of a parallelogram (called a *parallelepiped*). (See Figures 11.3–4.)

That is, if the unit cube is $I = (0, 1] \times (0, 1] \times \cdots \times (0, 1]$, then the corresponding parallelepiped is $G(I) = \{$vectors Gy when $y \in I\}$. You may check from the definition of matrix multiplication that the edges of the parallelepiped are the columns of G. You should recall the following fact from matrix algebra:

Proposition. *The volume* $V[G(I)] = |\det(G)|$, *the absolute value of the determinant of* G.

We call that volume $J_g = V(GI) = |\det(G)|$ the *Jacobian* of the affine transformation g. The one fact most people can remember about determinants is how to compute the 2-by-2 case: $\det \begin{pmatrix} a & b \\ c & d \end{pmatrix} = ad - bc$. If $\det(G) = 0$, we say that G is *singular*; this means that our cube was flattened by g into a lower-dimensional subspace. We will exclude that possibility. Again from multivariable calculus, we learn the following:

Proposition. *Let a transformation* $X = g(Y) = a + GY$ *be nonsingular. Then*

$$\int_{g(A)} f(X)dX = \int_A f[g(Y)]J_g dY,$$

where $g(A) = \{$*vectors* $g(y)$ *when* $y \in A\}$. [We see that the Jacobian corrects for the "stretching" of the region over which we are integrating by the transformation].

If f_X is the multivariate density of X, the proposition immediately tells us what the density f_Y of Y must be. Since $P(Y \in A) = P(X = g(Y) \in g(A))$, then $\int_{g(A)} f_X(X)dX = \int_A f_Y(Y)dY$. But the proposition tells us what the connection is:

Theorem (affine change of variables in a density). *Let* $X = g(Y) = a + GY$ *be nonsingular. Then* $f_Y(Y) = f_X[g(Y)]J_g$.

As an exercise, you should show that this does what we expect (see 10.4.1) for the one-variable transformation $X = a + bY$.

11.4.2 Dirichlet Densities

In our order-statistic example, $\det G = 1 \times 1 - 1 \times 0 = 1$; the area scale $J_g = 1$ does not change. Our guess about the density in Section 3.4 happened to be correct. We say that this particular change of variables is *volume-preserving*. Now let us write down the density of $P = (P_1, P_2, \cdots, P_k)^T$ where P_i is the n_ith order statistic of a uniform sample of n on $(0, 1)$, and $1 \leq n_1 < \cdots < n_i < \cdots < n_k \leq n$. You should discover as an exercise, by doing repeatedly what we did in Section 3.2, the following:

Proposition. *The joint distribution of the* $\{n_i\}$*th order statistics from a Uniform*$(0, 1)$ *sample of* n *is*

$$f(p) = \frac{n!}{(n_1 - 1)! \prod_{i=2}^{k}(n_i - n_{i-1} - 1)!(n - n_k)!} p_1^{n_1-1}$$

$$\times \prod_{i=2}^{k}(p_i - p_{i-1})^{n_i-n_{i-1}-1}(1 - p_k)^{n-n_k},$$

where $0 < p_1 < p_2 < \cdots < p_k < 1$.

Once again, we can clean up this expression by a transformation that singles out the divisions of the unit interval: $X_1 = P_1, X_2 = P_2 - P_1, \ldots, X_i = P_i - P_{i-1}, \ldots, X_{k+1} = 1 - P_k$. Notice that $X_i \geq 0$. Now add all the expressions together, and watch each of the P's cancel out in turn to get $\sum_{i=1}^{k+1} X_i = 1$. In order to find the density of $\mathbf{X}$, we will need to compute the Jacobian, and in order to compute the Jacobian, we need to solve for the P's. Solving for the P's one at a time, we get $P_1 = X_1, P_2 = X_1 + X_2$, and generally $P_i = \sum_{j=1}^{i} X_j$. Notice that we do not need X_{k+1}; the transformation is really k variables to k variables. Then the transformation is $\mathbf{P} = \mathbf{GX}$, where

$$
\mathbf{G} = \begin{pmatrix} 1 & 0 & \cdots & 0 \\ 1 & 1 & 0 & \cdots \\ \vdots & & \ddots & \\ 1 & 1 & \cdots & 1 \end{pmatrix}.
$$

It has 1's on the diagonal and below the diagonal, and zeros above the diagonal. Now, you probably still have nightmares about how complicated it was to calculate determinants for a matrix bigger than 2 by 2. But (time to dig into the matrix algebra book again) there was an important exception: *Triangular* matrices, with all zeros below the diagonal, or all zeros above the diagonal, have a very simple determinant. It is just the product of the diagonal elements. In our problem, we just multiply the 1's together to get det $\mathbf{G} = 1$. Our transformation is volume-preserving once again. The density has the simple form

$$
f(\mathbf{x}) = \frac{n!}{(n_1 - 1)! \prod_{i=2}^{k} (n_i - n_{i-1} - 1)!(n - n_k)!} x_1^{n_1 - 1} \prod_{i=2}^{k} x_i^{n_i - n_{i-1} - 1} x_{k+1}^{n - n_k}.
$$

Finally, we are ready to define the Dirichlet family of random vectors:

Definition. A random $(k + 1)$ vector $\mathbf{X}$ is **Dirichlet**(α) for $\alpha_i > 0$ if its sample space is $X_i > 0$ and $\sum_{i=1}^{k+1} X_i = 1$, and its density is

$$
f(\mathbf{x}) = \frac{\left[\left(\sum_{i=1}^{k+1} \alpha_i \right) - 1 \right]!}{\left(\prod_{i=1}^{k+1} (\alpha_i - 1)! \right)} \prod_{i=1}^{k+1} x_i^{\alpha_i - 1}.
$$

11.4.3 Some Properties of Dirichlet Variables

The vector $\mathbf{X}$ is really only a k-dimensional random vector because of the sum condition. (If you want to integrate it or graph it, get the dimension right by some such replacement as $X_{k+1} = 1 - \sum_{i=1}^{k} X_i$.) Notice that this big family includes the beta family as a special case: $X_1 = P, X_2 = 1 - P, \alpha_1 = \alpha$, and $\alpha_2 = \beta$. The beta family, of course, has reversal symmetry, which amounts to interchanging X_1 and X_2 and at the same time interchanging α_1 and α_2. The simple form of the Dirichlet density tells us that we have a complete **permutation** symmetry: We may interchange any X_i and X_j, so long as we interchange α_i and α_j at the same time, and the density is the same (just stare at the density formula).

To compute expectations, notice that X_1 is the αth order statistic from a uniform sample on $(0, 1)$ of size $(\sum_{i=1}^{k+1} \alpha_i) - 1$; therefore, it has a Beta$(\alpha_1, \sum_{i=2}^{k+1} \alpha_i)$ distribution. We therefore know its expectation and variance from (10.4.2). But we can interchange X_1 and any other X_i to get a general result:

Proposition. *Let* **X** *be Dirichlet*(α). *Then*

(i) $E(X_i) = (\alpha_i)/(\sum_{i=1}^{k+1} \alpha_j)$.

(ii) $Var(X_i) = (\alpha_i \sum_{j=1} \alpha_j)/((\sum_{i=1}^{k+1} \alpha_j)^2[(\sum_{i=1}^{k+1} \alpha_j) + 1])$.

We may as well try out our theory of multivariate expectation to find the covariance of two Dirichlet coordinates (you will do it another way as an exercise). We will use the formula $Cov(X_i, X_j) = E(X_i X_j) - E(X_i)E(X_j)$, which is still valid for our multivariate expectation, which we require only to be linear.

$$E(X_i X_j) = \int X_i X_j \frac{[(\sum_{l=1}^{k+1} \alpha_l) - 1]!}{\prod_{l=1}^{k+1}(\alpha_i - 1)!} \prod_{l=1}^{k+1} X_l^{\alpha_l - 1} dX$$

$$= \int \frac{[(\sum_{i=1}^{k+1} \alpha_i) - 1]!}{\prod_{i=1}^{k+1}(\alpha_i - 1)!} X_i^{\alpha_i} X_j^{\alpha_j} \prod_{i \neq i, j} X_l^{\alpha_l - 1} dX$$

$$= \frac{\alpha_i \alpha_j}{[(\sum_{l=1}^{k+1} \alpha_i) + 1](\sum_{i=1}^{k+1} \alpha_i)}$$

$$\times \int \frac{[(\sum_{l=1}^{k+1} \alpha_i) + 1]!}{\alpha_i! \alpha_j! \prod_{l \neq i,j}(\alpha_l - 1)!} X_i^{\alpha_i} X_j^{\alpha_j} \prod_{l \neq i, j} X_l^{\alpha_l - 1} dX.$$

The integrals are over the Dirichlet sample space, and the last integral is equal to 1, because it is of a Dirichlet density with two of the parameters increased by 1. Our expectation is just the constant in front, and after a little algebra (which you should do), we simplify the expression:

Proposition. *Let* X *be Dirichlet*(α). *Then*

$$Cov(X_i, X_j) = -\frac{\alpha_i \alpha_j}{(\sum_{l=1}^{k+1} \alpha_i)^2[(\sum_{l=1}^{k+1} \alpha_i) + 1]}.$$

The negative covariance says that the bigger one random piece of the interval is, the smaller any others tend to be.

We may apply this to the distribution of order statistics. If we are interested in the ith and jth uniform order statistics, $i < j$, the corresponding random intervals are Dirichlet$(i, j - i, n + 1 - j)$, with $P_i = X_1$ and $P_j = X_1 + X_2$. Then

$$Cov(P_i, P_j) = Cov(X_1, X_1 + X_2) = Var(X_1) + Cov(X_1, X_2)$$

using a result from Chapter 7 (see 7.5.4). You should check the following as an exercise:

Proposition. $Cov(P_i, P_j) = i(n + 1 - j)/((n + 1)^2(n + 2))$ *for* $i < j$.

We learn that uniform order statistics are always positively correlated, with the covariance larger the closer together they are in the ordered sample.

11.4.4   General Change of Variables

So far, our multivariate changes of variable have been rather simple affine transformations. We need sometimes to do changes of variable that are not so simple. For a possibly nonaffine transformation $X = g(Y)$, we need to learn about the density of Y from knowing the density of X. That is, for all events A, we need $P(Y \in A)$. But in general, the only thing we can be sure of is that $P(Y \in A) \leq P[X = g(Y) \in g(A)]$, because every vector Y on the left certainly corresponds to exactly one $X = g(Y)$ on the right. But it may not be an equality, because there may be Y's outside of A for which $g(Y) \in g(A)$. For example, if $g(y) = y^2$ and $A = (1, 2)$, then $g(A) = (1, 4)$. But $g(-1.5) = 2.25 \in g(A)$. This one-variable transformation is not monotone (see 10.4.1). We have to exclude these transformations:

Definition. A transformation $X = g(Y)$ is (**essentially**) **one-to-one** if you cannot find two vectors $Y_1 \neq Y_2$ for which $g(Y_1) = g(Y_2) = X$, except possibly for a set of X's whose total probability is zero.

We have disallowed all those cases $Y \notin A$ that worried us, because now we are sure that $g(Y) \notin g(A)$, unless they contribute nothing to probability calculations. Notice that an affine transformation is obviously one-to-one for G nonsingular; you learned in matrix algebra that such a G has an inverse. Then for any X there is only one $Y = G^{-1}(X - a)$ that is transformed to it.

Proposition. *If a transformation* $X = g(Y)$ *is essentially one-to-one, then*

$$P(Y \in A) = P[X = g(Y) \in g(A)].$$

Multivariate calculus also tells us how to do the change of variables as in Section 4.1 for many nonlinear cases. If the transformation is *differentiable*, then it is approximately an affine transformation near any given point y_0 : $g(y) \approx g(y_0) + G_g(y_0)(y - y_0)$. G is the matrix $G(y) = (\partial g_i(y)/\partial y_j)_{ij}$ of derivatives. The approximation means that if the Euclidean length of $y - y_0$ is small enough, then the Euclidean length of $g(y) - g(y_0)$ is much smaller still. Thus, the adjustment in the volume scale right at the point y_0 is given by $J_g(y_0) = |\det G(y_0)|$. Then the general multivariate calculus change-of-variables rule is this:

Proposition. *For a one-to-one differentiable transform* g, $\int_{g(A)} f(X)dX = \int_A f[g(Y)]J_g(Y)dY$.

As in the last section, this tells us exactly how to change variables in a density:

Theorem (multivariate change of variables in a density). *Let* $X = g(Y)$ *be essentially one-to-one and differentiable. Then* $f_Y(Y) = f_X[g(Y)]J_g(Y)$.

We have already seen one important example: In (10.5.3) we transformed an integral to polar coordinates. As an exercise, you should check that we could have accomplished the same thing with this theorem.

11.5   The Chi-Squared Distribution

11.5.1   Gammas Conditioned on Their Sum

We now want to use the technique of the last section to prove for a second time a proposition from Chapter 9 (see 9.5.5). Why? Each novel demonstration of an important result gives us more insight. Furthermore, as we shall see shortly, the new proof will make our result apply to many more cases.

Let X be Gamma(α) and Y be independently Gamma(β). We will discover once again the distribution of $Z = X + Y$, this time using transformation techniques. Since we have a two-dimensional vector, to use the Jacobian method we will need something in addition to Z (the vectors need to have the same dimension). We have many choices, but sneaking a look at Chapter 9 again, we choose $W = X/(X+Y)$, which is no longer linear. Notice that $Z > 0$ and $0 < W < 1$. Solve these for X and Y to get the transformation $X = ZW$ and $Y = Z(1 - W)$. How will this affect our density?

Then

$$G = \begin{pmatrix} \frac{\partial ZW}{\partial Z} & \frac{\partial ZW}{\partial W} \\ \frac{\partial Z(1-W)}{\partial Z} & \frac{\partial Z(1-W)}{\partial W} \end{pmatrix} = \begin{pmatrix} W & Z \\ 1 - W & -Z \end{pmatrix},$$

$$J = |\det G| = |-ZW - Z(1 - W)| = |-Z| = Z,$$

since Z is positive. It is *not* area-preserving.

Now we need the joint density of X and Y. For any two independent random variables, $F(x, y) = F_X(x)F_Y(y)$. Therefore,

$$f(x, y) = \frac{\partial^2 F(x, y)}{\partial x \partial y} = \frac{\partial F_X(x)}{\partial x} \frac{\partial F_Y(y)}{\partial y} = f_X(x)f_Y(y).$$

To say the same thing, $f_{Y|X}(y|x) = f_Y(y)$ for each value x. It works backwards as well: If the joint density is the product of the marginal densities, you can integrate twice to discover that the joint cumulative is the product of marginal cumulatives, and therefore the coordinates are independent.

For our problem, this becomes $f(x, y) = (x^{\alpha-1})/(\alpha-1)!e^{-x}(y^{\beta-1})/(\beta-1)!e^{-y}$. Substituting and including the Jacobian factor, we obtain

$$f(z, w) = z\frac{(zw)^{\alpha-1}}{(\alpha - 1)!}e^{-zw}\frac{[z(1 - w)]^{\beta-1}}{(\beta - 1)!}e^{-z(1-w)} = z^{\alpha+\beta-1}e^{-z}\frac{w^{\alpha-1}(1 - w)^{\beta-1}}{(\alpha - 1)!(\beta - 1)!}$$

after collecting terms. The z-terms are a gamma density except for a constant, $(\alpha + \beta - 1)!$, so we divide and multiply by that constant to get

$$f(z, w) = \frac{z^{\alpha+\beta-1}}{(\alpha + \beta - 1)!}e^{-z}\frac{(\alpha + \beta - 1)!}{(\alpha - 1)!(\beta - 1)!}w^{\alpha-1}(1 - w)^{\beta-1}.$$

It is a product of familiar densities. We conclude (no surprises here) that $Z = X+Y$ is Gamma($\alpha + \beta$), $W = X/(X + Y)$ is Beta(α, β), and the two are independent.

11.5.2   Squared Normal Variables

But this new turn on an old result will solve a fundamental problem for us. The simplest case of the chi-squared statistic from Chapter 8 (see 8.4.3) evaluated the closeness of fit of a contingency table to an independent Poisson model with means λ_i, $i = 1, \ldots, k$: $\chi^2 = \sum_{i=1}^{k}(X_i - \lambda_i)^2/\lambda_i$ where X_i is the random count in the ith cell. If the $\sqrt{\lambda_i}$ are all large, then we might reasonably expect that the standard normal approximations $Z_i = (X_i - \lambda_i)/\sqrt{\lambda_i}$ could be used (the continuity correction is controversial in this application; we shall leave it out). Then $\chi^2 = \sum_{i=1}^{k} Z_i^2$ and it is approximately a sum of squares of independent standard normal random variables. So what is its distribution?

First, let Z be standard normal and let $Y = Z^2$. We cannot quite use our transformation theory (see 10.4.1), because the square is neither nondecreasing nor nonincreasing. As an intermediate step, let $W = |Z|$. Its cumulative distribution function is

$$F_W(w) = P(W \le w) = P(-w \le Z \le w) = F_Z(w) - F_Z(-w).$$

Differentiating, we get $f(w) = f_Z(w) + f_Z(-w) = \frac{2}{\sqrt{2\pi}}e^{-w^2/2}$ on $W > 0$ (this is called a *half-normal* density). Now we are ready to apply the nondecreasing change of variables $Y = W^2$. The inverse transformation is $W = \sqrt{Y}$, whose derivative is $1/(2\sqrt{Y})$. Then its density is $f(y) = \frac{1}{\sqrt{2\pi}\sqrt{y}}e^{-y/2}$ for $y > 0$.

This density may not look familiar, but if you ignore the constant, you will see that it is y to a power $(-\frac{1}{2})$ times a negative exponential of y. It looks like a gamma density, but with a *fractional* shape parameter. If such a thing were possible, that parameter would be $\alpha - 1 = -\frac{1}{2}$, or $\alpha = \frac{1}{2}$. Then we also need a scale parameter $\beta = 2$.

11.5.3   Gamma Densities in General

Let us see what would happen in general if we allowed fractional gamma shape parameters. For one thing, we can no longer interpret a gamma random variable as the time to the αth Poisson event. But we can just define it in terms of the density:

Definition. A Gamma(α) random variable, with $\alpha > 0$ a real number, has density $f(x) = \frac{x^{\alpha-1}}{\Gamma(\alpha)}e^{-x}$ on $x > 0$, where $\Gamma(\alpha)$ is the constant that makes the density integrate to 1.

As usual, our big problem will be to evaluate the constant. First give it an (obvious) name and solve for it in the integration condition:

Definition. The **gamma function** is defined on $\alpha > 0$ by $\Gamma(\alpha) = \int_0^\infty x^{\alpha-1}e^{-x}dx$.

This integral cannot always be evaluated by elementary calculus, but we already know many values. By comparison with the old gamma density we get that $\Gamma(\alpha) = (\alpha - 1)!$ whenever α is a positive integer. Also, since $Y = \beta X$ is a Gamma(α, β) random variable, we get by change of variables that just as before, $f(y) = \frac{y^{\alpha-1}}{\beta^\alpha \Gamma(\alpha)}e^{-y/\beta}$. Now we see that, sure enough, the square of a standard normal variable is Gamma($\frac{1}{2}$, 2). And as gravy, we get the wonderful fact (first noticed by an English mathematician named Wallis in the seventeenth century) that $\Gamma(\frac{1}{2}) = \sqrt{\pi}$.

You should now stare at the last chapter, where we figured out that the gamma density was approximately normal for large α. You will see that nowhere did we require that α be an integer. Therefore, the theorem of the normal approximation to a gamma variable (10.5.5) is still true for (large) fractional α. But then Stirling's approximation also still works:

$$\Gamma(\alpha) \approx \sqrt{2\pi}(\alpha - 1)^{\alpha - \frac{1}{2}}e^{-\alpha + 1}$$

It is not traditional to write it this way; instead, we factor

$$(\alpha - 1)^{\alpha - 1/2} = \alpha^{\alpha - 1/2}(1 - \frac{1}{\alpha})^{\alpha - 1/2}.$$

But you should check that using the technique by which we derived the Poisson approximation to the binomial (see 6.4.1), we know that $(1 - \frac{1}{\alpha})^{\alpha - 1/2} \approx e^{-1}$ for large α. This cancels the 1 in the other exponent:

Proposition (Stirling's formula). $\Gamma(\alpha) \approx \sqrt{2\pi}\alpha^{\alpha - 1/2}e^{-\alpha}$ *for large α.*

When you try it, you will discover that the adjustment makes very little difference. So now we can calculate any large value of the gamma function, approximately.

One more step will allow us to calculate any value at all. Try to integrate the definition at $\alpha + 1$, by parts: $\Gamma(\alpha + 1) = \int_0^\infty x^\alpha e^{-x}dx = -x^\alpha e^{-x}|_0^\infty + \alpha \int_0^\infty x^{\alpha-1}e^{-x}dx$, where the parts are $u = x^a$ and $dv = e^{-x}dx$. You should check that the upper limit of the difference is zero, by L'Hospital's rule. We know the value of the integral on the right:

Theorem (gamma recursion). $\Gamma(\alpha + 1) = \alpha\Gamma(\alpha)$.

As an exercise, you will show how to use this to get the gamma function for α an integer (again). But the novelty is for half-integers: $\Gamma(\frac{3}{2}) = \Gamma\left(\frac{1}{2} + 1\right) = \frac{1}{2}\Gamma\left(\frac{1}{2}\right) = \sqrt{\pi}/2, \Gamma(\frac{5}{2}) = \Gamma\left(\frac{3}{2} + 1\right) = \frac{3}{2}\Gamma\left(\frac{3}{2}\right) = 3\sqrt{\pi}/4$, and in general,

$$\Gamma\left(\frac{2k + 1}{2}\right) = \frac{(2k - 1)(2k - 3)\cdots 3 \cdot 1}{2^k}\sqrt{\pi}.$$

So now we have covered half-integers as well as integers.

None of this seems to help with small, fractional α, such as $\Gamma(0.2)$. But notice that even for α small, if n is large enough Stirling's approximation will calculate $\Gamma(\alpha + n)$ reasonably accurately. Then from the recursion, $\Gamma(\alpha+n) = \prod_{i=1}^n (\alpha+i-1)\Gamma(\alpha)$. Nothing stops us from using this backwards: $\Gamma(\alpha) = \Gamma(\alpha + n)/\prod_{i=1}^n(\alpha + i - 1)$. Then we approximate $\Gamma(\alpha + n)$, with ever better accuracy for as large an n as we have time to compute; and we have a good approximation to $\Gamma(\alpha)$.

Example. $\Gamma(0.2) = \dfrac{\Gamma(10.2)}{0.2 \cdot 1.2 \cdot 2.2 \cdot 3.2 \cdot 4.2 \cdot 5.2 \cdot 6.2 \cdot 7.2 \cdot 8.2 \cdot 9.2} \approx 4.55$, with more accuracy if we are willing to divide more times. Since Stirling's approximation to $\Gamma(10)$ is within 1% of the correct value, it is a reasonable guess that so is our answer.

Theorem (recursive calculation of the gamma function). *For any α that is not a nonpositive integer, $\Gamma(\alpha) = \lim_{n\to\infty}(\sqrt{2\pi}(n+\alpha)^{n+\alpha-1/2}e^{-n-\alpha})/(\prod_{i=1}^{n}(\alpha+i-1))$.*

We needed the gamma function so we would know the constant in the gamma density. What we really need, though, is a cumulative distribution function from which to get probabilities. In general, we cannot evaluate the integral of the density. But there is really no problem; our old formula for the Gamma(α) cumulative still works, as you will check as an exercise: $F(x) = \sum_{y=a}^{\infty} \frac{x^y}{\Gamma(y+1)}e^{-x}$, where a gamma function has replaced the factorial. But we do not really have to compute a gamma function for each of enough terms for this computation to become reasonably accurate. Use the gamma recursion and factor out the constant terms:

Proposition. *For X a Gamma(α) variable,*

$$F(x) = \frac{x^{\alpha-1}}{\Gamma(\alpha)}e^{-x} \sum_{y=1}^{\infty} \frac{x^y}{\prod_{x=0}^{y-1}(\alpha+z)} = f(x) \sum_{y=1}^{\infty} \frac{x^y}{\prod_{x=0}^{y-1}(\alpha+z)}.$$

11.5.4   Chi-Squared Variables

By now you have probably forgotten why we cared about gamma variables with fractional shape parameter in the first place. It was because a chi-squared statistic for a k-independent Poisson cells contingency table model was approximately a sum of k squares of independent standard normal variables. Now we know that each square is a Gamma($\frac{1}{2}$, 2). But a sum of independent gammas with the same scale parameter is still a gamma; the new proof in this section works for the fractional case. We have discovered an enormously useful result:

Theorem (chi-squared distribution). *The sum of squares of k independent standard normal random variables is a Gamma($k/2$, 2) random variable.*

This is so widely used that it has been given a name:

Definition. A Gamma($k/2$, 2) random variable is called a **chi-squared variable on k degrees of freedom**.

We have its first application:

Proposition. *Let X be a vector of k counts with independently Poisson(λ) distributions, with each $\sqrt{\lambda_i}$ large. Then $\chi^2 = \sum_{i=1}^{k}(X_i - \lambda_i)^2/\lambda_i$ has approximately a chi-squared distribution on k degrees of freedom.*

Example. In Chapter 8 (see 8.4.3) we compared one year's tornadoes with the historical rates in three states, getting a chi-squared of 4.0247 on 3 degrees of freedom. This is larger than the 3.0 we would expect. Is it surprisingly large? We compute the probability that a chi-squared random variable on 3 degrees of freedom would be 4.0247, or perhaps even larger. $P(\chi^2 \geq 4.0247) = 1 - F(4.0247)$. We transform to a scale parameter of $1 : 4.0247/2 = 2.0124$. Our probability is 0.2588 after 10 terms in our series (you should check my calculation). It seems that this year was no worse for tornadoes than we might often have found by accident.

We will find later that this approximate distribution works as well for other models for contingency tables. Furthermore, it works only slightly less well for G-squared.

11.5.5 Beta Variables in General

Now that we know what fractional factorials mean, we can generalize other families as well.

Definition. A Beta(α, β) random variable with real parameters $\alpha, \beta > 0$ and sample space $0 < X < 1$ has density $f(x) = \frac{\Gamma(\alpha+\beta)}{\Gamma(\alpha)\Gamma(\beta)} x^{\alpha-1}(1 - x)^{\beta-1}$.

Example. In Chapter 4 we figured out that the density of the horizontal position of a dart hit on a unit-radius circular dart board was $f(x) = \frac{2}{\pi}\sqrt{1 - x^2}$ on $-1 < X < 1$ (see 4.8.1). Transform it to $0 < Y < 1$ by the change of variables $Y = (1+X)/2$; you should check that we get

$$f(y) = \frac{8}{\pi}\sqrt{y(1 - y)} = \frac{\Gamma(3)}{\Gamma(3/2)\Gamma(3/2)} y^{3/2-1}(1 - y)^{3/2-1}.$$

It turns out to be a Beta($\frac{3}{2}, \frac{3}{2}$) random variable.

How do we know that all these new beta random variables actually exist? Our theorem that says that a beta is a certain ratio of gammas, which now does not need to have integer parameters, allows us to construct them and says that we have the right density. See the next chapter for applications.

11.6   Bayesian Inference in Continuous Families

11.6.1 Bayes's Theorem Revisited.

Multivariate discrete random variables (see Chapter 7.8) turned out to be a useful tool for inference about an unknown parameter. We treated such a parameter as a random variable with a particular *prior* distribution, to reflect our uncertainty about it. We then imagined our experiment as involving the conditional distribution of the observed data, given possible values of the parameter. Finally, after doing the experiment and making those observations, we used Bayes's theorem to construct the conditional, *posterior*, distribution of the unknown parameter, given the values of the data we found. Reasonable conclusions then followed easily. We called this style of statistical reasoning *Bayesian* inference.

The example (see 7.8.2) involved n, an unknown binomial sample size. Since n takes on discrete values, we gave it a discrete prior (in fact, Poisson) random variable. Lately, though, most of our parameters have been continuous (we just figured out how to make α, the gamma shape parameter, continuous). This suggests that we would prefer to give them continuous prior distributions. It seems that we need a form of Bayes's theorem for densities.

Let random vectors $\mathbf{X}$, $\mathbf{Y}$ have joint density $f(\mathbf{x}, \mathbf{y})$. We know that $f(\mathbf{x}, \mathbf{y}) = f_{\mathbf{X}}(\mathbf{x}) f_{\mathbf{Y}|\mathbf{X}}(\mathbf{y}|\mathbf{x}) = f_{\mathbf{Y}}(\mathbf{y}) f_{\mathbf{X}|\mathbf{Y}}(\mathbf{x}|\mathbf{y})$. Clearing the denominator, we get an old friend:

Proposition (Bayes's theorem). $f_{\mathbf{X}|\mathbf{Y}}(\mathbf{x}|\mathbf{y}) = (f_{\mathbf{X}}(\mathbf{x}) f_{\mathbf{Y}|\mathbf{X}}(\mathbf{y}|\mathbf{x}))/(f_{\mathbf{Y}}(\mathbf{y}))$ *whenever* $f_{\mathbf{Y}}(\mathbf{y}) \neq 0$.

Now we write the density of a vector of observations $\mathbf{x}$ that depend on a value of the vector of parameters θ as $f_{\mathbf{X}|\theta}(\mathbf{x}|\theta)$. Since we do not know what θ is, we describe our uncertainty by letting it be a random variable with prior density $f_\theta(\theta)$. Bayes's theorem gives us a formula for the posterior density $f_{\theta|x}(\theta|x) = (f_\theta(\theta) f_{X|\theta}(x|\theta))/f_X(x)$.

Of course, we shall have to compute that marginal distribution of $\mathbf{X}$; but we can do this by integrating the basic relationship $f_{\mathbf{X}}(\mathbf{x}) f_{\mathbf{Y}|\mathbf{X}}(\mathbf{y}|\mathbf{x}) = f(\mathbf{x}, \mathbf{y}) = f_{\mathbf{Y}}(\mathbf{y}) f_{\mathbf{X}|\mathbf{Y}}(\mathbf{x}|\mathbf{y})$ with respect to $\mathbf{Y}$:

$$f_{\mathbf{X}}(\mathbf{x}) \int f_{\mathbf{Y}|\mathbf{X}}(\mathbf{Y}|\mathbf{x}) d\mathbf{Y} = \int f(\mathbf{x}, \mathbf{Y}) d\mathbf{Y} = \int f_{\mathbf{Y}}(\mathbf{y}) f_{\mathbf{X}|\mathbf{Y}} d\mathbf{Y}.$$

That first integral simply integrates a conditional density over all values of the random vector $\mathbf{Y}$; therefore, its value is 1; the second is the marginal density of $\mathbf{X}$.

Proposition. *For a random vector* $\mathbf{X}$, $\mathbf{Y}$ *with density* $f(\mathbf{x}, \mathbf{y})$,

$$f_{\mathbf{X}}(\mathbf{x}) = \int f(\mathbf{x}, \mathbf{Y}) d\mathbf{Y} = \int f_{\mathbf{Y}}(\mathbf{Y}) f_{\mathbf{X}|\mathbf{Y}}(\mathbf{x}|\mathbf{Y}) d\mathbf{Y}.$$

Now apply the proposition to our expression for the posterior density of parameters, to get $f_{\theta|\mathbf{X}}(\theta|\mathbf{x}) = (f_\theta(\theta) f_{\mathbf{X}|\theta}(\mathbf{x}|\theta))/(\int f_\theta(\theta) f_{\mathbf{X}|\theta}(\mathbf{x}|\theta) d\theta)$. You see that the denominator is simply the constant necessary to turn the numerator into a density for the random vector θ; sometimes we will be able to guess that constant without further effort. You should look at the parallel structure between this formula and the discrete one in (7.8.1).

11.6.2   *Application to Gamma Observations*

We have already considered the problem of inference about unknown parameters when the data are believed to come from a gamma random variable (see Chapter 9.6 and 10.7.3). Let us see whether Bayesian inference answers some interesting questions in this case. A Gamma(α, β) observation will have the density, conditioned on the parameters, $f(x|\alpha, \beta) = x^{\alpha-1}/(\beta^\alpha \Gamma(\alpha)) e^{-x/\beta}$. To keep the difficulties manageable, assume that α is known and a constant in all that follows. We need to assume a prior density for β that reflects our ignorance; the theory of the last section will apply to any prior density whose sample space makes sense for β. In practice, you may discover, most forms of prior density force you to do very difficult calculus. Therefore, we will look for forms that make life easy. To do this, stare at the gamma density, and try to imagine that β is the random variable and x is known. It seems as if β would itself be something like a gamma variate, except that it is consistently in the denominator instead of the numerator. This suggests that we might give β an *inverse gamma* prior (see Exercise 10.13). That is, $1/\beta$ would be thought of as a gamma variable.

Specifically, let the prior distribution for β be Inverse Gamma(γ, δ), so that its prior density is $\frac{1}{\beta^{\gamma+1}\delta^\gamma\Gamma(\gamma)}e^{-1/\delta\beta}$ for $\beta > 0$. (The hard part to remember will be that β is the variable here.) Then the numerator part of the posterior density is

$$\frac{1}{\beta^{\gamma+1}\delta^\gamma\Gamma(\gamma)}e^{-1/\delta\beta}\frac{x^{\alpha-1}}{\beta^\alpha\Gamma(\alpha)}e^{-x/\beta} = \frac{x^{\alpha-1}}{\beta^{\gamma+\alpha+1}\delta^\gamma\Gamma(\gamma)\Gamma(\alpha)}e^{-(x+1/\delta)/\beta}.$$

Again, the denominator would just be that constant that makes this function integrate to 1; so let us ignore all the constant pieces and look only at the part that varies with β : $1/\beta^{\gamma+\alpha+1}e^{-(x+1/\delta)/\beta}$. We conclude that the posterior density for β must be inverse gamma with shape parameter $\alpha + \gamma$ and scale parameter $\frac{1}{x+1/\delta}$.

We chose the prior so intelligently that the posterior distribution of β was in the same family of random variables as the prior was. Only the parameters changed, in simple ways. When we can do this, we say that the prior $f(\theta)$ is the *natural conjugate prior* of the family of distributions assumed to describe the observations $f(x|\theta)$. It is by no means always reasonable to do this; but if it is, it makes life so much easier that we leap at the chance.

One way we noticed in (7.8.2) that we could extract useful conclusions from that posterior was to exploit the fact that the expected value of a variable was the value that had the smallest mean squared error. Again checking Exercise 10.13, $\hat\beta = E(\beta) = (x + 1/\delta)/(\alpha + \gamma - 1)$. If we let $\gamma = 1$ (so β has an inverse negative exponential prior) and let δ be very large, the estimate approaches the method-of-moments estimate x/α. The variance of β provides us with the mean-squared error of this estimate: $\text{Var}(\beta) = (x + 1/\delta)^2/[(\alpha + \gamma - 1)^2(\alpha + \gamma - 2)]$.

Example. Return to the mouse lifetime data yet again (see 9.6.2) where we assumed that $\alpha = 5$ but wished to estimate β. Give β an inverse gamma prior distribution, with the shape parameter $\gamma = 1$ suggested above. The standard textbook suggested a β of 60 days. Unfortunately, our prior cannot match that on average because, you should notice, its expectation is infinite. Let us instead give it a *median* prior of 60 days. Since the median of a negative exponential(δ) random variable is $\delta \log(2)$, we can set the median of our prior to $60 = \frac{1}{\delta \log(2)}$. Thus let $1/\delta = 41.59$. Now, we assume the grand total lifetime of 7742 to have shape parameter $n\alpha = 20 \times 5 = 100$. The expected posterior estimate is then $\hat\beta = E(\beta) = \frac{7742+41.59}{100} = 77.84$ days. This estimate has a variance of 61.2, and so a standard error (square root of its variance) of 7.82 days.

A 99% Bayes interval is easy to build; it is just a range of β values that have a probability of 99% under the posterior distribution (in this continuous case we can do it exactly). That posterior distribution was inverse gamma; but that is no problem. We construct a 99% interval in a gamma random variable with the same parameters, then invert the end points. The (computer-assisted) computation gives P$[T \leq 0.009779|\text{Gamma}(100, 0.0001285)] = 0.005$ and P$[T \geq 0.016397|\text{Gamma}(100, 0.0001285)] = 0.005$. The 99% Bayes interval is then $60.99 \leq \beta \leq 102.25$. Notice that this is very close (within half a day) of the exact 99% *confidence* interval.

Expressing this analysis for the posterior expectation in terms of the sample of n measurements, we get $\hat{\beta} = E(\beta) = (\sum_{i=1}^{n} x_i + 1/\delta)/(n\alpha + \gamma - 1) = (\bar{x} + 1/n\delta)/(\alpha + (\gamma - 1)/n)$. Then for any choice of prior parameters, the Bayes estimate gets ever closer to the standard estimate $\hat{\beta} = \bar{x}/\alpha$ as the sample size n gets larger. This is a common observation with Bayesian analysis; it is called the *swamping* of the prior. Many statisticians worry that Bayesian methods are influenced by the statistician's opinion, as reflected in the prior distribution. But Bayesians can answer that if you have enough data, the prior will not matter much.

11.7 Two Normal Random Variables

11.7.1 Approximating Conditional Variables

Start again with the fact that for T and S independently Gamma(α) and Gamma(β), we have $U = T + S$ is Gamma($\alpha + \beta$). Notice that we have yet another consequence. For α and β both large, T is approximately Normal(α, α), and S is approximately Normal(β, β). But also U is approximately Normal($\alpha + \beta, \alpha + \beta$); furthermore, if we take parameters big enough, the approximations are as good as we wish. It looks as if normal random variables, with possibly very different centers and scales, have some sort of nice addition property, as gamma and Poisson variables do. If X and Y are any independent normal random variables, for what constants a and b will it turn out that $Z = aX + bY$ is normal?

We also found that $V = T/(T + S)$ is independently Beta(α, β). But that says that knowing the particular value $T + S = u$ has no effect on the distribution of $V = T/u$. Therefore, the conditional distribution of T, given $T + S = u$, is a constant multiple u of a Beta(α, β) random variable. This says that a large-parameter approximation to a beta distribution is a normal variable conditioned on another normal. Is it possible that beta variables are sometimes approximately normal? We will decide both these questions at once.

11.7.2 Linear Combinations of Normal Variables

We know from the definition that we can always change the means and variances of our variables to anything we want later; so let X, Y be independently standard normal. Their joint density is

$$f(x, y) = \frac{1}{\sqrt{2\pi}} e^{-\frac{1}{2}x^2} \frac{1}{\sqrt{2\pi}} e^{-\frac{1}{2}y^2} = \frac{1}{2\pi} e^{-\frac{1}{2}(x^2 + y^2)}.$$

We want the density of $Z = aX + bY$; the Jacobian method says that we need a second transformed variable to make it one-to-one. There are many choices, but you will see that in this case the algebra is easiest if the new variable is *uncorrelated* with the old one. A simple choice with this property is $W = bX - aY$. As an exercise, check that indeed Z and W have a covariance of 0. Now carry out the algebra of solving our equations for X and Y:

$$X = \frac{a}{a^2 + b^2}Z + \frac{b}{a^2 + b^2}W,$$

$$Y = \frac{b}{a^2 + b^2}Z - \frac{a}{a^2 + b^2}W.$$

The coefficients give us the **G** matrix; you should check that $J = |\det \mathbf{G}| = \frac{1}{a^2+b^2}$. Substituting in the exponent of the density, $-\frac{1}{2(a^2+b^2)^2}[(az+bw)^2 + (bz-aw)^2] = \frac{(z^2+w^2)}{2(a^2+b^2)}$. The cross-terms in zw canceled out, because of our bright choice to make z and w uncorrelated. We get a density

$$f(z, w) = \frac{1}{2\pi(a^2 + b^2)}e^{-(z^2+w^2)/[2(a^2+b^2)]}$$

$$= \frac{1}{\sqrt{2\pi(a^2 + b^2)}}e^{-z^2/[2(a^2+b^2)]}\frac{1}{\sqrt{2\pi(a^2 + b^2)}}e^{-w^2/[2(a^2+b^2)]},$$

which factors because of the missing cross-term. We recall that when the density factors, Z and W must be independent. Furthermore, we can read off that they are each independently Normal$(0, a^2 + b^2)$.

A general case is now easy. Let X be Normal(μ, σ^2) and Y be independently Normal(ν, τ^2). Now let us ask what the distribution of $Z = aX + bY + c$ is, for a, b, c constant. We know by definition that $X = \mu + S\sigma$ and $Y = \nu + T\tau$ for S and T independent standard normal. Then we need the distribution of $Z = a\sigma S + b\tau T + a\mu + b\nu + c$. We know from the last paragraph that $a\sigma S + b\tau T$ has a Normal$(0, a^2\sigma^2 + b^2\tau^2)$ distribution; finally we get that Z is Normal$(a\mu+b\nu+c, a^2\sigma^2+b^2\tau^2)$. Indeed, any linear combination of independent normal variables also has a normal distribution.

Theorem (linear combinations of normal variables). *Let X be Normal(μ, σ^2) and Y be independently Normal(ν, τ^2).*

(i) $Z = aX + bY + c$ is Normal$(a\mu + b\nu + c, a^2\sigma^2 + b^2\tau^2)$.

(ii) *Any two linear combinations of normal variables that are uncorrelated are also independent.*

This has a very useful consequence: Let $X_1, \ldots, X_n$ be an independent random sample of Normal(μ, σ^2) variables. Now consider its sample mean $\bar{X} = \frac{1}{n}\sum_{i=1}^{n} X_i$; after applying the theorem $n - 1$ times, introducing each of the observations in turn, we get the following:

Proposition.

(i) *For an i.i.d. random sample of n Normal(μ, σ^2) random variables, $\bar{X}$ is Normal$(\mu, \sigma^2/n)$.*

(ii) $P(\bar{X} - z_{\alpha/2}\frac{\sigma}{\sqrt{n}} \le \mu \le \bar{X} + z_{\alpha/2}\frac{\sigma}{\sqrt{n}}) = 1 - \alpha$.

This reassures us about what we conjectured earlier (see 10.7.3): What was approximately true for sample means with large n from gamma and Poisson random variables is exactly true if the observations are normal in the first place.

11.7.3   Conditional Normal Variables

We have also asked what might happen to one of our normal random variables when we learn about some linear function that gives us partial information about it. It will be general enough to ask, if Z and W are independently standard normal, and we learn that $aZ + bW = c$, then what will the conditional distribution of Z become? Following the procedure of the last section, let $S = aZ + bW$ and $T = bZ - aW$. Then S and T turned out to be independently normal with means 0 and variances $a^2 + b^2$. Now invert the transformation: $Z = a/(a^2 + b^2)S + b/(a^2 + b^2)T$. But our information says that $S = c$, and that has no effect on our knowledge of T, which is independent of it. Therefore, conditionally on $aZ + bW = S = c$, we learn that $Z = ac/(a^2 + b^2) + b/(a^2 + b^2)T$ is normal with mean $ac/(a^2 + b^2)$ and variance (after squaring the coefficient on T) $b^2/(a^2 + b^2)$.

As an exercise, we will let you translate this result into one about general normal variables, using $X = \mu + \sigma Z$ and $Y = \nu + \tau W$.

Proposition. *Let X be Normal(μ, σ^2) and Y be independently Normal(ν, τ^2). Then the distribution of X given that $dX + eY = f$ is*

$$\text{Normal}\left[\mu + \frac{d\sigma^2(f - d\mu - e\nu)}{d^2\sigma^2 + e^2\tau^2}, \frac{e^2\tau^2}{d^2\sigma^2 + e^2\tau^2}\sigma^2\right].$$

11.7.4   Approximating a Beta Variable.

This last result was addressed to our question about the normal approximation to a gamma variable T, conditioned on $T + S = u$, when the parameters α and β of the T and S are large. Using the normal approximations to each variable (see 10.5.5) and our calculation above, we find that T is conditionally approximately normal with mean $\alpha + [\alpha(u - \alpha - \beta)]/(\alpha + \beta) = (\alpha u)/(\alpha + \beta)$ and variance $(\alpha\beta)/(\alpha + \beta)$. Now $V = T/u$ we know to be Beta(α, β); and further, we know that V is approximately normal with mean $\alpha/(\alpha + \beta)$ and variance $(\alpha\beta)/[u^2(\alpha + \beta)]$. This last is silly: The distribution of V does not depend on u at all; so no reasonable approximation should. With high probability u would be relatively near its mean $\alpha + \beta$; so we may use that value, variance $(\alpha\beta)/(\alpha + \beta)^3$. This is so close to the actual beta variance $(\alpha\beta)/[(\alpha + \beta)^2(\alpha + \beta + 1)]$ for α and β large that we will now be happy to use that in our final result:

Theorem (normal approximation to the beta). *Let V be Beta(α, β). Then V is approximately* Normal$[\alpha/(\alpha + \beta), \alpha\beta/[(\alpha + \beta)^2(\alpha + \beta + 1)]]$ *for $\sqrt{\alpha}$ and $\sqrt{\beta}$ large.*

Example. Someone has proposed a new Uniform(0, 1) random number generator. As a quick test of whether it has a bias toward values too large or too small, you generate 101 random numbers, sort them, and pick out the median. If it is a good random number generator, then that median m, the middle order statistic, will have a Beta(51, 51) distribution. This is a good candidate for a normal approximation, Normal(0.5, 0.002427).

We shall do a two-sided hypothesis test with null hypothesis that the mean is in fact 0.5, with significance level 0.05. We noted earlier (see 10.7.1) that for Z standard normal, $P(-1.96 \le Z \le 1.96) = 0.95$. Using our normal approximation to m, we find that we should announce that the generator has peculiar behavior if $m > 0.59656$ or $m < 0.40344$. We will be blowing the whistle on a sound method with approximate probability only 0.05. (Using beta–binomial duality (see 9.5.3) it may be checked that the actual coverage probability of this normal interval is 0.9505—very close, and even a little conservative.)

It might be noticed that the errors in the approximation that came from the Gamma(α) and Gamma(β) part are of opposite signs; so in practice errors partially cancel out. In such a case, the normal approximation turns out to be even better than we expect. The effect is strongest, and the errors smallest, in cases like this where the two parameters are equal.

11.8 Normal Approximations to the Binomial and Negative Binomial Families

11.8.1 Binomial Variables with Large Variance

We can use the same conditioning technique to discover when a binomial random variable is close to a normal distribution. We just use the dual fact to the one about gammas and betas: If X is Poisson(λ) and Y is independently Poisson(μ), then conditional on the information that $X + Y = n$, X is B(n, p), where $p = \frac{\lambda}{\lambda+\mu}$ (see 7.4.2). But for large λ and μ, the Poisson variables are approximately normal. Our conditioning result for normal random variables then gives us (since $\lambda = np$ and $\mu = n(1 - p)$) the following:

Theorem (normal approximation to the binomial). *For X a B(n, p) random variable with $\sqrt{n(1 - p)}$ and $\sqrt{np}$ both large, $F_X(x) \approx F_Z\left(\frac{x+.5-np}{\sqrt{np(1-p)}}\right)$ where Z is standard normal.*

We have used a continuity correction here, for the same reason we did in the Poisson case (see 10.6.2). Our two conditions can be replaced by requiring $\sqrt{np(1 - p)}$, the standard deviation of X, to be large, since it is only slightly smaller than the smaller of our two check values.

Historically, this is the first normal approximation result, because deMoivre discovered the normal distribution while trying to compute binomial probabilities approximately.

Example. I toss what I believe to be a fair coin 100 times; so that I believe the number of heads X is B(100, 0.5). Should I be surprised when I get only 43 heads? One way to measure the surprise is by computing $F(43)$; but that is an unpleasant calculation. Instead I will compute the standard normal cumulative of $(43.5 - 50)/5 = -1.3$, which is 0.0968. It looks as if this is not really all that surprisingly low a total. (The exact answer is 0.0967; very close indeed.)

11.8.2   Negative Binomial Variables with Small Coefficient of Variation

By positive–negative duality (see 6.3.3), we of course have a normal approximation to the negative binomial family as well. Let X be NB(k, p). Then

$$F(x) = P[X \le x|\text{NB}(k, p)] = P[X \le x|\text{B}(k + x, p)].$$

But then we can use the normal approximation

$$F_X(x) \approx F_Z\big[[x - (k + x)p]/[\sqrt{(k + x)p(1 - p)}]\big]$$

(ignoring the continuity correction for a moment) whenever $\sqrt{(k + x)(1 - p)}$ and $\sqrt{(k + x)p}$ are large. To adjust this to the familiar standardized form, we want to know when we can replace x by its average value $kp/(1 - p)$. The relative error in this replacement is small with probability near 1 (by our law of large numbers) whenever this average is large compared to the standard deviation $\sqrt{kp}/(1 - p)$. That happens when $\sqrt{kp}$ is large. (The (small) ratio of the standard deviation to the mean of a positive random variable is called the *coefficient of variation*.) Then

$$\frac{x - (k + x)p}{\sqrt{(k + x)p(1 - p)}} \approx \frac{x(1 - p) - kp}{\sqrt{kp}} = \frac{x - kp/(1 - p)}{\sqrt{kp}/(1 - p)}.$$

The conditions become that $\sqrt{k}$ and $\sqrt{kp}/(1 - p)$ must be large. But we already have had to require that $\sqrt{kp}$ be large, and that would force both of these to be large. Now put back the continuity correction to get our approximation:

Theorem (normal approximation to the negative binomial).   *If X is NB(k, p) with $\sqrt{kp}$ large, and Z is standard normal, then*

$$F_X(x) \approx F_Z\big[[x + 0.5 - kp/(1 - p)]/[\sqrt{kp}/(1 - p)]\big].$$

Example. As in (6.2.5), we are seeking $k = 20$ left-handed people, given that they are a proportion $1 - p = 0.1$ of the population. What is the probability that we will interview at least 240 right-handed people along the way? It seems that $\sqrt{kp} = 4.24$ is a bit larger than 1, so a normal approximation is worth trying.

$$P(X \ge 240) \approx P\left[Z \ge \frac{239 + 0.5 - kp/(1 - p)}{\sqrt{kp}/(1 - p)}\right] = P(Z \ge 1.4024) = 0.0804$$

(A computer-assisted exact calculation gives 0.0884.)

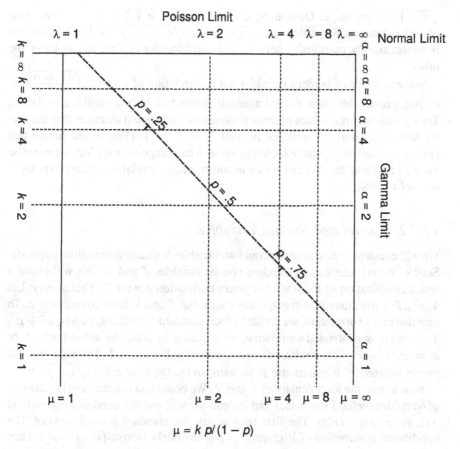

FIGURE 11.5. Negative binomial(k, p) parameter space

11.9 The Bivariate Normal Family

11.9.1 Approximating Two Order Statistics

At the beginning of the chapter, we began with the fact that uniform order statistics have beta distributions; then we found the joint distributions of two uniform order statistics. Later on, we used conditioning to discover that as the parameters get large, a beta random variable is approximately normal. This raises immediately the suggestion that two uniform order statistics may sometimes behave like such correlated linear combinations of normal variables. When might that happen?

If we are looking again at the ith order statistic P and jth order statistic $Q(j > i)$ from a uniform sample of n, then we know that for $\sqrt{i}$ and $\sqrt{n + 1 - i}$ large, P has approximately a normal distribution. But we noted in Section 3 that Q is, conditionally on $P = p$, a uniform $(j - i)$th order statistic from a sample of $n - i$ on the interval $(p, 1)$. Therefore, it was a rescaled Beta($j - i, n + 1 - j$) variable. This conditional distribution is approximately normal if $\sqrt{j - i}$ and

$\sqrt{n+1-j}$ are large. Obviously, $\sqrt{n+1-i} > \sqrt{n+1-j}$, so for any pair of order statistics for which $\sqrt{i}$, $\sqrt{j-i}$, and $\sqrt{n+1-j}$ are large, each one is approximately marginally normal, and conditionally normal if you know the other.

Since this pair of random variables has a correlation of $\rho_{PQ} = \sqrt{\frac{i(n+1-j)}{j(n+1-i)}}$ (exercise), presumably their normal approximations will have a similar correlation. Trying various large values of these parameters, you should discover that the correlations can range all over the interval $0 < \rho < 1$. (You might notice that correlations between jointly *Dirichlet* variates can range over all sorts of *negative* values.) Just how flexible can we be in constructing correlations between pairs of *normal* variables?

11.9.2   Correlated Normal Variables

We will construct a pair of normal random variables with *any* correlation we please. Start with two independent standard normal variables Z and W. We will create a linear combination of these with arbitrary correlation ρ with Z. This is easy: Let $V = \rho Z + a W$; then you may quickly check that Z and V have covariance, ρ. To turn that into a correlation, we divide by their standard deviations, 1 and $\sqrt{a^2 + \rho^2}$. To preserve our intended correlation, we choose a to make the second term 1, or $a = \pm\sqrt{1 - \rho^2}$. This works, of course, only if $-1 < \rho < 1$. The sign will not matter because W is symmetric about zero; so finally, $V = \rho Z + \sqrt{1 - \rho^2} W$.

Now to find the joint density of Z and V. We could just use the general change-of-variables method (exercise), but instead we will use the conditioning method $f(z, v) = f(z) f(v|z)$. The first term is just the standard normal density. The conditional distribution of V given $Z = z$ is obviously Normal$(\rho z, 1 - \rho^2)$. Then we can write the density

$$f(z, v) = \frac{1}{\sqrt{2\pi}} e^{-z^2/2} \frac{1}{\sqrt{2\pi}\sqrt{1-\rho^2}} e^{-(v-\rho z)^2/(2(1-\rho^2))}.$$

Combining the exponents,

$$z^2 + \frac{(v - \rho z)^2}{(1-\rho^2)} = \frac{z^2(1-\rho^2) + (v^2 - 2v\rho z + \rho^2 z^2)}{(1-\rho^2)} = \frac{z^2 - 2v\rho z + v^2}{(1-\rho^2)}.$$

We conclude that the joint density is as follows:

Proposition. *The joint normal density of two variables with means* 0*, variances* 1*, and correlation* ρ *is* $f(z, v) = \frac{1}{2\pi\sqrt{1-\rho^2}} e^{-(z^2 - 2v\rho z + v^2)/(2(1-\rho^2))}$.

(see Figure 11.6) It is symmetric in z and v, as we might have hoped.

Now, to modify our variables to have any mean and variance we want, let $X = \mu + Z\sigma$ and $Y = v + V\tau$. You should remind yourself from properties of the correlation in Chapter 7 (see 7.5.5) that these transformations have no effect on ρ. We will let you transform the density to one in X and Y, to get a density so complicated that no one should ever have to memorize it. Nevertheless, this

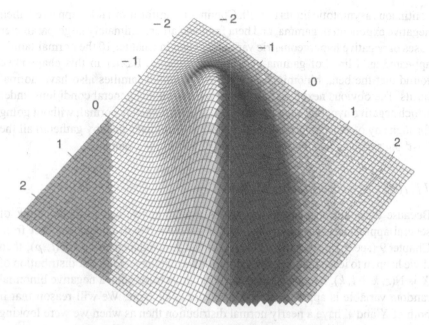

FIGURE 11.6. Bivariate normal density with $\rho = 0.7$

is called the *bivariate normal family*, with five parameters μ, ν, σ, τ, and ρ. We already know an application:

Proposition. *The ith and jth order statistics $(i < j)$ from a uniform sample of n with $\sqrt{i}$, $\sqrt{j - i}$, and $\sqrt{n + 1 - j}$ large has approximately a bivariate normal distribution with the same means, variances, and correlation.*

As an exercise, we will let you show that a (discrete) trinomial random vector, under conditions you will discover, has approximately a bivariate normal distribution.

11.10 The Negative Hypergeometric Family Revisited*

11.10.1 Family Relationships

We will close this chapter by bringing to a nice, logical conclusion one of the recurring themes of much of the book. In Chapter 5, we introduced the negative hypergeometric family of random variables. We found that Bernoulli variables and discrete uniform variables are simple special cases of this family and that the hypergeometric family was closely related to it. In Chapter 6 we found that as our parameters got large, negative hypergeometric variables became geometric, negative binomial, binomial, or Poisson random variables. These, of course, are all discrete random variables. But then in Chapter 9 we found that by rescaling our random variable (so that the discrete steps got close together), we could get continuous asymptotic limits as well. Continuous uniform variables appeared, then negative exponential, gamma, and beta families; all are ultimately large-parameter cases of negative hypergeometric variables. Then in Chapter 10 the normal family appeared as a limit of gamma and Poisson variates. Earlier in this chapter we found that the beta, binomial, and negative binomial families also have normal limits. The obvious next question is, can we find simple general conditions under which negative hypergeometric variables are approximately normal, without going through any of those intermediate families? We will have thereby gathered all the far-flung relatives together at grandmother's house for a reunion.

11.10.2 Asymptotic Normality

Because there are so many relationships available, you can probably think of several approaches we might take to this problem. I will choose the fact from Chapter 9 (see 9.5.5) that if X is NB(k, p) and Y is independently NB(l, p), then if we happen to learn that $X + Y = z$, it follows that the *conditional* distribution of X is N(z, $k + l$, k). But we learned earlier in this chapter that a negative binomial random variable is approximately normal for $\sqrt{kp}$ large. We will reason that if both of X and Y have a nearly normal distribution then as when we were looking at beta variables, their conditional distributions must as well.

In terms of our usual parameter names, we are interested in an N(W, B, b) variable. By the result mentioned above, we can realize it by conditioning on a sum of independent NB(b, p) and NB($B - b$, p) variables. Notice that the conditioning argument works no matter what value of p we use; but the fact that negative binomials converge in distribution to normality depends on the fact that they are, with high probability, relatively near their expectations. To ensure this, let us choose p to be the value that puts the total on which we condition, $W = X + Y$, precisely at its expectation: $W = E(X + Y) = B\frac{p}{1-p}$. (Recall that $X + Y$ is NB($k + l$, p).) Then $p = \frac{W}{W+B}$. Thus our conditions for normality of each component are that $\sqrt{\frac{Wb}{W+B}}$ and $\sqrt{\frac{W(B-b)}{W+B}}$ are large.

Theorem (normal approximation to negative hypergeometric probabilities). *Let X be an* N(W, B, b) *random variable for which* $\sqrt{\frac{Wb}{W+B}}$ *and* $\sqrt{\frac{W(B-b)}{W+B}}$ *are large. Then if F_Z is the standard normal cumulative distribution,*

$$F_X(x) \approx F_Z\left(\frac{x + 0.5 - Wb/(B+1)}{\sqrt{\frac{Wb(B-b+1)(W+B+1)}{B+2}}/(B+1)}\right).$$

Example. According to school medical records, 98 of the 240 students in a dormitory have type O positive blood. One night, after the records office has closed, there is a nearby medical emergency that puts enormous strain on the blood bank and requires large amounts of O positive blood. A medical technician decides that she must go from room to room in the dormitory until she has found 45 O positive students. What is the probability that she will have to ask no more than 110 people to donate blood?

Clearly, the number of people asked who are not O positive will be an $N(142, 98, 45)$ random variable, and we want to evaluate $F(65)$. This would involve quite a bit of computation. The conditions for a reasonably accurate normal approximation are that 5.16 and 5.6 be large compared to 1; and by our modest standards, they are. We compute $F_X(65) \approx F_Z(0.1036) = 0.541$.

11.11 Summary

The first family of continuous random vectors we studied was the joint distribution of several uniform order statistics; for example, the ith and jth of a sample of n have *joint density* $f(p, q) = \frac{n!}{(i-1)!(j-i-1)!(n-j)!} p^{i-1}(q - p)^{j-i-1}(1 - q)^{n-j}$ (3.2). The gaps between these order statistics made up the *Dirichlet* family, with density $f(\mathbf{x}) = [(\sum_{i=1}^{k+1} \alpha_i) - 1]! / [\prod_{i=1}^{k+1}(\alpha_i - 1)!] \prod_{i=1}^{k+1} x_i^{\alpha_i - 1}$ (4.3). In order to move from family to family, we needed the *multivariate transformation* formula for densities $f_Y(\mathbf{Y}) = f_X[\mathbf{g}(\mathbf{Y})] J_g(\mathbf{Y})$ for $\mathbf{g}$ a *one-to-one* function.

Sums of squares of standard normal variates then formed the *chi-squared* family of random variables, though these turned out to be gamma variables with fractional shape parameters and density $f(y) = \frac{y^{\alpha-1}}{\beta^\alpha \Gamma(\alpha)} e^{-y/\beta}$ (5.6). To handle such problems in general, we needed the factorial of a real number, the *gamma function* $\Gamma(\alpha) = \int_0^\infty x^{\alpha-1} e^{-x} dx$ (5.3), which further allowed us to define beta densities with fractional parameters:

$$f(x) = \frac{\Gamma(\alpha + \beta)}{\Gamma(\alpha)\Gamma(\beta)} x^{\alpha-1}(1 - x)^{\beta-1} \quad (5.7).$$

A continuous version of conditional distributions led us to a continuous extension of Bayes's theorem $f_{\theta|X}(\theta|\mathbf{x}) = [f_\theta(\theta) f_{X|\theta}(\mathbf{x}|\theta)] / [\int f_\theta(\theta) f_{x|\theta}(\mathbf{x}|\theta) d\theta]$ (6.1). This made Bayesian inference on parameters with a continuous range of values, such as a gamma scale parameter, possible (6.2).

The marginal normal approximation to large-parameter gamma variables led us, by conditioning both sides, to a normal approximation for large-parameter beta variables (7.4). The discrete dual to this argument led to a classical fact that high-variance binomial random variables (8.1) (and low *coefficient of variation* negative binomials (8.2)) are approximately normal. We then developed a *bivariate normal* density function

$$f(z, v) = \frac{1}{2\pi \sqrt{1 - \rho^2}} e^{-(z^2 - 2v\rho z + v^2)/(2(1-\rho^2))},$$

where ρ is the correlation between Z and V. Now our earlier results about joint distributions of uniform order statistics led us to conditions for their asymptotic joint normality (9.2). Finally, we used a conditioning argument to establish when negative hypergeometric variables are asymptotically normal (10.2).

11.12   Exercises

1. Solve Exercise 18 from Chapter 7 (about commissions on luxury cars) by the expectations-of-conditional-expectations method from Section 2 and check that your result is the same as you got before.

2. Write down the expectations-of-conditional-expectations expression for $E[g(X, Y, Z)]$ for a general trivariate random vector.

3. Show that our general definition of $E[g(X, Y)]$ makes it a positive linear operator.

4. Of the 5 friends in the Boston Marathon, what is the probability that the second-best was ahead of no more than 67% of the field and the fourth-best was ahead of no more than 28% of the field? (Be careful: Our definition of order statistics numbers people from the bottom, but race-finishers are counted from the winner down.)

5. For an absolutely continuous random vector X, Y with density $f(X, Y)$, show using Fubini's theorem that $E[g(X, Y)] = E_X\{E_{Y|X}[g(X, Y)|X]\} = E_Y\{E_{X|Y}[g(X, Y)|Y]\}$ when the two expectations exist.

6. For the joint distribution of two order statistics $i < j$ from a uniform $(0, 1)$ distribution, show that you can get the density from the cumulative distribution function by partial differentiation.

7. Let X and Y have the joint density $3360x^2(y - x)(1 - y)^3$ on $0 < x < y < 1$.

   a. Check that this is an order-statistic density.

   b. Find the joint density of $Z = \log X$ and $W = \log Y$ by two one-variable changes of variable as in Chapter 10 (see 10.4.1). Be sure to write down the sample space of your new vector.

8. a. Find the Jacobian for the transformation in Exercise 7(b).

   b. Find the Jacobian for the polar coordinate transformation of Chapter 10 (see 10.5.3).

9. Let X and Y be the market shares of two competing natural gas companies. On theoretical grounds, you propose a joint density

$$f(x, y) = 24x(1 - x - y) \text{ on } x, y > 0 \text{ and } x + y < 1.$$

   a. Compute $P(X + Y < 0.4)$.

   b. Find $E(Y)$.

10. Compute the *correlation* between uniform order statistics P_i and P_j.

11. Let X and Y have the joint density $[\cos(x + y) + \cos(x - y)]/8$ with sample space the square $(-\pi/2, \pi/2) \times (-\pi/2, \pi/2)$. Are X and Y independent?

12. Verify, using a result from Chapter 6 (see 6.4.1), that $(1 - \frac{1}{\alpha})^{\alpha - 1/2} \approx e^{-1}$ for large α.

13. In Exercise 17 from Chapter 9, estimate the average time between accidents β by a Bayesian analysis, by giving β a prior distribution that is Inverse Gamma$(1, 20)$.

   a. Compute $\hat{\beta}$, the posterior expectation of β. What is its standard error?

b. Construct a 95% Bayes interval for β.

14. If X and Y are uncorrelated, prove that $Z = aX + bY$ and $W = bX - aY$ are also uncorrelated for any constants a and b.

15. In Exercise 1 from Chapter 1, assume that the measurements of the mass ratios are from a normal distribution, with unknown mean, but from your experience with the measuring device they have a standard deviation of 0.0005. Construct a 95% confidence interval for the unknown mean.

16. The *third quartile* of a sample of $n = 4m + 1$ is the $i = (3m + 1)$st-order statistic. (It is an estimate of the 75th percentile, because three times as many observations are smaller than are larger.) Let $m = 10$ in a Uniform(0, 1) sample.

   a. What are the mean and standard deviation of the sample third quartile?
   b. What is the probability that the sample third quartile will exceed 0.8?
   c. Redo (b) using the normal approximation, and compare.

17. A census reports that 40% of children in your neighborhood school have blue eyes. A randomly assigned kindergarten class has 22 children in it.

   a. Compute the probability that 6 or fewer will have blue eyes.
   b. Redo (a) using the normal approximation, and compare.

18. You like to play a simple Scratcho lottery game that pays off 20% of the time. You decide that you will keep playing once a day until you have lost 100 times.

   a. What is the probability that you will have played at least 130 times?
   b. Redo (a) using the normal approximation, and compare.

19. Find the general formula for the correlation between two Dirichlet coordinates.

20. For Z and W independently standard normal, find the joint density of Z and $V = \rho Z + \sqrt{1 - \rho^2} W$ by the Jacobian change of variables method.

21. Let X and Y be bivariate normal, where X is marginally Normal(μ, σ^2), Y is marginally Normal(v, τ^2), and the correlation between the two is ρ. Find their joint density.

22. Let X and Y have the joint density $\frac{1}{16\pi} e^{-\frac{1}{128}(25x^2 - 12xy + 4y^2 - 174x + 52y + 313)}$ on the entire plane. What are the means, standard deviations, and correlation of X and Y?

23. In the O positive blood drive example (see Section 10.2), compute the exact probability of at most 100 students being asked; compare it to the normal approximation.

11.13   Supplementary Exercises

24. Write down the trivariate cumulative distribution of the order statistics $i < j < k$ from a Uniform(0, 1) sample of n.

25. Show that a transformation that is increasing is one-to-one. Then show that a two-variable transformation that is increasing in each variable (when you hold the other constant) is one-to-one.

26. Derive the joint density of k uniform (0, 1) order statistics from a sample of n, by the method of Section 3.

27. Derive the formula for the covariance of two Dirichlet coordinates in another way, the one that we used to find the multinomial covariance:

   a. Point out the simple reason that $X_i + X_j$ is itself a Dirichlet coordinate with parameter α_i, α_j.

   b. Find $\text{Var}(X_i + X_j)$; then use the variance-of-a-linear-combination formula to solve for $\text{Cov}(X_i, X_j)$.

28. Let X and Y have independently the Fisher–Tippet density $e^{-x}e^{-e^{-x}}$. Find the density of $Z = X - Y$.

29. Let Z and W be independent standard normal random variables. Find the density of $T = Z^2 + W^2$ by a polar coordinates change of variables (see Exercise 8). Did you get what you should have gotten from the results of Section 5.2?

30. Compute $\Gamma(14.5)$. Now compute it approximately using Stirling's formula, and compare.

31. Compute $\Gamma(0.75)$ approximately, using at least 20 recursive steps. Roughly how accurate do you believe your computation to be?

32. Derive the formula for a gamma cumulative

$$F(x) = \frac{x^{\alpha-1}}{\Gamma(\alpha)}e^{-x}\sum_{y=1}^{\infty}\frac{x^y}{\prod_{z=0}^{y-1}(\alpha+z)}$$

by starting with $F(x) = \frac{1}{\Gamma(\alpha)}\int_0^x t^{\alpha-1}e^{-t}dt$ and integrating by parts repeatedly, in such a way that the power of x always increases. Now show that the infinite series always converges.

33. Verify that $P(\chi^2 \geq 4.0247) = 0.2588$ using the series in Exercise 32, where we have 3 degrees of freedom.

34. Use the same formula to approximate $P(G^2 \geq 6.76)$ and $P(G^2 \geq 3.13)$, where each has one degree of freedom, in the analysis of the smoking survey in Chapter 8 (see 8.5.3). Interpret your results in words.

35. A *Rayleigh* random variable with parameters $a > 0$ and $b \geq 0$ has cumulative distribution function $F(t) = P(T \leq t) = 1 - e^{-at-bt^2}$ for $T > 0$.

   a. Find the *hazard rate* function (failure rate function) $z(t) = \frac{f(t)}{1-F(t)}$ for the Rayleigh family. Does it show increasing hazard or decreasing hazard over time?

b. A brand of incandescent light bulb has lifetime a Rayleigh random variable in months, with $a = 0.2$ and $b = 0.02$. What is the probability that of 58 bulbs in a building, at least will 36 will still be burning after 3 months? (You may use an appropriate approximate method.)

36. Find conditions under which a trinomial random vector has a bivariate normal approximation.

37. Prove again the normal approximation to the binomial family, this time using normal approximation to the beta family, and beta-binomial duality.

38. Find conditions under which a hypergeometric random variable $H(W + B, W, n)$ is approximately normally distributed. (This fact, discovered by the French mathematician Laplace in the late 18th century, was one of the first hints of just how widely useful normal random variables were.)

39. Assume that we have a collection of n independent observations $X_1, \ldots, X_n$ from a Normal(μ, σ^2), where σ is known but μ is unknown. Let the prior distribution for μ be Normal(ν, τ^2).

    a. Find the posterior distribution of μ, once the X's have been observed. Have we chosen a natural conjugate prior here?
    b. Write down the posterior expectation and variance of μ.
    c. In Exercise 15, compute the posterior expectation of the mass ratio and its standard error, if $\nu = 80$ and $\tau = 2$. Construct its 95% Bayes interval.

40. Solve the Buffon needle problem (see Chapter 4.1) to find the probability that a needle intersects a stripe boundary. **Hint:** Make the center of the needle a uniform random variable on the interval that is the width of a single stripe. Then make the angle the needle makes to the stripe a uniform random variable over possible angles, independent of the position of the center.

CHAPTER 12

Sampling Statistics
for the Linear Model

12.1 Introduction

In Chapter 8 we learned about probability models for contingency tables: that
we could estimate the parameters by the principle of maximum likelihood, and
that we could evaluate the fit of the models using the G-squared and chi-squared
statistics. But this leaves a big hole in our understanding. The very first models
we studied—linear models for rectangular layouts and regression models for measurement experiments, which are among the practically most important statistical
tools—we could estimate by the principle of least squares. But we have no way to
evaluate how well these models fit the data, no probability models to tell us how
much of what we think we see could really be due to chance. In this chapter we
will remedy this deficiency.

We will explore some probability models for errors in linear models and use
these to find the probability distributions of test statistics and statistics of fit for
our models. We will explore the absolutely continuous version of the method of
maximum likelihood as a general method of estimating unknown model parameters. Finally, we will discover that in many circumstances, least-squares estimates
and maximum likelihood estimates are in a certain sense the best, or nearly the
best, estimates we can make.

Time to Review

Chapter 2
Chapter 8, Section 2
Matrix inversion
Taking derivatives under an integral sign

12.2   Spherical Errors

12.2.1   A Probability Model for Errors

We want to discover some useful probability models for the noise, or lack of fit, in linear models for measurements (such as ANOVA and regression models) that we studied in Chapters 1 and 2. We called the noise the residual, or error, associated with each measurement. For example, in a simple linear regression model our predictions were $\hat{y}_i = \mu + (x_i - \bar{x})b$. Then in all such models the residuals are $e_i = y_i - \hat{y}_i$, where y_i are the measured experimental responses. To make this a probability model, the vector of residuals $e = (e_1, e_2, \ldots, e_n)^T$ must be thought of (before we do the experiment) as an n-dimensional random vector (usually continuous).

What random vector law should we use to describe our errors? This problem was not too difficult in the case of contingency tables, because we could very often see from the design of the experiment that, for example, a multinomial model with independent rows and columns was appropriate. But with measurement models, a description of an experiment in which the numbers we get are variable, usually for reasons that are not known, does not help us much when we want to conclude that the errors follow some specific random vector law. We will start by putting some strong restrictions on what sorts of random vectors we will consider, and see what we learn from that.

If we are going to use the method of least squares, which considers only Euclidean distance from the model predictions to the observations, to estimate our parameters, we are assuming that what is important is the *length* of the residual vector, and not its *direction*. This led us in Chapter 2 (see 2.5.4) to propose an assumption of spherical errors, which we can now restate in probability language:

Definition. A random vector is said to be **spherically distributed about the origin** if the probability of an event is entirely determined by the lengths of its vectors (and not by their directions).

In one dimension, this just says that the probabilities of intervals greater than zero are always the same as the probabilities of the reversed interval less than zero. That is, for any constant a, $P(X > a) = P(X < -a)$. If X has a density, this is the same as saying $f(x) = f(-x)$; when we encountered this before, we called it *symmetric about zero*.

In two dimensions we have *circular* symmetry—if you rotate the plane with axis at the origin, the appearance of the density does not change (see Figure 12.1).

If our spherically symmetric random vector has a density, it must depend only on the length of the vector x (or, for simplicity, the squared length); therefore, we can always find a function g such that $f(\mathbf{x}) = g(\mathbf{x}^T\mathbf{x})$. If the density can be written this way, and its sample space is also spherically symmetric, we know that it is symmetric about the origin. We will see an example shortly.

Now let us assume that in some linear model, the residuals have a spherical distribution about the origin. This is at least a plausible way to require that all

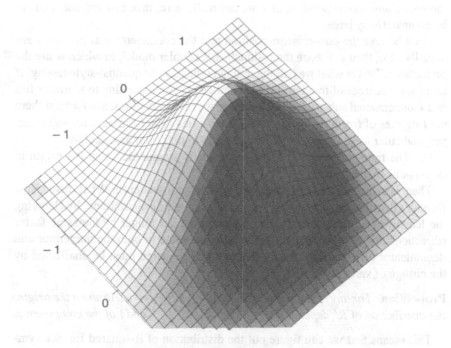

FIGURE 12.1. A circularly symmetric density

repetitions be subject to similar amounts of error. In an exercise, you will notice that it also requires that different observations be *uncorrelated*. However, it still does not say exactly what the random vector is; and a number of different laws are still possible.

12.2.2   Statistics of Fit for the Error Model

Let us see what consequences the assumption has for our statistics R^2 (see 2.5.3) and F. Remember that we informally justified our definition of F from spherical errors; so there is hope that we may learn something about it. You may have noticed long ago that even though we interpreted them in a very different way, R^2 and F may be calculated from each other. So we will proceed to see what we can learn about R^2, and work backwards from there to F. Remember that $R^2 = \frac{SSR}{SSR+SSE}$ for a simple regression problem, but let us come up with what it would mean in general. Imagine that we have, by least squares, decomposed the observation vector into $y = w + x + z$, where w, x, and z are orthogonal to each other. Let w be the prediction vector for parameters that we are willing to say may not be zero. Let x be the prediction vector for factors and effects that are controversial—our null hypothesis is that they might be equal to zero. Then let z be the residual vector from the entire model. Comparing this to the ANOVA for a one-way layout (see 2.5.2),

$\hat{\mu}$ is $\mathbf{w}$, $\hat{\mathbf{b}}$ is $\mathbf{x}$, and $\hat{\mathbf{e}}$ is $\mathbf{z}$, where the controversy is over whether the b's for the levels of the factor are really zero. Our test statistic is then $R^2 = (\mathbf{x}^T\mathbf{x})/(\mathbf{x}^T\mathbf{x} + \mathbf{z}^T\mathbf{z})$. If the coefficients corresponding to x are not really zero, then this statistic tends to be comparatively large.

If we believe the conservative position that the coefficients that give us $\mathbf{x}$ are actually zero, then $\mathbf{x} + \mathbf{z}$ are the residuals in a simpler model, in which $\mathbf{w}$ are the predictions. This is what we call the null hypothesis in frequentist-style testing. If there are k degrees of freedom for the coefficients corresponding to $\mathbf{x}$, then $\mathbf{x}$ lies in a k-dimensional subspace of the vector space of observations. Similarly, if there are l degrees of freedom for residuals $\mathbf{z}$, then $\mathbf{z}$ lies in an l-dimensional subspace, perpendicular to all the $\mathbf{x}$'s. Now we are ready to make our assumption:

H_0: The random vectors $\mathbf{x} + \mathbf{z}$ are spherically symmetric about the origin in $(k + l)$-dimensional space.

There is an immediate and somewhat amazing consequence: What if we transform this distribution to *any* other spherically symmetric distribution, by changing the lengths of all the vectors? Then $\mathbf{x}$ and $\mathbf{z}$ both get longer by the same factor (directions do not matter), and that factor is the same in both the numerator and denominator of R-squared. It cancels out, and R-squared must be unaffected by the change of variables.

Proposition. *For any random vector* $\mathbf{x} + \mathbf{z}$ *spherically symmetric about the origin, the distribution of* R^2 *depends only on the dimensions k and l of the components.*

This means that we can figure out the distribution of R-squared for *any* symmetric error law, and that will always be its distribution. Let us see whether we can do this. Therefore we will look for an easy-to-work-with density that looks like $g(\mathbf{x}^T\mathbf{x})$.

12.3   Normal Error Models

12.3.1   Independence Models for Errors

The assumption of spherical symmetry is certainly consistent with the method of least squares, but unlike some of our discrete models, we as yet have not connected it with what actually happens in experiments.

When we take measurements repeatedly in order to make them more accurate, it is usual to carefully reconstruct each repeated trial so that it is the same as the others, and so that it is not influenced at all by previous trials. Therefore, for example, we very carefully clean our test tubes so that no contamination remains and then do the next experiment. Mathematically, we pursue the ideal of mutual *independence* of observations; as in a Bernoulli process. Now, the spherical distribution assumption does capture our ideal that the errors be comparable from trial to trial—each try corresponds to a different direction, and all directions are equivalent. Furthermore, we figured out that spherical observations are uncorrelated. But we know that this does not make them independent. For example, if they all lie on the surface of a hypersphere of radius 1, and if you know all but the last coordinate, then you know a great deal about *it*, too: $|x_n| = \sqrt{1 - x_1^2 - x_2^2 - \cdots - x_{n-1}^2}$. There is no independence in such a case.

But it turns out that there is a possible distribution of measurement errors that both meets the assumption of independence of observations and the assumption of spherical errors. (In fact, though we will not prove it, there is only one.) Let each residual e_i be independently Normal$(0, \sigma^2)$. Then, since the densities multiply, we obtain

$$f(\mathbf{e}) = \prod_{i=1}^{n} \frac{1}{\sqrt{2\pi}\sigma} e^{-e_i^2/(2\sigma^2)} = \frac{1}{(2\pi)^{n/2}\sigma^2} e^{-\sum_{i=1}^{n} e_i^2/(2\sigma^2)} = \frac{1}{((2\pi)^{n/2})\sigma^n} e^{-\mathbf{e}^T\mathbf{e}/(2\sigma^2)}.$$

Thus, because of the fact that a normal density is the exponential of a squared coordinate, a vector of independent normals with mean zero has a density of the form $g(\mathbf{e}^T\mathbf{e})$. The density depends only on the squared length $\mathbf{e}^T\mathbf{e}$ of the error vector.

Proposition. *A vector of independent Normal$(0, \sigma^2)$ observations is spherically distributed about the origin.*

12.3.2 Distribution of R-squared

If we can find the distribution of a normal R^2, then this will solve our entire problem. Earlier, we had imagined an orthogonal decomposition of our residual vector $\mathbf{e} = \mathbf{x} + \mathbf{z}$. Then $\mathbf{e}^T\mathbf{e} = \mathbf{x}^T\mathbf{x} + \mathbf{z}^T\mathbf{z}$. That lets us rewrite the density

$$f(\mathbf{e}) = \frac{1}{(2\pi)^{k+l/2}\sigma^{k+l}} e^{-\mathbf{e}^T\mathbf{e}/(2\sigma^2)} = \frac{1}{(2\pi)^{(k+l)/2}\sigma^{k+l}} e^{-(\mathbf{x}^T\mathbf{x}+\mathbf{z}^T\mathbf{z})/(2\sigma^2)}$$

$$= \frac{1}{(2\pi)^{k/2}\sigma^k} e^{-\mathbf{x}^T\mathbf{x}/(2\sigma^2)} \frac{1}{(2\pi)^{l/2}\sigma^l} e^{-\mathbf{z}^T\mathbf{z}/(2\sigma^2)}.$$

We have discovered that $\mathbf{x}$ and $\mathbf{z}$ are independent of each other as random vectors. Furthermore, $\mathbf{x}^T\mathbf{x}$ and $\mathbf{z}^T\mathbf{z}$ are independent chi-squared random variables on k and l degrees of freedom, multiplied by σ^2 because they each have the joint densities of independent normal sums of squares (see 2.5.4 and 11.5.4). Therefore, they are gamma variables. But then we know that $R^2 = (\mathbf{x}^T\mathbf{x})/(\mathbf{x}^T\mathbf{x} + \mathbf{z}^T\mathbf{z})$ we know must follow a Beta$(k/2, l/2)$ law (see 9.5.5). Of course, now we know that this must be true for any spherically symmetric law:

Theorem. *Let $\mathbf{x}$ be a k-dimensional random vector and $\mathbf{z}$ an l-dimensional random vector orthogonal to it. Assume that $\mathbf{x} + \mathbf{z}$ is spherically distributed about the origin. Then $R^2 = (\mathbf{x}^T\mathbf{x})/(\mathbf{x}^T\mathbf{x} + \mathbf{z}^T\mathbf{z})$ has a Beta$(k/2, l/2)$ distribution.*

Example. In Chapters 1 and 2 we looked at a third-grade arithmetic test and at the effects of curriculum and gender on scores. It was plausible to conclude that curriculum mattered; but was there any improvement in our understanding if we went from there to a model that included gender and gender–curriculum interaction? The sum of squares for those two effects was 18.0 on 2 degrees of freedom, and the residual sum of squares was 774.8 on 16 degrees of freedom (see 2.8.1). The R-squared is then 0.0227; if these effects do not matter, and errors obey spherical symmetry, then this is a Beta$(1, 8)$ random variable. We can calculate the probability that it would be at least that large by chance, using the beta–binomial duality: $P(R^2 \geq 0.0227) = P[X < 1|B(8, 0.0227)] = 0.832$. The statistic is certainly not surprisingly large, so we have no evidence that gender contributes anything to the observed scores.

This example was chosen to have even numbers of degrees of freedom, so that the beta parameters were whole numbers and we could use duality to compute them. In an exercise, you will see ways to compute beta probabilities in some other cases.

We defined the F-statistic to be centered at 1 if the parameters are really zero. But we can turn the traditional F into a beta, and vice versa: In ourcurrent notation, $F_{k,l} = (\mathbf{x}^T\mathbf{x}/k)/(\mathbf{z}^T\mathbf{z}/l)$. After some easy algebra (which you should check), $R^2 = 1/(1 + l/(kF_{k,l}))$. We actually carried out this change of variables in Chapter 10 (using whole-number parameters; see 10.4.1):

Proposition. *Let Y be an $F_{k,l}$ statistic from errors spherically distributed about the origin. Then its density is*

$$f(y) = \Gamma(k/2 + l/2)/[\Gamma(k/2)\Gamma(l/2)]k^{k/2}l^{l/2}[y^{l/2-1}/(l + ky)^{k/2+l/2}]$$

on $y > 0$.

We do not need any new computing methods, because we can always change these into betas.

12.3.3   Elementary Errors

Independent normal error models have been the usual way to analyze linear measurement models ever since they were first proposed by Gauss in 1816. The traditional justification has been somewhat different from ours, and it is called the *theory of elementary errors*. We try to imagine in detail where measurement inaccuracies come from. For example, let there be many small independent sources of error for each trial, and let each of them either increase the measurement by a tiny amount δ or decrease it by that same amount, with probability 0.5 each. If there are m such sources, then the total error is $e_i = \delta X - m/2$, where X is $B(m, 0.5)$. We also know that for $\sqrt{m}/2$ large, X is approximately Normal$(m/2, m/4)$. But then e_i is approximately Normal$(0, \delta^2 m/4)$. If all other trials work similarly, then our test statistics will have approximately the distribution we have derived in this chapter.

Over the last 250 years, ever more sophisticated versions of the theory of elementary errors have been derived, called *central limit theorems*. The individual sources do not have to be Bernoulli, nor discrete, nor have exactly the same distributions (so long as they each have variance small enough to make negligible contribution to the total), nor completely independent; the total errors still are approximately normal for enough sources. You will study a result in that direction in the next chapter. Such facts, along with 150 years of practical experience with normal linear models, leave us with considerable confidence in their reasonableness.

12.4 Maximum Likelihood Estimation in Continuous Models

12.4.1 Continuous Likelihoods

Now that we have a plausible probabilistic model for ANOVA and regression, we might see what good estimates of the parameters would consist of. We would like something like the criterion of maximum likelihood from discrete model theory (see 8.2.2). The likelihood was the probability of the observations given a parameter value, but of course we no longer have probability mass functions. The (continuous) likelihood ratio test (see 9.7.2) suggested, at least for one variable, that the density should perhaps play the same role as the mass function in defining a likelihood.

We will once again use a Bayes's theorem argument to motivate a likelihood, which, recall, was a sort of probability that a parameter value was correct. For absolutely continuous random variables, we of course need the new form of the theorem from the last chapter (see 11.6.1). One possible prior distribution (which says that we are very ignorant) is to let θ be *uniform* on some large set A that includes all plausible values, whose volume is $V(A) < \infty$. Then on A the density is $f(\theta) = \frac{1}{V(A)}$. (For example, if the parameter were something like a binomial probability p, then an obvious A is $0 < p < 1$.) Now our version of Bayes's theorem says that the posterior density of θ after we have obtained our observations $\mathbf{x}$ is

$$f(\theta|\mathbf{x}) = \frac{f(\mathbf{x}|\theta)/V(A)}{\int_A f(\mathbf{x}|\theta)d\theta/V(A)} = \frac{f(\mathbf{x}|\theta)}{\int_A f(\mathbf{x}|\theta)d\theta}$$

over the entire large set A. But this just says that the posterior density for θ (what we know about how relatively probable various values are after we do our experiment) is proportional to $f(\mathbf{x}|\theta)$. We shall therefore define the likelihood of that vector as follows:

Definition.

(i) The **likelihood** of a parameter vector θ given an absolutely continuous observation vector $\mathbf{x}$ is $L(\theta|\mathbf{x}) = f(\mathbf{x}|\theta)$.

(ii) Its maximum likelihood estimate $\hat{\theta}$ is a value of the parameter vector that makes the likelihood as large as possible.

Finding maximum likelihood estimates for parameters in absolutely continuous families will be much like the same process in discrete families (see 8.2.2). For example, let each of n observations x_i be from a Gamma(α, β) random variable, with α known. As in (11.6.2), we wish to estimate the unknown β. The joint density of these independent observations is obtained by multiplication (see 11.5.1):

$$f(\mathbf{x}|\beta) = \prod_{i=1}^{n} f(x_i|\beta) = \prod_{i=1}^{n} \frac{x_i^{\alpha-1}}{\beta^\alpha \Gamma(\alpha)} e^{-x_i/\beta} = L(\beta|\mathbf{x}).$$

As in Chapter 8, taking the log-likelihood simplifies things considerably:

$$\log L(\beta|\mathbf{x}) = l(\beta|\mathbf{x}) = -n \log \Gamma(\alpha) + (\alpha - 1) \sum_{i=1}^{n} \log x_i - n\alpha \log \beta - \frac{1}{\beta} \sum_{i=1}^{n} x_i.$$

To find the β that makes it largest, differentiate with respect to β and set the result equal to zero: $0 = \partial l(\beta|\mathbf{x})/(\partial \beta) = -(n\alpha)/\beta + 1/\beta^2 \sum_{i=1}^{n} x_i$. Solving, we get a maximum likelihood estimate $\hat{\beta} = \frac{\bar{x}}{\alpha}$. In this case, we get the standard, method-of-moments estimate, which in (11.6.2) was seen to be the large-sample limit of a certain Bayes's estimate.

You will discover that life is not so simple when you try in an exercise to estimate an unknown gamma shape parameter α.

12.4.2   Maximum Likelihood with Normal Errors

A model with independent normal errors is a set of predictions $\hat{y}_i$ of experimental measurements y_i, where the residuals $y_i - \hat{y}_i$ are independently Normal$(0, \sigma^2)$. The likelihood is $L(\theta|\mathbf{y}) = 1/[(2\pi)^{n/2}\sigma^n]e^{-1/(2\sigma^2)\sum_{i=1}^{n}(y_i-\hat{y}_i)^2}$, where the parameters θ are whatever information you need to determine the predictions. As before, we instead maximize the log likelihood

$$l(\theta|\mathbf{y}) = \log L(\theta|\mathbf{y}) = -\frac{n}{2}\log(2\pi) - n\log\sigma - \frac{1}{2\sigma^2}\sum_{i=1}^{n}(y_i - \hat{y}_i)^2.$$

But θ comes into this expression only through the predictions $\hat{y}_i$, so maximizing the likelihood is equivalent to minimizing $\sum_{i=1}^{n}(y_i - \hat{y}_i)^2$. We have seen this before; it is the criterion of least squares. So far, maximum likelihood tells us nothing new.

Proposition. *A prediction model $\hat{y}$ for a dependent vector y with errors $y - \hat{y}$ distributed independently Normal$(0, \sigma^2)$ has maximum likelihood estimates for its parameters equal to the least-squares estimates.*

Now let us estimate the other parameter, σ, by maximum likelihood. Differentiate with respect to the unknown parameter to get $0 = (\partial l)/(\partial \sigma) = -(n/\sigma) + (1/\sigma^3)\sum_{i=1}^{n}(y_i - \hat{y}_i)^2$. Solving, we obtain $\hat{\sigma}^2 = \frac{1}{n}\sum_{i=1}^{n}(y_i - \hat{y}_i)^2$. Interestingly, this is just the MSE, except that we divide by n instead of by the degrees of freedom for error.

If you are going to use this as an estimate of the unknown quantity σ^2, you will want some idea of how good it is. If we knew all the parameters in our least-squares prediction model, then the predictions $\hat{y}_i$ would be known in advance. Then the random variables $(y_i - \hat{y}_i)/\sigma$ are independently standard normal, and then their sum of squares is $\sum_{i=1}^{n}(y_i - \hat{y}_i)^2/\sigma^2$. This last is, of course, a chi-squared random variable on n degrees of freedom, which is also a Gamma($n/2$, 2). Our maximum-likelihood estimator $\hat{\sigma}^2 = \frac{1}{n}\sum_{i=1}^{n}(y_i - \hat{y}_i)^2$ is just a multiple of this (multiply by σ^2 and divide by n), and so a Gamma($n/2$, $2\sigma/n$) random variable. Recall what we know about gamma variables, we compute $E(\hat{\sigma}^2) = (n/2) \cdot ((2\sigma^2)/n) = \sigma^2$. That is, the estimate is correct on average; it is *unbiased*. (Actually, by the argument used in (7.6.4), this is always true for independent and constant-variance errors, whether they are normal or not.) To get some idea of how uncertain it is, we compute $\text{Var}(\hat{\sigma}^2) = (n/2) \cdot [(2\sigma^2)/n]^2 = 2\sigma^4/n$. As when we estimated μ using the sample mean, we find the variance decreasing in proportion to the sample size. The standard deviation of $\hat{\sigma}^2$, which we call the *standard error* of the estimate, is then $(\sqrt{2\sigma^2})/\sqrt{n}$.

There is a chicken-or-egg problem with our accuracy estimates from the last paragraph: We have to *know* the value of σ^2 in order to calculate the standard error of its estimate. But we would want to estimate it only if it were *unknown*. In practice, we usually commit the obvious cheat here: Just insert the estimate $\hat{\sigma}^2$ in the formula for the standard error, $\sqrt{2}\hat{\sigma}^2/\sqrt{n}$, and hope it is accurate enough to be useful.

Example. In the collection of measurements of the speed of light by Michelson in Chapter 1 (see 1.2.1), we might imagine (for want of knowing any better) that his errors were independently Normal(0, σ^2). Oddly enough, we do know what the predictions should have been, $\hat{y}_i = 299,710.5$, which is (as this book is being written) believed to be, to high accuracy, the true speed of light in air. Therefore, the mean squared error we computed in Chapter 2, $\hat{\sigma}^2 = 12,586$, is a maximum likelihood estimate of σ^2. Its standard error is approximately 3,711.

12.4.3 Unbiased Variance Estimates

Of course, as the example suggests, there is an even more fundamental difficulty with estimating σ^2. In the real world of regression and ANOVA problems, we hardly ever know the parameters in advance; we have to estimate them, and therefore the predictions $\hat{y}_i$, by least squares. That should obviously mess up the distribution of the residuals $\mathbf{e} = \mathbf{y} - \hat{\mathbf{y}}$, because now $\hat{\mathbf{y}}$ is estimated from the data and is full of random errors. We need some notation for the "correct" predictions that would have been made if you had known the correct values of the parameters in one-, two-, three-way ANOVA or your single or multiple linear regression problem, or whatever; call the vector $\hat{\mathbf{y}}^C$. The assumption of normal errors says that $\mathbf{e}^C = \mathbf{y} - \hat{\mathbf{y}}^C$ consists of n independent Normal(0, σ^2) coordinates. Now, when we made our predictions by least squares, we found in each case that the vector $\hat{\mathbf{y}}$ of predictions was at right angles to $\mathbf{e} = \mathbf{y} - \hat{\mathbf{y}}$ (because by ordinary geometry that made the length of the residual vector as short as possible, review Chapter 2). But $\hat{\mathbf{y}}$ may lie anywhere in a k-dimensional subspace of the data space, determined by a linear model with k free parameters (for example, a one-way layout with k levels of the treatment). But $\hat{\mathbf{y}}^C$, because it is calculated from a model of the same form, lies in that same k-dimensional subspace. Therefore, so does $\mathbf{x} = \hat{\mathbf{y}} - \hat{\mathbf{y}}^C$, the vector of errors in the predictions. Then $\mathbf{x}$ is perpendicular to $\mathbf{e}$. This gives us a Pythagorean theorem $(\mathbf{e}^C)^T \mathbf{e}^C = \mathbf{x}^T \mathbf{x} + \mathbf{e}^T \mathbf{e}$.

One interesting consequence is that the estimated residuals $\mathbf{e}$ are always shorter than the residuals from the correct model $\mathbf{e}^C$ (the hypotenuse is always the longest side of a right triangle). This means that if we use our maximum likelihood estimator $\hat{\sigma}^2$ on the estimated residuals, we will tend to underestimate the true value of $\hat{\sigma}$. Let us quantify that underestimate. Remember that in Section 2 we discovered that we could factor the likelihood

$$f(\mathbf{e}^C) = \frac{1}{(2\pi)^{n/2}\sigma^2}e^{-\mathbf{e}^{C^T}\mathbf{e}^C/(2\sigma^2)} = \frac{1}{(2\pi)^{n/2}\sigma^n}e^{-(\mathbf{x}^T\mathbf{x}+\mathbf{e}^T\mathbf{e})/(2\sigma^2)}$$

$$= \frac{1}{(2\pi)^{k/2}\sigma^k}e^{-\mathbf{x}^T\mathbf{x}/(2\sigma^2)}\frac{1}{(2\pi)^{(n-k)/2}\sigma^{n-k}}e^{-\mathbf{e}^T\mathbf{e}/(2\sigma^2)}.$$

Because $\mathbf{e}$ is perpendicular to the vectors $\mathbf{x}$ in k dimensions, it may be anywhere in an $(n-k)$-dimensional subspace. Just as we reasoned before, $\mathbf{e}^T\mathbf{e} = \sum_{i=1}^{n}(y_i-\hat{y}_i)^2$ is σ^2 times a chi-squared random variable, but now on $n-k$ degrees of freedom. But then $E[\sum_{i=1}^{n}(y_i - \hat{y}_i)^2] = \sigma^2(n - k)$, and we conclude that $E(MSE) = E[\frac{1}{n-k}\sum_{i=1}^{n}(y_i - \hat{y}_i)^2] = \sigma^2$. Therefore, an adjusted version of the maximum likelihood estimator, $\hat{\sigma}^2 = \frac{1}{n-k}\sum_{i=1}^{n}(y_i-\hat{y}_i)^2$, is unbiased for σ^2. (As an exercise, and inspired by the argument in (7.6.4), you should check that this is always true, whether those independent errors with common variance are normal or not.) It is more common to use this estimate in practice than it is to use the simple maximum likelihood estimate. We will let you discover its variance and standard error.

Notice that if we have values we believe in (or wish to hypothesize) for the parameters of the model, then we have an alternative unbiased estimator $\hat{\sigma}^2 = \frac{1}{k}\mathbf{x}^T\mathbf{x} = \frac{1}{k}\sum_{i=1}^{n}(\hat{y}_i - \hat{y}_i^C)^2$. Its distribution is independent of the previous one. A ratio of the two unbiased estimates of σ^2 has the F-distribution from the last section; it is, of course, a test of whether or not you really should believe your hypothesis about the parameters. In exercises you will use this principle to carry out tests of some such hypotheses.

12.5   The G-Squared Statistic

12.5.1   When the Variance Is Known

You will remember (see Chapter 8.6) that we could compare two models estimated by maximum likelihood in which one was a special case of the other. We took the ratio of the likelihoods of the bigger model to the smaller model and computed twice its logarithm to get the G-squared statistic. This makes sense for comparing normal linear models as well. For example, let the question be, as in the last section, whether the treatments really have no effect, that is, whether all $b_j = 0$. Then the residuals from that simple model are, in the notation of the last section, $y_i - \hat{y}_i^C = y_i - \bar{y} = e_i^C$. If you are not willing to make that assumption, then you compute your usual least-squares (and therefore maximum likelihood) estimates

$\hat{b}_j$. The residuals from this more sophisticated model are $y_i - \hat{y}_i = y_i - \bar{y} - \hat{b}_j = e_i$, and the x-vector consists of the estimates $\hat{b}_j$.

Let us make another of those absurdly unrealistic assumptions: that we have independent normal errors, and that we *know* the error variance σ^2. It is hard to imagine why that would ever be true, but it will help us see how the G-squared statistic works. Under the more sophisticated model, which allows for treatment effects, the likelihood of all the observations is $L(\mu, \mathbf{b}|\mathbf{y}) = 1/[(2\pi)^{n/2}\sigma^n]e^{-\mathbf{e}^T\mathbf{e}/(2\sigma^2)}$. The likelihood of the simpler, no effects, model is $L(\mu, |\mathbf{y}) = \frac{1}{(2\pi)^{n/2}\sigma^n}e^{-\mathbf{e}^{C^T}\mathbf{e}^C/(2\sigma^2)} = \frac{1}{(2\pi)^{n/2}\sigma^n}e^{-(\mathbf{x}^T\mathbf{x}+\mathbf{e}^T\mathbf{e})/(2\sigma^2)}$. When we take the ratio of these two likelihoods, the constants cancel out. Then we take the logarithm, which removes the exponential, and multiply by two, to get our statistic

$$G^2(\mathbf{b}) = (\mathbf{x}^T\mathbf{x} + \mathbf{e}^T\mathbf{e})/\sigma^2 - \mathbf{e}^T\mathbf{e}/\sigma^2 = \mathbf{x}^T\mathbf{x}/\sigma^2.$$

Notice that this has *exactly* a chi-squared distribution on $k-1$ degrees of freedom. This is interesting, because in Chapter 8 and (11.5.4), working with our very different discrete models for contingency tables, we got *approximate* chi-squared distributions, also on degrees of freedom corresponding to the number of free parameters. You will learn in advanced courses that G-squared statistics are at least roughly chi-squared distributed for tests of a great variety sorts of models.

12.5.2   When the Variance Is Unknown

In the real world, of course, we usually have to estimate σ^2, a different estimate in the numerator and the denominator. These are the maximum likelihood estimates. In the numerator, treatment-effects, model $\hat{\sigma}^2 = \mathbf{e}^T\mathbf{e}/n$. Then the likelihood becomes

$$L(\mu, \mathbf{b}, \sigma|\mathbf{y}) = \frac{1}{(2\pi)^{n/2}(\mathbf{e}^T\mathbf{e}/n)^{n/2}}e^{-n/2};$$

the data have canceled out of the exponent. Similarly, for the denominator model,

$$L(\mu, \sigma|\mathbf{y}) = \frac{1}{(2\pi)^{n/2}(\mathbf{e}^{C^T}\mathbf{e}^C/n)^{n/2}}e^{-n/2}.$$

Twice the logarithm of the ratio gives us $G^2(\mathbf{b}) = n\log[(\mathbf{x}^T\mathbf{x} + \mathbf{e}^T\mathbf{e})/\mathbf{e}^T\mathbf{e}] = n\log(1 + \frac{\mathbf{x}^T\mathbf{x}}{\mathbf{e}^T\mathbf{e}})$. In earlier sections, our test for zero treatment effects was

$$F_{(k-1),(n-k)} = \frac{\mathbf{x}^T\mathbf{x}/(k-1)}{\mathbf{e}^T\mathbf{e}/(n-k)} = \frac{n-k}{k-1}\frac{\mathbf{x}^T\mathbf{x}}{\mathbf{e}^T\mathbf{e}}.$$

It seems that G-squared is an increasing function of F—we may use either measure of fit we want, and we will come to the same conclusion.

Often, we can say more. Ideally, we have many data points for each parameter we are estimating (for example, many measurements per treatment level). Then $n - k$ is large compared to $k - 1$. But then, the expected value of $\mathbf{e}^T\mathbf{e}$ is large compared to that of $\mathbf{x}^T\mathbf{x}$. Then, typically, $G^2(\mathbf{b}) = n\log(1 + \frac{\mathbf{x}^T\mathbf{x}}{\mathbf{e}^T\mathbf{e}}) \approx n\frac{\mathbf{x}^T\mathbf{x}}{\mathbf{e}^T\mathbf{e}}$, by the basic results about approximation to logarithms in Chapter 3 (see 3.5.1). In this case $\frac{n-k}{n} \approx 1$:

Proposition. *For a test of $k - 1$ free parameters in a normal model with $n - k$ error degrees of freedom, with n large compared to $k - 1$ and SSE large compared to SST,* $G^2 \approx (k - 1)F$.

But notice also that as n gets large for fixed k, $\hat{\sigma}^2 = \mathbf{e}^T\mathbf{e}/(n - k)$ converges in probability to σ^2. (Review (7.7.2), and notice its expected value and variance). Then $G^2(b) \approx (\mathbf{x}^T\mathbf{x})/\sigma^2$; and we have discovered once again that G-squared is *approximately* chi-squared on the appropriate numbers of degrees of freedom.

12.6   General Linear Models

12.6.1   Matrix Form

You have probably noticed that we have not yet tackled the issue of how accurate the parameter estimates are that we obtained by least-squares. It will turn out to be a reasonably easy problem to solve, once we get used to a drastic change in notation. We will find ourselves doing what we did in Chapter 2, but instead of using summation or vector notation, we will use *matrix* notation.

To illustrate, let us look at one of the more complicated models we studied in Chapter 1, the multiple linear regression model with two independent variables (see 1.6.2):

$$\hat{y}_i = \mu + (x_{i1} - \bar{x}_1)b_1 + (x_{i2} - \bar{x}_2)b_2.$$

We can write the individual predictions as a vector (inner) product

$$\bar{y}_i = \begin{pmatrix} 1 & x_{i1} - \bar{x}_1 & x_{i2} - \bar{x}_2 \end{pmatrix} \begin{pmatrix} \mu \\ b_1 \\ b_2 \end{pmatrix}.$$

An advantage of this way of writing is that we have separated the things that we knew before we made our experimental observations (the row vector with the x's) from the presumably unknown parameters μ and the b's that we shall have to estimate from the observations (the column vector). Of course, we want to predict a number of y's, so that we will have a (column) vector of $\hat{y}$'s. The obvious way to do that is to stack the row vectors into a matrix:

$$\begin{pmatrix} \hat{y}_1 \\ \hat{y}_2 \\ \vdots \\ \hat{y}_n \end{pmatrix} = \begin{pmatrix} 1 & x_{11} - \bar{x}_1 & x_{12} - \bar{x}_2 \\ 1 & x_{21} - \bar{x}_1 & x_{22} - \bar{x}_2 \\ \vdots & \vdots & \vdots \\ 1 & x_{n1} - \bar{x}_1 & x_{n2} - \bar{x}_2 \end{pmatrix} \begin{pmatrix} \mu \\ b_1 \\ b_2 \end{pmatrix}.$$

You should check that this is still a matrix product. We denote the matrix by some letter like X, and call it the *design* matrix. The vector of parameters we call something like **b**. Then we can write the whole prediction process $\hat{\mathbf{y}} = \mathbf{Xb}$.

Of course, we can write any of our linear regression models this way: The simple proportionality model has only one column for X and one entry in **b**. The simple linear regression model has two each; and other multiple regression and multilinear models have more than three columns and three parameters. The notation is so useful that we will give it a name. First, it is usual to express it in terms of the residuals $y - \hat{y} = e$.

Definition. A **general linear model** is $y = Xb + e$, where y is a vector of n experimental observations, X is an $n \times p$ **design** matrix of (known) experimental settings, **b** is a p-vector of (unknown) parameters, and **e** is an n-vector of residuals (errors).

In our two-independent-variable example, $p = 3$.

This notation is so flexible that it may be applied to many problems other than ordinary regression problems, in fact, to all the linear measurement models from the first chapter. For example, consider a one-way layout with 3 levels of the treatment and 2 observations at each level. We will start with the uncentered model, with parameters the individual treatment means μ. The trick will be to write, for, say, the second observation of the third treatment, $\hat{y}_{23} = \mu_3 = 0 \cdot \mu_1 + 0 \cdot \mu_2 + 1 \cdot \mu_3$. We have managed to construct another inner product involving all the parameters, even though two of them were redundant. Turn this, as above, into a matrix product

$$\begin{pmatrix} \hat{y}_{11} \\ \hat{y}_{21} \\ \hat{y}_{12} \\ \hat{y}_{22} \\ \hat{y}_{13} \\ \hat{y}_{23} \end{pmatrix} = \begin{pmatrix} 1 & 0 & 0 \\ 1 & 0 & 0 \\ 0 & 1 & 0 \\ 0 & 1 & 0 \\ 0 & 0 & 1 \\ 0 & 0 & 1 \end{pmatrix} \begin{pmatrix} \mu_1 \\ \mu_2 \\ \mu_3 \end{pmatrix}.$$

We have succeeded in putting it in general linear model form. Those 1's and 0's in the X-matrix are called *dummy variables*—they are not numerical settings, they just tell us into which treatment group the subject fell.

12.6.2   Centered Form

We, of course, preferred the centered form of our linear models, which must have a different matrix representation because its parameters are different. As in the regression case, the first column, the μ-column, just consists of a 1 for each observation. But what does the other centering condition $\frac{1}{n}\sum_{\text{all } i} b_j = 0$ mean in the new notation? It is hard to tell, because of course we do not know the b_j. Looking back at the centering conditions in the regression case, $0 - \frac{1}{n}\sum_{i=1}^{k}(x_{ij} - \bar{x}_j)b_j = \frac{b_j}{n}\sum_{i=1}^{k}(x_{ij} \quad \bar{x}_j)$, we see that we can guarantee them without knowing the regression coefficients, if only that last sum is zero (which it is). The same reasoning will work in ANOVA models if we adjust our 1's and 0's so that they average zero. That can be done in exactly the same way: Subtract the column average from each entry in the X matrix x_{ij} to get $x_{ij} - \bar{x}_j$. Then, just as in Chapter 1, each column average will be zero. In our example, each column average is $\frac{1}{3}$, so the new matrix expression is

$$
\begin{pmatrix} \bar{y}_{11} \\ \bar{y}_{21} \\ \bar{y}_{12} \\ \bar{y}_{22} \\ \bar{y}_{13} \\ \bar{y}_{23} \end{pmatrix} = \begin{pmatrix} 1 & 2/3 & -1/3 & -1/3 \\ 1 & 2/3 & -1/3 & -1/3 \\ 1 & -1/3 & 2/3 & -1/3 \\ 1 & -1/3 & 2/3 & -1/3 \\ 1 & -1/3 & -1/3 & 2/3 \\ 1 & -1/3 & -1/3 & 2/3 \end{pmatrix} \begin{pmatrix} \mu \\ b_1 \\ b_2 \\ b_3 \end{pmatrix}.
$$

Generally in the one-way layout the correction will be n_j/n. As exercises, we will let you write out the centered matrix models for other cases of ANOVA. For now, we will make this a definition and see later what advantages it offers.

Definition. The design matrix X is **centered** if the first column consists of 1's and the average of each later column is 0.

We know that we can compute the b_k coefficient once we have estimated the others; so we could still do the predictions in this model if we simply leave off the last column, and the last row of the coefficient vector. Later we will ask when we would wish to do that.

12.6.3 Least-Squares Estimates

Now for the fun part: we can estimate the general linear model by least squares, in all its many cases, by an easy application of the method from Chapter 2. We want that vector **b** that makes the squared length of the residual vector $\mathbf{e}^T\mathbf{e} = (\mathbf{y} - \mathbf{Xb})^T(\mathbf{y} - \mathbf{Xb})$ as small as possible. For *any* value **c** of the coefficient vector,

$$
\begin{aligned}
(\mathbf{y} - \mathbf{Xc})^T(\mathbf{y} - \mathbf{Xc}) &= [\mathbf{y} - \mathbf{Xb} + \mathbf{X(b} - \mathbf{c})]^T[\mathbf{y} - \mathbf{Xb} + \mathbf{X(b} - \mathbf{c})] \\
&= (\mathbf{y} - \mathbf{Xb})^T(\mathbf{y} - \mathbf{Xb}) + (\mathbf{b} - \mathbf{c})^T\mathbf{X}^T(\mathbf{y} - \mathbf{Xb}) \\
&\quad + (\mathbf{y} - \mathbf{Xb})^T\mathbf{X(b} - \mathbf{c}) + (\mathbf{b} - \mathbf{c})^T\mathbf{X}^T\mathbf{X(b} - \mathbf{c}).
\end{aligned}
$$

We did this by applying the familiar rules of algebra, but recall two special features of matrix multiplication: that $\mathbf{AB}$ does not necessarily equal $\mathbf{BA}$, but always $(\mathbf{AB})^T = \mathbf{B}^T\mathbf{A}^T$. This is the same argument as the one used repeatedly in Chapter 2. Notice that by the rule for transposes, the two middle terms are transposes of each other, but they are scalars (1×1 matrices), and so equal. We can get rid of both of them if we choose **b** such that $\mathbf{X}^T(\mathbf{y} - \mathbf{Xb}) = 0$, that is, if **b** is a solution of $\mathbf{X}^T\mathbf{y} = \mathbf{X}^T\mathbf{Xb}$. If you will decode the matrix algebra, this is a linear system of p equations in p unknowns; it will be the general version of our normal equations. Finally, the last term in our expansion is the squared length of the vector $\mathbf{X(b} - \mathbf{c})$; which is at least 0. We summarize these in a familiar form:

Theorem (least-squares estimates for the linear model). *For the prediction model* $\hat{\mathbf{y}} = \mathbf{Xb}$, *let* **b** *be a solution of the normal equation* $\mathbf{X}^T\mathbf{y} = \mathbf{X}^T\mathbf{Xb}$; *then*

(i) *for any vector* **c**, $(\mathbf{y} - \mathbf{Xc})^T(\mathbf{y} - \mathbf{Xc}) = (\mathbf{y} - \mathbf{Xb})^T(\mathbf{y} - \mathbf{Xb}) + (\mathbf{b} - \mathbf{c})^T\mathbf{X}^T\mathbf{X(b} - \mathbf{c})$; *therefore*

(ii) **b** *is a least-squares estimate; and in particular, letting* $\mathbf{c} = 0$, *we have the decomposition of the sum of squares*

(iii) $\mathbf{y}^T\mathbf{y} = \mathbf{b}^T\mathbf{X}^T\mathbf{X}\mathbf{b} + (\mathbf{y} - \mathbf{X}\mathbf{b})^T(\mathbf{y} - \mathbf{X}\mathbf{b})$.

As exercises, we will let you check that the various estimates from Chapter 2 were all special cases of this theorem. For example, simple linear regression ($p = 2$ columns) leads to the matrix equation (as you should check)

$$\mathbf{X}^T\mathbf{y} = \begin{pmatrix} \sum_{i=1}^n y_i \\ \sum_{i=1}^n (x_i - \bar{x})y_i \end{pmatrix} = (\mathbf{X}^T\mathbf{X})\mathbf{b} = \begin{pmatrix} n & 0 \\ 0 & \sum_{i=1}^n (x_i - \bar{x})^2 \end{pmatrix} \begin{pmatrix} \mu \\ \mathbf{b} \end{pmatrix}.$$

As an exercise, you should multiply this out and solve to see that the estimates are the same ones we got before. It will have been easy because of these zeros, which are there because the model is centered.

Often we can use matrix notation to write down the solution of the normal equations. If the $p \times p$ matrix $\mathbf{X}^T\mathbf{X}$ is *nonsingular*, then we can compute its inverse matrix $(\mathbf{X}^T\mathbf{X})^{-1}$ (time to review more matrix algebra). This is a matrix that has the property $(\mathbf{X}^T\mathbf{X})^{-1}\mathbf{X}^T\mathbf{X} = \mathbf{I}$, where $\mathbf{I}$, the ($p \times p$) identity matrix, has 1's on the diagonal and zeros elsewhere. Therefore, for any matrix $\mathbf{A}$ with p rows, $\mathbf{I}\mathbf{A} = \mathbf{A}$. Our method will be to multiply both sides of the normal equation by the inverse matrix to get $(\mathbf{X}^T\mathbf{X})^{-1}\mathbf{X}^T\mathbf{y} = (\mathbf{X}^T\mathbf{X})^{-1}\mathbf{X}^T\mathbf{X}\mathbf{b} = \mathbf{I}\mathbf{b}$. This gives us the explicit matrix equation $\hat{\mathbf{b}} = (\mathbf{X}^T\mathbf{X})^{-1}\mathbf{X}^T\mathbf{y}$ for the least-squares estimate.

This equation is less useful than you may think. For one thing, it is usually easier just to solve the normal equations than to compute the inverse. For another, very often $\mathbf{X}^T\mathbf{X}$ is singular, and therefore has no inverse (though you can still solve the equations). For example, the uncentered version of the one-way layout gives a singular $\mathbf{X}^T\mathbf{X}$ matrix. If we throw out the last column of $\mathbf{X}$, then $\mathbf{X}^T\mathbf{X}$ it can be inverted (exercise). Generally, you can pare down any design matrix until it has the minimum number of parameters; then $\mathbf{X}^T\mathbf{X}$ may be inverted.

12.6.4 Homoscedastic Errors

To talk about the accuracy of our estimates, we will need to imagine that the observation vector is a random vector. Earlier in the chapter we had made two somewhat different assumptions about the nature of the residuals: (1) that they have spherical symmetry and (2) that they are independent of one another and have the same marginal distributions. We then noted that independent normal errors with constant variance met both our criteria. To discuss accuracy, we will make yet a third assumption. No doubt you noticed in Chapters 2 and 7 how similar the theories of least squares and variances are (there had to be some reason for calling s^2 a *sample* variance). To exploit this, we will assume in this section that our residuals each have expectation 0 and the same finite variance, and that they are uncorrelated. There is an impressive word for this:

Definition. A residual vector **e** is *homoscedastic* if for all i, $E(e_i) = 0$, $\text{Var}(e_i) = \sigma^2$, and for all $i \neq j$, $\text{Cov}(e_i, e_j) = 0$.

This condition will let covariance terms drop out of variances of linear combinations, which will keep things simple. We have already noted the connections with the earlier assumptions: If either spherically symmetric or independent identically distributed residuals with expectation 0 have finite variances, then they are homoscedastic. (You will discover in exercises that it does not always work the other way.) Then, obviously, independent normal residuals are also homoscedastic.

For an experiment believed to have homoscedastic residuals, take the average value of our least-squares estimate

$$E(\hat{b}) = E[(X^TX)^{-1}X^Ty] = E[(X^TX)^{-1}X^T(Xb + e)].$$

Multiplying by a matrix of constants is just taking linear combinations; furthermore, we assumed $E(e) = 0$. Therefore, $E(\hat{b}) = (X^TX)^{-1}X^TXb = b$; a least-squares estimate of b is unbiased.

Now for the variances of the estimates. Actually, we are going to extract the *covariance* matrix of the random vector $\hat{b}$, which will include the covariances between the estimates of different parameters. The covariance matrix of the residual vector is, because it is homoscedastic, $\hat{b}$; where I is the $n \times n$ identity matrix. But since the predictions Xb are constants, $Var(y) = Var(Xb + e) = Var(e) = \sigma^2 I$. We want $Var(b) = E[(\hat{b} - b)(\hat{b} - b)^T]$. We know $\hat{b} = (X^TX)^{-1}X^Ty$. Back in Chapter 7 we derived a theorem that said that for any random vector Z and any matrix A, then the covariance matrix $Var(AZ) = AVar(Z)A^T$ (see 7.6.2).

We now calculate that covariance matrix:

$$Var[\hat{b}] = (X^TX)^{-1}X^TVar(y)X(X^TX)^{-1}$$
$$= \sigma^2(X^TX)^{-1}X^TIX(X^TX)^{-1} = \sigma^2(X^TX)^{-1}$$

after the identity matrices drop out. This is what we wanted.

Theorem (variance of least-squares estimates). *For the least-squares estimate of a homoscedastic linear model with nonsingular design matrix,* $Var(\hat{b}) = \sigma^2(X^TX)^{-1}$.

For example, in the simple linear regression model,

$$(X^TX)^{-1} = \begin{pmatrix} 1/n & 0 \\ 0 & 1/\sum_{i=1}^n (x_i - \bar{x})^2 \end{pmatrix}.$$

Therefore, $Var(\bar{\mu}) = \sigma^2/n$ (which we essentially discovered in Chapter 7) and $Var(\hat{b}) = \sigma^2/\sum_{i=1}^n (x_i - \bar{x})^2$. In addition, we learn that the two estimates are uncorrelated (another consequence of centering). In practice, of course, we do not know σ^2, but a sensible estimate of it using the MSE will let us plausibly approximate these quantities.

Example. In Chapter 2 we studied how the proportion of British children p taken to the doctor might be predicted from their ages a by $\hat{p} = \mu + (a - \bar{a})b$ (see 2.6.1). From the MSE we get $\hat{\sigma}^2 = 92.485$. We also found that $\sum_{i=1}^n (a_i - \bar{a})^2 = 280$. Our theorem then tells us that $Var(\bar{\mu}) = 92.485/15 = 6.1657$ and $Var(\hat{b}) = 92.485/280 = 0.3303$, approximately. Turning these into standard errors, we can express our conclusion as $\bar{\mu} = 56.33$ with error ± 2.48, and $\hat{b} = -1.5714$ with error ± 0.5747.

12.6.5 Linear Combinations of Parameters

Uncertainty in one parameter estimate may not be exactly what we want to know. For example, we may have done our estimates for the parameters μ and b_i for the centered one-way layout. What if we wanted instead to know the level mean $\mu_i = \mu + b_i$? As we know, we do not have to recalculate the estimate; it will still be least squares, because that depends only on making residuals small, and the residuals are still the same. But do we have to recalculate the variances? The new parameter is a linear combination of the old parameters, and we know from Chapter 7 how to compute variances of linear combinations (see 7.6.1). Let $\mathbf{a}$ be the vector of coefficients, and we want to know how good an estimate of $\mathbf{a}^T\mathbf{b}$ the obvious estimate $\mathbf{a}^T\hat{\mathbf{b}}$ might be. First, $E(\mathbf{a}^T\hat{\mathbf{b}}) = \mathbf{a}^T E(\hat{\mathbf{b}}) = \mathbf{a}^T\mathbf{b}$; the new estimate is also unbiased (we should have known that, because we knew that it was least squares). The variance is $\text{Var}(\mathbf{a}^T\hat{\mathbf{b}}) = \mathbf{a}^T\text{Var}(\hat{\mathbf{b}})\mathbf{a}$.

Proposition. *Given the least-squares estimate of a nonsingular linear model with homoscedastic errors and a linear combination* $\mathbf{a}^T\mathbf{b}$ *of its parameters,*

$$\text{Var}(\mathbf{a}^T\hat{\mathbf{b}}) = \sigma^2 \mathbf{a}^T(X^TX)^{-1}\mathbf{a}.$$

This may seem like an awfully special result, but it has a very important application. Our original reason for being interested in regression was to be able to predict the results of experiments we have not done yet, at other numerical settings of the independent variables. For example, in simple linear regression we might want to try the value x, so we predict $\hat{y} = \hat{\mu} + (x - \bar{x})\hat{b}$. This is a linear combination of our parameters, so it is unbiased: $E(\hat{y}) = \mu + (x - \bar{x})b$. Since the model is centered, our two parameters are uncorrelated, so

$$\text{Var}(\hat{y}) = \text{Var}(\hat{\mu}) + (x - \bar{x})^2\text{Var}(\hat{b}) = \sigma^2 \left(\frac{1}{n} + \frac{(x - \bar{x})^2}{\sum_{i=1}^{n}(x_i - \bar{x})^2} \right).$$

As usual, in practice we have to estimate this quantity by inserting an estimate of the residual variance.

Example (*cont.*). In our British doctor visit problem, we predicted that children 9.5 years old would be taken to the doctor 52.6% of the times they fell sick. The variance of that prediction is approximately $92.485(1/15 + 6.25/280) - 8.23$; the standard error is therefore something like plus or minus 2.87%.

Our formula quantifies our warning against using regression for extrapolation (see 1.5.1): If x is fairly close to $\bar{x}$, the variance does not increase very fast; but for x far away from $\bar{x}$, it increases by the square of the difference, and gets worse fast.

12.7   How Good Are Our Estimates?

12.7.1   Unbiased Linear Estimates

So we know all about least-squares estimates in the linear model, and we know that for normal errors they are also maximum likelihood. On various intuitive grounds, such estimates seem like a good idea. But assuming that our model is reasonable, are they really the best estimates we can make? One sensible way to measure that would be to ask whether the mean squared error of our $\hat{\mathbf{b}}$ is the smallest for any estimation method.

Actually, it is not. Imagine there was a statistician who had extrasensory perception, so that he always guessed our unknown parameters $\mathbf{b}$ correctly. His "estimate," with zero mean squared error, would certainly be better than ours.

But you say that my example is unrealistic. Okay, try this variation: Another statistician has a boyfriend at a lab that has done an experiment similar to yours, and she has heard about their preliminary results. You have not, because they have not been published yet. She computes an estimate from your data, then adjusts it a little in the direction of the result she got from her friend. Generally, her mean squared error is less than yours, because she in effect uses more data. What she has done may well have been quite reasonable, but we want to do our comparison without taking this alternative into account. After all, we can never be sure of having useful outside information, and even if we do there is the open question of how reliable it is.

We will try to disallow such possibilities by insisting that our estimate be unbiased (correct on average)—hers is not, because it will be biased toward the number she has overheard (at a time when the lab's experiment is complete but ours is not). Furthermore, we will assume that all our estimates are fixed *linear* combinations, only of our own data vector $\mathbf{y}$. The least-squares estimate we know meets both these restrictions. So let us rephrase our question as, Among all unbiased estimates of a nonsingular linear model that are linear combinations of the data, how good is the method of least squares?

Are there any such linear unbiased estimates other than the least-squares case? Yes. Imagine that you have an ANOVA model, with several replicates in each cell. Now reanalyze the problem, but use only *some*, not all, of your observations from each cell. You can still get all the parameter estimates by least squares, and the estimates are a linear combination of the observations that you did use. But then the estimates are a linear combination of all the data, with coefficients zero at the observations you did not use. So there are indeed alternative, unbiased linear estimates.

12.7.2   Gauss–Markov Theorem

Remember that the least-squares estimate is the linear combination $\hat{\mathbf{b}} = (\mathbf{X}^{\mathsf{T}}\mathbf{X})^{-1}\mathbf{X}^{\mathsf{T}}y$. Now let any other linear, unbiased estimate be $\tilde{\mathbf{b}} = \mathbf{A}y$ for a $p \times n$ matrix $\mathbf{A}$. Because it is unbiased, we know that $\mathrm{E}(\tilde{\mathbf{b}} - \hat{\mathbf{b}}) = \mathbf{b} - \mathbf{b} = 0$. But by factoring

$$0 = E(\tilde{b} - \hat{b}) = [A - (X^TX)^{-1}X^T]E(y) = [A - (X^TX)^{-1}X^T]Xb.$$

We will call the difference between the two estimating matrices $D = A - (X^TX)^{-1}X^T$. Then our condition for unbiasedness says that $E(Dy) = DXb = 0$, for any vector b (since we do not know b). The only possible conclusion is that DX is a $p \times p$ matrix of zeros, which fact we will write $DX = 0$. Again from Chapter 7, the covariance matrix of our new estimates is $\text{Var}(\tilde{b}) = A\text{Var}(y)A^T = \sigma^2AA^T$. But

$$AA^T = [(X^TX)^{-1}X^T + D][(X^TX)^{-1}X^T + D]^T$$
$$= (X^TX)^{-1} + DX(X^TX)^{-1} + (X^TX)^{-1}X^TD^T + DD^T.$$

We just noticed that DX (and its transpose X^TD^T) are zero, which gets rid of the two middle terms:

Proposition. *If we have an unbiased linear estimate* $\tilde{b} = Ay$, *then* $\text{Var}(\tilde{b}) = \text{Var}(\hat{b}) + \sigma^2DD^T$, *where* $\hat{b}$ *is the least-squares estimate and* $D = A - (X^TX)^{-1}X^T$.

Now, for any linear combination a of the coefficients, we use the theorem from Chapter 7 to get $\text{Var}(a^T\tilde{b}) = \text{Var}(a^T\hat{b}) + \sigma^2a^T(DD^T)a$. The matrix product in the second term is just the squared length of the vector D^Ta, which tells us that the second term is at least zero (this ought to remind you of the theorem of Pythagoras yet again):

Theorem (Gauss–Markov). *For a nonsingular linear model with homoscedastic errors, the least-squares estimate of any linear combination of the parameters is the linear unbiased estimate with the smallest variance (and so, mean squared error).*

So least squares not only leads to the best predictions of the observations, it leads to the best unbiased linear estimates of the constants (possibly important laws of nature) in our model. In our example of estimation by using only part of replicated data, we conclude that the mean squared error would be larger. Apparently, throwing away data always hurts us.

12.8   The Information Inequality

12.8.1   The Score Estimator

We know now how good least-squares estimates are—in a certain sense, the best in their class. Can we say similarly general things about how good maximum likelihood estimates are? Unfortunately, because they are the solution of a maximum problem, we usually cannot write them in some neat algebraic form. The most general approach to computing them involves Newton's method (see 8.8.1), which iteratively improves estimates of a parameter by computing

$$\theta^{(1)} = \theta^{(0)} - \left[\frac{\partial^2 l(\theta^{(0)}|\mathbf{x})}{\partial \theta^2}\right]^{-1} \frac{\partial l(\theta^{(0)}|\mathbf{x})}{\partial \theta}.$$

Here $l(\theta|x)$ is the log-likelihood for the value θ of the parameter, provided by the observations x. Since Newton's method is just as useful for absolutely continuous as for discrete models, in this section l will refer to either sort.

Our tactic will be to approximate iterative maximum likelihood estimates by similar but simpler estimates. One such estimate, sometimes used on real problems, is a *single-step* method: If we somehow get a pretty good estimate $\theta^{(0)}$, then $\theta^{(1)}$ obtained by one Newton iteration is usually a very good approximation to a maximum likelihood estimate.

Example. Let us fit the proportional regression model $\hat{y} = bx$ (see 2.3.2), where the errors are independently Normal$(0, \sigma^2)$ and σ is known. Then the log-likelihood for b from observations (x_i, y_i) is

$$l(b) = -\frac{n}{2}\log(2\pi) - n\log\sigma - \frac{1}{2\sigma^2}\sum_{i=1}^{n}(y_i - x_i b)^2.$$

Then $\frac{\partial l(b)}{\partial b} = \frac{1}{\sigma^2}\sum_{i=1}^{n} x_i(y_i - x_i b)$ and $\frac{\partial^2 l(b)}{\partial b^2} = -\frac{1}{\sigma^2}\sum_{i=1}^{n} x_i^2$. Our Newton step is

$$b^{(1)} = b^{(0)} + \frac{\frac{1}{\sigma^2}\sum_{i=1}^{n} x_i(y_i - x_i b^{(0)})}{\frac{1}{\sigma^2}\sum_{i=1}^{n} x_i^2} = b^{(0)} + \frac{\sum_{i=1}^{n} x_i y_i}{\sum_{i=1}^{n} x_i^2} - b^{(0)}.$$

The first guess canceled out (so in this case it did not matter what it was), and in a single step we got the exact least-squares solution.

Of course, this is the best possible behavior for a single step method. We need to further approximate our single-step method so we can learn about its behavior for other models. The second derivative, because it is in the denominator, does the most to make understanding our problem hard. In the example, it is *constant* (x's are the experimental settings), but we cannot count on that happening always. We shall hope that the second derivative does not vary much by chance and replace it by its *average* value. Furthermore, we will assume that the starting guess was good enough that it does not hurt much to replace it by the true value θ:

Definition. The **Fisher information** for a parameter θ is

$$I_\theta = E\left(-\frac{\partial^2 l(\theta|X)}{\partial \theta^2}\right).$$

(Yes, this is the R. A. Fisher of the tea-tasting experiment in Chapter 3.1.) In the regression example, the information is $I_b = \frac{1}{\sigma^2}\sum_{i=1}^{n} x_i^2$; in the single parameter logistic regression model (see 8.8.2), the second derivative is again constant, so $I_b = \sum_{i=1}^{n} x_i^2 p_i(1 - p_i)$.

The first-derivative piece of our Newton step always depends on the data, but it is in the numerator and so is easier to work with. Again, assume that our first guess was good enough that we may approximately replace it by the true value of the parameter θ. Then we make the following definition:

Definition. The **score statistic** is $s_\theta(\mathbf{x}) = [\partial l(\theta|\mathbf{x})]/(\partial\theta)$.

It will be our approximation to the nonconstant part of a single Newton step toward the maximum likelihood. So now we are ready to state that approximate maximum likelihood estimator for θ. We need a good first guess $\theta^{(0)}$; but what could be better than the true value θ? So our estimate is as follows:

Definition. The **score estimator** of θ is $\hat{\theta} = \theta + I_\theta^{-1} s_\theta(\mathbf{x})$.

12.8.2 How Good Is It?

I hope that this proposal bothers you, because it is of very little practical value. What use is an estimator that assumes that you know in advance the value you are trying to estimate? But as it turns out, we will learn some things about maximum likelihood and other methods by discovering simple properties of this score estimate. First, to find its expected value we need the expected value of the only variable part, the score statistic. Our result will apply both to discrete and to absolutely continuous variables; we will carry out our argument with densities and let you carry out similar argument with probability mass functions as an exercise:

$$E[s_\theta(\mathbf{X})] = E[\frac{\partial l(\theta|\mathbf{X})}{\partial\theta}] = \int \frac{\partial l(\theta|\mathbf{x})}{\partial\theta} f(\mathbf{x}|\theta)d\mathbf{x} = \int \frac{\partial \log f(\mathbf{x}|\theta)}{\partial\theta} f(\mathbf{x}|\theta)d\mathbf{x}$$

We are using the convention that the single integral sign with no explicit limits means integrate over the entire sample space of the multidimensional sample $\mathbf{x}$. Then using what we know about the derivative of a logarithm,

$$E[s_\theta(\mathbf{X})] = \int \frac{[\partial f(\mathbf{x}|\theta)]/(\partial\theta)}{f(\mathbf{x}|\theta)} f(\mathbf{x}|\theta)d\mathbf{x} = \int \frac{\partial f(\mathbf{x}|\theta)}{\partial\theta}d\mathbf{x}.$$

You may remember from calculus that often we can differentiate under the integral sign: $\frac{\partial}{\partial y} \int g(x, y)dx = \int [\partial g(x, y)]/(\partial y)dx$ is true whenever both integrals exist and the limits of integration do not depend on y. (If this last condition is not true, the left hand integral has a y at more than one place; and then the derivative is more complicated.) So let us assume that the score statistic has an expectation, and that the sample space of our family does not depend on θ (true, for example, of normal families). Then

$$E[s_\theta(\mathbf{X})] = \int \frac{\partial f(\mathbf{x}|\theta)}{\partial\theta}d\mathbf{x} = \frac{\partial}{\partial\theta} \int f(\mathbf{x}|\theta)d\mathbf{x} = \frac{\partial 1}{\partial\theta} = 0$$

because any density integrates to 1. But then for the score estimator, $E[\hat{\theta}] = \theta$; it is unbiased. Since it is only roughly equal to the score estimator, a maximum likelihood estimator is not necessarily unbiased; but our argument suggests that it is often nearly so.

To get the variance of the score estimator, again the only hard part is the variance of the score statistic:

$$\text{Var}[s_\theta(\mathbf{X})] = \text{E}\left\{\left[\frac{\partial l(\theta|\mathbf{x})}{\partial \theta}\right]^2\right\} = \int \left[\frac{\partial \log f(\mathbf{x}|\theta)}{\partial \theta}\right]^2 f(\mathbf{x}|\theta)\, d\mathbf{x}$$

$$= \int \frac{[[\partial f(\mathbf{x}|\theta)]/(\partial\theta)]^2}{f(\mathbf{x}|\theta)}\, d\mathbf{x}.$$

Although it is not yet obvious, this is connected with the Fisher information:

$$I_\theta = \text{E}\left(-\frac{\partial^2 l(\theta|\mathbf{X})}{\partial\theta^2}\right) = -\int \frac{\partial^2 \log f(\mathbf{x}|\theta)}{\partial\theta^2} f(\mathbf{x}|\theta)\, d\mathbf{x}$$

$$= \int \left\{-\frac{[\partial^2 f(\mathbf{x}|\theta)]/(\partial\theta^2)}{f(\mathbf{x}|\theta)} + \frac{[[\partial f(\mathbf{x}|\theta)]/(\partial\theta)]^2}{f(\mathbf{x}|\theta)^2}\right\} f(\mathbf{x}|\theta)\, d\mathbf{x}$$

$$= -\int \frac{\partial^2 f(\mathbf{x}|\theta)}{\partial\theta^2}\, d\mathbf{x} + \int \frac{[[\partial f(\mathbf{x}|\theta)]/(\partial\theta)]^2}{f(\mathbf{x}|\theta)}\, d\mathbf{x}.$$

By the same reasoning we used for the expectation, in the first integral $\int [\partial^2 f(\mathbf{x}|\theta)]/(\partial\theta^2) d\mathbf{x} = \frac{\partial^2}{\partial\theta^2}\int f(\mathbf{x}|\theta) d\mathbf{x} = 0$. We are left with the attractively simple $\text{Var}[s_\theta(\mathbf{X})] = I_\theta$. Now compute $\text{Var}[\hat{\theta}] = I_\theta^{-2} I_\theta$:

Proposition. *In a family that has a Fisher information for θ and whose sample space does not depend on θ, we have for the score estimator* $\text{Var}[\hat{\theta}] = I_\theta^{-1}$.

Our goal was to approximate the behavior of a maximum likelihood estimator. All we have presented is a plausibility argument that it might be similar to our (impractical) score estimator. Proof that in fact, the inverse Fisher information is often asymptotically equivalent to the variance of a maximum likelihood estimate shall have to await a more advanced course. Nevertheless, we have learned enough that you will not be surprised when I tell you that in many complicated estimation problems in which the variance cannot be computed exactly, this formula is used as an approximate variance of a maximum likelihood estimate.

For example, in the single-parameter logistic regression problem, we found that $I_b = \sum_{i=1}^n x_i^2 p_i(1 - p_i)$. Of course, this is still not computable, because the p_i depend on the true value of the parameter; but we have had to estimate them in the course of our calculations. Therefore, it is common to use $\left(\sum_{i=1}^n x_i^2 \hat{p}_i(1 - \hat{p}_i)\right)^{-1}$ as an approximate variance for $\hat{b}$. In the dose response example in (8.8.2), this suggests that the standard error to the estimated slope of 0.087 is about 0.024.

12.8.3   The Information Inequality

Now that we have a crude idea of the quality of a maximum likelihood estimate, we need to have something like a Gauss–Markov theorem to compare it to other estimators. Once again, it will be much easier to compare the score estimator to other estimators first. Let $g(\mathbf{x})$ be any unbiased estimator for θ, so that $\text{E}[g(\mathbf{X})] = \theta$.

To compare it to the score estimator, let us find the correlation of the two, $\rho_{g\hat{\theta}}$. But since $\hat{\theta}$ is just a linear function of the score statistic, $\rho_{g\hat{\theta}} = \rho_{gs_\theta}$, since a correlation is unaffected by scale and location changes. The numerator of this second correlation is $\text{Cov}[g(\mathbf{X}), s_\theta(\mathbf{X})] = E[g(\mathbf{X})s_\theta(\mathbf{X})] - E[g(\mathbf{X})]E[s_\theta(\mathbf{X})]$; but we know that $E[s_\theta(\mathbf{X})] = 0$. Thus

$$\text{Cov}[g(\mathbf{X}), s_\theta(\mathbf{X})] = \int g(\mathbf{x}) \frac{\partial \log f(\mathbf{x}|\theta)}{\partial \theta} f(\mathbf{x}|\theta) d\mathbf{x} = \int g(\mathbf{x}) \frac{\partial f(\mathbf{x}|\theta)}{\partial \theta} d\mathbf{x}.$$

If we are in one of those situations where we can differentiate under the integral sign, then

$$\int g(\mathbf{x}) \frac{\partial f(\mathbf{x}|\theta)}{\partial \theta} d\mathbf{x} = \frac{\partial}{\partial \theta} \int g(\mathbf{x}) f(\mathbf{x}|\theta) d\mathbf{x} = \frac{\partial}{\partial \theta} E[g(\mathbf{X})] = \frac{\partial \theta}{\partial \theta} = 1.$$

This is a delightfully simple result: When can we get away with it? Again, we need a random vector whose sample space does not depend on θ; then we need both integrals to exist. The second integral, which is just $E[g(\mathbf{X})]$; we already know the value of because we assumed that g was unbiased. The first integral, the covariance, exists if $\text{Var}(g)$ and $\text{Var}(s_\theta)$ both do, by the Cauchy–Schwarz inequality. We summarize:

Proposition. *Let $\mathbf{X}$ be a discrete or absolutely continuous random vector whose distribution depends on a parameter θ but whose sample space does not; and assume that its Fisher information exists. Let $g(\mathbf{x})$ be unbiased for θ and have a variance. Then $\text{Cov}[g(\mathbf{X}), s_\theta(\mathbf{X})] = 1$.*

You might notice that the result works in reverse as well: If a statistic has covariance 1 with the score statistic, it must be unbiased. This condition is another way of looking at unbiasedness.

Recall our fundamental inequality about the correlation, $\rho_{gs_\theta}^2 \leq 1$, we now know that $\rho_{gs_\theta}^2 = 1/(\text{Var}[g(\mathbf{X})]\text{Var}(s_\theta(\mathbf{X}))) = 1/(\text{Var}[g(\mathbf{X})]I_\theta) \leq 1$. Moving this variance out of the denominator, we have discovered a fundamental fact about unbiased statistics:

Theorem (the information inequality). *Let $\mathbf{X}$ be a discrete or absolutely continuous random vector whose distribution depends on a parameter θ but whose sample space does not. Let $g(\mathbf{x})$ be unbiased for θ. Then if the two quantities exist, $\text{Var}[g(\mathbf{X})] \geq I_\theta^{-1} = \text{Var}(\hat{\theta})$, where $\hat{\theta}$ is the score estimator.*

This important fact (in some older books known as the Cramér–Rao bound) says that our impractical score estimator is the ideal unbiased estimator; any other will be more variable. Our demonstration says that to be unbiased, g has to follow $\hat{\theta}$ around; but it may also have some extra variability of its own. We see why the maximum likelihood estimator is popular; it is a practical method that often approximates the ideal method.

12.8.4 MVUE Statistics

We noticed that in the one-parameter regression case with normal errors, the maximum likelihood (also least-squares) estimator was the same as the score estimator. Therefore, with normal errors, least squares is not only the best linear unbiased estimator, it is the best of all unbiased estimators, in the sense of mean squared error.

Definition. Let $g(\mathbf{x})$ be an unbiased estimator for a parameter θ that does not depend on the true value of θ. Then if $\text{Var}[g(\mathbf{X})] = I_\theta^{-1}$, so that its variance meets the lower bound given in the information inequality, we call g a **minimum variance unbiased estimator** (MVUE) of θ.

Any time we discover that we have such an estimate, we tend to have a high opinion of it. Fortunately, there are several other important examples besides least-squares estimates in the normal error model.

Theorem (common MVUE statistics). *Let $x = x_1, \dots, x_n$ be a random sample*

(i) *If x is from a binomial $B(m, p)$ random variable, $\hat{p} = \bar{x}/m$ is the MVUE;*
(ii) *If x is from a Poisson(λ) random variable, $\hat{\lambda} = \bar{x}$ is the MVUE;*
(iii) *If x is from a Normal(μ, σ^2) random variable with μ known, $\hat{\sigma}^2 = \frac{1}{n} \sum_{i=1}^{n} (x_i - \mu)^2$ is the MVUE;*
(iv) *If x is from a Gamma(α, β) random variable with α known, $\hat{\beta} = \frac{\bar{x}}{\alpha}$ is the MVUE.*

You will prove these results in exercises. More advanced courses will tackle the issue of how good estimates are for other families and parameters.

12.9   Summary

For linear models, we showed that the assumption of *spherically distributed* errors determines the probability distribution of our test statistics, R-squared (which then has a beta distribution) and F (2.2). Then we saw that the simplest probability model that leads to spherical errors is one in which residuals are independently normal, $f(\mathbf{e}) = 1/[(2\pi)^{n/2}\sigma^n]e^{-\mathbf{e}^T\mathbf{e}/(2\sigma^2)}$ (3.1). We then showed that the principle of maximum likelihood (for models with densities, $L(\theta|\mathbf{x}) = f(\mathbf{x}|\theta)$ (4.1)) for this class of models leads to least-squares estimates (4.2). The associated G-squared statistics are equivalent to F-statistics (5.2). Then we defined the *general linear model*, in matrix notation $\mathbf{y} = \mathbf{Xb} + \mathbf{e}$ (6.1) and derived its least-squares normal equations $\mathbf{X}^T\mathbf{y} = \mathbf{X}^T\mathbf{Xb}$ (6.3). On the assumption of *homoscedastic* errors (constant variance σ and uncorrelated), we derived an expression for the variance of our parameter estimates, $\text{Var}(\hat{\mathbf{b}}) = \sigma^2(\mathbf{X}^T\mathbf{X})^{-1}$ (6.4). We then discovered the *Gauss–Markov* theorem, which says that among all linear unbiased estimates, least-squares estimates have the smallest variance (7.2). Finally, the *information inequality* told us that there is a limit to how good any unbiased estimate can be,

Var$[g(X)] \geq I_\theta^{-1}$, where the *Fisher information* for θ is $I_\theta = E(\partial^2 l(\theta | X)/(\partial \theta^2))$ (8.3). It was observed that often maximum likelihood estimates come close to that limit. Estimates that actually achieve the lower limit are called *minimum variance unbiased* estimates (MVUE) (8.4).

12.10   Exercises

1. Determine whether the following densities are spherically distributed about the origin, and explain why or why not:

   a. $\dfrac{1}{\pi(1 + x^2)}$.

   b. $e^x e^{-e^x}$.

   c. $\dfrac{1}{\sqrt{2\pi}} e^{-x^2/2}$.

2. Demonstrate that the bivariate standard normal random vector with density $\frac{1}{2\pi} e^{-x^2/2 - y^2/2}$ is spherically symmetric about the origin, by showing that it is unchanged when we perform any rotational change of variables.

3. In Exercise 6 of Chapter 1:

   a. Compute the F-statistic that contrasts the combined treatment effects in the additive model (2 degrees of freedom) to the error sum of squares for the same model (6 degrees of freedom).

   b. Under the hypothesis of no treatment effect (and spherically symmetric errors), compute the probability that your F might by accident exceed the calculated value. What do you conclude about the additive model?

4. Our new absolutely continuous version of the method of maximum likelihood can help us with other than normal estimation problems. Let our random sample of n observations be taken independently from a member of the negative exponential family with density $\frac{1}{\beta} e^{-x/\beta}$ on $X > 0$. Find the maximum likelihood estimate for β.

5. In the linear model with independent normal errors, find the variance and standard error of the unbiased estimator $\hat{\sigma}^2 = \frac{1}{n-k} \sum_{i=1}^{n} (y_i - \hat{y}_i)^2$.

6. Calculate the G-squared statistic for Exercise 3, and see how well the approximation to the F-statistic works in this case.

7. Write out the design matrix X for the model of Exercise 21 in Chapter 2.

8. For Exercise 6 in Chapter 1:

   a. Write out the design matrix X for the additive model. You will be able to do this by simply writing out the columns for exercise and diet separately.

   b. Write out the design matrix for the centered, full model.

9. a. Calculate the inverses of the $X^T X$ matrices in Exercise 8.

   b. Solve the matrix forms of the normal equations for the observations of Exercise 6, Chapter 1. Check that they coincide with the original answers.

10. Write out the uncentered design matrix for a one-way layout with four levels and two observations per level. Now delete the last column, form the $\mathbf{X}^T\mathbf{X}$ matrix, and compute its inverse. Form the matrix normal equations and show that they indeed give standard estimates for the parameters.

11. In the model for aflatoxin in peanuts estimated in Exercise 14 of Chapter 2, estimate the standard errors of the parameters. Now estimate the standard error of the prediction of percent of bad peanuts at 50 parts per billion aflatoxin.

12. In Exercise 10, invent a linear estimate of the parameters A that is a least-squares estimate but uses only every second observation. Find its difference matrix $\mathbf{D}$ from the usual estimation matrix, and show that $\mathbf{DX} = \mathbf{0}$, so that your estimate is unbiased. Now compute $\mathbf{DD}^T$ and interpret it as increases in the variances of the estimates.

13. Find the Fisher information for

   a. β in a sample of n from a Negative Exponential(β) random variable.
   b. the standard deviation σ in a sample of n from a Normal$(0, \sigma^2)$ random variable.
   c. p in an NB(k, p) random variable.

14. Show that for a *discrete* family whose sample space does not depend on its parameter θ, for the score statistic $s_\theta(\mathbf{x})$, $E[s_\theta(\mathbf{X})] = 0$.

15. Derive our formula for the variance of the score estimator for a *discrete* family whose sample space does not depend on θ.

16. Prove the information inequality for a *discrete* family whose sample space does not depend on θ.

17. Show that the estimates of binomial p and Poisson λ are indeed MVUE:

   a. by showing that they are unbiased and have the minimum variance prescribed by the information inequality.
   b. by showing that they are the same as the score estimator.

18. Show that the estimates of normal σ^2 and gamma β are indeed MVUE:

   a. by showing they are unbiased and have the minimum variance prescribed by the information inequality.
   b. by showing that they are the same as the score estimator.

12.11   Supplementary Exercises

19. Show that any random vector spherically distributed about the origin whose coordinates have a finite variance must have any pair of coordinates uncorrelated.
   Hint: Show that $Z = (X_i + X_j)/\sqrt{2}$ has the same distribution as X_i does. Now compute its variance.

20. You may have noticed that our method of computing beta (and therefore F) probabilities exactly works only if both degrees of freedom are even. Let us tackle some other cases:

   a. Compute the cumulative distribution function for a Beta($n/2$, 1) random variable.

   b. Compute the cumulative distribution function for a Beta($n/2$, m) random variable by repeated integration by parts to get a finite series solution. Raise the half-integer power and lower the integer power until you reduce the problem to (a).

21. In Chapter 2, Exercise 10, you computed an F-statistic for the difference in DBH levels of 2 groups, on 1 and 22 degrees of freedom. Compute the probability that you would exceed this value by chance if in fact, psychotic state does not matter and the errors are spherically distributed. What do you conclude about levels of DBH in this group of patients?

22. If you assume that a sample $x_1, \ldots, x_n$ is from the Normal(μ, σ^2) distribution, notice that $z = (\bar{x} - \mu)/(\sigma/\sqrt{n})$ is Normal(0, 1). It might then be the test statistic for the null hypothesis that the expectation is really μ. But in the real world, we rarely know σ. In 1900, W. S. Gosset proposed the t-statistic $t = (\bar{x} - \mu)/(s/\sqrt{n})$ to test this hypothesis. Find its density, by the following method:

   a. First show that its square, $n(\bar{x} - \mu)^2/x^2$, has an F-distribution on 1 and $n - 1$ degrees of freedom.

   b. Then by change of variables find the density of $|t|$. Finally, extend to negative values by symmetry to obtain the density of t. (This is called a t-density on $n - 1$ degrees of freedom.)

23. In the mass-ratio data of Chapter 1, Exercise 1, assume that the observations are normal, and take the null hypothesis that they have a true average of 81.3. Use a t-statistic to test the one-sided alternative that the mass ratio is really greater than 81.3, at the 0.05 level.

24. Assume that we have two samples $x_{11}, \ldots, x_{1n_1}$ and $x_{21}, \ldots, x_{2n_2}$. Then the **two-sample t-statistic** for testing whether their means are different is

$$t = \frac{\bar{x}_1 - \bar{x}_2}{\sqrt{\frac{(n_1-1)s_1^2+(n_2-1)s_2^2}{n_1+n_2-2}}\sqrt{\frac{1}{n_1} + \frac{1}{n_2}}},$$

where s_1^2 and x_2^2 are sample variances of the individual samples.

Show that if the observations are independently Normal(μ, σ^2), then t^2 follows an F-distribution on 1 and $n_1 + n_2 - 2$ degrees of freedom. Therefore, t is a t-statistic as in Exercise 22 on $n_1 + n_2 - 2$ degrees of freedom. **Hint**: Notice that our problem is a one-way layout with $k = 2$ levels of treatment.

25. You want to try out a fertilizer called TomatoGro, so you apply it to seven tomato plants, while not using it on five other plants, chosen at random. Yields in pounds of tomatoes are, for the fertilized plants, 25, 16, 21, 28, 19,

19, and 22; for the others, 14, 20, 21, 12, 13. Use the t-test of Exercise 24 to test the (one-sided) hypothesis that TomatoGro improved yields (as opposed to the null hypothesis that the yields are the same).

26. In the data from Exercise 5.10, test whether allergic people tend to have higher histamine levels, using a t-test and a 0.01 significance level.

27. A Laplace location-and-scale model is a density that looks like $1/(2\sigma)e^{-|x-\mu|/\sigma}$. Assume that we have random sample of n observations from such a density. Find the maximum likelihood estimates for μ and σ.
    Hint: After you have worked on the problem for a while, look back at Exercise 19 from Chapter 2.

28. Consider a bivariate random vector (X, Y) with $p(1, 1) = p(-1, 1) = p(1, -1) = p(-1, -1) = p(0, 0) = 0.2$. Show that it is homoscedastic. Now show that it is **not** spherically distributed about the origin, and that X and Y are **not** independent.

29. In the Laplace model of Exercise 27, assume that σ is known:

    a. Compute the Fisher information for μ. (Use the equivalent first derivative formula. The likelihood has no second derivative.)

    b. Approximate the variance of the maximum likelihood estimate of μ (which turned out to be the sample median).

    c. Show that the sample mean $\bar{x}$ is an unbiased estimate of μ. Now compute its variance, and notice how much larger it is than your answer to (b) (illustrating the information inequality).

CHAPTER 13

Representing Distributions

13.1 Introduction

We have described some of the many useful families of random variables using their mass functions, density functions, and, sometimes, cumulative distribution functions. Often, though, these functions are not as helpful as we might hope in solving certain problems that require us to work with random variables. For example, in random sampling we are interested in the sums of independent observations. If we are lucky, as with Poisson and gamma variables, these sums stay in the family; if not, as with a uniform variable, properties of these sums may be difficult to extract.

In this chapter we will discover and exploit several new ways of representing our random variables, which will facilitate the solution of some important problems. These will include finding limits of sequences of variables, efficient estimation of parameters, and better probability approximations.

Time to Review

Power series and Taylor series
Chapter 8, Section 2.2.

13.2   Probability Generating Functions

13.2.1   Compounding Distributions

In biological or geographic studies, we might be interested in the number of off-spring an individual would be expected to have over a lifetime. One possible simple model might be a Poisson(λ) random variable, with mean known from our knowledge of many similar individuals. But what if I wanted to know something about the distribution of how many grandchildren, or great-grandchildren, that individual will have?

If we are willing to assume that the children all have, independently of each other and of their parents, a Poisson(μ) number of children of their own, then we could compute, by division into cases (see 4.6.2),

$$P(x \text{ grandchildren}) = \sum_{y=0}^{\infty} P(y \text{ children})P(x \text{ grandchildren}|y \text{ children}).$$

We are lucky that we know that the conditional number of grandchildren is Poisson($y\mu$), because it is just a sum of y independent Poisson variables. Therefore,

$$P(x \text{ grandchildren}) = \sum_{y=0}^{\infty} \frac{\lambda^y}{y!} e^{-\lambda} \frac{(y\mu)^x}{x!} e^{-y\mu},$$

with no obvious significant algebraic simplifications available. The situation gets worse with more complicated distributions than the Poisson, and much worse for great-grandchildren, and so forth. Incidentally, the number of grandchildren, or later descendants, in the Poisson case is called a **compound Poisson** random variable. Generating random variables in stages like this is generally called *compounding*.

13.2.2   The P.G.F. Representation

We shall tackle compounding problems in general by first solving a much simpler compounding problem. Let X be any discrete random variable whose sample space is a subset of the nonnegative integers $\{0, 1, 2, \ldots\}$. All variables that may be interpreted as random counts are therefore included; we will call these *counting* variables. Let us now compound X with independent Bernoulli(q) variables. That is, after we have had X successes, we do X additional independent Bernoulli trials. The answer to one particular question about this experiment will turn out to be most important: How likely are we to observe all successes, or no failures, at the second generation?

Example. The number of children born to each family in a certain community is roughly negative binomial NB(2, 0.5). In this group, 40% of children graduate from college. Assuming that a child's progress depends neither on family size nor on the success of any siblings, what is the probability that all a family's children will graduate from college?

The answer, of course, will depend on the probability q for each success. We will give it a name:

Definition. Let X be a random variable such that its sample space S is a subset of the nonnegative integers. Then let there be X independent Bernoulli(q) trials. The probability that all trials are successes we will call the **probability generating function (p.g.f.)** for X:

$$\pi_X(q) = P[\text{all successes} \mid X \text{ independent Bernoulli}(q) \text{ trials}].$$

This function, one of our new representations of the random variable X, will turn out to carry information about X in a very convenient form.

First, let us check that we can compute π_X in some important cases. The easiest one of all would be if X were itself Bernoulli(p). Then we would have a trial with probability $1 - p$ of failure, in which case there are no further trials, and so certainly no failures. If X is a success with probability p, then there is a further independent trial with probability q of success. Therefore, the total probability of no failures at the second stage is $\pi_X(q) = 1 - p + pq$.

The next simplest case would involve X a Geometric(p) random variable. Imagine two simultaneous sequences of Bernoulli trials. The first is the one with probability p of success that will determine X when the first failure appears. The second, with probability q of success, is the compound stage of the parallel trial. All the second-stage trials will have been a success if at the first trial where either trial fails, the failure is in the Bernoulli(p) trial. Therefore, $\pi_X(q) = P(p \text{ trial fails} \mid \text{either } p \text{ or } q \text{ fails}) = \frac{1-p}{1-pq}$ by a simple conditional probability calculation.

To find the probability generating functions of more complicated random variables, we need only a simple principle. Let X and Y be independent counts, and consider the random variable $Z = X + Y$. For all Z independent Bernoulli(q) trials at the second stage to be successes, first X of them must be successes, then a further independent Y of them must be. The probabilities that independent events both happen just multiply. We have established the following important result:

Proposition (p.g.f. of a sum). *For X and Y independent counting variables, and $Z = X + Y$,*

$$\pi_Z(q) = \pi_X(q)\pi_Y(q).$$

But recall that a binomial B(n, p) variable is a sum of n independent Bernoulli(p) variables; and a negative binomial NB(k, p) variable is a sum of k independent Geometric(p) variables.

Proposition.

(i) *For X a B(n, p) variable, $\pi_X(q) = (1 - p + pq)^n$.*

(ii) *For X an NB(k, p) variable, $\pi_X(q) = \left(\frac{1-p}{1-pq}\right)^k$.*

Example (*cont.*).   In the problem of children graduating from college,

$$P(\text{all graduate}) = \pi_X(0.4) = \left(\frac{1-0.5}{1-0.5 \times 0.4}\right)^2 = 0.391.$$

13.2.3   The P.G.F. As an Expectation

We suggested earlier that the probability generating function has more information hidden in it than just the answer to problems of this simple form. For example, use the division into cases theorem to write

$$\pi_X(q) = P(\text{all successes}) = \sum_{x=0}^{\infty} P(X = x) P[\text{all of } x \text{ Bernoulli}(q) \text{ trials succeed}].$$

This formula may be written compactly:

Proposition. $\pi_X(q) = \sum_{x=0}^{\infty} p(x)q^x = E(q^X).$

We have discovered a connection among the probability generating function, the mass function, and certain expectations.

If X is Poisson(λ), this tells us that $\pi_X(q) = \sum_{x=0}^{\infty}(\lambda^x/x!)e^{-\lambda}q^x = \sum_{x=0}^{\infty}((\lambda q)^x/x!)e^{-\lambda}$. We can now use the familiar device of converting the sum into a sum of Poisson(λq) probabilities by multiplying and dividing by $e^{\lambda q}$: $\pi_X(q) = \sum_{x=0}^{\infty}(\lambda^x/x!)e^{-\lambda}q^x = e^{-\lambda+\lambda q}\sum_{x=0}^{\lambda}((\lambda q)^x/x!)e^{-\lambda q}$. Since the probabilities sum to 1, we have a simple expression.

Proposition.   *For X Poisson(λ), $\pi_X(q) = e^{-\lambda(1-q)}$.*

We can do this reasoning in reverse. Recall how power series work from calculus, notice that the series $\pi_X(q) = \sum_{x=0}^{\infty} p(x)q^x$ can always be made smaller than a geometric series in q, since obviously $p(x) \leq 1$. Therefore, the series converges for $-1 < q < 1$ (even though so far, we do not know of any use for negative values of q). We conclude that there is only one possible power series for $\pi_X(q)$. This has the very important implication that if we know π and can expand it in a power series, the coefficients must be the probability mass function $p(x)$.

Example.   You send out two meter readers; one will go to 3 houses before lunch and read the meter for anyone who is home (60% of houses have someone at home). The other will just keep reading meters until the yard has a large dog in it (25% of houses have a large dog) and then will stop for lunch. What is the probability that exactly 4 meters will be read before lunch?

The number of houses read by the first reader is distributed as a binomial B(3, 0.6). The probability generating function is therefore $(0.4+0.6q)^3 = 0.064 + 0.288q + 0.432q^2 + 0.216q^3$ (so the probability that she will have read 2 houses is 0.432). For the second reader, houses are Geometric(0.75), so the probability generating function is $\frac{0.25}{1-0.75q} = 0.25(1+0.75q+0.563q^2+0.462q^3+0.316q^4+\cdots)$ by the usual formula for the sum of a geometric series, or for geometric probabilities.

By the proposition about sums of independent variables, we get the p.g.f. of the total of meters read by algebraic multiplication:

$$(0.064 + 0.288q + 0.432q^2 + 0.216q^3)$$
$$\times\, 0.25(1 + 0.75q + 0.563q^2 + 0.462q^3 + 0.316q^4 + \cdots)$$
$$= 0.016 + 0.084q + 0.171q^2 + 0.183q^3 + 0.140q^4 + \cdots.$$

You should check my arithmetic. We conclude that the probability that exactly 4 meters will be read is 0.14.

Please do not imagine that this procedure is magical. Before you ever got to this chapter, you would have solved such a problem as follows: Let X be the number read by the first reader, and Y the number read by the second. Then

$$P(X + Y = 4) = P(X = 0, Y = 4) + P(X = 1, Y = 3)$$
$$+\, P(X = 2, Y = 2) + P(X = 3, Y = 1).$$

You should verify that these are exactly the calculations you did to compute the coefficient of q^4 above. The algebra is so far mainly a memory aid for complicated calculations.

You should now review from calculus the fact that if a function f has a power series expansion near some point (in this case, 0), then it may be written as a *Taylor's series*:

$$f(x) = f(0) + xf'(0) + \frac{x^2}{2}f''(0) + \cdots = \sum_{i=0}^{\infty} \frac{x^i}{i!} f^{(i)}(0),$$

where $f^{(i)}$ means the ith derivative of f. Therefore, a probability generating function, since it has a unique power series, has coefficients that may be evaluated by taking derivatives:

Theorem (generating probabilities). $p(x) = (\pi_X^{(x)}(0))/x!$.

Example. In the meter reading problem, $\pi_X(q) = (0.4 + 0.6q)^3 \frac{0.25}{1-0.75q}$. Differentiate twice to get

$$\pi_X''(q) = 6 \times 6^2(0.4 + 0.6q)\frac{0.25}{1 - 0.75q}$$
$$+\, 6 \times 0.6 \times 0.75(0.4 + 0.6q)^2\frac{0.25}{(1 - 0.75q)^2}$$
$$+\, 2 \times 0.75^2(0.4 + 0.6q)^3\frac{0.25}{(1 - 0.75q)^3}.$$

Therefore, $p(2) = \pi_X''(0)/2 = (0.216 + 0.108 + 0.018)/2 = 0.171$.

13.2.4 Applications to Compound Variables

We motivated this whole section by the problem of compounding counting variables, and so far we have dealt only with the very special case of Bernoulli compounds. But now consider a counting variable X, and compound it with a variable Y; that is, after an experiment with X successes, perform X further independent experiments each with outcome Y_i distributed like Y, and total the successes $Z = \sum_{i=1}^{X} Y_i$. The probability generating function for Z will turn out to be easy to calculate.

Stating the following easy argument abstractly makes it sound obscure, so we will state a very concrete version of it. A *successful(q)* grandparent we will define as one all of whose grandchildren are successful, where each grandchild independently has probability q of success. In a similar way, a *successful(r)* parent is one all of whose children are successful, with probability r of success for each child. Therefore, a successful grandparent is one all of whose children are themselves successful parents. We conclude that a successful grandparent is a successful parent, with the probability r of success of each child equal to the probability that all its own children (who are the grandchildren) succeed.

We restate this as a general principle about compounding. Let X be the random variable for the number of children at the first generation, and let Y be the number of children in each second-generation family. Then $r = \pi_Y(q)$ is the probability that each first-generation child will be a successful parent. But then the probability that the original parent will be a successful parent is $\pi_X(r)$, and the probability that it will be a successful grandparent is the same thing, $\pi_X(r) = \pi_X[\pi_Y(q)]$.

Theorem (fundamental theorem of branching processes). *Let a counting variable X be compounded by a counting variable Y. Let the total count at the second generation be $Z = \sum_{i=1}^{X} Y_i$. Then $\pi_Z(q) = \pi_X[\pi_Y(q)]$.*

Experiments like these are called *branching* processes, of course, because we are studying family trees.

Now let me prove this important theorem again, in a shorter but abstract way: $\pi_Z(q) = E(q^Z) = E(q^{\sum_{i=1}^{X} Y_i}) = E(\prod_{i=1}^{X} q^{Y_i})$. Now (see 11.2.1) write this

$$\pi_Z(q) = E_X\left[\prod_{i=1}^{X} E_{Y_i|X}(q^{Y_i}|X)\right] = E_X\left[\prod_{i=1}^{X} E_{Y_i|X}(q^{Y_i}|X)\right] = E_X\left[\prod_{i=1}^{X} E_{Y_i}(q^{Y_i})\right],$$

where the last step follows just because of the independence of generations. Finally,

$$\pi_Z(q) = E_X\left[\prod_{i=1}^{X} \pi_Y(q)\right] = E\left[\pi_Y(q)^X\right] = \pi_X\left[\pi_Y(q)\right].$$

When we compound a Poisson(λ) variable with a Poisson(μ) variable, as at the beginning of this section, the probability generating function is now trivial to compute: $\pi_Z(q) = e^{-\lambda[1 - e^{-\mu(1-q)}]}$. We may use Taylor's theorem now to find any probabilities we want.

Example. If each generation has Poisson(2) children, let us find the probability that an individual will have 3 grandchildren. After a little calculus (which you should check) we find that $\pi_X''(q) = 4[2 + 4e^{-2(1-q)}]e^{-2(1-q)-2[1-e^{-2(1-q)}]}$. Then $p(2) = \frac{\pi_X''(0)}{2} = 0.122$.

13.2.5 Factorial Moments

We now look more closely at what happens when we differentiate a p.g.f. Using the expectation form, we have $\pi'_X(q) = E(q^X)' = E(Xq^{X-1})$. We remind ourselves that we are allowed to differentiate inside the expectation (a sum) inside the radius of convergence $-1 < q < 1$. But now, instead of evaluating at 0, let $q = 1$: $\pi'_X(1) = E(X)$. This will work whenever the corresponding series converges, which is, of course, when we have an expectation at all (so long as we think of it as a *one-sided* derivative, measuring the rate of change as q approaches 1 from below). Sometimes this is a reasonable way of discovering the expectation of a counting variable; you will compute some you already know as exercises.

For compound variables, since $\pi_Z(q) = \pi_X[\pi_Y(q)]$, we have that $\pi'_Z(q) = \pi'_X[\pi_Y(q)]\pi'_Y(q)$ by the chain rule. Then to substitute $q = 1$, recall first that $\pi_Y(1) = 1$, from the definition of a p.g.f. We have $\pi'_Z(1) = \pi'_X(1)\pi'_Y(1)$.

Proposition. *If Z is a compound of X and Y, then* $E(Z) = E(X)E(Y)$.

You will discover a less sophisticated proof of this fact in the exercises.

Let us differentiate a p.g.f. once more, to get

$$\pi''_X(q) = E(q^X)'' = E[X(X-1)q^{X-1}].$$

Then whenever the expectation exists, $\pi''_X(1) = E[X(X-1)]$. This should look familiar: When we were using the inductive method to calculate variances, this expression turned out to be easy to evaluate for many of our families of discrete random variables (see 6.6.3). Now we see that it is somehow a fundamental expectation for any counting variable. Repeating this process yields $\pi_X^{(k)}(1) = E[X(X-1)\cdots(X-k+1)]$. We shall give such expressions a name, using permutation notation:

Definition. The *k*th **factorial moment** of X, denoted by f_X^k, is equal to $E[(x)_k]$.

This suggests that we are interested in a Taylor series approximation for π again, but now the series is to be expanded near $q = 1$. We will assume that π has an $(m+1)$-term Taylor series there; that is, $\pi_X(q) \approx 1 + \sum_{k=1}^{m} \frac{f_X^k}{k!}(q-1)^k$. For example, when X is Poisson(λ), we know that $\pi_X(q) = e^{-\lambda(1-q)} = 1 + \sum_{k=1}^{\infty} \frac{\lambda^k}{k!}(q-1)^k$ from calculus. Therefore, $f_X^k = \lambda^k$. (Of course, this is quite easy to evaluate directly.)

We may use factorial moments to get other expectations. For example, we noted in Chapter 6.3 that we could calculate $\text{Var}(X) = f_X^2 + f_X^1(1 - f_X^1)$. Since we have already calculated some derivatives of a compound Poisson p.g.f,

$$\pi''_Y(q) = \lambda\mu\left[\mu + \lambda\mu e^{-\mu(1-q)}\right]e^{-\mu(1-q)-\lambda[1-e^{-\mu(1-q)}]},$$

we just substitute $q = 1$ to get $f_X^2 = \lambda\mu(\mu+\lambda\mu)$. Therefore, $\text{Var}(Z) = \lambda\mu(\mu+1)$.

For the general case of a compound of X and Y, we compute

$$\pi''_Z(q) = \{\pi_X[\pi_Y(q)]\}'' = \{\pi'_X[\pi_Y(q)]\pi'_Y(q)\}'$$

$$= \pi''_X[\pi_Y(q)][\pi'_Y(q)]^2 + \pi'_X[\pi_Y(q)]\pi''_Y(q).$$

Now substitute $q = 1$ to get $f_Z^2 = f_X^2 [f_Y^1]^2 + f_X^1 f_Y^2$. Each term here can be written in terms of variances and expectations. You should do so. When you rearrange these you will get the following general formula:

Proposition. *If Z is a compound of X and Y, Then* $\text{Var}(Z) = \text{Var}(X)\text{E}(Y)^2 + \text{Var}(Y)\text{E}(X)$.

Unlike the earlier expectation formula, this is not intuitively obvious. You will discover another proof in the exercises.

Example. In the case where children and grandchildren were each Poisson(2), then an individual, of course, has an average of 4 grandchildren. But now we know that the variance is 12.

13.2.6   Comparison with Geometric Variables

The probability generating function is directly related to the solution of another very simple probability problem. Recall the definition of a Geometric(q) random variable. It is the number of Bernoulli(q) successes in a row that precede the first failure. But the p.g.f. of a counting variable X we described as being the probability that we had X successes in a row, where each try is Bernoulli(q). Therefore, the p.g.f. is the probability of the second-stage, parallel Bernoulli sequence, the geometric variable, passing the first.

Proposition. *Given an independent Geometric(q) variable Y, the probability generating function of X is* $\pi_X(q) = \text{P}[Y \geq X \mid Y \text{ is Geometric}(q)]$.

Example. You and a friend are each shooting at decoys in adjacent booths at the county fair. You hit 90% of the time, and your friend hits 95% of the time. You will each shoot until you miss. What is the probability that you will last as long as your friend, or longer?

The number of hits for each of you is geometric, with probabilities 0.9 and 0.95. Your friend's p.g.f is $\frac{0.05}{1-0.95q}$. By the proposition above, let $q = 0.9$ for your probability of hitting, and we see that you will keep shooting at least as long as your friend with probability 0.345.

13.3   Moment Generating Functions

13.3.1   Comparison with Exponential Variables

You may now be wondering whether there is a representation of continuous random variables as useful as the p.g.f. is for discrete counting variables. We will first discover a representative for variables continuous on $X \geq 0$, using as a clue the last problem, involving comparison with geometric variables. The random variable that corresponds to a geometric random variable will be its asymptotic limit for p large, a Negative Exponential(β) variable. This suggests a class of problems:

Example. It is estimated that a dangerously large meteor strikes the earth unpredictably once every 2000 years. The space agencies of the world get together to build an automatic satellite to warn us of such an approaching meteor. They want it to last as long as possible without maintenance, because wars, depressions, and antitechnological fads on earth might prevent the satellite from being maintained. Therefore, they create a very durable control computer that should have a lifetime that is negative exponential with mean survival 700 years. They then place 2 identical backup computers to go on-line after the previous one has failed, so that the warning system will fail only when all three computers have failed. What is the probability that this system will detect the next large meteor to approach the earth?

We can formulate this as follows: The satellite's life X is Gamma($\alpha = 3, \beta = 700$). The time Y to the next meteor is Negative Exponential($\beta = 2000$). The problem is then $P(Y < X)$. It will turn out to be convenient to express the parameter of the negative exponential variable in this problem in terms of the Poisson rate of events λ, so that $\lambda = 1/\beta$ (see 9.4.1).

Definition. Let X be any random variable with sample space contained in $X \geq 0$ (a **positive** random variable). Now let Y be independently Negative Exponential($\lambda = t$), where t may be any positive number. The **moment generating function** (m.g.f.) of X is then $m_X(t) = P(Y \geq X)$.

Notice the similarity to the proposition above that says that the p.g.f. is the probability that a geometric variable is larger than a given counting variable.

In the example above, the answer will be $1 - m_X(0.0005)$. We may as well first solve for an easier moment generating function. Let X be negative exponential(β), while Y is negative exponential(γ). Then by a little ingenuity, X/β and Y/γ are independently standard negative exponential; and clearly,

$$P(Y \geq X) = P\left[Y/\gamma \geq \frac{\beta}{\gamma}(X/\beta)\right] = P\left(\frac{Y/\gamma}{Y/\gamma + X/\beta} \geq \frac{\beta}{\gamma}\frac{X/\beta}{Y/\gamma + X/\beta}\right)$$

$$= P\left[\frac{Y/\gamma}{Y/\gamma + X/\beta} \geq \frac{\beta}{\gamma}\left(1 - \frac{Y/\gamma}{Y/\gamma + X/\beta}\right)\right]$$

$$= P\left(\frac{Y/\gamma}{Y/\gamma + X/\beta} \geq \frac{\beta/\gamma}{1 + \beta/\gamma} = \frac{\beta}{\gamma + \beta}\right).$$

But I expressed it this way because we discovered a long time ago (see 9.5.5) that $U = (Y/\gamma)/((Y/\gamma) + (X/\beta))$ is a Beta(1, 1) variable, which, you should notice, is Uniform(0, 1). Therefore, $P(Y \geq X) = 1 - \beta/(\gamma + \beta) = \gamma/(\gamma + \beta) = 1/(1 + (\beta/\gamma))$. We recall also that the t in the definition is $1/\gamma$, and we have calculated the m.g.f.:

Proposition. *If X is negative exponential(β), then its moment generating function is $m_X(t) = 1/(1 + \beta t)$.*

We will be able to extend this to the gamma variable we needed in the meteor-defense example by noticing the following important fact: Let X and Z be independent positive random variables, and let $W = X + Z$. As usual, Y will be our competing negative exponential variable. Then $P(Y \geq W) = P(Y \geq X + Z)$, and we will interpret the latter as saying that Y lasts past X, and then lasts a further amount of time Z. But by the memoryless property of a Poisson process (see 9.3.3), the chances of surviving a further Z are independent of having survived past X (we simply restart the clock after a time X has passed). Therefore, $P(Y \geq W) = P(Y \geq X)P(Y \geq W)$.

Theorem (m.g.f. of a sum). *For X and Z independent positive variables and $W = X + Z$, $m_W(t) =$ we have $m_X(t)m_Z(t)$.*

Now recall that if T is Gamma(α, β) for α an integer, we may write $T = \sum_{i=1}^{\infty} T_i$ where the T_i are each independently Negative Exponential(β) (see 9.3.4). Then using the product formula above repeatedly gets us the m.g.f. of T:

Proposition. *If T is Gamma(α, β) with α an integer, then $m_T(t) = (1 + \beta t)^{-\alpha}$.*

Example (*cont.*). In the problem of the meteor-defense satellite, the satellite's life X has m.g.f. $m_X(t) = (1 + 700t)^{-3}$. We conclude that the probability that it will be around when the next 2000-year meteor arrives is $1 - \left(1 + \frac{700}{2000}\right)^{-3} = 0.594$.

13.3.2 The M.G.F. as an Expectation

Let us look at the general form of the calculation one must do to find the moment generating function of a variable. The technique we will use transforms a probability into an expectation using *indicator* functions χ_A (see 11.3.1), which are 1 when the outcome is in the event A and 0 everywhere else. Recall that $E(\chi_A) = P(A)$. Then $P(Y \geq x) = E(\chi_{Y \geq x}) = E_X[E_{Y|X}(\chi_{Y \geq x} \mid X)]$. The inner expectation is a probability: $P(Y \geq X) = E[P(Y \geq X \mid X)]$. Now, $P(Y \geq X \mid X = x) = e^{-xt}$, using the cumulative distribution for a Negative Exponential($\lambda = t$) variable. Therefore, $P(Y \geq X) = E(e^{-Xt})$.

Proposition. *For a positive random variable X, $m_X(t) = E(e^{-Xt})$.*

For reasons we shall see later, this is often used as the *definition* of a moment generating function. (Another common convention that you may encounter says that the moment generating function should be $E(e^{Xt})$. You can easily switch to this convention by changing an occasional sign in the formulas in the rest of this chapter. Our version is, of course, easier to interpret in problems like the one about satellite maintenance.)

These counting variables we studied in Section 2 are positive random variables, of course, so they have moment generating functions as well. Since $\pi_X(q) = E(q^X)$, we may just substitute e^{-t} for q in our earlier calculations:

Proposition.

(i) *For X distributed* B(n, p), $m_X(t) = (1 - p + pe^{-t})^n$.

(ii) *For X distributed* NB(k, p), $m_X(t) = \left((1 - p)/(1 - pe^{-t})\right)^k$.

(iii) *For X distributed* Poisson(λ), $m_X(t) = e^{-\lambda(1 - e^{-t})}$.

For counting variables, the m.g.f. representation is less useful than the p.g.f., but it will soon have some applications.

13.3.3 Moments

The p.g.f. was connected with factorial moments; something similar will happen with the m.g.f. Taking its derivative with respect to t, we get $m'_X(t) = E(-Xe^{-Xt})$ whenever that expectation exists. Taking its value at 0, $E(X) = -m'_X(0)$. As with the p.g.f., this derivative, one-sided for t approaching 0 from above, will exist whenever the corresponding expectation does (since the exponential only makes it smaller). For example, in the gamma case, $m'_T(t) = -\alpha\beta(1+\beta t)^{-\alpha-1}$. Substituting 0, we reproduce the standard result $E(X) = \alpha\beta$.

Taking another derivative, we find that $m''_X(0) = E(X^2)$, which would allow us to compute the variance (though almost always there is some easier way to do the same thing). We have a name for expectations of powers: $E(X^k) = m_X^k$ is the *kth moment of* X. If the moments exist, we may compute them by repeated differentiation:

Proposition.

(i) $m_X^k = (-1)^k m_X^{(k)}(0)$.

(ii) *If* $m_X(t)$ *has an* $(m + 1)$-*term Taylor series near 0, then*

$$m_X(t) \approx 1 + \sum_{k=1}^{m} (-1)^k \frac{t^k}{k!} m_X^k$$

for small t.

This last expression explains why m_X is called the moment generating function.

13.4   Limits of Generating Functions

13.4.1 Poisson Limits

One of the most important applications of probability generating functions is to asymptotic approximations of probabilities. For example, since a binomial family of random variables converges in distribution to the Poisson as $np^2 \to 0$ (see 6.4.3), let us see what happens to its p.g.f.: $\pi_X(q) = (1 - p + pq)^n$. A proposition we proved in (6.4.1) established that $(1 - p + pq)^n = [1 - p(1-q)]^n \approx e^{-np(1-q)}$ whenever $(np^2(1-q)^2)/(1 - p(1-q))$ was close to zero. But when $np^2 \to 0$, this last ratio clearly also approaches zero. We have shown that under these conditions the binomial p.g.f.s converge to a Poisson p.g.f. Notice that our argument is a bit simpler than the one we used in Chapter 6 to establish the Poisson approximation to the binomial.

But have we really shown anything important? Just because we can approximate a p.g.f., does that mean we have also approximated a distribution? Fortunately, it does, as a consequence of the nice property of probability generating functions that the power series in q always converges for $-1 < q < 1$. You should review from advanced calculus the following fact: If a sequence of functions with power series about 0 with at least some constant positive radius of convergence (here, 1) converges to another function with at least the same minimum radius of convergence, then the powers series for those functions also converge to the series for the other function. So the terms in our power series, proportional to the probability mass function, also converge to Poisson probabilities. We have established a general result:

Theorem (limits of p.g.f.s). *If a series of counting random variables X_1, X_2, X_3, ... have probability generating functions that converge to the p.g.f. of a random variable Y on $-1 < q < 1$, then the X_1, X_2, X_3, ... converge in distribution to Y.*

So we have an advanced, and often rather easy to use, method of discovering asymptotic limits for counting variables. You will get a chance to apply it further in the exercises.

13.4.2   Law of Large Numbers

We might hope that we could do limit arguments for other families of random variables with similar ease, perhaps using the m.g.f. (since it applies to a larger class of random variables). For example, you will remember that we showed earlier (see 7.7.3) that sample means from i.i.d. random variables that have a variance converge in probability to the true expectation (we called it a law of large numbers). We already know that if $Y = X_1 + X_2 + X_3 + \cdots + X_n$, then $m_Y(t) = m_X(t)^n$. Now we need to see what will happen when we divide Y by n to get $\bar{X}$.

Proposition. *If $Y = aX$, then $m_Y(t) = m_X(at)$.*

You should check this as an easy exercise. We conclude that $m_{\bar{X}}(t) = m_X(t/n)^n$. To make this relationship easier to work with, we will take its logarithm.

Definition. For a random variable X, its **cumulant generating function** is $\kappa_X(t) = \log m_X(t)$.

You should check the following obvious properties:

Proposition.
   (i) $\kappa_X(0) = 0$.
   (ii) *If X and Y are independent, and $Z = X + Y$, then $\kappa_Z(t) = \kappa_X(t) + \kappa_Y(t)$.*
   (iii) $\kappa'_X(0) = -\mathrm{E}(X)$ *when the expectation exists.*

Therefore, $\kappa_{\bar{X}}(t) = n\kappa_X(t/n)$. But then

$$\lim_{n \to \infty} \kappa_{\bar{X}}(t) = \lim_{n \to \infty} t\frac{\kappa_X(t/n)}{t/n} = t\kappa'_X(0) = -t\mu,$$

by the definition of a derivative. Therefore, $\lim_{n \to \infty} m_{\bar{X}}(t) = e^{-\mu t}$.

To see what this means, let us invent the peculiar concept of a random variable that does not vary: If the numerical result X of an experiment is always the constant a (so $p(a) = 1$), then we will call X a random variable *degenerate* at a. Then, of course, $m_X(t) = e^{-at}$. Our calculation above established that the m.g.f. of $\bar{X}$ converges, as the sample size grows, to the m.g.f. of a variable degenerate at the true mean μ. Since we will have several results of this form, let us give this phenomenon a name:

Definition. If each of a sequence of random variables $X_1, X_2, X_3, \ldots$ has a moment generating function defined at least in a fixed interval including 0, and Y is another random variable with moment generating function defined at least in that same interval, then if $\lim_{i \to \infty} m_{X_i}(t) = m_Y(t)$ for each t in that interval, we say that $X_1, X_2, X_3, \cdots$ **converge in m.g.f.** to Y.

Proposition (law of large numbers). *If X has a moment generating function defined in an interval including 0, and $E(X) = \mu$ exists, then the means $\bar{X}$ computed from i.i.d. samples of n converge in m.g.f. to a random variable degenerate at μ.*

This sounds like an advance over the law of large numbers from (7.7.3), because there we required the X's to have finite variances. No such requirement appears here. However, even though moment generating functions do solve some problems for us, as we saw earlier, it will still probably seem to you that knowing that we have convergence in m.g.f. is a fact of little obvious value. You will discover in advanced courses that when it makes sense, convergence in m.g.f. means that we also have convergence in distribution. Therefore, the existence of a variance really will turn out to be unnecessary in laws of large numbers.

13.4.3 Normal Limits

We are inspired to hope that some such simple limit argument might tell us when a family of random variables converges in m.g.f. to a *normal* variable, presumably using moment generating functions. Unfortunately, we do not even know what the m.g.f. of a normal random variable might mean, because so far we only have a definition for positive random variables. Normal random variables, of course, are defined for all real numbers.

Now our expectation formula for the m.g.f. comes into its own, as the definition of a more general moment generating function:

Definition. Let X be a random variable. Then its moment generating function is $m_X(t) = E(e^{Xt})$ whenever this function is defined in an interval including $t = 0$.

(Again, you may run into an alternative definition with no minus sign.) We know, of course, that this new definition is exactly the same as the old, for any positive random variable. The new approach will have one serious limitation: You should verify as an exercise that there are some important random variables that do not have moment generating functions.

In the standard normal case $E(e^{-Zt}) = \frac{1}{\sqrt{2\pi}} \int_{-\infty}^{\infty} e^{-Zt - Z^2/2} dZ$. Use a method you may remember from algebra, and complete the square in the exponent:

$$-Zt - Z^2/2 = -\frac{1}{2}(Z^2 + 2Zt + t^2 - t^2) = \frac{t^2}{2} - \frac{1}{2}(Z + t)^2,$$

so that $E(e^{-Zt}) = e^{t^2/2} \int_{-\infty}^{\infty} \frac{1}{\sqrt{2\pi}} e^{-(Z+t)^2/2} dZ$. The integral is of a Normal$(-t, 1)$ density over its sample space, which is, of course, equal to 1.

Proposition. *For Z standard normal, $m_Z(t) = e^{t^2/2}$.*

Since we know how to expand this in a Taylor series, we can use the general result above to conclude that $m_Z(t) = 1 + \sum_{k=1}^{\infty} (-1)^k (t^2/k!) m_Z^k = 1 + \sum_{l=1}^{\infty} (t^{2l}/2^l l!)$. The coefficients of the same power have to be the same on both sides, so we learn immediately that for k odd, $m_Z^k = 0$. For the even powers of t, $(m_Z^{2l}/(2l)!) = (1/2^l l)$, so that $(2l)!/2^l l! = (2l - 1) \times (2l - 3)x \cdots \times 3 \times 1$. For example, the 6th moment of a standard normal variable is $3 \times 5 = 15$.

When we found a normal limit of the gamma family, we first had to standardize our variable by subtracting its mean and dividing by its standard deviation. We need to find out what effect that has on the m.g.f.; so please check the following result as an exercise:

Proposition. *If $Y = X + b$, then $m_Y(t) = e^{-bt} m_X(t)$.*

Since a Gamma(α) variable has $m_T(t) = (1 + t)^{-\alpha}$, let us standardize to mean 0 and variance 1 by $Z = \frac{T - \alpha}{\sqrt{\alpha}}$. Then by our transformation rules,

$$m_Z(t) = e^{\sqrt{\alpha} t}(1 + t/\sqrt{\alpha})^{-\alpha}.$$

Turn this into a cumulant generating function: $\kappa_Z(t) \approx \sqrt{\alpha} t - \alpha \log(1 + t/\sqrt{\alpha})$. Now use the quadratic approximation to the logarithm (see 10.5.2) to get that

$$\kappa_Z(t) \approx \sqrt{\alpha} t - \alpha(t/\sqrt{\alpha} - t^2/2\alpha) = t^2/2$$

whenever $t^3/(3\sqrt{\alpha})$ and $t^3/[3(t + \sqrt{\alpha})]$ are small. For a fixed value of t this says that $m_Z(t) \approx e^{t^2/2}$ for $3\sqrt{\alpha}$ large. Thus, the standardized version of a gamma random variable converges in m.g.f. to a standard normal variable as $\sqrt{\alpha}$ goes to infinity.

In your exercises, you will apply the same approach to check when other families (for which we have a convenient form of m.g.f.) converge to normality. In each case, the conclusion will be that it happens just when we already knew that they converged in distribution. Thus we have more informal evidence for my claim that the two concepts of convergence are often the same. You may notice that the argument for convergence to normality seems usually to be somewhat easier for m.g.f.s.

13.4.4   A Central Limit Theorem

We will exploit the relative simplicity of m.g.f. limit arguments to discover a much more general normal limit. Consider an i.i.d. random sample $X_1, X_2, \ldots, X_n$, where each observation has mean μ, but also positive and finite variance σ^2. We know that to look for a normal limit, we will need eventually to standardize, so let us do it now, by $Z_i = (X_i - \mu)/\sigma$. If each observation has a moment generating function, we know that so will each Z; call it $m_Z(t)$. Then the sample mean of the Z's has m.g.f. $m_{\bar{Z}}(t) = m_Z(t/n)^n$. But the sample mean, though still with expectation zero, now has variance $1/n$. We standardize again by $Z_n = \sqrt{n}\bar{Z}$, so that $\bar{Z}_n$ has mean 0 and variance 1. Now we have $m_{Z_n}(t) = m_Z(t/\sqrt{n})^n$. Our goal will be to check when the limit of this for large n is the normal m.g.f.

We remember that if we take the second derivative of a moment generating function, we get $m_X''(t) = E(X^2 e^{-tX})$. Since we are working with variables for which this moment exists, it follows that this expectation exists, at least for t not negative. Therefore, $m_X''(0) = E(X^2)$, where the derivative is one-sided from the positive direction. Since we know that powers are involved, we will again find it handy to work with the cumulant generating function $\kappa_X(t) = \log m_X(t)$. Taking two derivatives, we have $\kappa_X''(t) = (m_X''(t))/(m_X(t)) - (m_X'(t)^2)/(m_X(t)^2)$. Recall that $m_X(0) = 1$ and $m_X'(0) = -E(X)$, we discover a wonderful fact:

Proposition. $\kappa_X''(0) = \text{Var}(X)$.

Now we know that $\kappa_{Z_n}(t) = n\kappa_Z(t/\sqrt{n})$. Here is where we hope to take a limit as n goes to infinity. The existence of a one-sided second derivative says that $\lim_{a\to 0^+}(f(a) - f(0) - af'(0))/a^2 = (f''(0))/2$ (check this as an exercise). Since Z has mean zero and variance 1, we have $\kappa_Z(0) = 0$, $\kappa_Z'(0) = 0$, and $\kappa_Z''(0) = 1$. The limit says that $\lim_{a\to 0^+}(\kappa_Z(a)/a^2) = 1/2$. For a fixed positive t, let $a = t/\sqrt{n}$; this becomes $\lim_{n\to\infty}(\kappa_Z(t/\sqrt{n}))/(t^2/n) = (1/2)$. Then $\lim_{n\to\infty}\kappa_{Z_n}(t) = t^2/2$. Comparing this to our cumulant generating function for Z_n, we discover that $\lim_{n\to\infty}\kappa_{Z_n}(t) = t^2/2$. Going back to the moment generating function, $\lim_{n\to\infty}m_{Z_n}(t) = e^{t^2/2}$, which is, of course, the m.g.f. of a standard normal variable.

We got to Z_n by first standardizing our X's, then taking the sample mean, then standardizing again. We could just as well have taken the sample mean of the X's, then standardized all at once.

Theorem (central limit theorem). *Let a random sample $X_1, X_2, \ldots, X_n$ consist of i.i.d. variables with mean μ and variance σ^2. Assume further that they have an m.g.f. in a common interval that includes 0. Then the standardized sample mean $Z_n = (\bar{X} - \mu)/(\sigma/\sqrt{n})$ converges in m.g.f. to a standard normal distribution.*

This is an example of a class of wonderful results that were some of the greatest achievements of early twentieth-century mathematical statistics. Notice how amazingly general it is: We get a normal distribution asymptotically for the means of observations about which we have assumed almost nothing. In (12.3.3) we noted that such facts lead people to have great confidence in statistical models that have normal errors, because they are approximately correct in so many cases. But because normal approximations are so important, it will be worthwhile to stop and comment on some limitations of the above central limit theorem.

(1) We are talking only about convergence in m.g.f. The fact that this is almost the same as convergence in distribution still has to wait for more advanced courses.

(2) The theorem applies only to distributions that have a moment generating function. I hope that you will take some of those more advanced mathematical statistics courses, because there you will learn that this restriction is completely unnecessary. (One way to discover this will be by using a representation closely related to the m.g.f., called a *characteristic function*).

(3) Unlike our normal limits in the special families obtained in earlier chapters, we do not have even rough estimates of the sizes of errors we make by using the normal approximation. This is because our theorem is so general that nothing is always true about the size of the errors. However, our earlier estimates should give you useful qualitative information about how quickly convergence to normality happens in common cases.

13.5   Exponential Families

13.5.1   Natural Exponential Forms

You may have noticed that many of the approximation and asymptotic argument we have seen have involved first taking the logarithm of a probability mass function, or a density function, or more recently a p.g.f. or m.g.f. Let us look a little deeper into what is going on when we do that. For example, consider the binomial $B(n, p)$ mass function $p(x) = \binom{n}{x} p^x (1 - p)^{n-x}$. You should remember that in (8.2.2) we took the logarithm of this to get the log-likelihood

$$l(p|x) = \log p(x) = \log \binom{n}{x} + x \log \frac{p}{(1 - p)} + n \log(1 - p).$$

Our purpose there was to find a good estimate for p, which is more often the unknown parameter in binomial problems. At that time, we pointed out the different roles the various quantities x, n, and p played in the three terms. Then in the several likelihoods studied in Chapter 8 (see 8.5.1), we called the middle term the *core*, because of the fundamental role it plays in estimation and inference.

This structure for log-likelihoods appears as well in superficially very different random variables. Starting with a Gamma(α, β) density $f(t) = t^{\alpha-1}/(\beta^\alpha \Gamma(\alpha)) e^{-t/\beta}$, we take its logarithm $\log f(t) = [(\alpha - 1) \log t - \log \Gamma(\alpha)] - t/\beta - \alpha \log \beta$. We have written this in the form of the log-likelihood $l(\beta \mid t)$, for the case where α was known and we were trying to estimate β (see 12.4.1). Now, the first term involves data t and the known parameter, the second is linear in t and involves the unknown parameter, and the third is a function of the unknown (and the known) parameter. The pattern is the same as in the binomial case.

Our goal will be to see how much we can learn about random variables purely from the fact that their likelihoods can be written in this pattern. First, let us simplify things by ignoring the known parameters, since they will not change in the course of carrying out and analyzing the experiment.

Definition. A family of random variables X indexed by a parameter θ is called a **natural exponential family** if the logarithm of its mass function or density (that is, its log-likelihood) may be written $l(\theta \mid x) = A(x) + xB(\theta) - C(\theta)$ and if its sample space does not depend on θ.

We call the family exponential because the density or mass function looks like e to the power of this simple expression. You should check as easy exercises that a number of our favorite families may be written in this form; note that we have already established that a binomial family with a given n and parametrized by p can; as well as a gamma with given α and parametrized by β.

Several of the things we commonly do when we want to estimate parameters turn out to be easy in a natural exponential family. If we take an i.i.d. sample $\mathbf{x}$ of n observations from such a family, the likelihoods multiply, so that the log-likelihoods add: $l(\theta \mid \mathbf{x}) = \sum_{i=1}^{n} A(x_i) + B(\theta) \sum_{i=1}^{n} x_i - nC(\theta)$. If we want the maximum likelihood estimate of θ, we take the derivative with respect to θ to get $\partial l(\theta \mid \mathbf{x})/(\partial\theta) = B'(\theta) \sum_{i=1}^{n} x_i - nC'(\theta)$. Setting this expression equal to zero and rearranging, we find that we must solve $\bar{x} = C'(\theta)/(B'(\theta))$. You may remember that the information from the sample that is needed to estimate a parameter by maximum likelihood is called its *sufficient statistic* (see 8.5.1). We conclude that the sample mean is a sufficient statistic for the parameter in any natural exponential family.

13.5.2 Expectations

Assume for the moment that X is absolutely continuous. Since $f(x|\theta) = e^{l(\theta|x)}$, then $\partial f(x|\theta)/(\partial\theta) = [xB'(\theta) - C'(\theta)]f(x|\theta)$. Now integrate both sides over the sample space of X to get $\int \partial f(X|\theta)/(\partial\theta)dX = \mathrm{E}[XB'(\theta) - C'(\theta)]$. But the sample space of X was assumed not to depend on θ, so we may differentiate under the integral sign, and $\int \partial f(X|\theta)/(\partial\theta)\,dX = \partial \int f(X|\theta)dx/(\partial\theta) = \partial 1/(\partial\theta) = 0$, because densities always integrate to 1. We conclude that $\mathrm{E}(X) = C'(\theta)/(B'(\theta))$. As an exercise you should check that the same thing is true if X is discrete, using mass functions and summations. In another exercise, you will use the same approach to find a general formula for $\mathrm{Var}(X)$.

The second derivative of the log-likelihood, $(\partial^2 l(\theta|x))/(\partial\theta^2) = xB''(\theta) - C''(\theta)$, now gives us the Fisher information

$$I_\theta = \mathrm{E}\left[-\frac{\partial^2 l(\theta|x)}{\partial\theta^2}\right] = \frac{C''(\theta)B'(\theta) - C'(\theta)B''(\theta)}{B'(\theta)}.$$

You should check that the Fisher information from a sample of n is just n times this expression. You will remember from Chapter 12 (see 12.8.2) that the inverse of this Fisher information is roughly the variance of the maximum likelihood estimate, and it is the best variance you can hope to get from an unbiased estimate (see 12.8.3).

Proposition. *For X in a natural exponential family*

(i) $\mathrm{E}(X) = \dfrac{C'(\theta)}{B'(\theta)}.$

(ii) $\text{Var}(X) = \dfrac{C''(\theta)B'(\theta) - C'(\theta)B''(\theta)}{B'(\theta)^3}$.

(iii) $I_\theta = \dfrac{C''(\theta)B'(\theta) - C'(\theta)B''(\theta)}{B'(\theta)}$.

13.5.3 Natural Parameters

Long ago, you may have noticed that the particular way we parametrized the families we have studied was somewhat arbitrary. For example, we used β, the mean time between Poisson events, rather than $\lambda = 1/\beta$, the rate of events, to scale the gamma family. Staring at our exponential form for the binomial family,

$$l(p|x) = \log \binom{n}{x} + x \log \frac{p}{(1-p)} + n \log(1 - p),$$

we notice an important quantity, the logit $l = \log \frac{p}{(1-p)}$ (see 1.7.3), in the middle term. Since every logit corresponds to a p, and every value of p to a logit, we feel free to let l be the parameter in this exponential family: $l(l|x) = \log \binom{n}{x} + xl - n \log(1 + e^l)$. We can make this simplification in any of our natural exponential families:

Definition. If a natural exponential family has a log-likelihood written $l(\eta|x) = H(x) + x\eta - K(\eta)$, then η is called its **natural** parameter.

Proposition. *For a natural exponential family in natural parameter form:*

(i) $E(X) = K'(\eta)$.
(ii) $\text{Var}(X) = K''(\eta)$.
(iii) $I_\eta = K''(\eta)$.

These follow immediately from the previous proposition.

A connection between moments and derivatives should have reminded you of moment generating functions. If X is absolutely continuous, let us compute

$$E(e^{-Xt}) = \int e^{-Xt} e^{H(X) + X\eta - K(\eta)} dX = e^{K(\eta-t) - K(\eta)} \int e^{H(X) + X(\eta-t) - K(\eta-t)} dX,$$

where we have multiplied and divided by $e^{K(\eta-t)}$ in order to leave a density under the integral sign (which therefore integrates to one). You should check as an exercise that you get the same result for discrete families.

Proposition. *For a natural exponential family in natural parameter form*

(i) $m_X(t) = E^{K(\eta-t) - K(\eta)}$.
(ii) $\kappa_X(t) = K(\eta - t) - K(\eta)$ *for any t for which $\eta - t$ is a possible value of the parameter.*

Thus, there is a close connection between the m.g.f. representation and the exponential family representation, for those random variables that have such things.

13.5.4 MVUE Statistics

On the other hand, the main parameter of interest may be related to the mean of the random variable. For example, in the gamma case, $E(X) = \alpha\beta$; since α is known, the obvious estimate of β is just $\hat{\beta} = \bar{X}/\alpha$. In a similar way, in the binomial case we have used $\hat{p} = X/n$. To see what will happen to natural families when the form of the parameter we care about is just the expectation of X, let $E(X) = \theta = C'(\theta)/(B'(\theta))$. First of all, this will mean that $\hat{\theta} = \bar{X}$ is an unbiased estimate of θ. We find that $C'(\theta) = \theta B'(\theta)$; and, by differentiating, that $C''(\theta) = \theta B''(\theta) + B'(\theta)$. Use these to get rid of C in the expression in Section 5.1 for the variance, to obtain $C''(\theta)B'(\theta) - C'(\theta)B''(\theta) = B'(\theta)^2$. Therefore, $\mathrm{Var}(X) = 1/B'(\theta)$ and $I_\theta = B'(\theta)$. We use the information inequality (see 12.8.2) to conclude that X and therefore $\bar{X}$ are MVUE for of θ. None of this reasoning is affected by taking a linear function of θ.

Theorem. *In a natural exponential family with a parameter θ, so that $E(aX+b) = \theta$ for some constants a and b, $\hat{\theta} = a\bar{X} + b$ is an MVUE statistic for θ.*

You should notice that several of the examples in (12.8.4) are summarized under this theorem. Unfortunately, these examples represent most of those to which this result applies.

13.5.5 Other Sufficient Statistics

We have so far treated these as one-parameter families, usually by assuming that other parameters are known. For example, you can easily check that the MVUE statistic for a normal mean is just $\hat{\mu} = \bar{X}$. In the analysis, we assumed that the standard deviation σ was known, though clearly that is unnecessary here because $\bar{X}$ does not depend on σ. (On the other hand, knowledge of α is clearly necessary to compute $\hat{\beta} = \bar{X}/\alpha$ in the gamma case.)

If instead we assume the normal mean known, we may write the log-likelihood for σ^2: $l(\sigma^2 \mid x) = -(x - \mu)^2/(2\sigma^2) - \frac{1}{2}\log(2\pi\sigma^2)$. With a little imagination, you can see the parallel to the exponential form, if only we were willing to treat $(X - \mu)^2$ as the random variable instead of X. Let us allow for this possibility.

Definition. A family of random variables X with parameter θ whose sample space does not depend on θ is called an **exponential family** if its log-likelihood may be written $l(\theta|x) = A(x) + T(x)B(\theta) - C(\theta)$.

Then a natural exponential family is just a special kind of exponential family, in which $T(x) = x$. Taking an independent random sample of n observations, we get

$$l(\theta|x) = \sum_{i=1}^{n} A(x_i) + \sum_{i=1}^{n} T(x)B(\theta) - nC(\theta).$$

If we let $\bar{T} = \frac{1}{n}\sum_{i=1}^{n} T(x)$, then the maximum likelihood estimate for θ is gotten by solving the equation $\bar{T} = C'(\theta)/(B'(\theta))$. This must be the *sufficient statistic* for estimating θ. For example, the maximum likelihood estimate for the normal variance is $\hat{\sigma}^2 = \bar{T} = \frac{1}{n} \sum_{i=1}^{n} (x_i - \mu)^2$ (which we already know to be MVUE).

It turns out, though, that we have very little more work to do to learn as much about these families as we knew about natural families. Just perform the change of variables $T = T(X)$. The Jacobian can involve only T and the known parameters, and not θ, so it becomes part of the new $A(t)$ term. The result is that we have a natural parametric family in T, with the same B and C terms. Therefore, our previous work gets us the expectation and variance of T, and exactly the same Fisher information for θ. The new families may still be thought of as having a natural parameter, $\eta = B(\theta)$. For example, it seems that the natural scale parameter in the normal family is $-1/(2\sigma^2)$. (Notice, though, that this line of argument does *not* get us the m.g.f., $m_X(t)$.) Finally, we can generalize the main theorem of the last section:

Theorem. *In an exponential family with a parameter θ and sufficient statistic $T(X)$, so that $E(aT(X) + b) = \theta$ for some constants a and b, $\hat{\theta} = a\bar{T} + b$ is an MVUE statistic for θ.*

In a later course, you should learn that much more is true: Any unbiased estimate of θ that is a function only of $\bar{T}$ (and not just a linear function) is an MVUE statistic.

Treating more than one parameter as unknown can still lead to a suggestive way of writing the likelihood. In the Normal(μ, σ^2) case, we can expand the square in the exponent and rearrange to get $l(\mu, \sigma^2|x) = x(\mu/\sigma^2) - x^2(1/(2\sigma^2)) - \frac{1}{2}[(\mu^2/\sigma^2) + \log(2\pi\sigma^2)]$. This seems to have more than one middle term, but otherwise looks similar to the exponential family form. A k-parameter exponential family generally looks like $l(\boldsymbol{\theta}|x) = A(x) + \sum_{j=1}^{k} T_j(x)B_j(\boldsymbol{\theta}) - C(\boldsymbol{\theta})$, where $\boldsymbol{\theta}$ is a k-vector of parameters. We see that from a sample of n there is now a k-vector $\mathbf{T}$ of sufficient statistics $\bar{T}_j = \frac{1}{n} \sum_{i=1}^{n} T_j(x_i)$ that are necessary to solve the simultaneous equations required by maximum likelihood. In a later course you should learn to what extent the results of this section can be extended to the problem of estimating more than one parameter at once.

13.6   The Rao–Blackwell Method

13.6.1   Conditional Improvement

We started out in Chapter 8 (see 8.5.1) with the idea that sufficient statistics are those data summaries (in that case, row and column sums) that turned out to be what we needed to know in order to calculate maximum likelihood estimates of the parameters. Then in the last chapter (see 12.8.3) we noted that in a certain sense, maximum likelihood estimates are close to the best estimates we can get. Now we know that for many of our favorite families, the sufficient statistic seems to be the quantity $T(\mathbf{X})$ that appears in the middle of the exponent. It is time to ask, when can we use this sufficient statistic to get good estimates of parameters, if linear functions will not suffice?

Example. In many semitropical climates it is not certain to freeze every year. Years without a freeze have unusually severe insect problems the following summer. What is the probability that we will not observe a freeze next year? For the last 10 years, we have had 1, 1, 0, 2, 1, 3, 0, 1, 1, 1 freezes.

Let us model the as independent Poisson(λ) observations, with λ the average number of freezes per year. The maximum likelihood estimate is, of course, $\hat{\lambda} = \bar{x} = 1.1$ freezes per year. But the question is about $P(X = 0) = p(0) = e^{-\lambda}$, whose maximum likelihood estimate is just $P(X = 0) = p(0) = e^{-\hat{\lambda}} = e^{-\bar{x}}$ (since changing the way we write the parameter does not change the likelihood of the observations).

The maximum likelihood estimate, though, lacks one perhaps-desirable feature; let us take its expectation. The sum X of n Poisson(λ), observations is Poisson($n\lambda$); and we want $E(e^{-\bar{x}}) = E(e^{-X/n})$. Never work too hard; this is its m.g.f. with $t = 1/n$. Therefore, $E(e^{-\bar{x}}) = e^{-n\lambda(1-e^{-1/n})}$. Since this does not equal $e^{-\lambda}$, our maximum likelihood statistic (though it may be quite good) is *biased*. You will check in an exercise that it is asymptotically unbiased as n gets large, but is nevertheless a consistent overestimate on average.

Of course, we could have estimated $p(0)$ in the obvious way, by calculating the proportion of times we observed zero: $\hat{p}(0) = \frac{\|x_i=0\|}{n}$. This is just an observed proportion, so it is *unbiased*: its expectation is the true proportion of zeros. Notice that its variance is $e^{-\lambda}(1 - e^{-\lambda})/n$. We know of a way to see how good this is: The information inequality gives the best possible variance. To use it, we redefine our parameter as $\theta = e^{-\lambda}$. Then the log-likelihood becomes $l = -\sum_{i=1}^{n} \log(x_i!) + \sum_{i=1}^{n} x_i \log(-\log\theta) + n\log\theta$. As an exercise, calculate its Fisher information $I_\theta = -n/(\theta^2 \log\theta)$. Therefore, the lowest limit on possible variances of unbiased estimators is $-\theta^2 \log\theta/n = \lambda e^{-2\lambda}/n$. This is definitely smaller than $e^{-\lambda}(1 - e^{-\lambda})/n$; for example, if $\lambda = 1$, the simple proportion estimator has a relative efficiency of 58%. Real improvement may be possible here.

To find better estimators, we will reason as follows: The ideal estimator (the score estimator) was inspired by maximum likelihood, which uses only the sufficient statistic $T(\mathbf{x})$ in its calculations. Assume that we have an unbiased estimator $S(\mathbf{x})$ of the function $g(\theta)$ that we care about. That is, $E[S(\mathbf{X})] = g(\theta)$. If we knew only that minimum amount of information $T(\mathbf{x})$, what would we guess S to be? A reasonable conjecture would, as usual, be the average over all possible values S might be, for the given value of the sufficient statistic: $S^*(t) = E[S(\mathbf{X})|T(\mathbf{X}) = t]$. Leaving aside for the moment whether we can actually calculate this expectation, let us pretend that $S^* = S^*[T(\mathbf{x})]$ might actually be an estimator. To figure out its properties, we will consider T and S to be themselves random variables. Then $E(S^*) = E_T[E_{S|T}(S|T)] = E(S) = g(\theta)$ (see 11.2.2); therefore, this statistic S^* is unbiased as an estimator of $g(\theta)$. Now compute the variance of S by conditioning on possible values of T (see 7.5.3):

$$\text{Var}(S) = E_T[\text{Var}_{S|T}(S|T)] + \text{Var}_T[E_{S|T}(S|T)] = E_T[\text{Var}_{S|T}(S|T)] + \text{Var}[S^*].$$

We learn that S^* is an unbiased estimator of $g(\theta)$, whose variance is guaranteed to be smaller than that of S (since that first term is positive if S is not just a function of T). Therefore, S^* is a better estimator than S.

In a sample of n Poisson observations, we had an unbiased statistic $\hat{p}$ that was just the sample proportion of zero counts. The sufficient statistic is just $x = \sum_{i=1}^{n} x_i$, which of course has a Poisson($n\lambda$) distribution. To find the conditional expectation of $\hat{p}$, conditioned on this sum, we will use the method of indicators: let $\hat{p} = \frac{1}{n} \sum_{i=1}^{n} Y_i$, where

$$Y_i = \begin{cases} 1 & x_i = 0, \\ 0 & \text{if not.} \end{cases}$$

Then we want $E(Y_i|x) = P(x_i = 0|x)$. But a single Poisson count conditioned on a total Poisson count is binomial (see 7.4.2), with x trials and $p = \frac{\lambda}{n\lambda} = \frac{1}{n}$. The probability that the probability count is zero is then $(1 - 1/n)^x$. Finally, S^* is just the average of n of these, or the same probability.

Example (*cont.*).   In the annual freeze data, $n = 10$ and $x = 11$, so $S^* = 0.9^{11} = 0.3138$. Contrast this probability of not freezing with the simple unbiased estimate 0.2 and the (biased) maximum likelihood estimate 0.3329. This latter is larger, as predicted, and pretty close to our new estimate.

I will leave it to you as an exercise to check directly that the new statistic is unbiased. Furthermore, you should verify that its variance is $e^{-2\lambda}(e^{\lambda/2} - 1)$. This is indeed always larger than the information inequality lower limit, but it quickly converges to it as n grows. If, for example, $\lambda = 1$ and $n = 10$, we find that the new estimate is only 5% less efficient than the theoretical ideal.

13.6.2 Sufficient Statistics

I hope that something bothers you about the foregoing analysis: We could have "improved" S by conditioning on any statistic T. The miraculous thing that happened in our example is that the result S^* does not require you to know the true parameter λ in order to calculate it (recall the score estimator from (12.8.1)). We will generalize our concept of a sufficient statistic to that situation.

Definition. A **sufficient statistic** for a parameter θ in a family is a function $T(\mathbf{x})$ of the data such that the distribution of X conditioned on the value of T does not depend on θ.

Then our procedure for improving an unbiased S for $g(\theta)$ gives an unbiased statistic S^* called its **Rao–Blackwell statistic**, which has lower variance. It is generally true that the sufficient statistics we have seen in our exponential families meet this condition; in certain cases, we are prepared to prove it.

Proposition. *In a random sample from a natural exponential family, $\bar{x}$ is a sufficient statistic.*

PROOF.   First, look at a discrete family, and consider any value of $\bar{X} = \bar{x}$ that has positive probability. Then

$$p(\mathbf{X}|\bar{X} = \bar{x}) = [e^{\sum_{i=1}^{n} a(X_i)+n\bar{x}B(\theta)-nC(\theta)}]/[\sum_{\hat{Y}=\bar{x}} e^{\sum_{i=1}^{n} a(Y_i)+n\bar{x}B(\theta)-nC(\theta)}].$$

But because of the exponential form, all references to θ cancel out, and the definition of a sufficient statistic is met.

For an absolutely continuous family, we will need to integrate over the event in n-dimensional space that meets the condition $\bar{X} = \bar{x}$. To facilitate this, define, for example, the change of variables $T_1 = X_1, T_2 = X_1 + X_2, \ldots, T_n = \sum_{i=1}^{n} X_i$. Then to invert this, we just use $X_1 = T_1$ and $X_i = T_i - T_{i-1}$. This affine change of variables has a matrix that is triangular and has a determinant of one, so it is volume preserving (see 11.4.1). Then we can calculate a conditional density

$$f(\mathbf{X}|\bar{X} = \bar{x}) = \frac{e^{\sum_{i=1}^{n} a(X_i)+n\bar{x}B(\theta)-nC(\theta)}}{\int_{\hat{Y}=\bar{x}} e^{\sum_{i=1}^{n} a(Y_i)+n\bar{x}B(\theta)-nC(\theta)}} = \frac{e^{\sum_{i=1}^{n} a(X_i)+n\bar{x}B(\theta)-nC(\theta)}}{\int_{T_n=n\bar{x}} e^{\sum_{i=1}^{n} A(T_i-T_{i-1})+n\bar{x}B(\theta)-nC(\theta)}}.$$

The second integral makes perfect sense: It is just the marginal density of T_n, integrated over all the other T's and evaluated at the specified value. One again, all references to θ cancel out of the exponents, and we have a sufficient statistic.

It may strike you that we have proved less than we could have. It seems likely that the same thing is true for a general exponential family with the possibly nonlinear sufficient statistic $\sum_{i=1}^{n} T(x_i)$. For discrete families, this is true by the same proof we just showed. For the absolutely continuous case, we would have to first carry out the nonlinear transformation to a natural exponential family in the random variable T; then everything works as before. In advanced courses, you will study the generalization of this result to several parameters.

In more advanced courses you should discover that usually S^* is not merely better, but actually the MVUE for $g(\theta)$. It will turn out there that its key feature is that it is a function of the sufficient statistic T only.

13.7   Exponential Tilting

13.7.1   Tail Probability Approximation

For the last three sections, we have been looking at attractive and general ways to tackle some problems that we had been doing earlier in a more case-by-case way. However, we have not yet really solved many new problems using the representation techniques of this chapter. In this section we will show that indeed, m.g.f.s and exponential families can be used to do a better job with some very practical probability calculations.

Our first application of the normal distribution was to approximate gamma probability calculations in the case where α is large (see 10.5.5). At that time, though, we noted that as the value of the random variable moved away from the mean (so that the cube of the standard score grew), the approximation was less trustworthy.

Example. Using the chi-squared statistic with 32 degrees of freedom to evaluate the goodness-of-fit in an independence model for a contingency table (see 8.4.1), we get a value of 37.2. We believe that its probabilities should be well approximated by a Gamma($\alpha = 16$, $\beta = 2$) random variable (see 11.5.4), so after a lengthy series calculation, we get a p-value $P(\chi^2 \geq 37.2 | 32 \, \text{d.f.}) = 0.242$. A normal approximation, using the mean 32 and standard deviation 8, yields $P(\chi^2 \geq 37.2 | 32 \, \text{d.f.}) \approx P(Z \geq \frac{37.2-32}{8} = 0.65) = 0.258$, which is fairly good.

The important practical applications, though, are to large values of the test statistic when the probabilities are so low that we are tempted to doubt a null hypothesis. In a contingency table with the same shape, imagine that we instead got a value of 53.4 for chi-squared. Now our laborious calculation gets $P(\chi^2 \geq 53.4 | 32 \, \text{d.f.}) = 0.0102$. The normal approximation becomes

$$P(\chi^2 \geq 53.4 | 32 \, \text{d.f.}) \approx P\left(Z \geq \frac{53.4 - 32}{8} = 2.68\right) = 0.00374,$$

which is almost three times smaller. Such a large error could make a difference in our decision about our independence hypothesis.

You should observe in exercises that many other of our favorite normal approximations have the same problem: They are good near the mean, but deteriorate badly in the tails. This is not just characteristic of normal approximation: Poisson approximation shows the same phenomenon.

Example. A manufacturer of hay-fever medication advertises that an irritating side effect (temporary lycanthropy) occurs in only 5% of users. You do a study of 200 representative users; 14 show the syndrome. Should you be surprised?

The number of victims should be binomial $B(200, 0.05)$; we compute $P[X \geq 14 | B(200, 0.05)] = 0.130$. Using the Poisson($np = 10$) approximation, $P[X \geq 14 | B(200, 0.05)] = 0.136$; this is satisfactory for many purposes.

On the other hand, what if 22 users have problems? Now $P[X \geq 22 | B(200, 0.05)] = 0.000481$; the Poisson approximation now gets $P[X \geq 22 | B(200, 0.05)] = 0.00070$. This is in error by almost 50%.

It would be nice if somehow the approximation could be made in the neighborhood of the values of interest, rather than in the center of the distribution.

13.7.2 Tilting a Random Variable

We will come up with a transformation that moves our density or mass function to the neighborhood in which it is needed. Consider a natural exponential log-likelihood with natural parameter value η: $H(x) + x\eta - K(\eta)$. If we want to move our parameter from η to a new point, which we will write $\eta - t$, the new likelihood is $H(x) + x(\eta - t) - K(\eta - t)$. Now notice how the density or mass function has been changed: We have multiplied it by e^{-xt} and divided by $e^{K(\eta-t)-K(\eta)}$. This latter expression is, of course, $m_X(t)$. We call this transformation (exponential) *tilting* of the distribution.

For example, in a normal distribution with μ the free parameter, $\eta = \mu/\sigma^2$. Then to tilt the distribution from mean μ to mean v, the tilting factor is $t = (\mu - v)/\sigma^2$. The variance does not change.

The natural parameter for the Poisson family is $\log \lambda$; so to tilt to a new mean κ, we just use a tilting factor $t = \log \lambda - \log \kappa$. This tilt has a new interpretation as $t = \log \frac{\lambda}{k}$; if we call the ratio between the means $\kappa/\lambda = q$, then $e^{-t} = q$. Then we find ourselves multiplying the mass function by $e^{-xt} = q^x$ and dividing by the constant $m_x(- \log q) = \pi_X(q)$, the probability generating function. Thus, we have two ways of looking at exponential tilting, according to whether X has a useful p.g.f.

Actually, we can apply exponential tilting to move the mean of any density that has a moment generating function. Just multiply by e^{-xt} and divide by $m_X(t)$, as before. For example, if we have started with a density $f(x)$, then

$$f_t(x) = \frac{e^{-xt} f(x)}{m_X(t)} = \frac{e^{-xt} f(x)}{E(e^{-Xt})} = \frac{e^{-xt} f(x)}{\int e^{-Xt} f(x)dX}$$

will be the version tilted by t. That last equality should make it obvious that our choice of denominator is exactly right to make the tilted density integrate to 1. As you may verify, the same transformation can be applied to discrete mass functions.

But where have we moved the mean to? Remember that the mean of f was, in terms of the log of the m.g.f., the cumulant generating function, $E(X) = -\kappa'_X(0)$, and $\text{Var}(X) = \kappa''_X(0)$. Now compute the m.g.f. of the tilted variable (call it X_t):

$$m_{X_t}(s) = \frac{\int e^{-Xs} e^{-Xt} f(X)dX}{m_X(t)} = \frac{\int e^{-X(s+t)} f(X)dX}{m_X(t)} = \frac{m_X(s + t)}{m_X(t)}.$$

Its cumulant generating function is then $\kappa_{X_t}(s) = \kappa_X(s + t) - \kappa_{X_t}(t)$. From this we extract the mean and variance of the tilted distribution: $E(X_t) = -\kappa'_X(t)$ and $\text{Var}(X_t) = \kappa''_X(t)$. If we want to tilt to a particular new mean v, we just solve the equation $v = -\kappa'_X(t)$ for the appropriate tilt t.

Example. Let X be Uniform(0, 1). Then you should check as an exercise that $m_X(t) = (1 - e^{-t})/t$, and the density of X_t is $(te^{-xt})/(1 - e^{-t})$. Taking derivatives of the log m.g.f. gives us that $E(X_t) = (1/t) - e^{-t}/(1 - e^{-t})$ and $\text{Var}(X_t) = 1/t^2 - e^{-t}/(1 - e^{-t})^2$.

13.7.3   Normal Tail Approximation

Our strategy will be to find a normal approximation to a tail probability $P(X \geq y)$ (where, for example, y is much larger than the mean μ of X) by first tilting to an X_t whose mean is y. As noted, we choose t by solving $y = -\kappa'_X(t)$ for t.

Then we approximate the tilted distribution by a normal distribution with mean y and variance $\kappa_X''(t)$. To illustrate this better, we will use a common convention that the standard normal density is $\phi(z)$ and the standard normal tail probability is $P(Z \geq z) = \Phi(z)$. Then

$$f_t(x) = \frac{(e^{-xt} f(x))}{(m_X(t))} \approx \frac{1}{\sqrt{\kappa''(t)}} \phi\left(\frac{(x-y)}{\sqrt{\kappa''(t)}}\right).$$

But our goal was to approximate f, so we solve for it to find

$$f(x) \approx \frac{(m_X(t)e^{xt})}{\sqrt{\kappa''(t)}} \phi\left(\frac{(x-y)}{\sqrt{\kappa''(t)}}\right).$$

As complicated as this looks, we discover something nice: The exponential factor will tilt the normal density back by a degree $-t$, which we know how to do from the previous section. We solve for the tilted mean, $v = \mu + \sigma^2 t = y + \kappa''(t)t$. Then

$$f(x) \approx \frac{(m_X(t)e^{yt+\kappa''(t)t^2/2})}{\sqrt{\kappa''(t)}} \phi\left(\frac{(x-y-\kappa''(t)t)}{\sqrt{\kappa''(t)}}\right)$$

(the exponential factor is just $m_Z(-t)$ for Z the normal approximation at y, by which we had to divide and then multiply). Now integrate to get a tail probability:

Theorem. *For a random variable X with m.g.f. $m_X(t) = e^{\kappa(t)}$ that is twice differentiable, let y be a value of interest and t be the solution of $y = -\kappa'(t)$. Then the (first-order) saddle point approximation is*

$$P(X \geq y) = \int_y^\infty f(X)dX \approx m_X(t)e^{yt+(t^2/2)\kappa''(t)} \Phi\left(-\sqrt{\kappa''(t)}t\right).$$

This is an example of a very old mathematical method called a *saddle-point* approximation.

Let us see what this does for a Gamma(α) random variable. The m.g.f. of the gamma is $m_T(t) = (1-t)^{-\alpha}$. We compute $\kappa_T(t) = -\alpha \log(1+t)$, $\kappa_T'(t) = \frac{-\alpha}{1+t}$, and $\kappa_T''(t) = \frac{\alpha}{(1+t)^2}$. To find the tilt point, we must solve $y = -\kappa_T'(t) = \frac{\alpha}{1+t}$; then $t = -\frac{y-\alpha}{y}$, $\kappa_T''(t) = y^2/\alpha$, and $m_T(t) = (y/\alpha)^\alpha$. The general formula for our saddle-point approximation for a Gamma(α) tail probability is then

$$P(X \geq y) \approx (y/\alpha)^\alpha e^{-(y-\alpha)+(y-\alpha)^2/2\alpha} \Phi\left(\frac{y-\alpha}{\sqrt{\alpha}}\right).$$

Example. Let us apply this formula to the chi-squared tail approximation in Section 7.1. Removing the scale factor, our tail approximation problem is

$$P(T \geq 26.7|\text{Gamma}(16)) \approx (26.7/16)^{16} e^{-(10.7)+(10.7)^2/32} \Phi\left(\frac{10.7}{4}\right) = 0.0109.$$

This is now within about 7% of the correct answer 0.0102. This is approximately as accurate as our earlier example of a y near the mean.

We can see how accurate this technique is going to be by noticing that gamma random variables are themselves natural exponential families in β, for fixed α. Therefore, when we tilted our distribution, only β changed. A normal approximation, now at the mean, has relative error at most about $\frac{1}{3\sqrt{\alpha}}$. The saddle point approximation will stay approximately equally good no matter how far out in the tail we go.

In exercises, you will see that saddle-point approximation works in a number of other cases.

13.7.4 Poisson Tail Approximations

It should be equally straightforward to use exponential tilting to approximate any distribution with an m.g.f. by any natural exponential family. For example, let us see whether we can get improved Poisson approximations in the tails of counting distributions.

Consider a counting random variable X with mass function $p(x)$. We wish to compute $P(X \geq y)$ for some value y much larger than the mean, in a case where we would like to use Poisson approximation. Our examples suggest that this approximation soon breaks down, even if it worked fairly well near the mean. Therefore, we will tilt our distribution until its mean is y. Following a suggestion in Section 7.2, we will use a tilting factor $q = e^{-t}$, where q is the variable in a probability generating function:

$$p_q(x) = \frac{q^x p(x)}{\sum_{Z=0}^{\infty} q^Z p(Z)} = \frac{q^x p(x)}{E(q^Z)} = \frac{q^x p(x)}{\pi_X(q)}.$$

Once again, you can see from this way of writing it that $\sum_{X=0}^{\infty} p_q(X) = 1$, so it really is a probability mass function.

To work with this new random variable X, we will first extract its p.g.f.:

$$\pi_{X_q}(r) = \frac{\sum_{X=0}^{\infty} f^X q^X p(X)}{\pi_X(q)} = \frac{\sum_{X=0}^{\infty}(rq)^X p(X)}{\pi_X(q)} = \frac{\pi_X(rq)}{\pi_X(q)}.$$

Just as with the m.g.f., the logarithm of the p.g.f. will be very useful here.

Definition. The **discrete cumulant generating function** of a counting variable X is $k_X(q) = \log \pi_X(q)$.

Proposition.
  (i) $k_X(1) = 0$.
  (ii) *If X and Y are independent, and $Z = X + Y$, then $k_Z(q) = k_X(q) + k_Y(q)$.*
  (iii) $k_X'(1) = E(X)$.

You should check these as (very easy) exercises. Using the p.g.f. equation above and taking logs, we get for our tilted variable $k_{X_q}(r) = k_X(rq) - k_X(q)$. Its mean is then $E(X_q) = k_{X_q}'(1) = q k_X'(q)$.

To tilt the mean to the prespecified value y, we then solve for q in the equation $y = q k_X'(q)$. Now approximate the mass function of X_q by a Poisson(y) mass function. To make our formulas compact, we will invent a notation for a Poisson mass function $\rho_\lambda(x) = (\lambda^x/x!)e^{-\lambda}$ and tail probability $R_\lambda(x) = P(X \geq x) = \sum_{X=x}^\infty \rho_\lambda(X)$. We write $p_q(x) = q^x p(x)/(\pi_X(q)) \approx p_\lambda(x)$, then solve for the original mass function $p(x) \approx \pi_X(q)(1/q)^x \rho_y(x)$. Just as in the last section, we notice that this amounts to tilting the Poisson mass function (back) by a factor $1/q$.

The Poisson(y) p.g.f. is $\pi_X(1/q) = e^{-y(1-1/q)}$; we know that we will have to multiply by this to cancel the denominator. The mean we tilt back to is y/q. Thus $p(x) \approx \pi_X(q)e^{-y(1-1/q)}\rho_{y/q}(x)$.

Theorem. *If X is a counting variable and y is a value of interest, let q be the solution of $y = qk'_X(q)$. Then a tilted Poisson approximation is given by*

$$P(X \geq y) \approx \pi_X(q)e^{-y(1-1/q)}R_{y/q}(x).$$

In the example at the beginning of this section, we were interested in approximations to binomial random variables in the case of small p. To apply the tilting technique, we need $\pi_X(q) = (1-p+pq)^n$, its logarithm $k_X(q) = n\log(1-p+pq)$, and $k'_X(q) = \frac{np}{1-p+pq}$. To find the tilting factor, we solve $qk'_X(q) = \frac{npq}{1-p+pq} = y$ for q, getting $q = \frac{(1-p)y}{p(n-y)}$. Substituting this in the formula above, we obtain $P(X \geq y) \approx \left[\frac{n(1-p)}{n-y}\right]^n e^{-(y-np)/(1-p)}R_{p(n-y)/(1-p)}(x).$

Example. In the example in Section 7.1, the probability of at least 22 sufferers is approximately $P[X \geq 22 \mid B(200, 0.05)] \approx 0.000455$, which is much closer to the correct value than the simple Poisson method.

To get some idea of how good this approximation should be, remember that a binomial random variable is also a natural exponential family in p. Therefore, when we tilt to a mean of y, we create a B(n, y/n) random variable. The Poisson approximation is good when y/n is small, which for the case of a large y is larger than p; therefore, the approximation worsens as y gets larger (though not so fast as the simple approximation would). This suggests, though, that if we find ourselves interested in binomial left tail approximations, y smaller than np, then the approximations may work even better than near the mean. You will try this in exercises.

13.7.5   Small-Sample Asymptotics

Saddle-point approximations and tilted approximations are the fundamental methods of a growing modern class of probability calculations, called *small-sample asymptotic* methods. In advanced courses, you will learn that by studying higher derivatives of κ and k (higher *cumulants*), probability approximations based on what are called *Edgeworth* or *Charlier* series corrections can achieve very high accuracy near the mean. Then we can tilt the point of approximation out to a value in the tails and achieve similarly high accuracy. This is a lively area of statistical research and a very promising area for future practical applications.

13.8   Summary

If counting variables are thought of as enumerating descendants, then they may be *compounded* by looking at descendants after several generations (2.1). A valuable tool for studying compounds is the *probability generating function* $\pi_X(q) = \sum_{x=0}^{\infty} p(x)q^x = E(q^X)$ (2.3). Then If Z is a compound of X and Y, $\pi_Z(q) = [\pi_X[\pi_Y(q)]]$ (2.4). This p.g.f. may be expanded in a power series in either the mass function $p(x) = \frac{\pi_X^{(x)}(0)}{x!}$ or the *factorial moments* $E[(X)_k] = f_X^k$ of X, since $\pi_X(q) \approx 1 + \sum_{k=1}^{m} \frac{f_X^k}{k!}(q-1)^k$ (2.5).

For continuous random variables, we may define a similar representation, called the *moment generating function* $m_X(t) = E(e^{-Xt})$ (3.2). The p.g.f. and the m.g.f. may be used to construct remarkably general asymptotic arguments, leading to new laws of large numbers and to a *central limit theorem*, which says that in a certain sense sample means with a variance are asymptotically normal (4.4).

Many of our favorite families of random variables have a particularly simple form for their log-likelihood $l(\theta|x) = A(x) + xB(\theta) - C(\theta)$. We call these *natural exponential families*; there is a simple connection between their log-likelihood, their m.g.f., and their Fisher information (5.1). In important cases, we saw that the parameters of these families have MVUE statistics (5.4). In some cases we can use sufficient statistics and the *Rao–Blackwell* process to improve parameter estimates (6.2).

Normal and Poisson approximation far from the mean of a distribution (in the *tails*) may be significantly improved by exploiting the natural exponential form and the m.g.f. and p.g.f. The basic idea is to move the natural parameter by *exponential tilting* so that the approximation is once again performed near the mean (7.2). In the normal case, a simple version says

$$P(X \geq y) = \int_y^{\infty} f(X)dx \approx m_X(t)e^{yt + (t^2/2)\kappa''(t)}\Phi\left(-\sqrt{\kappa''(t)}t\right),$$

where κ is the *cumulant generating function* and Φ is the standard normal tail probability (7.3). In the Poisson case, our simple version is $P(X \geq y) \approx \pi_X(q)e^{-y(1-1/q)}R_{y/q}(x)$, where R is a Poisson tail probability (7.4). These methods are called *saddle-point* or *small-sample asymptotic* methods.

13.9   Exercises

1. Find the probability generating function for a discrete uniform $(0, \dots, W)$ random variable.
2. There are known to be 20 nesting pairs of doves in a certain small preserve. One of these pairs is of the extremely rare barred dove species. Once egg-laying season is finished, you proceed to search out dove nests until you find and mark for protection the barred dove nest. Along the way, at the behest of

a colleague, you will test one egg from each regular dove nest for viability. Historically, 90% of eggs were viable.

What is the probability that all the eggs tested will be viable?

3. Typically, 60% of premed students from your university will be asked for an interview at a medical school. Of those interviewed, 75% will be offered admission. If there are 22 premeds in the senior class, what is the probability that all of those who get an interview will also be offered admission?

4. In Exercise 45 of Chapter 6, we defined a logarithmic(p) random variable. Find its probability generating function.

5. A **Gambler's Ruin** random variable, $G(n, p)$ is the number of Bernoulli trials X until there have been n more successes than failures. The sample space is $X = \{n, n+2, n+4, n+6, \ldots\}$. It has probability generating function

$$\pi_X(q) = \left[\frac{1 - \sqrt{1 - 4p(1-p)q^2}}{2(1-p)q} \right]^n .$$

You show up in Atlantic City one evening with $500 in your pocket. Each night, you play a double-or-nothing game that either earns you an additional $100, or takes $100 that you already had. The probability of winning is 0.45. After each night of gambling, you spend the morning on the beach. When you go broke, you will leave for home in the afternoon. Five percent of mornings this time of year are rainy. What is the probability that none of your mornings on the beach are ruined by rain?

6. Explain why a $G(n, p)$ variable is defined only for every second value of $X(n, n+2, \ldots)$.

7. Fifty percent of physics majors and 35% of math majors go on to graduate school. 3 of your friends are physics majors and 4 are math majors. What is the probability that 3 of these friends will go on to graduate school? Solve this problem twice, by direct algebra and by differentiation, to see that you get the same thing.

8. If each generation of a family has NB(2,0.5) children, what is the probability that an individual will have 5 grandchildren?

9. Find the mean and variance of a $G(n, p)$ random variable.

10. Find the mean and variance of the number of grandchildren in Exercise 8.

11. For discrete uniform $(0, \ldots, W)$ random variables, find the kurtosis $E\{[X - E(X)]^4\}$ using probability generating functions.

12. Derive the moment generating function of a (continuous) Uniform$(0, \mu)$ random variable.

13. A professor has promised that a pop quiz will be given in class with no warning sometime anytime in the next 90 days (for simplicity treat its occurrence as a continuous random variable). You believe that the event of getting the flu this year is a negative exponential random variable with mean $\beta = 200$ days. What is the probability that you will get the flu before the pop quiz (so you are sure to be behind when it happens)?

14. Very complex systems that are not maintained until failure, so that they gradually drift out of adjustment, then fail when they are finally working too poorly to function, have times to failure that may be modeled by an **inverse Gaussian**(a, b) random variable X. The quantity $a > 0$ measures how fast the system tends to get out of adjustment, and $b > 0$ measures how tolerant the system is of bad adjustment. The moment generating function of X is then
$$m_X(t) = e^{b(a-\sqrt{a^2+2t})}.$$
Find $E(X)$ for **any** a, b. Now, if X is life in months, how long would you expect a pump on the Alaska pipeline to work on average if $a = 0.2$ and $b = 3$?

15. Show that a Cauchy random variable (with density $f(x) = \frac{1}{\pi(1+x^2)}$ and sample space $-\infty < X < \infty$) has no moment generating function.

16. Use moment generating functions to compute the skewness $E\{[X - E(X)]^3\}$ of a Gamma(α, β) random variable.

17. Use a limit of p.g.f.s argument to establish when a family of negative binomial random variables converges in distribution to a Poisson random variable.

18. Prove that

    a. if $Y = aX$ then $m_Y(t) = m_X(at)$.
    b. if $Y = X + b$, then $m_Y(t) = e^{-bt}m_X(t)$.

19. The battery in your smoke alarm is known to have a life of an average of 60 days with a standard deviation of 50 days (but a very long shelf life when not in use). You know nothing more about the shape of its distribution. You buy a box of twelve such batteries, intending to replace each battery in turn as it gives out. Local fire statistics suggest that your next fire serious enough to trigger an alarm might occur as a negative exponential event with mean $\beta = 800$ days. Use our version of the central limit theorem to estimate the probability that you will have a fire before your box of batteries is used up.

20. Write the NB(k, p) family as a natural exponential family in p. What is its Fisher information for p? What is its natural parameter? What is its Fisher information for its natural parameter?

21. Show that any natural exponential family is a *monotone likelihood ratio* family in terms of its natural parameter (see 9.7.2).

22. Write the Beta(α, β) family as an exponential family in α (with β assumed known). What is its natural parameter? What is the sufficient statistic, if we had an i.i.d. sample of n?

23. You are given a sample of n observations from a Poisson(λ) family and the maximum likelihood estimate of $P(X = 0) = p(0) = e^{-\lambda}$.

    a. Show that its expectation is always larger than the correct value; but
    b. it is asymptotically unbiased; that is, the limit of the expectation as n gets large is correct.

24. In a sample of n from a Poisson family with parameter $\theta = e^{-\lambda}$, compute its Fisher information (see 12.8.1).

25. In (10.5.7) we studied a box of 24 fuses, each with a negative exponential life of $\beta = 5$ days. We would now like to know the probability that the box lasts at least 183 days. The exact answer is 0.011, but you should try it by normal approximation.

    a. Calculate the ordinary normal approximation (fairly poor, because you are in the far tail).
    b. Calculate the saddle-point approximation.

26. In Exercise 25, calculate the probability that the box will last at most 60 days (exact answer 0.00147)

    a. by ordinary normal approximation.
    b. by saddle-point approximation.

27. If 10% of Americans are left-handed, find the probability that a sample of 75 will have at least 18 lefties (exact answer 0.000348)

    a. by the usual Poisson approximation.
    b. by Poisson tilting.

28. In Exercise 27, find the probability of 2 or fewer lefties (exact answer 0.0161)

    a. by the usual Poisson approximation.
    b. by Poisson tilting.

13.10 Supplementary Exercises

29. Find the probability mass function of a Gambler's Ruin $G(n, p)$ random variable. (You may have to review the general binomial theorem from calculus.) In Exercise 5, what is the probability that you will stay 13 days at Atlantic City?

30. For Z a compound of X and Y, prove that $E(Z) = E(X)E(Y)$ without using p.g.f.s. (**Hint:** Apply the proposition in (7.5.2) that says that $E[g(X, Y)] = E_X\{E_{Y|X}[g(X, Y)|X]\}$.)

31. Prove again the proposition in Section 2.5 about the variance of a compound, using the proposition about conditional decomposition of variance (see 7.5.3). **Hint**: try conditioning on X.

32. A maintenance engineer drops in on the pump station in Exercise 14 on average once a year (12 months), but his schedule is so complicated by weather and problems at other stations that his arrival follows a negative exponential law. If the pump is in adjustment when he leaves, what is the probability that it will be working when he returns?

33. Compute the moment generating function of a Beta(2, 2) random variable.

34. Redo Exercise 6.26 using the method of moment generating functions.

35. A *logistic* random variable has density $f(x) = e^x/(1 + e^x)^2$ on the entire real line. Find its m.g.f.

36. Show that t-statistics on n degrees of freedom (see Exercise 12.22)

    a. converge in distribution to a standard normal distribution as n goes to infinity (you might approximate the log-density by a quadratic).

    b. never have a moment generating function.

37. When the cumulant generating function has a power series approximation at 0, $\kappa_X(t) \approx \sum_{i=1}^{k} \frac{\kappa_X^i}{i!}(-t)^i$, the *cumulants* of a random variable X are defined as the constants $\kappa_X^i = (-1)^i \kappa_X^{(i)}(0)$.
Write the 3rd and 4th cumulants, κ_X^3 and κ_X^4, as functions of the moments of X.

38. Find general expressions for the cumulants of

    a. a Normal(μ, σ^2) variable.

    b. a Poisson(λ) variable.

    c. a Gamma(α) variable.

39. Show that if X_i is NB$(\alpha, 1 - p_i)$ and $p_i \to 0$ then $T_i = X_i p_i/(1 - p_i)$ converge in m.g.f. to Gamma(α).

40. Discover under what conditions on its parameters an inverse Gaussian (see Exercise 14) family of random variables will converge in m.g.f. to a normal distribution.

41. Let η be the natural parameter of a family, and let $\theta = g(\eta)$ be a monotone, differentiable transformed parameter for the family. Find the Fisher information for θ in terms of the Fisher information for η.

42. Show by a direct calculation that for a sample of n from a Poisson(λ) family, our new estimator of $\theta = e^{-\lambda}$, which is $(1 - 1/n)^x$, is unbiased. Also, show that its variance is $e^{-2\lambda}(e^{\lambda/n} - 1)$.

43. Let $x_x, \ldots, x_n$ be an i.i.d. sample from a Negative Exponential(β) random variable, with β unknown. We are going to look for unbiased estimates of the cumulative distribution function $F(x) = 1 - e^{-x/\beta}$ for a fixed value of x.

    a. Show that the empirical cumulative distribution function $F_n(x) = |\{x_i \le x\}|/n$ is unbiased for $F(x)$ (in fact, for any random variable at all).

    b. Apply the Rao–Blackwell method to find a better unbiased estimate of $F(x)$. **Hint:** $\bar{x}$ is the sufficient statistic for β. Now show that each $x_i/(n\bar{x})$ has a Beta distribution.

44. Derive the saddle-point approximation for B(n, p) tail probabilities in case $np(1 - p)$ is large. (You should use continuity correction.)

45. If 35% of Americans have the gene to be a super taster, we would like to know the probability that of a random sample of 125, we get at least 60 super tasters

    a. by normal approximation; or

    b. by saddle-point approximation (the exact answer is 0.00187).

46. In the early days of computers, standard normal random variables were sometimes approximately obtained by summing 12 independent Uniform(0, 1) variables, and subtracting 6. Approximate the probability that this would get at least 3.5

   a. by ordinary normal probabilities.

   b. by the saddle-point method.

47. The factor that multiplies the normal tail probability in the saddle-point formula is $m_X(t)e^{yt+(t^2/2)\kappa''(t)}$. Show that this is equal to the (easier to remember) expression $e^{\int_0^t (t^2/2)\kappa'''(s)ds}$ if the third derivative of κ exists. **Hint:** Integrate twice by parts.

48. Derive the Poisson tilting approximation for NB(k, p) random variables with kp^2 small.

49. Eight percent of Americans are allergic to penicillin. A medical study needs 50 people who are **not**. In the course of collecting our nonallergic sample, approximate the probability of finding at least 12 who are allergic (exact answer 0.00338)

   a. by the usual Poisson approximation.

   b. by Poisson tilting.

Hints and Solutions
to Certain Exercises

Warning: This section may be bad for your academic health. In order to get any benefit from doing the exercises, do not look at the hints until you have given the problem an honest try and are stuck. More importantly, never look at the partial solutions below until you have finished that exercise yourself. Then use the answer only to check that you have not made a mistake. "Working backwards" from a solution will never teach you crucial problem-solving skills.

Chapter 1

1. **b.** $\bar{x} = 81.3008$.
2. **b.** $\hat{\mu}_n = \bar{x}_n = 0.01617$.
   c. $\hat{b}_p = 0.00472$.
4. **Hint**: Average both sides of the equation $\mu_i = \mu + b_i$ over all observations. Using the hint, you obtain $\mu = \frac{1}{n} \sum_{i=1}^{k} n_i b_i$.
5. **Hint**: Solve for the last observation at each level; then $x_{in_i} = n_i \hat{\mu}_i - \sum_{j=1}^{n_i-1} x_{ij}$, so the last observation is determined.
6. **b.** $\hat{\mu} = 7.33$, $\hat{b}_d = 1.33$, $\hat{c}_{ex} = 1.83$.
   c. $\hat{d}_{ex,d} = 0.25$.
7. $\frac{1}{n} \sum_{i=1}^{l} \sum_{j=1}^{m} n_{ij} \hat{b}_i = \frac{1}{n} \sum_{i=1}^{l} n_{i\bullet}(\bar{x}_{i\bullet} - \bar{x}) = \bar{x} - \bar{x} = 0$.
8. $\frac{1}{n_{i\bullet}} \sum_{j=1}^{m} n_{ij} \hat{d}_{ij} = \frac{1}{mr} \sum_{j=1}^{m} r(x_{ij} - \bar{x}_{i\bullet} - \bar{x}_{\bullet j} + \bar{x}) = \bar{x}_{i\bullet} - \bar{x}_{i\bullet} - \bar{x} + \bar{x} = 0$.
9. **b.** $173.22.
10. **b.** 60.85 crimes per 1000 people.
11. expect 15.89 complaints in the first group.
13. **Hint**: When you sum expected counts over columns j, you should find that $x_{i\bullet}$ factors out.

14. The model is $\log(x_i) = \mu + b_i$; then, e.g. $3\mu = \log(43) + \log(35) + \log(22)$ because the b's sum to zero.
15. **a.** Imitate the argument in (7.3) using summation notation.
16. Do Exercise 15 first.
18. Do Exercise 17b, then interpolate.

Chapter 2

1. MSE $= 1.56$.
2. **c.** 0.14% bad peanuts.
3. The smaller side is 92% of the larger.
5. **a.** Sample variance $= 0.00000069$.
6. **Hint:** What does the principle of least squares say about a sample variance?
7. **Hint:** Write out the defining formula, then look at Exercise 4.
9. **Hint:** Express the orthogonality condition in summation notation, and try to simplify.
10. $F = 15.76$.
11. $F = 5.37$.
12. $K = 8.27$.
14. **b.** RMSE $= 0.0394\%$.
17. **b.** F(inter) $= 0.19$; F(exer) $= 3.94$.

Chapter 3

2. **Hint:** P(different) $= 1 - $ P(same).
3. Your list should have 70 items on it.
6. **Hint:** $50 \times 49 \times \cdots \times 39$.
7. P(5 spades) $= 0.000495$.
9. **Hint:** Let one marble in the urn be special. Now count separately the sets of k that include it and those that do not.
11. **Hint:** At each intersection on the way, I must choose either west or south, until I have taken each direction a certain number of times.
12. **Hint:** Section 3.4 has two suggestions.
14. **b. Hint:** Consider the three possible countries separately.
15. P(7|10) $= 0.117$.
16. P $= 0.06$.
19. P $= 0.000558$.
20. P(9 cameras) $= 0.06$.
21. P(7 cameras) $= 0.317$.
22. **Hint:** Noticing that $\binom{i}{1} = i$, use the result of Exercise 9 repeatedly, as in Pascal's triangle.
23. **Hint:** Use Exercise 9 repeatedly.

24. Hint: Use our inequality for the logarithm. But do not ignore the cases in which $x \leq -1$.

Chapter 4

3. Hint: What event is $A - (A - B)$?

5. Hint: A naive answer might be that one coin is certainly heads up, so the chance that the other is heads is $\frac{1}{2}$; this is wrong.

7. Hint: Look at the derivation of the multiplication rule for combinatorial probabilities (see 3.4.3) and translate it into our new notation.

8. Hint: if n outcomes are equally likely, what is the probability of each particular outcome?

11. $P(H|N) = 0.00105$.

12. b. $P(New|Out) = 0.571$.

13. a. Hint: If she has talked to more than six people, then the first six were all right-handed.

14. Hint: There are two reasons why it might be for sale on Sunday. It might have just arrived, or it might have been there Saturday but did not sell.

16. Hint: if x is any number in $(0, 1]$, then $y = 1/x$ is ≥ 1. What determines the value of the integer part of y?

17. Hint: Figure out what event is described by the expression
$$\cup_j A_j - \cup_i (\cup_j A_j - A_i).$$

20. Hint: Piece events together out of finite rectangles.

Chapter 5

1. b. A typical entry is $p(1) = 0.214$.

3. $P(5 \text{ older}|2 \text{ Sophs}) = 0.1107$.

4. Hint: Generalize the idea of a negative hypergeometric variable to more than two categories (three kinds of trees).

5. $P(3 \text{ or more}) = 0.097$.

9. p-value $= 0.0354$.

10. p-value $= 0.0403$.

13. Hint: You want the probability of a vertical strip whose left edge is at $x = -1$.

15. a. Hint: $F(x) = \int_0^x f(X) dX$.

16. Hint: The two calculations should be quite different but have exactly the same answer, illustrating positive–negative duality.

19. Hint: The calculations will be different, but the answers should be the same, illustrating black–white symmetry.

20. Hint: The fact that $F(8|N(32, 8, 5)) = 0.0574$ should reduce (but not eliminate) your arithmetic.

21. $E(X) = 2.3929$.

26. a. 179 sheep.
   b. Hint: Start with P(so many if the total is 80) = 0.1016 and P(if the total is 50) = 0.0147.

Chapter 6

1. b. $P \approx 0.076$.
2. $P = 0.0493$.
3. a. Hint: You need not use any formula you may happen to remember for the sum of a finite or infinite geometric series; use only the definition of F and the reasoning in (5.4.1).
5. b. $P \approx 0.334$.
6. b. Hint: What is past is past. Consider only future races and wrecks.
9. Hint: Reread the derivation of the birthday inequality in (3.5.3).
10. b. $P \approx 0.712$.
13. $P = 0.136$.
15. Hint: The equality between your answers illustrates positive–negative duality.
16. Hint: The equality illustrates black–white symmetry for the negative binomial family.
17. b. Hint: Treat each day's work as an independent experiment.
18. b. Hint: Notice that failing is a rare event.
20. Hint: There is a simple duality principle that drastically reduces your work.
22. Hint: You will need the expectation of a Geometric(p) variable.
23. a. $E([x - 3]^2) = 1.3375$.
25. Hint: Use the inductive method.
27. b. $\hat{p} = 0.585$.
28. Hint: In Exercise 3 you derived a helpful formula.
29. Hint: For a balanced die, $p(39) = 0.0145$ and $p(38) = 0.0108$.

Chapter 7

1. c. $P = 0.0125$.
2. b. $P = 0.286$.
5. b. Hint: It may be easier to reason it out than to compute with mass functions.
8. One entry is $F(5, 5) = 0.55$.
10. Hint: Our formula for the probability of a rectangle should inspire you.
14. $Cov(X, Y) = 0.6$.
17. $\sigma = 11.14$.
18. $\sigma = \$377.20$.
20. $\rho_{XY} = -0.041$.
22. $Var(\bar{x}) = 3$.

23. **Hint**: After a particular failure, how many successes will there be before the *next* failure?

25. **b.** $\sigma = \$316.66$.

28. **b.** $P(X \geq 81|B(96, 0.75)) = 0.0189$; $P(X \geq 81|B(97, 0.75)) = 0.0306$.

29. **a.** **Hint**: The posterior distribution of the number of bears is $48 + Z$, where Z is a Poisson random variable with mean 105.

Chapter 8

1. $R = 8.7$.
2. $R = 2.826$.
3. $\hat{p} = 0.897$.
4. **Hint**: Do not use calculus. Find the ratio between probabilities for successive larger integers W (or B), and note when it stops increasing and starts decreasing.
6. **Hint**: In order to force your estimates to sum to 1, substitute $p_k = 1 - \sum_{i=1}^{k-1} p_i$.
7. **Hint**: Move x to the denominator.
9. **Hint**: It helps to replace $p_{r\bullet} = 1 - p_{1\bullet} - \cdots - p_{(r-1)\bullet}$, where there are r rows; and similarly for columns.
10. **a.** G-squared $= 41.4$.
    b. Chi-squared $= 42.3$.
14. For example, the expected count of male, urban nonsmokers is 124.49.
16. $\lambda = 1.678$.
18. **c.** $P(\text{six}|25 \text{ mg}) = 0.353$.

Chapter 9

1. **b.** **Hint**: Be careful about what happens at $X = 0$.
2. The answer starts out 0, 0, 1, 4, 5, ...
3. **b.** $P = 0.713$.
5. **Hint**: Use the results in Chapter 6.8.
9. **b.** mode $= b\left(\frac{a-1}{a}\right)^{1/a}$.
10. **Hint**: As often happens, it is easier to find the maximum of the log of f.
14. $P - 0.577$.
16. $P = 0.474$.
17. **Hint**: A statistics program that calculates cumulative Poisson probabilities will help here.
18. **a.** **Hint**: The best size you can get turns out to be about 0.035.

Chapter 10

1. **Hint**: Your answer should be in the form of a table.
2. The 25th percentile is 0.82.
5. **Hint**: See Exercise 9.6
7. **b.** P = 0.612.
12. **Hint**: $E(1 + X)$ and $[(1 + X)^2]$ are easy to find.
14. $\text{Cov}(X, Y) = -690$.
17. $0.4028 \leq \log(1.5) \leq 0.4167$.
18. **Hint**: Move the exponent to the denominator.
19. **Hint**: Write them as integrals.
20. Your answer will be within one part in 300 of the exact value.
21. Exact probability is 0.1254 and approximate probability is 0.1262.
22. The seventh and eighth terms of the series in Section 5.6 should give your bounds.
24. **Hint**: This involves summing many terms, but if you sum them in a sensible order, you will find that the terms quickly become negligible and you have an answer accurate to 2 significant figures.
27. **c.** $P(39) \approx 0.0518$.
28. Coverage probability = 0.918.

Chapter 11

4. P = 0.278.
9. **a.** P = 0.1792.
11. **Hint**: Review addition formulas from trigonometry.
13. **b.** The lower bound is 12.34
15. The upper limit of the interval is 81.30117.
16. **b. Hint**: Remember Beta–Binomial duality (see 9.5.3). P = 0.182.
17. **b.** $P \approx 0.158$.
18. **a. Hint**: A computer would help here. P = 0.219.
22. **Hint**: Try to rearrange it so it looks like a bivariate normal density.

Chapter 12

2. **Hint**: These look like $z = x \cos\theta + y \sin\theta$ and $w = -x \sin\theta + y \cos\theta$ for a rotation through any angle θ.
3. **b. Hint**: This involves evaluating a very easy integral.
8. **Hint**: The column for the one degree of freedom for interaction should be proportional to the product of the two centered columns for the factors.
9. **a. Hint**: You may have to discard some redundant columns so that X^TX becomes nonsingular.

11. standard error of prediction $= 0.00744$.

13. **a.** $I_\beta = \frac{n}{\beta^2}$.

18. **Hint**: treat σ^2 as a parameter itself, not the square of σ.

Chapter 13

1. $\pi(q) = \dfrac{q(1 - q^W)}{(W + 1)(1 - q)}$.

2. **Hint**: Does Exercise 1 help?

3. $P = 0.028$.

7. $P = 0.297$.

8. $P = 0.0593$.

10. Var $= 24$.

13. **Hint**: Does Exercise 12 help?

14. $E(X) = 15$ months.

19. $P \approx 0.584$.

25. **b.** $P \approx 0.0117$.

27. **b.** $P \approx 0.000305$.

Index

Springer Texts in Statistics *(continued from page ii)*